Animal Welfare

2nd Edition

Edited by

Michael C. Appleby
World Society for the Protection of Animals (WSPA), UK

Joy A. Mench
University of California, Davis, USA

I. Anna S. Olsson
Institute for Molecular and Cell Biology, Portugal

and

Barry O. Hughes
Edinburgh, UK

www.cabi.org

CABI is a trading name of CAB International

CABI Head Office	CABI North American Office
Nosworthy Way	875 Massachusetts Avenue
Wallingford	7th Floor
Oxfordshire OX10 8DE	Cambridge, MA 02139
UK	USA
Tel: +44 (0)1491 832111	Tel: +1 617 395 4056
Fax: +44 (0)1491 833508	Fax: +1 617 354 6875
E-mail: cabi@cabi.org	E-mail: cabi-nao@cabi.org
Website: www.cabi.org	

A catalogue record for this book is available from the British Library, London, UK.

Library of Congress Cataloging-in-Publication Data

Animal welfare / edited by Michael Appleby . . . [et al.]. -- 2nd ed.
      p.; cm.
   ISBN 978-1-84593-659-4 (perm. paper)
1. Animal welfare. I. Appleby, Michael C. II. C.A.B. International. III. Title.
[DNLM: 1. Animal Welfare. HV 4708]

HV4711.A587 2011
179'.3--dc22

2010050005

ISBN-13: 978 1 84593 659 4
CABI South Asia Edition ISBN: 978 1 84593 867 3

Commissioning editor: Sarah Hulbert
Editorial assistant: Alexandra Lainsbury
Production editor: Tracy Head

Typeset by SPi, Pondicherry, India.
Printed and bound in the UK by Cambridge University Press, Cambridge.

Animal Welfare

2nd Edition

Contents

Contributors

Michael C. Appleby, World Society for the Protection of Animals, 5th floor, 222 Gray's Inn Road, London WC1X 8HB, UK. E-mail: michaelappleby@wspa-international.org.

Richard Bennett, School of Agriculture, Policy and Development, University of Reading, Reading RG6 6AR, UK. E-mail: r.m.bennett@reading.ac.uk.

Dominique Blache, School of Animal Biology, Faculty of Natural and Agricultural Sciences, University of Western Australia, 35 Stirling Highway, Crawley, WA 6009, Australia. E-mail: Dominique.Blache@uwa.edu.au.

Alain Boissy, Institut National de la Recherche Agronomique (INRA), UR 1213 Herbivores, Research Centre of Clermont-Ferrand – Theix, F-63122 Saint-Gènes Champanelle, France. E-mail: alain.boissy@clermont.inra.fr.

Xavier Boivin, Institut National de la Recherche Agronomique (INRA), UR 1213 Herbivores, Research Centre of Clermont-Ferrand – Theix, F-63122 Saint-Gènes-Champanelle, France. E-mail: xavier@clermont.inra.fr.

Charlotte C. Burn, The Royal Veterinary College, Hawkshead Lane, North Mymms, Hatfield, Hertfordshire AL9 7TA, UK. E-mail: cburn@rvc.ac.uk.

Andrew Butterworth, Division of Food Animal Science, University of Bristol, Langford House, Bristol BS40 5DU, UK. E-mail: andy.butterworth@bristol.ac.uk.

Michael S. Cockram, Sir James Dunn Animal Welfare Centre, Department of Health Management, Atlantic Veterinary College, University of Prince Edward Island, 550 University Avenue, Charlottetown, PEI C1A 4P3, Canada. E-mail: mcockram@upei.ca.

Richard B. D'Eath, Scottish Agricultural College (SAC), West Mains Road, Edinburgh EH9 3JG, UK. E-mail: Rick.DEath@sac.ac.uk.

Ian J.H. Duncan, Department of Animal and Poultry Science, University of Guelph, Guelph, Ontario N1G 2W1, Canada. E-mail: iduncan@uoguelph.ca.

Paul Flecknell, Institute of Neuroscience, Faculty of Medical Sciences, Newcastle University, Framlington Place, Newcastle upon Tyne NE2 4HH, UK. E-mail: p.a.flecknell@ncl.ac.uk.

David Fraser, Animal Welfare Program, Faculty of Land and Food Systems, and W. Maurice Young Centre for Applied Ethics, University of British Columbia, 2357 Main Mall, Vancouver V6T 1Z4, Canada. E-mail: david.fraser@ubc.ca.

Francisco Galindo, Department of Ethology and Wildlife, Faculty of Veterinary Medicine, National Autonomous University of Mexico (UNAM), Mexico City 04510, Mexico. E-mail: galindof@unam.mx.

Paul H. Hemsworth, Animal Welfare Science Centre, The Melbourne School of Land and Environment, University of Melbourne and the Department of Primary Industries (Victoria), Parkville, Victoria 3010, Australia. E-mail: phh@unimelb.edu.au.

Paul M. Hocking, The Roslin Institute and Royal (Dick) School of Veterinary Studies, University of Edinburgh, Easter Bush, Midlothian EH25 9RG, UK. E-mail: paul.hocking@roslin.ed.ac.uk.

Stella Maris Huertas, Facultad de Veterinaria, Universidad de la República (UDELAR), Lasplaces 1550 CP 11600, Montevideo, Uruguay. E-mail: stellamaris32@hotmail.com.

Barry O. Hughes, 19 Comiston Drive, Edinburgh EH10 5QR, UK. E-mail: barry.hughes@waitrose.com.

William T. Jackson, Flat 43, 4 Sanctuary Street, Borough, London SE1 1EA, UK. E-mail: DrWilliamJackson@ hotmail.co.uk.

Bryan Jones, Animal Behaviour and Welfare Consultant, 110 Blackford Avenue, Edinburgh EH9 3HH, UK. E-mail: bryanjones@abwc.wanadoo.co.uk.

Linda J. Keeling, Department of Animal Environment and Health, Swedish University of Agricultural Sciences, 750 07 Uppsala, Sweden. E-mail: linda.keeling@hmh.slu.se.

Joergen B. Kjaer, Institute for Animal Welfare and Animal Husbandry, Friedrich-Loeffler-Institut (FLI), Dörnbergstrasse 25-27, 29223 Celle, Germany. E-mail: Joergen.Kjaer@fli.bund.de.

Ute Knierim, Department of Farm Animal Behaviour and Husbandry, Universität Kassel, Witzenhausen D-37213, Germany. E-mail: knierim@wiz.uni-kassel.de.

Ilias Kyriazakis, University of Thessaly, PO Box 199, 43100, Karditsa, Greece and School of Agriculture, Food and Rural Development, Newcastle University, Newcastle upon Tyne NE1 7RU, UK. E-mail: ilias. kyriazakis@newcastle.ac.uk.

Shane K. Maloney, Physiology: Biomedical, Biomolecular and Chemical Science, The University of Western Australia, 35 Stirling Highway, Crawley, WA 6009, Australia. E-mail: Shane.Maloney@uwa.edu.au.

Georgia J. Mason, Department of Animal and Poultry Science, University of Guelph, Guelph, Ontario N1G 2W1, Canada. E-mail: gmason@uoguelph.ca.

Joy A. Mench, Department of Animal Science, University of California, Davis, One Shields Avenue, Davis, CA 95616, USA. E-mail: jamench@ucdavis.edu.

Mike Mendl, School of Veterinary Science, University of Bristol, Langford House, Bristol BS40 5DU, UK. E-mail: mike.mendl@bristol.ac.uk.

Ruth C. Newberry, Center for the Study of Animal Well-being, Washington State University, Pullman, WA 99164, USA. E-mail: rnewberry@wsu.edu.

Christine J. Nicol, School of Veterinary Science, University of Bristol, Langford House, Bristol BS40 5DU, UK. E-mail: C.J.Nicol@bristol.ac.uk.

Birte L. Nielsen, Faculty of Agricultural Sciences, Department of Animal Health and Bioscience, Aarhus University, Blichers Allé 20, PO Box 50, DK-8830 Tjele, Denmark. Present address: Institut National de la Recherche Agronomique (INRA), UR1197, Bâtiment 325, F-78350 Jouy-en-Josas, France. E-mail: birte. nielsen@jouy.inra.fr.

I. Anna S. Olsson, Institute for Molecular and Cell Biology, University of Porto, Rua do Campo Alegre, 823 4150-180 Porto, Portugal. E-mail: olsson@ibmc.up.pt.

Edmond A. Pajor, Department of Production Animal Health, Faculty of Veterinary Medicine, University of Calgary, Alberta T2N 4N1, Canada. E-mail: eapajor@ucalgary.ca.

Clare Palmer, Department of Philosophy, Texas A&M University, Bolton Hall, 4237 TAMU, College Station, TX 77843-4237, USA. E-mail: cpalmer@philosophy.tamu.edu.

Jeff Rushen, Agriculture and Agri-Food Canada, PO Box 1000 Agassiz, British Columbia V0M 1A0, Canada. E-mail: Jeff.Rushen@agr.gc.ca.

Peter Sandøe, Danish Centre for Bioethics and Risk Assessment, Faculty of Life Sciences, University of Copenhagen, Rolighedsvej 25, Frederiksberg C 1958, Denmark. E-mail: pes@life.ku.dk.

Marek Špinka, Department of Ethology, Institute of Animal Science, Pratelstvi 815, 104 00 Prague – Uhrineves, Czech Republic. E-mail: spinka@vuzv.cz.

Andreas Steiger, formerly Division of Animal Housing and Welfare, Vetsuisse Faculty, University of Bern, Bremgartenstrasse 109a, Postfach CH 3001, Bern, Switzerland. Current address: Breitenrain 64, CH 3032 Hinterkappelen, Switzerland. E-mail: steiger.andreas@bluewin.ch.

Claudia Terlouw, Institut National de la Recherche Agronomique (INRA), UR 1213 Herbivores, Research Centre of Clermont-Ferrand – Theix, F-63122 Saint-Gènes-Champanelle, France. E-mail: claudia.terlouw@clermont.inra.fr.

Paul Thompson, Michigan State University, 1611 Osborn Road, Lansing, MI 48915-1283, USA. E-mail: thomp649@msu.edu.

Bert Tolkamp, Scottish Agricultural College (SAC), West Mains Road, Edinburgh EH9 3JG, UK. E-mail: Bert.Tolkamp@sac.ac.uk.

Ignacio Viñuela-Fernández, Biological Services, University of Edinburgh, Chancellor's Building, 49 Little France Crescent, Edinburgh EH16 4SB, UK. E-mail: ignacio.vinuela-fernandez@ed.ac.uk.

Natalie K. Waran, Jeanne Marchig International Centre for Animal Welfare Education, Royal (Dick) School of Veterinary Studies, University of Edinburgh, Easter Bush, Midlothian EH25 9RG, UK. E-mail: nwaran@ed.ac.uk.

Daniel M. Weary, Animal Welfare Program, Faculty of Land and Food Systems, University of British Columbia, 2357 Main Mall, Vancouver, BC V6T 1Z4, Canada. E-mail: dan.weary@ubc.ca.

Françoise Wemelsfelder, Scottish Agricultural College (SAC), West Mains Road, Edinburgh EH9 3JG, UK. E-mail: francoise.wemelsfelder@sac.ac.uk.

Nadja Wielebnowski, Chicago Zoological Society, Brookfield Zoo, 3300 Golf Road, Brookfield, IL 60513, USA. E-mail: nadja.wielebnowski@czs.org.

Hanno Würbel, Animal Welfare and Ethology, Justus-Liebig-University of Giessen, 35392 Giessen, Germany. E-mail: hanno.wuerbel@vetmed.uni-giessen.de.

Introduction

MICHAEL C. APPLEBY, JOY A. MENCH, I. ANNA S. OLSSON
AND BARRY O. HUGHES

Concern for animal welfare continues to grow worldwide and is increasingly implemented in measures to safeguard and improve that welfare. In one sense that is surprising, given the burgeoning problems of the day, notably climate change and economic crises. Yet in another sense it is not, because animals are a hugely important part of our world. Achieving sustainability – which can be described as the primary challenge of humanity – involves ecological, economic and ethical responsibility, and all three of these 'Three Es' lead us to consider animals, because animal management has impacts on many other issues, such as hunger, poverty, disease control and environmental protection. In the first edition of this book (Appleby and Hughes, 1997), we quoted Mahatma Gandhi:

> The greatness of a nation and its moral progress can be judged by the way its animals are treated.

That ethical mandate remains, but the complementary ecological and economic imperatives are increasingly recognized, as emphasized by an Expert Group convened by the Food and Agriculture Organization of the United Nations (Fraser *et al.*, 2009, p. 1):

> The welfare of humans and the welfare of animals are closely linked. In many regions, a secure supply of food for people depends on the health and productivity of animals, and these in turn depend on the care and nutrition that animals receive. Many diseases of humans are derived from animals, and the prevention of these animal diseases is important for safeguarding human health. Roughly one billion people, including many of the world's poor, depend directly on animals for income, social status and security as well as food and clothing, and the welfare of their animals is essential for their livelihood. Moreover, positive relations with animals are an important source of comfort, social contact and cultural identification for many people.

There have been many changes in the subject since 1997, with more professionals involved, much new research completed and more academic courses taught around the world. Landmarks include:

- The decision by legislators to phase out battery cages for laying hens in Europe.
- The decision by production companies and retailers to phase out crates for pregnant sows on many farms in North America.
- The initiation of global animal welfare guidelines by the World Organisation for Animal Health (OIE) for farm, laboratory and companion animals.
- Legislation or proposals of legislation in many countries not previously regulating animal welfare matters, including China.
- The proposal for a Universal Declaration on Animal Welfare at the United Nations.

So we invited new contributors to help in updating or replacing previous chapters of the first edition of the book, and ensured breadth of coverage by involving two or more writers on each. The worldwide spread of the subject and the increased emphasis on non-farm animals are reflected in the authorship: we have 46 authors with varied expertise from 15 countries. Two completely new chapters have been added at opposite ends of the scale of human involvement with animals: hands-on practical approaches to assess and improve welfare (Chapter 12) and international issues (Chapter 19). There is a shift in emphasis away from history and towards current review, demonstrated by the fact that more than 60% of references cited have been published since the first edition. We trust that where readers found valuable material in the first edition which has now had to be excluded, the first edition will remain on their bookshelves.

As in the first edition, the terms animal welfare and animal well-being are regarded as synonyms and the former is generally preferred. Welfare is not 'all or nothing', but the state of an animal on a continuum, from poor to good. Humans are, of course, animals, but for convenience the word 'animals' is generally used to refer to non-human animals.

The book has five parts. Part I, Issues, introduces the background and philosophy of the subject. Part II, Problems, then covers problems for animal welfare, starting in Chapter 3 with consideration of the animal's interactions with its environment. This includes the important point that animals are not just passive in these interactions but also active. The following four chapters use categories similar to the UK Farm Animal Welfare Council's Five Freedoms (FAWC, 2009), although there is not a specific chapter on discomfort. Discomfort frequently occurs in association with other problems – or as a lesser form of such problems – and there have been few studies specifically on this topic.

Part III, Assessment, considers various approaches to the assessment of welfare. Welfare is not a simple, unitary variable, so it is not possible to measure it in the same way that mass or length can be measured. However, welfare can be assessed by considering its various aspects and the problems relating to them, as outlined previously. Part IV, Solutions, then emphasizes the ways in which such problems can be ameliorated and welfare enhanced. There is inevitably some overlap between the chapters in the book, for example between those considering problems and solutions, but this should aid complementary coverage.

Finally, Part V, Implementation, is concerned with the implementation of solutions to problems, the means by which solutions are put into practice. It concludes with coverage of how awareness and communication about animal welfare, particularly in relation to major issues such as transport and slaughter, are affecting animals on a global basis. To an even greater extent than might have been predicted in 1997, animal welfare is being addressed on the world stage.

Acknowledgements

We are very grateful to our co-authors and to other colleagues for discussion and assistance during the planning and completion of the book. We would also like to acknowledge the contribution of those authors in the first edition who do not reappear here, because we believe that 'Animal Welfare 1' made a useful contribution to the subject. We hope that 'Animal Welfare 2' will do the same.

Michael Appleby would like to dedicate his involvement in this book to his sons, Duncan and Andrew Appleby, for their enthusiasm and dedication to their vocations, and for helping to restore a similar zeal to their father by way of competition.

Joy Mench dedicates her contribution to her husband, Clive Watson, for his patience and support.

Anna Olsson wishes to dedicate her contribution to her father, Jan Olsson, for inspiring her interest in animal behaviour, and to Margareta Rundgren for encouraging her academic engagement in the topic.

Barry Hughes dedicates his contribution to his wife, Helen, for her patience and constant support. He would also like to recognize the seminal roles played by Ruth Harrison, whose book *Animal Machines* (1964) drew the subject of animal welfare to modern attention, and by David Wood-Gush, who recognized that it must rest on scientific foundations.

References

Appleby, M.C. and Hughes, B.O. (eds) (1997) *Animal Welfare.* CAB International, Wallingford, UK.

FAWC (Farm Animal Welfare Council) (2009) *Five Freedoms.* Available at: http://www.fawc.org.uk/freedoms.htm (accessed July 2010).

Fraser, D., Kharb, R.M., McCrindle, C., Mench, J., Paranhos da Costa, M., Promchan, K., Sundrum, A., Thornber, P., Whittington, P. and Song, W. (2009) *Capacity Building to Implement Good Animal Welfare Practices:* Report of the FAO Expert Meeting, FAO Headquarters (Rome), 30 September–3 October 2008. Available at: ftp://ftp.fao.org/docrep/fao/011/i0483e/i0483e00.pdf (accessed July 2010).

Harrison, R. (1964) *Animal Machines.* Vincent Stuart, London.

1 Animal Ethics

CLARE PALMER AND PETER SANDØE

Abstract

This chapter describes and discusses different views concerning our duties towards animals. First, we explain why it is necessary to engage in thinking about animal ethics and why it is not enough to rely on feelings alone. Secondly, we present and discuss five different kinds of views about the nature of our duties to animals. They are: contractarianism, utilitarianism, the animal rights view, contextual views and a respect for nature view. Finally, we briefly consider whether it is possible to combine elements from the views presented, and how to make up one's mind.

1.1 Introduction: the Need to Give Reasons for One's Ethical Views

This chapter describes and discusses different views on right and wrong in our dealings with animals. What might be right or wrong is not a factual question, and therefore cannot be settled by the same methods as those used in biology and other natural sciences. Some readers of this chapter may even wonder whether moral issues can be settled at all; rather, they may be seen as matters purely of feeling or taste. We shall suggest below that the latter position is problematic.

The primary focus of this book is to discuss *factual* issues relating to the way that animals are used and treated by humans. Until recently, ethics was seen as something that should be kept at arm's length from the fact-oriented science-based study of animal welfare; only once the facts are established would it be appropriate to discuss, from an ethical perspective, where to draw the line between what is acceptable and what is not.

But the link between factual knowledge and sound ethical judgement is not that simple – often the study of the facts relies on tacit ethical judgements. For example, studying the consequences for animal welfare of various ways of housing farm animals proceeds on the assumption that it is acceptable to use animals for food production as long as the animals do not suffer from bad welfare. Also, assessments of animal welfare rely on assumptions regarding what matters, ethically speaking, in our dealings with animals. Is it to avoid pain and other forms of suffering? Is it to give pleasure and other positive emotion? Or is it to allow animals to live natural lives? To be able to deal with such questions and to justify the tacit judgements underlying studies of animal welfare we need not only to know the facts but to engage with and be proficient in ethical thinking.

This chapter focuses on possible answers to basic questions about animal ethics: 'Do animals have moral standing in their own right? And if so, what kind of duties do we have towards them? Does it matter whether animals are wild or domesticated? Do we only have obligations to individual animals or also to species or populations of animals? How should we balance our duties to animals against other kinds of duties?' We do not attempt to answer these questions. Rather, we take a pluralist approach to animal ethics, presenting five diverse ethical positions, each with its own answers. We do not side with any of these views, but we encourage the reader to consider their strong points and why people have been drawn to them. Although we (as authors of this chapter) have our own views, we have tried to present the arguments in a balanced way (though we may not have always succeeded in concealing our sympathies).

However, we do take the view that it is important to adopt a reasoned approach to animal ethics, rather than one based on feelings alone. Reliance on feelings makes for difficulty in entering ethical debates, and in explaining to others why particular attitudes or practices are either problematic or beneficial. For animal professionals to be taken seriously by people who hold different views, they must show that they can comprehend the nature of disagreements about animal ethics. This entails understanding why people make the moral judgements that they do.

But what are moral judgements? They do not seem to be just statements of personal taste. The philosopher James Rachels (1993, p. 10) suggests:

If someone says 'I like coffee,' he does not need to have a reason – he is merely making a statement about himself, and nothing more. There is no such thing as 'rationally defending' one's like or dislike of coffee, and so there is no arguing about it. So long as he is accurately reporting his tastes, what he says must be true. Moreover, there is no implication that anyone should feel the same way; if everyone else in the world hates coffee, it doesn't matter. On the other hand, if someone says that something is *morally wrong*, he does need reasons, and if his reasons are sound, other people must acknowledge their force.

Here, Rachels points out the importance of being able to give *reasons* to justify our ethical views. A consequence of the requirement to give reasons is a requirement of *consistency*: if something provides a moral reason in one case it should also count as a reason in other, similar cases. We can see this process of reasoning by appeal to consistency in the following famous passage, first published in 1789, in which Jeremy Bentham argues that animals ought to be protected by the law (Bentham, 1989, pp. 25–26):

The day *may* come, when the rest of the animal creation may acquire those rights which never could have been withholden from them but by the hand of tyranny. The French have already discovered that the blackness of the skin is no reason why a human being should be abandoned without redress to the caprice of a tormentor. It may come one day to be recognized, that the number of the legs, the villosity of the skin, or the termination of the *os sacrum*, are reasons equally insufficient for abandoning a sensitive being to the same fate. What else is it that should trace the insuperable line? Is it the faculty of reason, or, perhaps, the faculty of discourse? But a full-grown horse or dog is beyond comparison a more rational, as well as a more conversible animal, than an infant of a day, or a week, or even a month, old. But suppose the case were otherwise what would it avail? The question is not, Can they *reason*? nor, Can they *talk*? but, Can they *suffer*?

Bentham asks the reader to consider on what grounds legal rights (for instance, legal protection against torture) are assigned to people. We now accept that factors such as skin colour are irrelevant to the possession of basic legal rights. But what, then,

is the relevant factor? One possible answer, Bentham suggests, is the ability to reason and to use language. So, it might be suggested that reason and language provide a basis for separating humans and animals, and for assigning legal rights to humans, and not to animals. But Bentham raises a number of questions about this kind of response. First, why think that reason and language are relevant to the generation of legal rights (any more than, say, skin colour)? Secondly, some animals *do* appear to have reasoning abilities. Thirdly, some animals are at least as reasonable as some people – as human infants, or those who have severe mental disabilities – so reason and language do not obviously provide the suggested firm dividing line between *all* people and *all* animals.

Bentham makes us consider whether it is possible to argue *consistently* that all humans should be treated in one way, and all animals in another. The first of the five views presented below maintains that we can, indeed, consistently distinguish morally between animals and humans.

1.2 Five Views About Humanity's Duties to Animals

Moral philosophers distinguish a number of types of ethical theory, and in principle any of these might underlie a person's views about the acceptable use of animals. Here, five prominent theoretical positions will be presented: contractarianism, utilitarianism, the animal rights view, contextual approaches, and respect for nature. These have direct implications for the ongoing debate over animal use.

1.2.1 Contractarianism

Why be moral? This is a central question in moral philosophy, and one to which the contractarian gives a straightforward answer: you should be moral because it is in your self-interest. Showing consideration to others is really for your own sake. Moral rules are conventions that best serve the self-interest of all members of society.

Contractarian morality as defined here (the term may also be used of other views that we do not discuss here) applies only to individuals who can 'contract in' to the moral community, so it is important to define who these members are. The philosopher Narveson (1983, p. 56) puts this as follows:

On the contract view of morality, morality is a sort of agreement among rational, independent, self-interested persons, persons who have something to gain from entering into such an agreement ... A major feature of this view of morality is that it explains why we have it and who is party to it. We have it for reasons of long-term self-interest, and parties to it include all and only those who have *both* of the following characteristics: 1) they stand to gain by subscribing to it, at least in the long run, compared with not doing so, and 2) they are *capable* of entering into (and keeping) an agreement ... Given these requirements, it will be clear why animals do not have rights. For there are evident shortcomings on both scores. On the one hand, humans have nothing generally to gain by voluntarily refraining from (for instance) killing animals or 'treating them as mere means'. And on the other, animals cannot generally make agreements with us anyway, even if we wanted to have them do so.

So, on this view, people are dependent on the respect and cooperation of other people. If someone treats fellow humans badly, he or she will be treated badly in return. In contrast, the animal community will not strike back if, for example, some of its members are used in painful experiments. So a person needs only to treat animals well enough for them to be fit for his or her own purposes.

Because non-human animals cannot enter into a contract or an agreement governing future conduct, so they cannot, according to the contractarian view, join the moral community. In this view, any kind of animal use may be permissible inasmuch as it brings human benefits, such as income, desirable food and new medical treatments.

That animals are not members of the moral community does not necessarily mean that their treatment is irrelevant from the contractarian perspective: if people *like* animals, for example, animal use can become important, because it is in a person's interests to get what he or she likes. But the contractarian view of animals is human centred; any protection animals have will always depend on, and be secondary to, human concerns. A further implication is that in this view, it would be likely that levels of protection would differ across animal species. As most people like cats and dogs more than they like rats and mice, causing distress to cats and dogs is likely to turn out to be a more serious problem than causing the same amount of distress to rats and mice.

This contractarian view accords with attitudes to animal treatment that are common in many societies. But it raises many problems. Is causing animals to suffer for a trivial reason really morally unproblematic, if no human being cares? After all, some *humans* – small children, for instance – also

cannot behave in reciprocal ways, or make contracts with other people. Would it be morally acceptable to eat or experiment on them, if other human contractors didn't object? Many people consider that it is immoral *as such* to cause another to suffer for little or no reason, whether one's victim is a human being or an animal. An ethical theory that captures this view is *utilitarianism*.

1.2.2 Utilitarianism

Utilitarian ethical theory provides probably the most well-known approach to animal ethics. Utilitarianism is *consequentialist* in form; that is, only *consequences* are important when making ethical decisions, and we should always aim at bringing about the best possible consequences. But what counts as the best possible consequences? Here, forms of utilitarianism diverge. One leading form – promoted by Jeremy Bentham – maintains that consequences should be measured in terms of *maximizing pleasures* and *minimizing pains*. If animals feel pain and pleasure, then they should be included in our calculations about what to do. Indeed, on this view, there is no reason to privilege human pain over animal pain. Pain is pain, wherever it occurs. So a certain kind of equality is very important; the pains of every being should be taken equally into account, whatever the species of the being concerned.

In recent animal ethics, this view has been most forcefully defended by the philosopher Peter Singer (1989, p. 152). Singer uses the language of *interests* in outlining his position. If a being can suffer, it has an interest in avoiding suffering, and its interests should be treated equally to the similar interests of other beings, whether they are human or not:

> I am urging that we extend to other species the basic principle of equality that most of us recognize should be extended to all members of our own species ... Jeremy Bentham incorporated the essential basis of moral equality into his utilitarian system of ethics in the formula: 'Each to count for one and none for more than one.' In other words, the interests of every being affected by an action are to be taken into account and given the same weight as the like interests of any other being ... The racist violates the principle of equality by giving greater weight to the interests of members of his own race, when there is a clash between their interests and the interests of those of another race. Similarly the speciesist allows the interests of his own species to override the greater interests of members of other species. The pattern is the same in each case.

For the utilitarian, what matters are the interests of those affected by our actions – not the race or the species of the beings who have the interests. The strongest interests should prevail no matter who has them. This view can have radical consequences. Take modern intensive livestock production. Broiler chickens and animals in confined feeding operations often suffer. Some basic interests of these animals are set aside so that production is efficient and meat is cheap. But for affluent individuals, cheap meat is not a basic interest. In a country such as Denmark, ordinary consumers only spend around 10% of their disposable income on food. If such consumers paid 30% or 50% more, and the extra money was used to improve the living conditions of the animals, this would mean an immense increase in welfare and a substantial reduction in suffering, without significantly decreasing human welfare. Therefore, according to the utilitarian view, we ought to make radical changes in the treatment of intensively farmed animals.

Indeed, Singer (1979, p. 152) argues that we should become vegetarians, because consumption of meat and other products from commercially reared animals creates animal suffering that is not outweighed by the human pleasure it generates. (There are also other utilitarian arguments in favour of reducing meat consumption based on the negative consequences for sentient beings from the effects of meat production on the environment, on climate and on resource use.)

However, utilitarianism does *not* endorse a principle that killing animals is wrong. Killing is certainly likely to be morally problematic for two reasons: it may cause suffering, and once a being is killed, it can no longer have positive experiences. So, killing *may* both increase suffering and reduce pleasure in the world; but it need not. As Singer says: 'It is not wrong to rear and kill it [an animal] for food, provided that it lives a pleasant life, and after being killed will be replaced by another animal which will lead a similarly pleasant life and would not have existed if the first animal had not been killed. This means that vegetarianism is not obligatory for those who can obtain meat from animals that they know to have been reared in this manner' (Singer, 1979, p. 153).

This utilitarian view on killing animals may give rise to worries that are animated by Michael Lockwood's (1979, p. 168) troublesome (fictional) case, Disposapup:

Many families, especially ones with young children, find that dogs are an asset when they are still playful puppies … but become an increasing liability as they grow into middle age, with an adult appetite but *sans* youthful allure. Moreover, there is always a problem of what to do with the animal when they go on holiday. It is often inconvenient or even impossible to take the dog with them, whereas friends tend to resent the imposition, and kennels are expensive and unreliable. Let us suppose that, inspired by Singer's article, people were to hit on the idea of having their pets painlessly put down at the start of each holiday (as some pet owners already do), acquiring new ones upon their return. Suppose, indeed, that a company grows up, 'Disposapup Ltd', which rears the animals, house-trains them, supplies them to any willing purchaser, takes them back, exterminates them and supplies replacements, on demand. It is clear, is it not, that there can, for Singer, be absolutely nothing directly wrong with such a practice. Every puppy has, we may assume, an extremely happy, albeit brief, life – and indeed, would not have existed at all but for the practice.

Some people may, after thinking a bit, accept that it is in principle acceptable to replace dogs in this way. However, they will then have to face a further, related problem: the apparent implication that we can painlessly kill humans, if we create new humans to replace them! This difficulty has led some utilitarians – including Singer himself – to make a further distinction, based on the possession of *self-consciousness*. Although it is difficult to define self-consciousness, some utilitarians have maintained that a self-conscious being is one that has a *preference* or a *desire* to go on living, and that the frustration of such basic desires is morally relevant. They argue that (either in addition to, or instead of) minimizing *pain*, we should minimize the *frustration of desires* in the world, especially the frustration of that most basic desire of a self-conscious creature – the desire to go on living.

Yet this does not really seem to solve the problem. For it sounds as though, in principle, it would be morally permissible to painlessly kill a self-conscious human if the human were replaced by another human who lives a better life than the first (to make up for the loss incurred by the killing) and who would not otherwise have existed. Admittedly, the utilitarian may argue that killing humans and animals has very different consequences. Killing humans usually has negative emotional and social effects on survivors in a way that killing animals does not. Then again, to say that the wrong in painlessly killing humans lies in effects on other people may reasonably be regarded as missing the point.

C. Palmer and P. Sandøe

Singer (1999) himself, in his book *Practical Ethics*, argues that the creation of a new desire to live cannot be weighed against the frustration of someone else's desire to go on living – that is, that preferences are not substitutable in this way. This, though, starts to move away from some of the fundamental calculative elements of utilitarianism because it suggests that there are some goods (such as a desire to go on living) that just cannot be compensated for by the creation of more of the same goods (more desires to live). In fact, this kind of view – that some harms are just unacceptable, whatever the ensuing benefits – is much more closely associated with a different approach to animal ethics: a rights view.

1.2.3 The animal rights view

We can think about rights in two senses: *legal* and *moral*. Legal rights are rights that are created and that exist within legal systems. Moral rights, though, are not created by the law, and those who argue from a moral rights-based perspective give a variety of different accounts of the origin of rights. One traditional – though now controversial – claim depends on the intuition that humans *naturally* have rights; to be a rights-holder is just part of what it is to be human.

Claims about rights are particularly important here for two reasons. First is the special force that rights language carries. Although the term 'rights' is sometimes loosely used just to mean having moral status (it is in this loose sense that Singer is sometimes called the 'father of animal rights'), philosophers generally understand rights in a more restricted sense. In this restricted sense, to say that a being has moral rights is to make a very strong claim that those rights should be protected or promoted. Indeed, sometimes possessing a right is described as having a 'trump card' – the kind of claim that wins over any competing claims. Second is the fact that some philosophers have extended the idea of moral rights beyond humans, arguing that animals also have moral rights. After all, such philosophers argue, it is not just being biologically human – a member of the species *Homo sapiens* – that gives a being rights. Rather, it must be the possession of particular *capacities* (such as sentience or self-awareness) that one has as a species member that underpins the rights of humans. But if it is *capacities*, not genes, on which rights possession is based, then perhaps some *animals* share the

relevant basic capacities, and should be thought of as having rights? It is this view that is adopted by animal rights advocates, most prominently by the philosopher Tom Regan (1984) in *The Case for Animal Rights*.

Regan (2007, p. 209) argues that all 'experiencing subjects of a life' should be thought of as possessing moral rights. An experiencing subject of a life is 'a conscious creature having an individual welfare that has importance to it whatever its usefulness to others'. Such beings 'want and prefer things, believe and feel things, recall and expect things'. They can undergo pleasure and pain, experience satisfaction and frustration, and have a sense of themselves as beings that persist over time. Such beings have, on Regan's account, *inherent value* of their own, based on their nature and capacities. They are not just instruments for someone else's use and benefit. Inherent value, Regan maintains, cannot be traded off, factored into calculations about consequences, or replaced. Creatures that possess it – and Regan argues that all mentally normal adult mammals fall into this category – have basic moral rights, including the right to life and to liberty. The evidence that infant mammals, birds, fish, reptiles and some invertebrates are experiencing subjects of a life is less clear. However, as we cannot be sure about their inner worlds, Regan (1984) argues that we should give them the benefit of the doubt in moral decision making, as they too may have inherent value.

Regan explicitly sets up his rights view in opposition to utilitarianism. Utilitarians, he maintains, are fundamentally mistaken in thinking that harming some beings to bring about good consequences for others is morally acceptable. On the contrary, he argues (Regan, 1985, p. 22) 'That would be to sanction the disrespectful treatment of the individual in the name of the social good, something the rights view will not – categorically will not – ever allow'. So utilitarian and rights views will, in some cases, diverge in practice – and they will always diverge in principle. Regan (2007, p. 210), for instance, comments on animal experimentation and commercial animal agriculture:

> The rights view is categorically abolitionist. Lab animals are not our tasters; we are not their kings. Because these animals are treated routinely, systematically as if their value were reducible to their usefulness to others, they are routinely, systematically treated with lack of respect, and thus are their rights routinely, systematically violated … As for

commercial animal agriculture, the rights view takes a similar abolitionist position. The fundamental moral wrong here is not that animals are kept in stressful close confinement or in isolation, or that their pain and suffering, their needs and preferences are ignored or discounted. All these *are* wrong, of course, but they are not the fundamental wrong. They are symptoms and effects of the deeper, systematic wrong that allows these animals to be viewed and treated as lacking independent value, as resources for us – as, indeed, a renewable resource.

Of course, sometimes the judgements of a utilitarian and a rights theorist about particular cases of experimentation or commercial animal agriculture will coincide; some animal experimentation, and most commercial animal agriculture as currently practised, should consistently be condemned by both. But the underlying reasons for these judgements differ. A utilitarian is primarily concerned about suffering or desire frustration in cases where the benefits derived do not seem to outweigh the costs. In contrast, a rights theorist is concerned about failing to respect animals' inherent value, and violating animals' rights, irrespective of potential good consequences. From the rights perspective, the utilitarian idea that the interest of an animal in continuing to live may be outweighed by conflicting interests, i.e. the combined interests of the future animal which will replace it and human interests in animal production, is morally abhorrent. So a rights view is abolitionist, whereas a utilitarian will ask questions about the benefits of any particular practice involving suffering to animals before coming to a view about its moral permissibility.

On Regan's rights view, killing – even where it is painless and another being is created – harms the being that is killed. Regan (1984) describes killing as harm by *deprivation* – an animal that is killed is deprived of all the goods that the rest of its life would otherwise have contained, even if its death is sudden and unanticipated. Indeed, to kill an experiencing subject of a life is to display ultimate disrespect, by destroying the animal's inherent value, and thus violating its rights.

An animal rights view such as Regan's – though providing a plausible alternative to utilitarianism – generates its own difficulties. One problem concerns how to handle rights conflicts. For example, it may be difficult to combine respect for the rights of all rodents with the aim of securing human health and welfare. If these 'pests' are not 'controlled' they

may pose a threat because they eat our food, and because they spread disease. It seems to be either them or us. What has the rights view to say about this?

Regan certainly thinks that we are entitled to self-defence. If I am attacked by a bear, for instance, I may kill the bear because this is a case of 'my life or the bear's'; and, Regan might suggest, there are probably ways of avoiding conflict over food and disease, by more systematically and efficiently separating rodents from our food supplies. But still, it is possible that if human lives really were at stake from threats presented by rodents to our basic resources, killing them would be morally permissible, even on a rights view, as a form of self-defence.

While humans and animals may sometimes be in *conflict* over resources, on other occasions humans deliberately *share* their resources with particular favoured animals. After all, some animals live, by invitation, alongside us, as family members. More than a third of US households include at least one dog; virtually a third include at least one cat (AVMA, 2007). What position does a rights view take with respect to pets?

Actually, there are different answers to this question. Some advocates of animal rights – such as Gary Francione (2000) – argue that pet keeping depends on the idea that pets are human property. As beings with rights should, most fundamentally, not be treated as human property, we should not keep pets. But on Regan's account, it is plausible that, in principle at least, one could live with a pet (perhaps 'companion animal' would be a better term here) without infringing its rights. After all, a pet is not necessarily being treated 'merely as a means' to the ends of the person with whom it is living, because one could live alongside an animal while respecting its inherent value.

However, in practice, pet keeping presents a number of challenges to a rights view. Animals kept as pets are frequently confined against their will, and often against their interests. Breeding practices may infringe on animal liberty, and the creation of some pedigree breeds generates animals unable to live healthy lives (albeit in shapes and sizes that are very appealing to people). Spaying and castration foreclose the sexual and reproductive freedom of animals and plausibly, in this view, constitute rights infringements (we would certainly think this in the human case, but, of course, it is very difficult to know what these freedoms might mean to animals). Many pet animals are fed carnivorous diets made out of *other* animals whose rights have been infringed (it is unclear whether all pets can flourish on a wholly vegetarian diet). The freedom to roam of some pets – cats in particular – may devastate wildlife; and although (as a cat is not a moral agent, as noted below) this does not raise direct issues of the infringements of the rights of individual wild animals, it is difficult to deny that human pet keepers are at least indirectly responsible for the predation of their pets. Yet confining a cat indoors may, in a rights view, deprive it of its right to liberty. For all these reasons, even though a rights view such as Regan's does not necessarily condemn the keeping of pets in principle, it is likely to be at least an uncomfortable fit with many common pet-keeping practices.

A rights view, then, allows for self-defence and, in principle, allows us to live alongside animals, provided that their rights are fully respected. What, though, about those animals that neither threaten us, nor live in our homes – wild animals that live their lives independently of us? What are our duties towards them? This issue is often thought to be problematic for rights views (and even more so for utilitarian views). For instance, does a rights view imply that humans should defend the rights of wild animals against wild animal predators? Should utilitarians promote wild animal rescue services, in case of storm or wildfire, to minimize suffering?

Regan (1984) argues that there is no duty to protect animals against threats from other animals. For, he maintains, rights only hold against *moral agents* – that is, those beings that can recognize and respect rights. Antelope do not have rights against lions, because lions are not moral agents; lions do not threaten their rights. So humans do not have to act to protect antelope against lions (though they should protect them against other *people*, because people are moral agents, and do threaten their rights). In Regan's view, humans also do not have duties to rescue wild animals, or at least not on the basis of their rights. Regan suggests that rights provide animals with protections against particular kinds of *interference* from moral agents (inflicting pain, constraining liberty); this does not mean that humans have duties to *assist* in cases where harm was not caused by a moral agent.

For utilitarians, though, this issue is more difficult, as utilitarians are concerned to minimize suffering or desire frustration, *whatever* its cause. This does sound like a mandate to act in the wild. One utilitarian response is to maintain that acting in the wild to relieve wild animal suffering is likely – over time – to cause more suffering than staying out of it, because such actions might disturb natural systems. But neither view here is unproblematic; Regan's rights view seems to have *too little* to say about any kind of assistance (including to distant suffering *people*), while a utilitarian view may imply *too much* human action in the wild.

To summarize so far: there is genuine moral disagreement between a utilitarian and a rights view in relation to animals, but there are also points on which both agree. For instance, both maintain that the *capacities* of individual animals are of primary importance in moral decision making (even though they differ on which capacities, exactly, are relevant). In order to decide how to act, we need to ask questions such as: Does this being have the capacity to feel pain? Does it have the relevant capacities to be an experiencing subject of a life? If the answer is positive, then – providing, in the case of utilitarianism, we have some idea of the possible consequences of our actions – we have almost complete guidance as to what to do. However, in contextual views, of the kind we shall now consider, this capacities-oriented approach is too narrow, and ignores a range of *other* important factors that are relevant to our ethical duties towards animals.

1.2.4 Contextual approaches

Several different positions can be grouped together as *contextual* approaches to animal ethics. These positions share the view that although animal capacities are not irrelevant to moral decision making, and may indeed be very important to it, these capacities are not enough, in themselves, to give comprehensive guidance about how we should act. Advocates of contextual views argue that the capacities focused on in utilitarian and rights views are very narrowly understood; that utilitarian and rights views give no real weight to the different *relations* that humans have with animals; that they have no substantial place for human emotions such as empathy; and that they barely discuss the special obligations that humans may have towards particular animals, based on prior commitments to them or prior interactions with them. We shall consider just two such contextual approaches here.

One kind of contextual approach emphasizes the role of what are sometimes called the moral emotions – such as sympathy, empathy and care – in all of our transactions with others, including animals. This view – as the philosopher of care Nel Noddings (1984, p. 149) maintains – certainly includes responding to animal (and human) suffering, but interprets this somewhat differently to utilitarian and rights views:

> Pain crosses the line between the species over a wide range. When a creature writhes or groans or pants frantically, we feel a sympathetic twinge in response to its manifestation of pain. With respect to this feeling, this pain, there does seem to be a transfer that arouses in us the induced feeling, 'I must do something'. Or, of course, the 'I must' may present itself negatively in the form, 'I must not do this thing'. The desire to prevent or relieve pain is a natural element of caring, and we betray our ethical selves when we ignore it or concoct rationalizations to act in defiance of it.

According to an ethics of care, what is wrong with causing suffering to animals is not primarily that suffering is increased (utilitarianism) or that it violates rights (an animal rights view) but that it demonstrates a lack of care, or inappropriate emotional response, in the person concerned. A view of this kind provides a basis for differentiating between what is owed to animals in different contexts that is not easily available to a rights or a utilitarian view. So, for instance, people usually develop deep emotional relations with their pets, making them sensitive to that particular animal's well-being. Because people care for their pets, they protect them from external threats, give them veterinary treatment, feed them, and – as we have already noted – frequently understand them to be 'family members'. This emotional closeness, though, does not (usually) extend to wild animals; where bonds of care and sympathy are much weaker. So on this account, even though two animals might have similar capacities, if human emotional relations to the animals differ, their ethical responsibilities will differ too.

An ethics of care, in this form, is controversial – in the human, as well as in the animal case. Critics have pointed out that this view implies that we have no, or very few, duties towards distant strangers (both in the human and the non-human case) because we do not know them personally, and so have not developed caring relations with them. In the animal case, this might suggest that, providing we made sure we never encountered animals heading for the slaughterhouse, we could eat them without any ethical concern. Many ethicists of care are unhappy with conclusions of this kind, and

have argued more recently that we can feel sympathy for the suffering of those we never encounter; that sympathy can be extended to strangers – including distant animals. This may not generate the intensity of obligation that we have towards those to whom we are close, but because even distant suffering generates responses of care and sympathy, distant sufferers are, none the less, of moral concern.

Alternative contextual views, however, shift the focus from human *emotions* to human *relations*, interpreting 'relations' to include much more than human emotional responses to particular animals. On this approach, for instance, humans have quite different relations – and hence moral obligations – to wild animals than to domestic ones. This is not primarily due to differing human emotional responses (though these may play some part). Rather, it is because humans are responsible for the very existence of domestic animals (unlike wild ones) and, additionally, through selective breeding, for their natures (frequently natures that render these animals dependent and vulnerable, in ways that wild animals are not). After all, we think that those who bring dependent and vulnerable human children into existence have a special responsibility to protect and provide for them; on this account, the same reasoning can be applied to animals. Alongside the creation of dependence and vulnerability through breeding and captivity, on this view other human actions also generate special obligations towards some animals. Suppose a population of animals has been displaced by human development, and is struggling to survive. Because humans have harmed these animals and increased their vulnerability, there is a *special obligation* to assist them. This kind of special obligation would not exist towards animals struggling owing to (say) natural drought or heavy snowfall. In summary then, this relational approach takes into account a variety of different factors, in particular human interactions with and causal responsibility for the situations of particular animals, before coming to a judgement about what obligations there might be in any particular context.

Of course, complications are generated by this view. One question concerns the way in which causal links are supposed to work here. Suppose someone dumps some kittens that they cannot sell. If I come across them, am I personally responsible for assisting them? The kittens were bred by a human and dumped by a human, of course; but am I,

C. Palmer and P. Sandøe

on this view, responsible for all the ills committed by other people? If human relations to animals are to be thought of as morally significant in the sense that this view implies, then a complex account of how to think through individual and collective moral responsibility for the actions of other individuals, and of groups of which one is merely one member, is required.

So while contextualist positions accept that the possession of particular capacities – such as the capacity to feel pain – provide a basis for moral status, unlike utilitarian and rights views they maintain that we need to know more than this before deciding how to act. However, utilitarian, rights and contextual views do all share one thing in common: a focus on animals as *individuals*. It is the capacities of individual animals, and/or our relations to individual animals, that provide guidance as to how we should treat them. An alternative approach, though, shifts the focus away from individual animals towards protecting what is understood to be *natural* and, in particular, to concern about groups perceived to be natural such as wild *species*.

1.2.5 Respect for nature

Moral concern about animals need not be based around the suffering, rights or well-being of particular individuals; it is also often expressed about the extinction of species. Indeed, such concern often extends to including the extinction of species of plants and insects, where suffering and the possession of rights is not an issue. The worry here is about the loss of a particular natural *form*, the species, that is manifested in each of the individual species members. Here, the value of animals lies in their membership of a valued species, not in their individual capacities.

Although some species are obviously useful or potentially useful to people (for instance as resources for medical research) and others are of high symbolic value (such as polar bears) this is not all that is at stake here. Some ethicists argue that a species has *value in itself*, and therefore should be protected (both from extinction, and from some kinds of 'meddling' in its genetic integrity). This kind of value – as Rolston (1989, pp. 252–255) maintains below – falls outside the 'individualist' frameworks that we have so far been considering, and is thus rejected by them. Rolston, for instance, makes this case:

Many will be uncomfortable with the view that we can have duties to a collection ... Singer asserts, 'Species as such are not conscious entities and so do not have interests above and beyond the interests of the individual animals that are members of the species.' Regan maintains, 'The rights view is a view about the moral rights of individuals. Species are not individuals, and the rights view does not recognize the moral rights of species to anything, including survival' ... But duties to a species are not duties to a class or a category, not to an aggregation of sentient interests, but to a lifeline. An ethic about species needs to see how the species *is* a bigger event than individual interests or sentience ... Thinking this way, the life the individual has is something passing through the individual as much as something it intrinsically possesses. The individual is subordinate to the species, not the other way round. The genetic set, in which is coded the *telos*, is as evidently a 'property' of the species as of the individual ... Defending a form of life, resisting death, regeneration that maintains a normative identity over time – all this is as true of species as of individuals. So what prevents duties arising at that level? The appropriate survival unit is the appropriate level of moral concern.

In Rolston's view, the extinction of a species is deplorable not just because of its consequences for the welfare of humans or animals but as something that is *in itself* bad. If the blue whale becomes extinct this is not, after all, a problem for animal welfare – individual whales, for example, do not suffer from becoming extinct. For Rolston, it reverses the correct order of things to say that loss of a species is bad because it is regretted by humans; rather, humans have duties to protect species – and regret their loss – because species are themselves valuable. Why is this? Rolston argues that a species is, in itself, rather like a living individual – a lifeline. A species comes into being, reproduces itself, and will eventually die, like any other living individual. Indeed, to push Rolston's argument a bit further, we can even think of a species as having interests distinct from those of its members. So, for instance, we could keep all the remaining individuals of a particular species in captivity in a zoo for captive breeding; this might produce welfare problems for all those individuals, but it might, none the less, be good for the *species*, allowing it to continue and perhaps to flourish in the future.

If the focus of this view is respect for nature, and in particular for natural species, what do those who hold such a view think about *domesticated* animals? The genetic make-up of domesticated animals has for countless generations been influenced, shaped and, more recently, in some cases at least, controlled by human beings. Animals created in these ways are not members of 'natural' species in the sense that Rolston describes. Indeed, many members of domesticated species would find life extremely difficult were they to be taken out of human-created environments and placed into natural ones; we might describe such animals as being artefactual as much as natural.

For this reason, some environmental ethicists regard domesticated animals as being less valuable than wild ones. The environmental ethicist J. Baird Callicott (1980, p. 53) has argued that domesticated animals are 'living artefacts … they constitute yet another extension of the works of man into the ecosystem'. Unlike wild animals, they are bred to 'docility, tractability, stupidity and dependency'. Indeed, not only do they lack the value of wild naturalness, they also threaten the very beings that do manifest such values, by overrunning their habitats.

One response to views of this kind has been to argue that the wild/domestic divide on which this position depends just cannot be so clearly divined. Domesticated animals are still related to wild animals; indeed, in Europe, some animals are now being bred for 'de-domestication', to fit back into natural landscapes as now extinct wild animal species, such as aurochs, once did. Some 'wild' animal species such as squirrels – wild in the sense that no one has tried to domesticate them – have evolved alongside humans over generations. Even the wildest of animal species are now likely to be shaped by human impacts, for instance by agricultural expansion and by climate change – impacts that will only intensify in the future. Other critics of this view – such as Stephen Budiansky (1999) – argue that domestication should be regarded much more positively than Callicott's view suggests, as a kind of 'win–win' contract, of benefit both to humans and to animal species themselves.

A further concern here may be with the kinds of processes that humans use in order to *change* animal species. Someone with a 'respect for nature' view might regard all human attempts to change animals as morally impermissible, creating artefacts that threaten the flourishing of wild animal species. However, others maintain that some processes by which humans adapt animals are more 'natural' and thus more morally acceptable than others. So, for instance, it is sometimes argued that

slow, selective breeding, as practised by farmers across the centuries, merely accentuates and guides changes that could have happened naturally. These traditional practices, it is maintained, are rather different from the fast-changing modern 'engineering' of animals by genetic modification and intensive breeding programmes, in which animals are adapted to suit narrow human purposes. In this view, selective breeding and domestication, in the traditional sense, are understood to be relatively natural and so morally permissible, while genetic modification and intensive breeding programmes are understood to be unnatural and morally impermissible.

Of course – as with the other four positions we have considered – positions based on respect for nature have been widely challenged. One challenge focuses on the difficulty of identifying what is and is not natural, given human embeddedness in and entanglement with nature. Another challenge asks why we should think that what is natural has some special value anyway. From a perspective centred on animal welfare, for instance, if a highly artificial process such as genetic modification could create animals resistant to certain painful diseases, then it would seem to be morally desirable because such a process could reduce animal suffering. Again, then, we see just how far different approaches to animal ethics can produce widely divergent views on how we should treat animals.

1.3 Combining Views and Decision Making

In this chapter, we have outlined a number of different approaches to animal ethics: contractarian views, utilitarian views, rights views, contextual views and views concerned for the protection of natural species. These different approaches certainly seem to give divergent answers to the questions raised at the beginning of the chapter: 'Do animals have moral standing in their own right? And if so, what kind of duties do we have towards them?' Must we, then, choose to adopt one of these approaches (or some other approach altogether) and reject all the others? Or are there ways of combining attractive elements from different approaches to create some kind of a 'hybrid view'?

Some kinds of hybridity do seem plausible. It is perfectly possible to be morally concerned about species extinction while also thinking that the well-being of individual species members is of moral significance. Also, frequently, both species-oriented and individualist views will recommend the same policies – protecting a species will usually protect individual members. But on occasions these two values will come apart – for instance, in the case where to protect an endangered species, sentient animals of another species would have to be culled. In cases of conflict of this kind, someone who held this kind of hybrid view would have to decide which ethical approach had priority.

Another kind of hybrid view might combine elements of a rights position with a kind of contextual view. So, for instance, one might argue that the capacities of animals give them basic rights protections. This may not tell us *everything* about our moral responsibilities towards all animals, though; our relations with particular animals (such as pets) might give us additional special obligations that we owe only to a few animals and not (as with respect for rights) to animals in general.

The ground looks rather fertile, then, for possible hybridization, especially if one view is taken as 'baseline' or given priority in a situation where the different approaches may conflict. But not all views hybridize well. The utilitarian aim, for best consequences, is in clear tension with the claim of rights theorists that there are some actions we should never carry out, however good the consequences. But even here some form of hybridization might work. For instance, it might be argued that there are certain things that may never be done to animals, no matter how beneficial the possible consequences – perhaps causing an animal to experience intense and unrelenting suffering. But – on this hybrid view – as long as we abstain from these absolutely impermissible actions, we can otherwise reason as a utilitarian would. So, for example, painless killing of animals or causing them mild distress or inconvenience may be allowed if sufficiently good consequences follow, even though severe and unrelenting pain should never be inflicted.

1.4 Conclusions

- Ethical decision making relating to animals is problematic, highly contested, and requires reasoned discussion. A number of competing positions exist. We have outlined five leading positions here:
  - The *contractarian view* only considers human interests. Individual humans belong to a human-only moral contract that benefits the individual human concerned, along with other collaborating fellow humans.
  - According to the *utilitarian view*, we should consider not just the interests of all affected humans, but also of all affected sentient beings. The aim should be to produce the best balance of good over bad, by maximizing the fulfilment of sentient interests.
  - In the *animal rights view* animals that are sentient and have high-level cognitive abilities have rights to life, liberty and respectful treatment. The rights of individuals cannot be overridden in order to benefit others.
  - In *contextual views*, a variety of factors, as well as the capacities of animals, are of moral significance; these include the emotional bonds between humans and animals and the special commitments that humans have made to particular animals.
  - Finally, in the *respect for nature view*, the protection of natural species, genetic integrity and some kinds of natural processes are thought to be of moral significance, and animals are valued as tokens of their species.
- These different theoretical approaches to animal ethics should not be understood as rigid and uncompromising. There are, for instance, ways in which they can hybridize with one another.
- These approaches can perhaps be thought of as lenses, each focusing on a different aspect of what might be ethically troubling about the treatment of animals by humans – their suffering, their instrumentalization, their vulnerability and dependence, their natural form.

Acknowledgements

The authors would like to acknowledge the efforts of Roger Crisp and Nils Holtug who were co-authors of the version of this chapter published in the first edition of the present book. Also, we would like to thank Mike Appleby and Anna Olsson for comments on an earlier draft of this chapter.

References

AVMA (American Veterinary Medical Association) (2007) U.S. pet ownership 2007. Available at: http://www. avma.org/reference/marketstats/ownership.asp (accessed 13 December 2010).

Bentham, J. ([1789] 1989) A utilitarian view. In: Regan, T. and Singer, P. (eds) *Animal Rights and Human Obligations*. Prentice Hall, Englewood Cliffs, New Jersey, pp. 25–26.

Budiansky, S. (1999) *The Covenant of the Wild*. Yale University Press, New Haven, Connecticut.

Callicott, J. B. (1980) Animal liberation: a triangular affair. *Environmental Ethics* 2, 311–338.

Francione, G. (2000) *Introduction to Animal Rights: Your Child or Your Dog?* Temple University Press, Philadelphia, Pennsylvania.

Lockwood, M. (1979) Singer on killing and the preference for life. *Inquiry* 22, 157–171.

Narveson, J. (1983) Animal rights revisited. In: Miller, H.B. and Williams, W.H. (eds) *Ethics and Animals*. Humana Press, Clifton, New Jersey, pp. 45–60.

Noddings, N. (1984) *Caring*. University of California Press, Berkeley, California.

Rachels, J. (1993) *The Elements of Moral Philosophy*, 2nd edn. McGraw Hill, New York.

Regan, T. (1984) *The Case for Animal Rights*. Routledge, London.

Regan, T. (1985) The case for animal rights. In: Singer, P. (ed.) *In Defense of Animals*. Blackwell, Oxford, pp. 13–26.

Regan, T. (2007) The case for animal rights. In: Lafollette, H. (ed.) *Ethics in Practice*, 3rd edn. Blackwell, Malden, Massachusetts, pp. 205–211.

Rolston, H. (1989) *Environmental Ethics: Duties To and Values In the Natural World*. Temple University Press, Philadelphia, Pennsylvania.

Singer, P. (1979) Killing humans and killing animals. *Inquiry* 22, 145–156.

Singer, P. (1989) All animals are equal. In: Regan, T. and Singer, P. (eds) *Animal Rights and Human Obligations*. Prentice Hall, Englewood Cliffs, New Jersey, pp. 148–162.

Singer, P. (1999) *Practical Ethics*, 2nd edn. Cambridge University Press, Cambridge, UK.

C. Palmer and P. Sandøe

2 Understanding Animal Welfare

LINDA J. KEELING, JEFF RUSHEN AND IAN J.H. DUNCAN

Abstract

In this chapter, we introduce some of the principal issues that have arisen in relation to scientific approaches to animal welfare, most of which are treated in more detail later in the book. Much of the apparent disagreement between people about animal welfare stems from mixing up scientific questions about the actual welfare of animals and ethical questions about how we ought to treat and care for animals. This chapter does not deal with these ethical questions but focuses on the science of animal welfare and on the different approaches taken in the past to understand what animal welfare is and how to assess it. We nevertheless put animal welfare in its social context by presenting a brief history and referring to some key events that have shaped the development of animal welfare science. We discuss the links between animal welfare and animal health, and the links between animal welfare and natural behaviour. Although these links seem self-evident, and the assessment of welfare based on these approaches often leads to the same conclusions, there are many considerations that are not usually taken into account. Examples include how difficult it is to define good health, or to compare the degree of suffering experienced by animals with different types of disease, illness and injury. Furthermore, even if behaviour has evolved because it contributes to the survival of animals under natural conditions, not all natural behaviour is desirable. Ultimately, we argue that as concern about the welfare of animals stems from the fact that they are sentient (capable of feelings), then feelings have to be a major part, perhaps the central part, of their welfare.

2.1 Introduction

Most people have quite strong views on how animals ought to be treated, but many appear not to appreciate the complexity of the issue. Admittedly, the area of animal welfare and animal protection is a complex one, but in reality it is no more (or less) difficult to understand than many other areas where science and society meet, even though the debate may be more emotionally loaded than usual. The term 'animal welfare' is used widely, and often loosely, in the media and by society in general, often in the context of discussion on politics, economics, trade, food production and generally what society 'ought' to do to protect animals. However, this chapter will focus on understanding the science of animal welfare and so we use the term 'animal welfare' in the context of animal welfare science.

In scientific use, the term 'animal welfare' refers to the actual state of an animal rather than to the ethical obligations that people have to care for animals. Thus, we consider 'welfare as a characteristic of an animal, not something that is given to it' (Broom, 1996) and that the 'term animal welfare describes the quality of an animal's life as it is experienced by an individual animal' (Bracke et al., 1999). Therefore, an animal in the wild suffering from disease or malnutrition can be said to have poor welfare even if people were not responsible for this and have no ethical obligations to care for that animal.

The scientific concept of animal welfare is still developing. It arose out of attempts by scientists to deal with the ethical concerns about the way that we treat animals. Thus, to understand the concept of animal welfare, we need to understand the concerns that people have about animals. David Fraser and his colleagues (Fraser et al., 1997; Fraser, 2008) have examined the most common statements made about farm animal welfare in an effort to clarify the most important concerns that people have. Three broad types of questions are typically raised about the effects of modern farming systems upon animal welfare: (i) Is the animal happy or is it suffering from pain or other undesirable emotions? (i.e. concerns about the animal's feelings or emotions); (ii) Is the animal healthy and producing well? (i.e. concerns about the animal's ability to function biologically); and (iii) Is the animal able to perform its normal behaviour and live a reasonably natural life? (i.e. concerns about the naturalness of how the animal lives). These three aspects of animal welfare are usually included in official definitions. For example, the World Organisation for Animal Health (OIE) defines an animal as having good welfare if it is: 'healthy, comfortable, well nourished, safe, able to express innate behaviour, and … is not suffering from unpleasant states such as pain, fear, and distress' (OIE, 2010). These aspects are important not only for farm animals, but also encompass most of the concerns about the welfare of zoo, laboratory and companion animals; they are examined in more detail in Sections 2.3, 2.4 and 2.5.

In reading the various discussions of animal welfare, one is often left with the impression that there is little agreement as to its best definition (Keeling, 2004). Much of the disagreement between people about the welfare of animals arises from the fact that different stakeholders tend to put emphasis on different aspects of animal welfare. For example, veterinarians and farmers generally focus on disease, injury, poor growth rates and reproductive problems. Medical researchers using animals concentrate on hygienic conditions and freedom from disease. In contrast, many members of the public focus upon whether the animals are suffering from unpleasant feelings, such as pain, fear or hunger. For people interested in animal welfare, especially consumers of organic products, a key concern is whether the animal is able to live a relatively natural life and express its natural behaviour. Some wrongly link a happy and healthy animal with a tasty and nutritious animal product. These different points of view often give the impression that there is no generally accepted definition of animal welfare (Fig. 2.1).

The recognition that animal welfare is multifaceted, with links to animal health, animal feelings and behaviour, has led to definitions of animal welfare that basically list the conditions under which animal welfare is good or bad. The best known example of such a definition is provided by

Fig. 2.1. Mistaking a part for the whole. Most readers are probably familiar with the parable of the blind men and the elephant. One, feeling only the leg, states that the elephant is like a tree; another feeling only the tail, believes it more like a piece of rope. This is the classic case of people mistaking a part of something for the whole. We suggest that much of the debate about animal welfare is equivalent, with different groups of people focusing on different aspects, and being unable to appreciate fully the multifaceted nature of animal welfare. Image from Miscellaneous Items in High Demand Collection, Prints & Photographs Division, Library of Congress LC-USZ62-134246.

the Five Freedoms, which define good welfare as freedom: (i) from hunger and thirst; (ii) from discomfort; (iii) from pain, injury or disease; (iv) from fear and distress; and (v) to express normal behaviour (FAWC, 2009). While such definitions have the advantage of being inclusive, they often lead to trade-offs between the different threats and opportunities to good animal welfare. For example, ample research has now shown that the traditional battery cage for laying hens frustrates the hen's motivation to perform nest-building behaviour, thus breaching the 'freedom' to perform normal behaviour. Some alternative systems that do allow this behaviour though, such as non-cage or free-range systems, tend to be associated with poorer health and increased injurious behaviour, such as feather pecking and cannibalism, thus breaching the freedom from pain, injury and disease (Blokhuis *et al.*, 2007). Without a concept of animal welfare that provides a 'common currency' it is very difficult to assess when and where a hen's overall welfare is best safeguarded. The absence of a scientific concept that allows us to balance the different threats to animal welfare is one of the most important weaknesses in animal welfare science at present.

However, we must not overestimate the differences between the different interested parties in what they consider to be good animal welfare. In practice, when having to make real-life decisions about whether or not animal welfare is good or bad, people are capable of coming to some consensus. For example, Whay *et al.* (2003) found considerable agreement between 'experts' (veterinarians and behavioural scientists) as to which of two dairy farms had the higher level of welfare when they were presented with a variety of information about the state of the cows on the two farms. There are also many other examples that show how consensus can be achieved on what constitutes good or poor welfare by using expert consultation methods (Anonymous, 2001; Hegelund and Sorenson, 2007; Leach *et al.*, 2008). The results show that with appropriate techniques, experts are capable of integrating a variety of information about the animals and their housing and management on any given farm to achieve a fair consensus on the level of welfare on that farm. Such techniques have heuristic value in making decisions about animal welfare, even in the absence of a clear definition of welfare. Of course there is always the possibility that all the 'experts' could be wrong, but this most often occurs when there is a lack of information rather than a difference of opinion about the meaning of animal welfare.

2.2 A Brief History of Animal Welfare

Concern about the welfare of animals is nothing new – pet owners, zoo managers, farmers and veterinarians have always been concerned about the condition of animals in their care and have tried to ensure that they are healthy and well nourished. There is little doubt that good health is an essential component of good welfare. However, the field of research that has become known as 'animal welfare science' stems largely from the public's concern about some modern farming techniques, especially the use of intensive husbandry (Fig. 2.2), the increased use of animals in medical research involving painful experiments and housing conditions that are designed more for the convenience of the human beings using or caring for the animals than for the animals themselves.

An important event that started this recent interest in farm animal welfare was the publication of *Animal Machines* by Ruth Harrison (1964) (Fig. 2.3). This book described many of the changes that had happened in agriculture over the previous decades; changes that were seen as 'unnatural' and contrary to what the public often assumed to be the norm for farming. It introduced the term 'factory farm' and focused on intensive farming practices, in particular the use of battery cages for laying hens, gestation crates for pigs and methods of raising veal calves (Fig. 2.4). These particular farming issues have tended to dominate both research and animal welfare legislation. Somewhat later, philosophers, such as Peter Singer (1975) and Tom Regan (1985), questioned this exploitation of animals for our own purposes. The issues they raised, especially the degree of suffering caused to animals by farming and research practices, and the apparently 'unnatural' conditions under which animals are kept, have shaped the legislative and other approaches that various countries have adopted in dealing with the issue of animal welfare.

In response to the writings of Ruth Harrison, the Brambell Committee was established by the UK government to 'Enquire into the welfare of animals kept under intensive husbandry systems'. The Brambell Report to the UK government was one of the most influential writings on animal welfare

Fig. 2.2. The apparent 'unnaturalness' of many modern farms is one factor that has provoked disquiet in the general public about the welfare of farm animals. For example, many people who are uninformed about the modern dairy industry assume that cows are kept outdoors, grazing on grass and in contact with their calves. In reality, in most industrialized countries, dairy cows are separated from their calves soon after birth, housed indoors often with no access to pasture, and eat a diet of grain rather than grass. There are welfare problems associated with such intensive housing systems, but there are also welfare problems associated with outdoor housing. The apparent 'naturalness' of a method of housing animals gives little information about its effect on the welfare of the animals.

(Brambell Committee, 1965). We have included quotes from this report in which the Committee expresses itself on the issue of feelings in animals, and on health and natural behaviour in the sections on these topics. The views expressed on animal welfare, and the particular issues and concerns that were examined by the Committee, had a great influence on the topics and nature of the research that was done in animal welfare and on the animal welfare legislation that was adopted in Europe. The effects of this Committee's report on subsequent legislative approaches to animal welfare, especially in Europe, have been well described in Veissier *et al.* (2008).

Governments of many European countries have responded to the public concerns about animal welfare by adopting legislation that prohibited certain practices. The Swedish and Swiss regulations were among the earliest and perhaps the most notable in explicitly dealing with the problems of behavioural deprivation and in laying down in detail which practices would no longer be permitted. For example, the *Swiss Animal Protection Ordinance* (Swiss Federal Council, 1978, amended 1998) states that animals 'shall be kept in such a manner as not to interfere with … their behaviour'. Similar animal welfare legislation was adopted in other European countries and formed the basis for subsequent European Union (EU) legislation.

The European Convention for the Protection of Animals Kept for Farming Purposes of 1978 focused upon the importance of avoiding suffering and ensuring that housing, nutrition and management systems should be appropriate to animals' 'physiological and ethological needs in accordance with … scientific knowledge' (Council of Europe, 1978, p. 15). These last requirements, especially the reference to 'ethological needs', precipitated considerable scientific research aimed at better understanding such needs and how these differ among species. However, the importance given to concepts associated with natural behaviour in legislation or official definitions of welfare can cause problems for science. For example, the OIE's definition of animal welfare (OIE, 2010) refers to 'innate behaviour' but there is no clear scientific definition of innateness (Mameli and Bateson, 2006). This topic is discussed in more detail in Section 2.5.

Surveys undertaken in the EU show that consumers often state that animal welfare issues are important to them in making purchasing decisions (e.g. European Commission, 2007). Animal welfare is an accepted part of product quality in many countries (Blokhuis *et al.*, 2008) and this has led to a proliferation of 'quality assurance' schemes that try to assure consumers that the products they buy are produced according to practices that respect animal welfare.

Fig. 2.3. A 'concerned member of the public'. Ruth Harrison (the author of *Animal Machines*, 1964) played a central role in the development of animal welfare science, not because she was a scientist, but because she was able to crystallize the growing public concern about the effects of modern farming systems on the welfare of the animals. The fact that research into animal welfare in large part involves responding to public concerns means that we cannot define animal welfare in a purely scientific fashion without referring to the ethical concerns that the public has about how we treat animals. A good biography of Ruth Harrison can be found in an article by van de Weerd and Sandilands (2008). Image from the library of Ruth Harrison.

Fig. 2.4. A 'crate' for veal calves (top), a stall for pregnant pigs (centre) and a battery cage for laying hens (bottom). These intensive forms of housing systems have been the subject of most attention in terms of the effect on the welfare of farm animals.

One of the earliest successful independent schemes is the Freedom Foods scheme of the RSPCA (Royal Society for the Prevention of Cruelty to Animals) in the UK. Quality assurance schemes have also become the most common way of dealing with farm animal welfare issues in North America (Mench, 2008), even though there is little control of how the claims being made actually relate to animal welfare. Because consumers are the drivers in this development, companies are attempting to satisfy the public perception of welfare, at least in Europe, by moving animals to large areas outdoors. This is happening even though there is no guarantee of good welfare by keeping animals outdoors. Increasingly, research is directed towards validating measures of animal welfare that can reliably and feasibly be used on farm, during transport and at slaughter, in order to verify such claims about animal welfare.

More recently, the globalization of food has led to increasing awareness that animal welfare issues are not restricted to modern intensive systems, but occur also in traditional farming systems that have changed little over decades, if not hundreds of years. Some recent initiatives by the Food and Agriculture Organization of the United Nations (FAO, 2009) are specifically targeted at increasing awareness about animal welfare in developing countries as a way of reducing poverty.

Animal welfare science grew initially out of concern about how animals were being treated and this is still the case 50 years later. Assessment of animal welfare in practice as part of enforcing legislation or in checking compliance with assurance schemes will be addressed in more detail in later chapters, as will economic aspects of animal welfare. For this reason, we now turn in this chapter to discussing animal welfare science from the animal's perspective.

2.3 Welfare and the Subjective Experience of Animals

As we have stated previously, the ethical concern that human beings have for animals is a result of the capacity of animals for subjective experience and especially of their capacity to suffer from pain or other aversive mental states, such as fear or boredom. Consequently, developing scientific methods to deal with the feelings or emotions of animals have played a major role in animal welfare science. When Ruth Harrison (1964, p. 3) criticized intensive livestock husbandry practices in *Animal Machines*, she was concerned about the feelings of the animals:

Today the exploitation has been taken to a degree which involves not only the elimination of all enjoyment, the frustration of almost all natural instincts, but its replacement with acute discomfort, boredom and the actual denial of health. It has been taken to a degree where the animal is not allowed to live before it dies.

A year later, the Brambell Committee (1965) also acknowledged that feelings were an important feature of welfare. In our view they were very far sighted in claiming that:

Welfare is a wide term that embraces both the physical and mental well-being of the animal. Any attempt to evaluate welfare, therefore, must take into account the scientific evidence available concerning the feelings of animals that can be derived from their structure and functions and also from their behaviour.

Interest in the subjective or emotional experiences of animals and their importance for animal welfare has a long history (Preece, 2007). In 1839, William Youatt, an English veterinarian who took a humanitarian approach to animal welfare, criticized many practices that he observed in 19th century England. The list will be uncannily familiar to 21st century readers: too early training of racehorses; steeple chasing; transport methods for newly born calves; the raising of veal calves; tail docking and ear cropping of dogs; using live bait for fishing; the dissection of living animals. In making these criticisms, Youatt always relied on scientific evidence. For example, when talking about slaughterhouse management, he strongly recommends the poleaxe for stunning animals before bleeding them to death, rather than cervical (neck) dislocation. He explains very clearly that, although the area below the cervical dislocation will be insensitive, there is probably some sensation in the area above it and this could lead to a reduction of welfare. In Youatt's words (1839, p 179), 'When we use the poll-axe, we come upon the very seat of sensation – *we crush all feeling*'. Some 50 years later, George John Romanes (1884), a prominent biologist and follower of Darwin, wrote that:

Pleasures and Pains must have been evolved as the subjective accompaniment of processes which are respectively beneficial or injurious to the organism, and so evolved for the purpose or to the end that the organism should seek the one and shun the other.

The more recent prominence given to the feelings or emotional states of animals and how these affect their welfare came particularly from the writings of Marian Dawkins and Ian Duncan (e.g. Duncan and Dawkins, 1983), who suggested that feelings play a major role in welfare. In fact, Duncan went further in suggesting that not only were feelings an important component of welfare, but that they might be the only thing that mattered (Duncan, 1996). For example, consider the case of an animal that is not able to eat enough to meet its biological requirements. While the deficiency of nutrients may be the primary factor that reduces the biological functioning of the animal, it is the subjective experience of this state by the animal, i.e. the feeling of hunger, that reduces the animal's welfare. Similarly, it is possible to separate the primary state of being ill, which reduces health, from the secondary subjective experience of feeling ill, which reduces welfare.

One advantage of seeing animal welfare as threatened only when the animal is suffering is that this does provide a common currency for ranking different threats to animal welfare. Thus, the question of whether a laying hen's welfare is reduced more by frustration at not being able to build a nest or by being attacked by other hens can be answered (in theory) by finding which causes more suffering. But is it sufficient to ensure that animals are not suffering as a result of how we treat them, or should we aim to provide them with opportunities to experience pleasure? Some have argued that it is the degree of pleasure that an animal obtains from performing a behaviour or obtaining a valued resource, rather than the amount of suffering caused by the inability to perform the behaviour or the absence of the resource, that determines its welfare, and that this should be considered the common currency (Cabanac, 1992; Spruijt *et al.*, 2001).

A commonly heard criticism of the emphasis placed on animals' sentience or emotions is that these are beyond the reach of science. A real breakthrough in the acceptance of feelings in scientific investigations came with the publication of Donald Griffin's book *The Question of Animal Awareness* (Griffin, 1975). Applied ethologists who had been struggling to answer questions raised by Ruth Harrison (1964) and the Brambell Committee (1965), now began to recognize that the feelings and emotions of animals are accessible to scientific investigation. Many recent publications show that an understanding of the subjective feelings of animals is now widely accepted to be amenable to scientific investigation (Duncan 2006; Brydges and Braithwaite, 2008; Dawkins, 2008; Mendl *et al.*, 2009; Reefmann *et al.* 2009). In particular, considerable progress has been made on understanding and measuring animal pain, and a large and rapidly developing scientific literature on pain assessment and prevention is now available for farm animals (for a review see Weary *et al.*, 2006). More recent research is beginning to examine pain in a much wider range of species, such as invertebrates (Elwood *et al.*, 2009). There has also been interest in finding signs that may indicate that the welfare of an animal is good rather than signs that its welfare is bad, and in investigating positive emotions (Boissy *et al.*, 2007).

2.4 Health, Production and Reproduction

It seems self-evident that good animal welfare requires that the animals be healthy, and that the occurrence of disease and injury will lead to poor welfare. Failure to treat disease adequately is widely condemned as breaching the basic tenets of animal care. For companion animals, most illnesses are treated fairly quickly, and the main ethical issue involves the question of the relative benefits of treating chronic illness, or in using treatments that are themselves painful, as opposed to immediate euthanasia. The relationship between animal welfare and animal health has been of most concern in relation to farm animals, where the cost of treating commonly occurring health problems in large numbers of animals can be prohibitive. Farm animals, both in traditional and modern housing systems, and under both intensive and extensive management, suffer from a variety of health problems, and the incidence of these problems can be surprisingly high. For example, on average, 25% of dairy cows in the USA suffer from lameness (Espejo *et al.*, 2006) and 75% of broiler chickens in Portugal suffer from foot-pad dermatitis (Gouveia *et al.*, 2009).

The importance of good health for good welfare is one of the least controversial aspects of the debate about animal welfare. Farmers readily appreciate the economic cost of disease and scientists have also long accepted the importance of good health for good animal welfare. The importance of disease was recognized as an essential part of good welfare by the Brambell Committee (1965), who wrote that:

> [A] principal cause of suffering in animals, as it is in men, is disease … we lay stress on the incidence of disease and on the guarantee that a sick animal will be quickly recognized and appropriately treated or slaughtered. Any given intensive system of husbandry may, or may not, be satisfactory in one or both of these respects. Some compare favourably with traditional methods with regard to disease, others compare unfavourably.

However, we should not assume that concern about disease typifies the welfare concerns only of veterinarians and farmers. Graphic descriptions of illness in modern intensive housing systems figured prominently among attempts to turn the public against 'factory farms'. For example, Singer (1975, pp. 104–105), in describing modern methods of housing broilers writes that:

> [Their] fast growth rate also causes crippling and deformities that force producers to kill an additional 1 to 2 percent of broiler chickens – and since only severe cases are culled, the number of birds suffering from deformities is bound to be higher … When the

birds must stand and sit on rotting, dirty ammonia-charged litter, they also suffer from ulcerated feet, breast blisters, and hock burns.

Measures of the occurrence of disease and injury have long been used as indicators of animal welfare and have figured prominently in assessments of farm animal welfare in many species. The main difficulty in using health measures to assess animal welfare lies though in judging the degree of suffering associated with different forms of illness or injury. As Dawkins (1998, p. 73) explains, a key question when dealing with symptoms of illness as a sign of poor welfare is:

> Are animals that can be shown to have these objectively measured symptoms consciously experiencing what humans would call suffering, if we were experiencing these same symptoms?

Wells *et al.* (1998, p. 3034) further point out that:

> [Concern about animal welfare] seems to be directed toward diseases leading to perceived suffering by animals ... Clinical lameness has been recognized as an animal welfare concern because of the sometimes obvious signs of pain in the affected cattle.

Although most attention has been paid to clinical diseases where the symptoms of illness are obvious, scientists have recently been interested in trying to assess animal welfare by examining subclinical physiological changes that may show that animals are under 'stress' or are at risk of becoming ill (Moberg, 2000). Moberg argues that an animal's responses to any challenge or stressor require utilization of the animal's biological resources (e.g. time or energy or nutrients) that would usually be used to support normal biological functioning, and that the severity of any stress can be measured in terms of the biological cost of the response. When the stress is sufficiently severe or prolonged (Moberg, 2000, p. 13):

> [This] results in distress when the stress response shifts sufficient resources to impair other biological functions. When this occurs the animal enters the pre-pathological state, is at risk of developing a pathological state and experiences distress.

Thus, pre-pathological changes, such as alterations in immune function, may lead to pathological changes such as lesions in the lungs or abomasums of veal calves. These types of lesions are increasingly being recorded at slaughter as a reliable indicator of the welfare state of the animal while it was alive. Such indicators can be used to differentiate between housing and management systems, or even between different animal producers.

Whereas the use of pathological indicators is generally accepted, there are many difficulties in using subclinical changes as indicators of reduced welfare (Rushen and de Passillé, 2009), most of which arise from the fact (Dawkins, 2006, p. 79) that:

> [These changes] can be difficult to interpret in welfare terms because many of these changes are part of the adaptive way in which an animal responds to its _ environment, and because apparently pleasurable activities, such as sex and hunting prey, can lead to similar changes to those that are apparently unpleasant, such as escaping a predator.

Furthermore, equating any sign of stress with reduced welfare can be particularly dangerous because (Moberg, 2000, p. 1):

> Stress is a part of life and is not inherently bad ... Our challenge is to differentiate between the little non-threatening stresses of life and those stresses that adversely affect an animal's welfare.

Hence, such subclinical physiological or immunological changes can only be used to assess animal welfare if they are directly associated with some suffering, or are likely to result in chronic changes that will threaten animal welfare in the future, for example by causing poor health.

It may seem evident that a farm animal's productivity will also reflect its welfare: a dairy cow with poor welfare will produce less milk, and a pig or broiler chicken in poor conditions will grow less well (see, for example, Curtis, 2007). Yet the relationship between the welfare of an animal and its productivity is not straightforward. High levels of productivity often result from specific practices, such as the use of growth enhancers (hormones, antibiotics, etc.). Furthermore, there is increasing evidence that genetic selection for fast growth in broiler chickens and pigs, and for high milk production in dairy cows, is associated with an increased occurrence of certain health problems (Rauw *et al.*, 1998). Clearly, the relationship between an animal's welfare and its productivity is controversial.

Likewise, it may seem evident that reproduction is a sign of good welfare and that failure to reproduce, for example as it occurs in some animals in zoos, is a sign of a welfare problem. However, this does not mean that reproductive success can always be used to assess welfare. When reproductive problems arise because of poor health, this is undoubtedly a sign of poor welfare, but differences among farms in reproductive failure could be due to many factors that are not related to the welfare of the animals, such as success at oestrus detection, effective artificial insemination strategies and general reproductive management. Indeed, reproductive success itself can be a major threat to the survival of individuals, a fact that is most apparent in wild animals. In discussing the use of measures of reproductive fitness to assess welfare, Dawkins (1998, p. 73) states that the main problem is that:

> Animals have been selected to reproduce, not just to live a long time as individuals. A consequence of gene level selection is that the health, longevity and even survival of individual animals may be reduced in the interests of gene propagation. A very obvious example of this is the considerable costs, including reduction in physical health, that occur in animals during the breeding season.

Barnard and Hurst (1996) provide more subtle examples from laboratory rodents of how increased reproductive success is sometimes achieved at the expense of the welfare of the individual animals: high testosterone levels, which are a necessity to achieve high reproductive success in male mice, are immunosuppressive and can increase the animal's susceptibility to disease.

In summary, changes in productivity or reproduction due to stress or poor health, and which involve the animals suffering in some way, can serve as indicators of changes in animal welfare, but the relationship between the welfare of animals and their productivity or their willingness or ability to reproduce is not a simple one.

2.5 Welfare, Natural Behaviour and the 'Nature' of Animals

It is the 'unnaturalness' of modern housing conditions and management practices, even with species that have been domesticated for several thousands of years, that is one of the greatest concerns of the public. The possibility that farm, laboratory and zoo animals are suffering because they cannot perform behaviour that they would normally perform in a more natural environment is one of the enduring concerns. The Brambell Committee drew particular attention to the problems of behavioural restriction and frustration that result from intensive indoor housing systems for farm animals. In the appendix to this report, Thorpe (an ethologist) wrote (Brambell Committee, 1965, p. 79):

> We must draw the line at conditions which completely suppress all or nearly all the natural, instinctive urges and behaviour patterns characteristic of actions … as found in the ancestral wild species and which have been little, if at all, bred out in the process of domestication.

As a result of this, some of the earlier research in farm animal welfare involved attempts to determine how much modern breeds of farm animals have retained the behavioural repertoire of their evolutionary ancestors (Jensen, 1989) (Fig. 2.5). Also, many of the improvements in the housing of zoo animals have come from developing enclosures that provide a more natural environment (Fig. 2.6).

It is very appealing then, when faced with uncertainty about the best way to keep animals, to say they should be kept in a natural environment. But what is natural for a domestic dog such as a chihuahua, a highly selected, artificially inseminated farm animal, or a genetically modified mouse? As David Fraser (2008, p. 174) says:

> Although there are many examples of animal welfare problems caused by artificial conditions, looking in 'natural' environments as a way to safeguard animal welfare raises serious difficulties. For a given species it may not be possible to identify a 'natural' environment nor, for practical reasons, to recreate it; and the problem is further compounded for animals with a long history of domestication.

For most animals, it seems society has already taken a pragmatic view. People want to be able to come close to wild animals at the zoo, have pets for companionship and so on. Some forms of restriction compared with the natural environment are therefore accepted. The question for animal welfare is the extent to which animals should be able to show natural behaviour, rather than the extent to which their environment should be natural. For example, populations of herring gulls (*Larus argentatus*) and lesser black-backed gulls (*Larus fuscus*) are very successful in Europe through living in very artificial environments. Rather than living by and from the sea, they forage in garbage dumps, nest on buildings and roost on playing fields, but these artificial environments allow the performance of natural behaviour (Duncan, 1995). The ubiquity of the concept of 'natural behaviour' suggests that it does capture some of the disquiet that modern farming systems provoke in many people, despite reservations by some scientists (Špinka, 2006; Broom, 2008).

However, there is a problem with the notion of encouraging natural behaviour, and that is that not all natural behaviour is desirable. Flight reactions

Fig. 2.5. An argument raised against some modern, intensive housing systems is that they do not allow animals to perform their natural behaviour. Critics of this point of view argue that modern farm animals have been subjected to artificial selection for many generations and have lost much of their natural behaviour. This has led to experiments in which modern breeds of farm animals have been released into semi-natural environments. Despite generations of being housed in farrowing crates, modern domestic pigs will show typical nest-building behaviour when given the opportunity, showing that they have retained much of their evolved behavioural repertoire.

Fig. 2.6. Knowledge of the natural behaviour of zoo animals in particular can provide useful guidelines to improve their housing. In the case of a primate, providing ropes or other appropriate structures will allow them to perform this behaviour, most likely improving their welfare, while also entertaining zoo visitors. Photo by Yezica Norling.

L.J. Keeling *et al.*

in response to real or perceived predators are adaptive in the wild, but under confined conditions this can lead to problems, for example panic reactions in poultry flocks that lead to birds suffocating under a pile of other birds (Hansen, 1976; Mills and Faure, 1990). Aggression and the establishment and maintenance of the social hierarchy is natural for many of our domesticated species, but there can be negative consequences for the low-ranked animals if there are limited resources or insufficient space to escape. For many species of farmed fish, those individuals that are more aggressive will catch more food from the feed dispensers and thus have better growth rates. None the less, this better growth for some individuals is achieved at the expense of other fish that not only grow less well, but are injured in the process of competing for food (Brännas et al., 2008). Infanticide is a natural behaviour of many primates, seen in many natural habitats (Hiraiwa-Hasegawa, 1988), but it is not a desirable behaviour to see in zoo settings, either for the welfare of the animals or the zoo visitors. Thus, natural behaviour, while enhancing the welfare of one individual can at the same time lead to a reduction in welfare for another individual in the same group.

To understand better the relationship between natural behaviour and animal welfare, we need also to understand better the evolutionary background of our domesticated animals. Dawkins (2006), in her paper entitled 'A user's guide to animal welfare science', promotes an evolutionary approach and says that:

Behavioural ecologists have a major role to play in understanding the mechanisms by which different species respond to threats to their fitness and thus in defining what constitutes 'welfare.'

But also that:

Paradoxically, it is evolved mechanisms for coping with anticipated threats to fitness … that causes more concern about welfare than the direct threats themselves.

So here lies the crux of the problem when using a natural behaviour approach to answer questions about animal welfare: what is natural is not necessarily appropriate in our artificial environments and what is appropriate in our artificial environments is not necessarily what comes naturally to our domesticated animals.

Owing to the difficulties with the concept of 'natural behaviour', scientists have refocused their attention on trying to understand which behaviour is important for the animal to be able to perform in order to have good welfare. This research has often been done using the term 'behavioural need'. For example, scientists proposed that in the absence of the environmental trigger, natural behaviour is important to the animal only if it is internally motivated and if the performance of the behaviour itself is important in providing the appropriate feedback to the animal. For example, dairy calves usually suck to get their milk, but on many modern farms calves are fed from a bucket and do not need to suck in order to consume their milk. Despite this, research has shown that the performance of sucking behaviour itself plays a role in satisfying the calf's feeding motivation, independently of the intake of milk (Rushen et al., 2008). Furthermore, if the calf is prevented from performing this behaviour in its normal context, it will be likely to continue to show the behaviour but in an abnormal fashion, for example, by sucking another calf (Fig. 2.7). Thus, the importance of a behaviour pattern for animal welfare depends on its source of motivation (see Chapters 7 and 9).

In addition, it is widely recognized that the emotions of animals play a major role in behavioural motivation. It has been suggested that negative feelings, such as pain, fear and frustration, occur most in 'need situations', that is, where immediate action by the animal is required. In contrast, positive feelings, such as pleasure and excitement, occur more in 'opportunity situations' and increase the likelihood that the animal will perform the behaviour, even where the fitness benefits are far in the future (Fraser and Duncan, 1998). Fear is an example of a negative feeling associated with motivating an animal to escape from a predator. Pleasure is an example of a positive feeling associated with the performance of play behaviour. In other words, feelings are part of the natural selection process that allows survival mechanisms to act. Some of this relationship is captured in one of few recent attempts to define natural behaviour (Bracke and Hopster, 2006):

Natural behaviour may be defined as behaviour that animals have a tendency to exhibit under natural conditions, because these behaviours are pleasurable and promote biological functioning.

The difficulties with the concept of natural behaviour do not mean that we cannot benefit

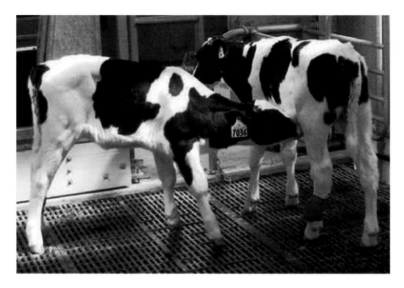

Fig. 2.7. A dairy calf cross-sucks another. A calf's natural behaviour to obtain its milk is to suck on its mother's udder. In modern farming, calves often drink milk from a bucket. However, calves will often continue to show sucking behaviour, but in an abnormal context. The fact that animals continue to show behaviours such as this, even when they are no longer 'necessary' to achieve their usual function, show that the performance of the behaviour is important to the animal.

from a better knowledge of the natural behaviour of domestic animals. Marek Špinka (2006) proposes three examples of the importance of natural behaviour to good farm animal welfare:

> First, it is often more efficient to allow animals to satisfy their own needs and achieve their goals than to address these needs and goals through technical means. Second, a larger class of natural behaviours is associated with positive affective experience, and thus their performance directly enhances animal welfare. Third, the performance of natural behaviour in its richness and complexity often brings long-term benefits for the animal, such as improved proficiency in coping with social and physical challenges.

In summary, while freedom to perform the whole repertoire of natural behaviour is not crucial for an animal's welfare, the opportunity to perform natural behaviour may be an effective way to improve its welfare in practice.

2.6 Conclusions

- Different groups of people put emphasis on different aspects of animal welfare. These different aspects are interconnected and revolve round the same key issues.
- The physiology and behaviour of animals have evolved to maximize health and fitness under natural conditions. Feelings have evolved as a flexible mechanism to motivate behaviour.
- Negative feelings motivate behaviour in situations where immediate action is required. Positive feelings motivate behaviour in situations where no immediate action is required but where there is a long-term benefit. There is growing acceptance that it is negative feelings that reduce welfare and positive feelings that improve it.
- Work on health is increasingly being directed towards preventing and understanding the impact of reduced health on an animal's welfare.
- Because of the central role played by feelings, it seems likely that future research will be increasingly directed towards cognition and emotions in animals.

References

Anonymous (2001) Scientists' assessment of the impact of housing and management on animal welfare. *Journal of Applied Animal Welfare Science* 4, 3–52.

Barnard, C.J. and Hurst, J.L. (1996) Welfare by design: the natural selection of welfare criteria. *Animal Welfare* 5, 405–433.

Blokhuis, H.J., Van Niekerk, T.F., Bessei, W., Elson, A., Guemene, D., Kjaer, J.B., Levrino, G.A.M., Nicol,

C.J., Tauson, R., Weeks, C.A. and De Weerd, H.A.V. (2007) The LayWel project: welfare implications of changes in production systems for laying hens. *World's Poultry Science Journal* 63, 101–114.

Blokhuis, H.J., Keeling, L.J., Gavinelli, A. and Serratosa, J. (2008) Animal welfare's impact on the food chain. *Trends in Food Science and Technology* 19 (Suppl. 11) S75–S83.

Boissy, A., Manteuffel, G., Jensen, M.B., Moe, R.O., Spruijt, B., Keeling, L.J., Winckler, C., Forkman, B., Dimitrov, I., Langbein, J., Bakken, M., Veissier, I. and Aubert, A. (2007) Assessment of positive emotions in animals to improve their welfare. *Physiology and Behaviour* 92, 375–397.

Bracke, M.B.M. and Hopster, H. (2006) Assessing the importance of natural behavior for animal welfare. *Journal of Agricultural and Environmental Ethics* 19, 77–89.

Bracke, M.B.M., Spruijt, B.M. and Metz, J.H.M. (1999) Overall animal welfare assessment reviewed. Part 1: Is it possible? *Netherlands Journal of Agricultural Science* 47, 279–291.

Brambell Committee (1965) *Report of the Technical Committee to Enquire into the Welfare of Animal Kept under Intensive Livestock Husbandry Systems.* Command Paper 2836, Her Majesty's Stationery Office, London.

Brännas, E. and Johnsson, J.I. (2008) Behaviour and welfare in farmed fish. In: Magnhagen, C., Braithwaite, V.A., Forsgren, E. and Kapoor, B.G. (eds) *Fish Behaviour.* Science Publishers, Enfield, New Hampshire, pp. 593–627.

Broom, D.M. (1996) Animal welfare defined in terms of attempts to cope with the environment. *Acta Agriculturae Scandinavica, Section A – Animal Science* Supplement 27, 22–28.

Broom, D.M. (2008) Consequences of biological engineering for resource allocation and welfare. In: Rauw, H.W. (ed.) *Resource Allocation Theory Applied to Farm Animal Production.* CAB International, Wallingford, UK, pp. 261–274.

Brydges, N.M. and Braithwaite V.A. (2008) Measuring animal welfare: what can cognition contribute? *Annual Review of Biomedical Sciences* 10, T91–T103.

Cabanac, M. (1992) Pleasure: the common currency. *Journal of Theoretical Biology* 155, 173–200.

Council of Europe (1978) European Convention for the Protection of Animals Kept for Farming Purposes. *Official Journal of the European Communities* No. L323 (17.11.78).

Curtis, S.E. (2007) Commentary: performance indicates animal state of being: a Cinderella axiom? *Professional Animal Scientist* 23, 573–583.

Dawkins, M.S. (1998) Evolution and animal welfare. *Quarterly Review of Biology* 73, 305–328.

Dawkins, M.S. (2006) A user's guide to animal welfare science. *Trends in Ecology and Evolution* 21, 77–82.

Dawkins, M.S. (2008) The science of animal suffering. *Ethology* 114, 937–945.

Duncan, I.J.H. (1995) D.G.M. Wood-Gush Memorial Lecture: an applied ethologist looks at the question "why?". *Applied Animal Behaviour Science* 44, 205–217.

Duncan, I.J.H. (1996) Animal welfare defined in terms of feelings. *Acta Agriculturae Scandinavica Section A – Animal Science* Supplement 27, 29–35.

Duncan, I.J.H. (2006) The changing concept of animal sentience. *Applied Animal Behaviour Science* 100, 11–19.

Duncan, I.J.H. and Dawkins, M.S. (1983) The problem of assessing 'well-being' and 'suffering' in farm animals. In: Smidt, D. (ed.) *Indicators Relevant to Farm Animal Welfare.* Martinus Nijhoff, The Hague, The Netherlands, pp. 13–24.

Elwood, R.W., Barr, S. and Patterson, L. (2009) Pain and stress in crustaceans? *Applied Animal Behaviour Science* 118, 128–136.

Espejo, L.A., Endres, M.I. and Salfer, J.A. (2006) Prevalence of lameness in high-producing Holstein cows housed in freestall barns in Minnesota. *Journal of Dairy Science* 89, 3052–3058.

European Commission (2007) *Attitudes of Consumers Towards the Welfare of Farmed Animals, Wave 2.* Special Eurobarometer [Report] 229(2)/Wave 64.4 – TNJ Opinion and Social. Brussels, Belgium/Luxembourg.

FAO (Food and Agriculture Organization of the United Nations) (2009) *Capacity Building to Implement Good Animal Welfare Practices,* Report of FAO Expert Meeting Rome 30 September to 3 October 2008. Rome, Italy.

FAWC (Farm Animal Welfare Council) (2009) *Five Freedoms.* Available at: http://www.fawc.org.uk/freedoms.htm (accessed July 2010).

Fraser, D. (2008) Toward a global perspective on farm animal welfare. *Applied Animal Behaviour Science* 113, 330–339.

Fraser, D. and Duncan, I.J.H. (1998) 'Pleasures', 'pains' and animal welfare: toward a natural history of affect. *Animal Welfare* 7, 383–396.

Fraser, D., Weary, D.M., Pajor, E.A., Milligan, B.N. (1997) A scientific conception of animal welfare that reflects ethical concerns. *Animal Welfare* 6, 187–205.

Gouveia, K.G., Vaz-Pires, P. and da Costa, P.M. (2009) Welfare assessment of broilers through examination of haematomas, foot-pad dermatitis, scratches and breast blisters at processing. *Animal Welfare* 18, 43–48.

Griffin, D.R. (1975) *Animal Minds.* University of Chicago Press, Chicago, Illinois.

Hansen, R.S. (1976) Nervousness and hysteria of mature female chickens. *Poultry Science* 55, 531–543.

Harrison, R. (1964) *Animal Machines.* Vincent Stuart, London.

Hegelund, L. and Sorensen, J.T. (2007) Measuring fearfulness of hens in commercial organic egg production. *Animal Welfare* 16, 169–171.

Hiraiwa-Hasegawa, M. (1988) Adaptive significance of infanticide in primates. *Trends in Ecology and Evolution* 3, 102–105.

Jensen, P. (1989) Nest site choice and nest building of free-ranging domestic pigs due to farrow. *Applied Animal Behaviour Science* 22, 13–21.

Keeling, L.J. (2004) Applying scientific advances to the welfare of farm animals: why it is getting more difficult. *Animal Welfare* 13, 187–191

Leach, M.C., Thornton, P.D. and Main, D.C.J. (2008) Identification of appropriate measures for the assessment of laboratory mouse welfare. *Animal Welfare* 17, 161–170.

Mameli, M. and Bateson, P. (2006) Innateness and the sciences. *Biology and Philosophy* 21, 155–188.

Mench, J.A. (2008) Farm animal welfare in the USA: farming practices, research, education, regulation, and assurance programs. *Applied Animal Behaviour Science* 113, 298–312.

Mendl, M., Burman, O.H.P., Parker, R.M.A. and Paul, E.S. (2009) Cognitive bias as an indicator of animal emotion and welfare: emerging evidence and underlying mechanisms. *Applied Animal Behaviour Science* 118, 161–181.

Mills, A.D. and Faure, J.-M. (1990) Panic and hysteria in domestic fowl: a review. In: Zayan, R. and Dantzer, R. (eds) *Social Stress in Domestic Animals*. Kluwer, Dordrecht, The Netherlands, pp. 248–272.

Moberg, G.P. (2000) Biological response to stress: implications for animal welfare. In: Moberg, G.P. and Mench, J.A. (eds) *The Biology of Animal Stress*. CAB International, Wallingford, UK, pp. 1–21.

OIE (World Organisation for Animal Health) (2010) Chapter 7.1: Introduction to the recommendations for animal welfare. Article 7.1.1. In: *Terrestrial Animal Health Code, Volume 1*. Available at: http://www.oie.int/eng/normes/mcode/en_chapitre_1.7.1.htm (accessed 14 December 2010).

Preece, R. (2007) Thoughts out of season on the history of animal ethics. *Society and Animals* 15, 365–378.

Rauw, W.M., Kanis, E., Noordhuizen-Stassen, E.N. and Grommers, F.J. (1998) Undesirable side effects of selection for high production efficiency in farm animals: a review. *Livestock Production Science* 56, 15–33.

Reefman, N., Wechsler, B. and Gygax, L. (2009) Behavioural and physiological assessment of positive and negative emotion in sheep. *Animal Behaviour* 78, 651–659.

Regan, T. (1985) *The Case for Animal Rights*. Routledge, London.

Romanes, G.J. (1884, reprinted 1969) *Mental Evolution in Animals*. AMS Press, New York.

Rushen, J. and de Passillé, A.M.B. (2009) The scientific basis of animal welfare indicators. In: Smulders, F.J.M. and Algers, B. (eds) The assessment and management of risks for the welfare of production animals. In: *Food Safety Assurance and Veterinary Public Health, Volume 5. Welfare of Production Animals: Assessment and Management of Risks. Part 3 – Management of Risks for the Welfare of Production Animals*. Wageningen Academic Publishers, Wageningen, The Netherlands, pp. 391–416.

Rushen, J., de Passillé, A.M., von Keyserlingk, M. and Weary, D.M. (2008) *The Welfare of Cattle*. Springer, Dordrecht, The Netherlands, p. 303.

Singer, P. (1975) *Animal Liberation*. Avon Books, New York.

Špinka, M. (2006) How important is natural behaviour in animal farming systems? *Applied Animal Behaviour Science* 100, 117–128.

Spruijt, B.M., van den Bos, R. and Pijlman, F.T.A. (2001) A concept of welfare based on reward evaluating mechanisms in the brain: anticipatory behaviour as an indicator for the state of reward systems. *Applied Animal Behaviour Science* 72, 145–171.

Swiss Federal Council (1978, amended 1998) *Swiss Animal Protection Ordinance*. Bern, Switzerland.

van de Weerd, H. and Sandilands, V. (2008) Bringing the issue of animal welfare to the public: a biography of Ruth Harrison (1920–2000). *Applied Animal Behaviour Science* 113, 404–410.

Veissier, I., Butterworth, A., Bock, B. and Roe, E. (2008) European approaches to ensure good animal welfare. *Applied Animal Behaviour Science* 113, 279–297.

Weary, D.M., Niel, L., Flower, F.C. and Fraser, D. (2006) Identifying and preventing pain in animals. *Applied Animal Behaviour Science* 100, 64–76.

Wells, S.J., Ott, S.L. and Seitzinger, A.H. (1998) Key health issues for dairy cattle – new and old. *Journal of Dairy Science* 81, 3029–3035.

Whay, H.R., Main, D.C.J., Green, L.E. and Webster, A.J.F. (2003) Assessment of the welfare of dairy cattle using animal-based measurements: direct observations and investigation of farm records. *The Veterinary Record* 153, 197–202.

Youatt, W. (1839) *The Obligation and Extent of Humanity to Brutes, Principally Considered with Reference to the Domesticated Animals*. Republished in 2004, edited, introduced and annotated by Rod Preece. Edwin Mellen Press, Lewiston, New York.

3 Environmental Challenge and Animal Agency

Marek Špinka and Françoise Wemelsfelder

Abstract

Challenges are there to be overcome – seen usually as problems to avoid rather than as opportunities to enjoy. However, for humans a life without challenge would be likely to be dull and boring, lacking the enthusiasm and satisfaction that come with individual development. Could this also be true for animals? This chapter looks at the positive value of engaging with environmental challenges for animal welfare, proposing that this value lies in an animal's expression of agency and the enhanced functional competence that it gains through this. It explores the different facets of agency, and provides more detailed discussion of key elements such as problem solving, exploration and play, as well as discussing responses to challenge and how an animal's welfare is affected if the animal is prevented from performing behaviours of this kind. The final sections of the chapter consider how monotonous, predictable, captive environments may lead to apathy and boredom, and prevent animals from experiencing a positive quality of life. Agency should be regarded as an integrative capacity that works across specific modules of organization and, as such, forms an important condition for an animal's overall well-being and health.

3.1 Introduction

Animals in the wild face many challenges. Predators, food shortage, social competition and intricacies, weather, illness – all these threaten their health and survival and hamper their reproductive efforts. For any species and any individual, the natural environment is complex and in an ever-fluid state as a result of fluctuating physical conditions and the actions of cohabiting living beings. Animals are adept at turning challenges into opportunities, sometimes through mastery in a specific task, sometimes through flexibility of response, yet such capabilities are often not enough to stay fit, survive and reproduce. Challenges are intrinsic and natural yet, if they are too severe or too many, animals fail to cope with them and their individual welfare deteriorates, frequently to the point of premature death.

By contrast, animals in captivity often live in simple, monotonous and predictable environments, where they are challenged infrequently or not at all. One might expect this to be an improvement: the animal can relax and get on with its life. But could it go too far the other way – might animals also suffer from a lack of environmental challenge? How much of a welfare problem is the fact that captive animals live in barren environments that give them very little

opportunity to engage actively with meeting the needs of their own life?

Before turning to this question, it is necessary to look at how wild-living animals deal with challenges. Some challenges come with little novelty but still demand a lot of attention, time or energy because they pose an important threat or opportunity. For instance, coping with seasonal temperature stress or defending a position in a stable dominance hierarchy are serious challenges, but the aspect of novelty may not play an important role in them. Animals can overcome these challenges by being masters in specific behavioural or physiological tasks and responses. Other challenges are highly unpredictable, such as predator attacks, sudden turmoils in social situations or random changes in food accessibility. Such challenges are much more likely to present animals with novel problems, for which they must be able to find new solutions. While challenges of the first type (such as social stress, hunger, illness) are dealt with in other chapters, in this chapter we focus on challenges induced by novelty. Our discussion of an animal's ability to deal with such challenges will centre around two key concepts, *competence* and *agency* (White, 1959). We see these terms as reflecting complementary aspects of the animal's engagement with novel challenges, and in this chapter we will explore the evidence and interpretations that support their relevance to the study of animal welfare.

We use the term *competence* to denote the whole array of cognitive and behavioural experience, tools and strategies that an animal possesses at any given moment to deal with novel challenges. How do animals acquire and enhance such competence? As for any biological trait, animals inherit genetic predispositions that feed into and support the different aspects of competence. These predispositions unfold through developmental maturation and sensory experience (Rogers, 2008), and by learning through interaction with the hard-and-real everyday striving for food, security, partners, social ties etc. However, if competence could be stepped up only as a result of reaction to external events, the process would be haphazard and risky. It may be too late to start looking for a solution when the challenge is already arriving. Therefore, it is also worth it for the animal to invest in the future, i.e. to expend time and energy in exposing itself to a degree of risk in order to have better chances at dealing with unexpected events later on. Many animals hoard food or build body reserves for lean times – but they also gather information and/or train their own abilities for later times and moments of need.

In this chapter, we label as *agency* the propensity of an animal to engage actively with the environment with the main purpose of gathering knowledge and enhancing its skills for future use. In other words, agency is the intrinsic tendency of animals to behave actively beyond the degree dictated by momentary needs, and to widen their range of competencies. In goal-oriented behavioural sequences such as foraging, mate seeking and predator avoidance, the animal will mostly use existing skills, whereas through agency-based patterns such as exploration and play, it is building novel competencies. In reality, however, these two aspects of a day's ongoing problem-solving activities are often intertwined, because features of the environment are rarely sharply divided into 'old' and 'new' challenges, and the animal's response will often be a mixture of reactive and proactive decision making. One could say that agency and competence reflect, respectively, the procedural (more proactive) and functional (more reactive) aspects of the animal's ability to prepare for, discern and resolve challenges that are meaningful to it in the context of its ecological niche (White, 1959).

3.2 Facets of Agency/Competence

What then should we think of as the main features of animal agency/competence? In this section, we list several major aspects of environmental challenge and link these aspects with complementary facets of animal agency/competence.

First, the environment is *rich* and *complex*. Most animal species live in complex ecosystems that offer a plethora of stimuli and their inter-contingencies. Only few of the apparent associations between environmental features reflect true causal links, and even fewer can be harnessed, but the utility of those few can be pivotal. For instance, food can often be discerned only through very subtle or highly complex cues. Therefore, animals need a highly developed ability of *associative learning*.

Secondly, elements of the natural environment, especially other organisms, often *resist* attempts to harness them. For example, plants and animals defend themselves from being eaten through structural and/or behavioural defences, such as hard shells, or poisonous skins or stings. Thus, animals need to learn how to overcome such hindrances effectively, and to develop good *operant or instrumental learning* capability.

Thirdly, the environment continually presents the animal with *new objects, situations* and *events*. It is important to learn more about the nature of these as soon as possible, and it is safer to do this when other challenges are not looming; therefore animals tend to react to novel objects with *inspective exploration*.

Fourthly, most wild animals live in an *open world*. That is, they live in an environment where there is a possibility to expand their horizon of knowledge and activity – a valuable resource could be hidden just behind the next bush or stone. Therefore, in addition to inspective exploration, *inquisitive exploration* is a fundamental element of the animal's competence.

Fifthly, from the animal's perspective, the environment is highly *probabilistic*. That is, because of the intrinsic variability of so many features of the environment, the same action by the animal works on some occasions but not on others. This probability may change in time and it may also contain a hidden regularity or combination of contingencies that the animal might be able to detect through more intense engagement. In order to allocate its time, energy and attention efficiently, the animal should be motivated to track the environment's stochasticity, i.e. it would have to be able to *assess uncertainty* and *update this information* regularly.

M. Špinka and F. Wemelsfelder

Sixthly, many sources of environmental variability, especially those generated by other animals, interfere with the animal's activities and this often leads to *lack or loss of control* over its own movements and actions. Therefore, animals possess not only species- and situation-specific skills but also a general behavioural, cognitive and emotional *flexibility*. One prominent way to train for the unexpected is through *play*.

Lastly, but not least, others living in the natural environment are also *knowledgeable*. Conspecifics and other organisms also gather and appraise information about the environment, and it is often faster, more precise and/or more efficient to use, share and combine this available knowledge than to rely solely on one's own experience and assessment. Therefore, animals are adept at *observational and social learning*, and at *communicating with others around them in general*.

The list of challenge types in this section does not aspire to be comprehensive but it shows that in order to live and reproduce in the natural world, animals need a large array of behavioural and cognitive activities, such as associative, operational/instrumental and social learning, information gathering and updating, flexibility in the face of atypical events and/or loss of control and, generally, a sophisticated capacity for communication. These facets continuously interact with and enhance each other; for instance, regular patrolling in order to update information can reveal novel features in the environment that stimulate exploration, which, in turn, may produce incentives to employ operational learning, e.g. on a potential food source. The concepts of agency and competence can thus best be regarded as denoting the animal's ability to integrate these various facets into effective, intelligent conduct that will optimize its survival and well-being. Such a view sits well with the growing tendency among scientists no longer to regard 'intelligence' as the prerogative of a select few so-called 'higher' animal species, but rather as a systemic characteristic of adaptively behaving organisms (e.g. Manrod *et al.*, 2008; Matzel and Kolata, 2010).

3.3 The Expression of Agency/ Competence in Problem Solving, Exploration and Play

In this section, we focus on problem solving, exploration and play as examples of prominent facets of agency/competence. These three facets illustrate well how agency and competence are intertwined, although functional competence-oriented aspects are more pronounced in problem solving, while procedural agency-based aspects prevail in play.

3.3.1 Problem solving

Problem solving comes into action when previously applied behavioural solutions no longer work to attain a goal such as obtaining food. The animal then switches back to appetitive behaviours and modifies them, but also engages 'off-line' higher levels of cognitive control where representations of the world beyond the current sensory input as well as memories of the animal's own past actions, successes and failures are stored (Toates, 2004). So problem solving is initially driven by an external situation, but it triggers cognitive processes that have many degrees of freedom and that may well continue beyond the instant when the animal solves the actual problem.

Harlow (1950) was one of the first to demonstrate that problem solving itself is intrinsically rewarding to animals when he showed that rhesus monkeys will manipulate and learn to open a complex six-step mechanical puzzle even when no explicit reward is given for either manipulating or solving the puzzle. Recent evidence for an intrinsic motivation for problem solving came from Langbein *et al.* (2009), who taught dwarf goats to discriminate between sets of visual shapes with water as a reward. When the goats were later presented both with freely available water and water attained through the cognitive task, the goats still oriented about one-third of their drinking activity towards the cognitive task. Such data illustrate that problem solving encompasses an element of active cognitive engagement that goes beyond the immediate problem at hand, and that animals continue to exercise even when the problem no longer exists.

This propensity appears to have longer term beneficial consequences for the animal's ability to cope with its environment. For instance, Bell *et al.* (2009) rearranged the spatial configuration of the living environment for laboratory rats frequently for several weeks and made it more complex every 10 days. Rats living in this dynamic, enriched and cognitively demanding space were subsequently shown to be faster learners in spatial memory and danger avoidance tasks than control rats. Another example is a study by Ernst *et al.* (2005), who provided individual pigs with an automated feeding system that summoned them to the feeding station by individually distinct acoustic stimuli. Pigs cognitively challenged in this way showed fewer aberrant behaviours in the home pen and less fear in a novel environment than conventionally fed pigs, indicating an improvement of their coping abilities (Puppe *et al.*, 2007).

3.3.2 Exploration

Exploration is a form of behaviour that appears to be directly aimed at gathering information (Archer and Birke, 1983; Wemelsfelder and Birke, 1997). If, for example, we observe a laboratory rat placed into an unfamiliar arena, or cattle entering a new pasture, what we are likely to see is the animals moving around inspecting all kinds of stimuli in their new surroundings. However, animals not only explore in response to new situations, but also go out and actively seek novel stimuli, which is often referred to as 'inquisitive exploration'. For instance, piglets prefer to visit places where they can expect novel objects to places where they will encounter familiar objects (Wood-Gush and Vestergaard, 1991). Thus, exploration has its own motivation that is expressed, for instance, in the strong rebound of explorative activity that can be observed when animals housed in impoverished conditions are presented with a novel object or situation (Stolba and Wood-Gush, 1981; Wood-Gush and Vestergaard, 1993).

The motivation for exploration (labelled 'curiosity drive' by Berlyne, 1960) probably evolved because animals need to reduce the environmental uncertainties that they are constantly faced with in the wild (Inglis, 1983, 2000; Dall *et al.*, 2005). Recent research has identified the neural mechanisms that underlie the motivation for exploration. For instance, Cohen *et al.* (2007) argue that the trade-off between the exploitation of known resources and the exploration of new alternatives is governed by a complex interplay of brain systems in which the forebrain cholinergic and adrenergic systems monitor the expected and unexpected forms of environmental uncertainty, medial frontal brain structures report about rewards and costs, and the locus coeruleus noradrenergic system integrates these inputs and shifts the behaviour either towards exploitation or exploration. Intensive research is also being pursued on many other aspects of exploration such as evolutionary

modelling (Dall *et al.*, 2005), or the social dimension of information gathering. Seppanen *et al.* (2007) for example, review evidence that animals explore the environment not only directly but also indirectly, through paying specific attention to cues and signals from both conspecific and heterospecific animals. The common theme of all such research is that active gathering of information brings animals crucial advantages for the future, and they are therefore well equipped and strongly motivated to engage in it.

3.3.3 Play

One of the defining features of play behaviour is that it is a spontaneous, intrinsically motivated activity that is being performed for its own sake rather than to achieve a consummatory goal such as to obtain food, escape a predator or gather information (Burghardt, 2005). It has, therefore, always been assumed that play has deferred, long-lasting positive effects on the development of young animals. More recently, it has been documented that play can also have immediate functions. For instance, domestic dogs seem to confirm their dominance relationships during play (Bauer and Smuts, 2007), and post-pubescent laboratory rats use social play initiation to maintain friendly relationships with the dominant male (Pellis and Pellis, 2009). Nevertheless, the delayed, lasting functions of play are still considered very important. It seems that what animals mainly learn and train for in play are not so much physical fitness/endurance (which fades away quickly; Byers and Walker, 1995), or specific skills such as prey catching in cats (Caro, 1980) or fighting proficiency in meerkats (*Suricata suricatta*) (Sharpe, 2005), as these seem to mature to full function even if play is prevented or reduced, but rather various kinds of general physical and/or psychological flexibility. This is underscored by the facts that the neurobiology of play is distinct from that of 'serious' adult types of behaviour such as social, sexual or aggressive behaviour (Vanderschuren *et al.*, 1997) and that play repertoires include many elements totally dissimilar to 'serious' behaviours (Petrů *et al.*, 2009).

Play has several features that channel it towards creating novelty: play elements are incomplete, exaggerated or awkward compared with elements used in 'serious' contexts; play elements follow each other in variable sequences; and many play

M. Špinka and F. Wemelsfelder

elements have a self-handicapping character, that is, they put the playing animal into unnecessary disadvantageous positions and situations where the animal loses control over its movements (Petrů et al., 2009). For instance, vigorous and variable head rotations, torso twists and body pirouettes are among the most widely occurring elements of play (Byers, 1984). Petrů et al. (2008) analysed the kinematics of play head rotations in Hanuman langurs (*Presbytis entellus*) and found that they include different, sometimes extreme positions of the head that follow each other in variable sequences; vision is most probably blurred during such rotations owing to the high angular velocities. Špinka et al. (2001) specifically suggested that one major and widely present function of play is to train for unexpected situations and mishaps, i.e. to practice in a 'relaxed field' how to handle, behaviourally and emotionally, situations where external forces kick the animal out of control and routine. In their book on rat play fighting, Pellis and Pellis (2009) conclude that juvenile play fighting enhances the experience of unpredictability and thus provides a perfect means by which to fine-tune emotional reactivity, and so to produce an animal that is capable of subtle and nuanced responses to novel and potentially dangerous situations. According to Pellis and Pellis, deprivation of play fighting during rat ontogeny does not take away specific social or cognitive skills but, rather, impairs the ability of the animals to calibrate their emotional response; hence, play-deprived animals are unable to apply their motor, social or cognitive skills effectively in challenging situations.

These examples from different fields of behavioural research indicate and support that animals possess a general skill to initiate, and persist in, interaction with their environment in a way that appears directly beneficial to a range of specific skills and their ability to cope with adverse, restrictive conditions.

3.4 Agency Responds to Appropriate Challenge

The examples of agency/competence in the preceding sections illustrate that an animal may be intrinsically motivated to engage with the environment, but still requires certain conditions to be met before it will do so. Not all types and levels of challenge elicit exploration or play equally – it seems that a moderate degree of challenge is most

likely to evoke a positive interactive response. To capture this, Hebb (1955) proposed a model in which he linked the occurrence of explorative behaviour to optimal levels of arousal, postulating that too little novelty would fail to arouse the animal's attention, whereas too much would startle or frighten it into a fear or stress response. This idea was further developed by Inglis (1983), who suggested that an animal prefers the greatest degree of discrepant input to occur when it is best able to assimilate that input. Watters (2009) takes this notion further in a zoo context, arguing that apparently paradoxically, zoo animals are most motivated to interact with types of enrichment that produce a pay-off that is uncertain, i.e. that is neither guaranteed nor highly improbable. For example, play behaviour tends to be stimulated by environments that are slippery or otherwise tricky to such a degree that full control over the animal's own movements becomes difficult, yet the risk of injury or serious mishap is small, such as is the case for shallow water, fresh snow, sloped terrains, thin flexible branches or swinging suspensions (Byers, 1977; Heinrich and Smolker 1998; Petrů et al., 2009).

Not only is agency best stimulated by intermediate levels of challenge, it also affects the animal's competence most positively at such levels. This notion is akin to the idea of 'eustress': the idea that moderately challenging environments can evoke a stress response in the animal which in the longer term has a positive effect on its survival and welfare (Langbein et al., 2004; Moncek et al., 2004). Summarizing such information, Meehan and Mench (2007, p. 248) propose the notion of 'appropriate challenge', defined as 'problems that may elicit frustration, but are potentially solvable or escapable through the application of cognitive and behavioural skills'. Fig. 3.1 illustrates this notion. If challenges are too strong for the skills of the animal, fear will freeze agency; if challenges are not up to the skills, boredom will result. An appropriate level of challenge stimulates agency and this engagement, in turn, enhances competence. The type and relative level of challenge may also influence the type of agency in which the animal will engage. For instance, relatively high levels of uncertainty or novelty may incite exploration, but as the confidence of the animal with the situation grows, exploration may give way to play (Špinka et al., 2001).

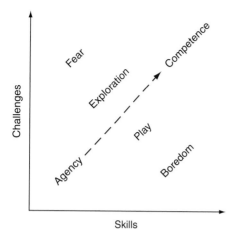

Fig. 3.1. Schematic depiction of the relationship between the strength of a challenge and the strength of an animal's skill.

3.5 The Importance of Agency/Competence for Welfare

Agency and competence are clearly important for the survival and reproductive success of wild animals. But do these abilities also play an important role for the welfare of captive animals whose survival and concomitant needs are mostly met by the captive environment? In the following paragraphs, we argue that there might be a threefold positive relevance of agency/competence for animal welfare. In Fig. 3.2a (and also in Fig. 3.2b, as discussed in Section 3.6), we illustrate our argument with the simplifying assumption that feelings (i.e. sentient experiences) are at the core of animal welfare (Duncan, 1996), although more complex approaches that also include health (Dawkins, 2008), naturalness (Appleby, 1999; Fraser and Weary, 2005) and indeed the wholeness of the animal (Wemelsfelder, 2007; see also Section 3.8) may be more appropriate.

First, there is growing evidence that expressions of agency are rewarding for animals independently of any functional outcome that they may have. We know, as argued above, that animals will engage in solving problems without any apparent form of external reward, and this suggests they may enjoy the process of learning itself (Harlow, 1950). For example, pigs, cattle, chickens and other species engage in what is known as 'contra-freeloading'; that is, they will make an effort to work for a reward even if the reward is also available freely (Inglis *et al.*, 1997; de Jonge *et al.*, 2008; Hessle *et al.*, 2008; Lindqvist and Jensen, 2008). Combining a cognitive task with a food reward in an otherwise barren environment can lead to overeating in rats and goats; this also supports the rewarding value of such tasks over and above obtaining food (Johnson *et al.*, 2004; Langbein *et al.*, 2009). Physiological evidence in support of such value comes from a study by Kalbe and Puppe (2010), who found that long-term cognitive enrichment for pigs in the form of an operant feeding system significantly affects gene expression of reward-sensitive cerebral receptors in the amygdalae of the animals concerned.

In addition, a number of studies indicate more directly that the process of problem solving affects an animal's mood. Hagen and Broom (2004) set five heifers (the experimental animals) the task of learning how to open a gate to gain access to a food reward, and matched these animals with five control heifers for whom the gate opened automatically the moment that the experimental animals had succeeded in opening the gate. Detailed comparison of fluctuations in the heart-rate and behavioural vigour of the two animal groups led the authors to suggest that progression of the problem-solving process in experimental animals was associated with raised arousal and agitation, and that this may in turn reflect an awareness by the animal of its progress in learning – in other words, an understanding of, and excitement about, 'getting there'. Whether this was mainly a positive or negative excitement (i.e. frustration or enjoyment) cannot be told from these quantitative data. A qualitative assessment approach addressing an animal's 'body language', such as developed by Wemelsfelder *et al.* (2001, 2009), may shed further light on the actual experience of such experimental animals. Langbein *et al.* (2004), in a study of instrumental learning in dwarf goats, also looked in detail at correlations between the learning process, learning success and physiological indicators such as heart rate and heart rate variability. Like Hagen and Broom (2004), they suggest that the observed response patterns reflect a process of understanding of, and gaining control over, the task – a process they interpret in terms of 'positive stress'. Yet this term still reflects an abstract scientific understanding; what precisely this means for the actual experience of the animals requires more direct qualitative investigation of their behavioural expression (Wemelsfelder, 2007).

M. Špinka and F. Wemelsfelder

Expressions of agency other than problem solving appear to be similarly self-rewarding. Play routinely tends to be considered as self-rewarding because it does not result in any obvious goal, and often emanates a sense of relaxed and intensive in-the-moment enjoyment (Fagen, 1992; Held and Špinka, 2011). Common ravens, for example, will fly upside down, slide down snowy slopes on their backs, play tug of war, or play 'pass the stick' in mid-air (Heinrich and Smolker, 1998); Siberian ibex kids may jump into the air from overhangs and perform two or three neck twists and heel kicks before landing (Byers, 1977); and domestic piglets can stimulate each other into a playing frenzy in which the whole litter sprints around barking excitedly (M. Špinka, personal observation). Such behavioural examples are complemented by evidence that the performance of play instigates an increase in brain opioid levels (Vanderschuren *et al.*, 1995).

Exploration, too, is positively valued by animals, even when the acquired information has no immediate use for feeding, sexual behaviour or other actions leading to consummatory rewards. Wood-Gush and Vestergaard (1991) demonstrated that piglets select an arena containing a novel object rather than one containing a familiar object, even though neither object was of any utility. Moreover, the piglets displayed an increase of locomotor play near the novel objects, indicating their positive experience from the exploration. Newberry (1999) showed that domestic chickens value the possibility of exploring novel objects of zero utility as much as the exploitation of non-essential resources such as peat moss or straw bale.

The second way in which agency may benefit welfare stems from the self-building nature of agency/competence. The idea is that because animals benefit from being as competent as possible, and because exercising agency engenders competence, animals will get 'drawn into' agency-based activity through positive competence–reinforcement loops (Fig. 3.2). Inglis *et al.* (2001) and Inglis and Langton (2006) showed that mathematical modelling of behaviour based on such starting points could indeed simulate empirically observed behaviour in, for example, studies of contrafreeloading and latent learning. Csikszentmihalyi (1992) is an author well known for his studies of the interaction between competence and happiness in humans, which resulted in what he calls a 'theory of flow'. This theory posits that a person's experienced happiness is a function of the interaction between perceived

challenge and skill levels. If the level of challenge is perceived to be higher than the level of skill, a person will try to learn new skills, while if perceived skills are greater than the challenge, he/she will seek more challenge. Thus, perceived challenge and skill chase each other, which, Csikszentmihalyi argues, leads to 'reorganization and growth in the order and complexity of consciousness' (Moneta and Csikszentmihalyi, 1996, p. 277), a process subjectively experienced as 'flow'. The greatest experience of 'flow', and the greatest associated happiness, arise when both perceived challenges and skills are high and the person is intensely engaged with what he or she is doing through focused attention and sustained concentration and activity. So for animals as well, opportunities to initiate and maintain meaningful cycles of behavioural and cognitive effort are likely to produce a similar feeling of 'flow', and thereby contribute significantly to their longer term welfare.

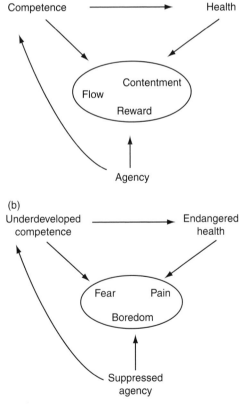

Fig. 3.2. Welfare effects of agency (a) expressed and (b) suppressed.

The third welfare effect of agency arises from the influence that increased competence potentially has on an animal's physical health and fitness, and thereby on positive feelings such as contentment. In the first place, increased levels of interaction and mobility may have direct physical benefits, such as stronger bones, stronger muscles, a stronger heart and higher physical endurance (Spangenberg *et al*, 2005, 2009; Schenck *et al.*, 2008), while improved sensorimotor coordination has been shown to lead to greater neural complexity and plasticity which, in turn, may enhance physical fitness and rehabilitation (Kleim and Jones, 2008). More generally though, through the skills and information that they acquire, active animals are likely to be more confident and perform better in fulfilling their daily needs – ending up better fed, better protected and socially better positioned than animals that are highly restrained. Evidence is growing that such general improvement of an animal's ability to cope can affect its health and fitness even in captive environments; for example, rats exposed to spatially demanding tasks are subject to lower mortality rates (Bell *et al.*, 2009), while pigs taught to discriminate between sounds to obtain food showed faster wound healing and better overall immunity than pigs not exposed to these challenges (Ernst *et al.*, 2006). So allowing an animal to exercise agency not only affects its immediate welfare, but is also likely to improve its longer term physical health and fitness.

3.6 The Consequences of Suppressed Agency/Competence in Restrictive Environments

What happens when animals adapted to deal with the vagaries of natural environments are born and raised in simple, restrictive captive environments? Does this affect their expression of agency/competence, and concomitantly, the richness of their behaviour and experience? Also, most importantly in the context of this book: what are the consequences of such behavioural and cognitive changes for their welfare?

Agency is intrinsic to the way animals behave, and animals will therefore engage with their environment even if it is unchallenging, restrictive and unresponsive. However, such environments will limit the frequency and diversity with which agency is expressed. For instance, enclosures are often much too small to allow animals to go out and look for novelty, and prevent the expression of inquisitive exploration; at best, an opportunity for inspective exploration may occasionally arrive. Equally, if social partners are few, of the wrong category, or absent altogether, an animal will not be able to engage in social play (Pellis and Pellis, 2009). With the animal's agency suppressed this way, its development of competence will soon hit a limit too (Fig. 3.3), and the animal's scope for interaction will remain limited and stagnant (Von Frijtag *et al.*, 2002).

There is considerable evidence indicating that animals housed in restrictive environments show reduced behavioural diversity both in their home pens (Gunn and Morton, 1995; Haskell *et al.*, 1996) and in response to novel objects (Beattie *et al.*, 2000; Wemelsfelder *et al.*, 2000; Meehan and Mench, 2002), and use a large proportion of their time lying down, sleeping or dozing (Gunn and Morton, 1995, Zanella *et al.*, 1996). They may also spend extended periods of time in motionless sitting or standing, often with drooping heads and ears, half-closed eyes, abnormally bent limbs, or pressed against a wall or stall division. Such passive postures have been characterized qualitatively as apathetic, helpless or depressed (Wood-Gush and Vestergaard, 1991; Martin, 2002). In addition, certain behaviour patterns appear to become less versatile, and more fixed and compulsive in their execution; stereotyped pacing in captive polar bears, for example, has been linked to the frustration of their freedom to patrol (Clubb and Mason, 2004), while food-related stereotypies in intensively

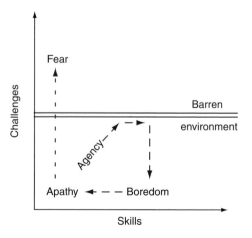

Fig. 3.3. Effects of suppressed agency on animals living in barren environments.

M. Špinka and F. Wemelsfelder

farmed animals are linked to, among other factors, the prevention of foraging (Mason and Mendl, 1997). Stereotyping animals may also show an overly aggressive and/or fearful reaction to novel or unexpected events (Broom, 1986).

Thus, captive animals may be physically mobile and respond to perceived stimuli, but the question is whether such activity is reflective of normal agency. Do the animals interact resourcefully and playfully with their environment, are they busy and absorbed in organizing their lives? The passive, unvaried and sometimes rigid nature of behaviour patterns that are, as discussed in the paragraph above, observed in captive animals, suggests this may not be the case. Wemelsfelder (2005) proposes that such characteristics, in their multifaceted complexity, reflect a chronic disruption of 'flow' in the organization of an animal's behaviour (cf. Csikszentmihalyi, 1992; see Section 3.5), and with that a potential suppression, or even dismantling, of its agency (Fig. 3.3). Animals are prevented from sustaining activities that they are motivated to perform, and so their engagement with the environment deteriorates, and the versatility and flow of their behaviour dries up.

What are the welfare consequences of such deterioration? First, the suppression of agency will directly affect an animal's emotional state in various ways. For a start, inability to express agency will deprive animals of experiencing the positive feelings that accompany agency (Fig. 3.2). In recent years, there has been increasing emphasis on the importance of the positive aspects of welfare for an animal's well-being, related for example to social interaction or exploration and play (Boissy et al., 2007; Napolitano et al., 2009). Welfare is no longer viewed solely in terms of functional health and the absence of suffering, but also in terms of positive experiences or, generally, good 'quality of life' (McMillan, 2005). In Section 3.5 the expression of agency was described as being associated with the positive interest and excitement of exploration and problem solving, the relaxation and enjoyment of play, and the contentment and happiness of experiencing sustained competence and 'flow'. Depriving animals of opportunities for agency would therefore deny them a source of naturally sustained positive experience. Clearly, more research is needed to investigate whether, and under what circumstances, animals experience states of pleasure, enjoyment or contentment.

In addition, various authors have suggested that the suppression of agency is also associated with an experience of boredom and, in the longer run, perhaps also with states of depression and/or helplessness. Glanzer (1958) argued that suboptimal levels of information processing induce boredom, but adds that animals growing up in impoverished environments may adjust to such conditions and not acutely suffer. Inglis (1983) postulates as well that animals may get used to discrepancies between expected and actual levels of novelty, and although initially bored, may eventually settle down. So one could argue that passivity in captive animals could be seen as an adaptive strategy or life history avenue designed for highly predictable environments. However, the question is to what extent such assertions correspond with actual observed behavioural processes in captive animals. Wemelsfelder (2005), as discussed above, argues that a multifaceted range of behavioural symptoms in captive animals reflects chronic disruption of behavioural 'flow', and that, as such, this should not be regarded as a form of functional adaptation. Rather, in analogy with human behavioural organization, such symptoms may be considered indicative of chronic boredom and depression, or general psychological atrophy (cf. Csikszentmihalyi, 1975; Harris, 2000). Early discussion of this condition in animals was provided by McFarland (1989), who characterized states of chronic behavioural deprivation and indecision as 'limbo'. In the more recent experimental literature, the term boredom is frequently used to interpret the observed effects of impoverished environments on animal welfare (e.g. Newberry, 1999; Ernst et al., 2005; Manteuffel et al., 2009). Therefore, if we can accept that through active engagement with their environment animals experience meaning and enjoyment in what they do, then there seems to be no reason why chronic disruption of such engagement should not be experienced as debilitating, boring or even depressingly dull.

A second way in which declining agency may affect welfare is through the consequences of underdeveloped competence, such as heightened fear and anxiety and compromised social coping. Not exercising competence may lead animals to regress in their ability to develop appropriate expectancies and act upon these; they become less well able to classify and evaluate perceived environmental stimuli, and will be less ready to deal with challenges once they arise. As a consequence, unexpected or novel events will startle and arouse animals more, and they will fail to exploit the novel information available. Laboratory animals, for example, may adjust poorly to experimental procedures which involve transport and handling (e.g. Rennie and Buchanan-Smith, 2006), while for farm animals, arrival at a slaughterhouse, with its crowded, noisy, unfamiliar conditions, may be particularly stressful (Deiss et al., 2009). Rats and pigs deprived of natural amounts of play during early periods of ontogeny have compromised ability to solve social conflicts and therefore experience more prolonged and/or more intense fights (Pellis and Pellis, 2009; Newberry et al., 2000). Thus, rather than living in quiescent adjustment to their barren surroundings, animals from impoverished backgrounds may be overwhelmed by events when they arise, fail to cope and experience intense fear or anxiety (Von Frijtag et al., 2002; Chaloupková et al., 2007).

A third potential implication of declining agency for welfare lies in health-related effects. As described at the end of Section 3.5, animals living permanently in impoverished surroundings may heal from injuries less well (Ernst et al., 2006), and are therefore likely to experience more pain. Farm animals living in more restrictive systems generally suffer from a higher incidence of painful production-related diseases, such as mastitis and lameness in dairy cows (Laven et al., 2008a,b), or leg problems in pigs (Kilbride et al., 2009a,b). To what extent these deleterious health effects are indeed a result of the effects of impoverished agency remains to be established.

3.7 Is the Expression of Agency/Competence the Same in All Animals?

The ability of animals to act competently in a challenging environment does not imply that animals always tackle challenges in innovative or complicated ways. In everyday tasks, animals use well-proven strategies, long-adopted behavioural routines, simple rules of thumb and direct cues from the environment whenever possible (e.g. McLinn and Stephens, 2006). For instance, after dwelling in a locality for some time, the process of motor learning trains the animal to move quickly and efficiently through the environment, with little further cognitive processing (Stamps, 1995). Also, animal species, populations and individuals often specialize on a limited diet, even though other types of food in the environment are available and potentially equally rewarding (Tosh et al., 2009). So many everyday tasks are performed in a skilled, routine way, and yet the mastery of routine is complemented by the astuteness and flexibility of competence (when challenges arise) and by the self-driven dynamism of agency.

Notwithstanding the fundamental value of agency, its expression can differ to considerable extent between species, age groups, gender categories and individuals. In other words, the quantitative balance between application of routines and the utilization of competence- and agency-related behaviours varies across species, animals and situations. For instance, among non-human primate species, the proportion of time devoted to social play ranges between 1 and 22% (Lewis and Barton, 2006), a variation that correlates closely with differences in relative volume of the amygdala and hypothalamus, two brain regions involved in the organization of social and emotional responsiveness. In polygynous and promiscuous mammal species, males engage more frequently in social play than females, whereas in monogamous species there is no difference between the sexes (Chau et al., 2008). Even in closely related species, the amount and complexity of play can differ considerably, such as between the Norway rat and domestic mice (Pellis and Pellis, 2009), or between kaka and kea parrots (Diamond and Bond, 2004). Such variations can be due to the different demands of species niches. For instance, migratory garden warblers explore more over a wide area, perhaps because they need to find food quickly during short stays on stopover sites, while the related resident Sardinian warblers are much less keen to explore widely (Mettke-Hofmann and Gwinner, 2004). However the amount of everyday expressions of agency is not in any obvious way related to the apparent intelligence of a species. For instance, wild-living orang-utans are renowned for their tendency to spend several days on one tree with abundant fruits just sitting, eating and sleeping (Lhota, personal communication), while some reptiles, such as monitor lizards, may be agile learners when it comes to novel ways for acquiring food (Manrod et al., 2008).

Ontogenetic variation in agency can also be prominent. During the later phases of ontogeny, expressions of agency should theoretically decrease because the shortening remainder of lifespan diminishes the value of future competence in relation to

M. Špinka and F. Wemelsfelder

the current costs of agency. Indeed, declining levels of exploratory behaviour after the prepubertal peak have been documented in rats and mice (Arakawa, 2007). Some facets of agency are mostly expressed during a well-defined period of ontogeny. For instance, play has a typical inverted U-shaped ontogenetic occurrence with the highest levels during the juvenile period in mice, rats, cats and pigs (Byers and Walker, 1995; Blackshaw *et al.*, 1997).

Furthermore, the expression of agency varies considerably within age and gender categories of the same species. It has long been recognized that individual animals respond to challenges in different ways, and this fact has been the focus of intensive study over the past two decades, often under headings such as 'individual difference', 'coping style', or 'temperament' (Croft *et al.*, 2009; Jones and Godin, 2010), although where such studies include the tendency of animals to explore, play and be sociable, researchers are more likely to speak of 'personality dimensions' (Svartberg *et al.*, 2005; Smith and Blumstein, 2008). From such differences, it follows that restrictive environments and their inhibiting effect on agency may affect some individuals more than others, and will certainly affect different individuals in different ways. Research with captive orang-utans for example, has indicated that animals scoring highly on 'extraversion' and 'agreeableness' personality factors and low on 'neuroticism' factors also score highly on a subjective well-being questionnaire (Weiss *et al.*, 2006).

Finally, even within one individual animal, the propensity to explore, play or patrol may vary with different locations, seasons, times of day or even moods. Theoretically, an animal should engage in such behaviours if and when this is likely to provide benefits that are greater than the costs, relative to the other things that it could be doing (Dall *et al.*, 2005). In times of nutritional hardship, many mammalian species will dramatically reduce their engagement in play, focusing instead on essential foraging and maintenance behaviours (Baldwin and Baldwin, 1976; Muller-Schwarze *et al.*, 1982; Stone, 2008). It has been suggested that, in this light, play may be considered a 'luxury behaviour' (Lawrence, 1987). While there may be clear prioritization of life-sustaining and reproductive behaviours over play, and possibly over other types of agency in many instances, agency-type behaviours are often reinstated at an early occasion, indicating

that they do have long-term fitness benefits. Moreover, agency-type behaviours, including play, may actually increase during tense times, such as pre-feeding periods or crowded indoor housing (Palagi *et al.*, 2006; Tacconi and Palagi, 2009), because they enhance competence through, for instance, relieving tension or signalling friendly intentions. The term 'luxury behaviour' is also misleading from a welfare point of view because it disregards how systematically important agency-type behaviours are for the positive side of animal welfare (Held and Špinka, 2011).

As we argue throughout this chapter, the continuous opportunity to engage in agency and thus to enhance competence is fundamental for the welfare of any animal. However, the variability in quantity and quality of agency expression that we have just discussed indicates that how specifically welfare will be compromised as a result of the suppression of agency in captive environments depends very much on the species and category of animals.

3.8 The Integrated Nature of Agency and Competence – What Does It Mean?

We have discussed different expressions, functions and benefits of agency and competence. However, it is important to note that in the end agency is an integrative capacity that works across specific modules of organization. The neuroscience literature also agrees that there are common neuro-motivational pathways connecting specific modules and integrating emerging information into anticipatory, flexible, reward-sensitive patterns of behaviour (Van der Harst and Spruijt, 2007). Yet, in the neuroscience literature that discusses such integrative systems, one seldom finds references to agency and/or competence. 'Integration' is not necessarily conceived as something actively *done by* the animal, but rather as something that *happens in* the animal, a systemic feature of behavioural organization which can at best be regarded as possibly a 'higher-order' neural state (e.g. Sterelny, 2001). This may perhaps seem a small semantic disparity, but philosophically it lies at the heart of what it means to talk about agency. In the final section of this chapter we will, therefore, briefly touch on the philosophical debate surrounding this concept.

Philosophical discussions of agency tend to centre around the question to what extent, if at all, humans and animals can be considered 'do-ers', that is, 'authors' of their own conduct (Hyman and

Steward, 2004; McFarland and Hediger, 2009). Traditionally, in this context, agency is seen as yet another signifier of the human–animal divide: humans behave intentionally, with insight and foresight, and hence can be held responsible for their conduct, whereas animals behave instinctively, blindly and cannot. In recent times, however, this ground has begun to shift. Animal intentionality, the question of whether or not animals act 'knowingly', is the subject of a rapidly growing field of research covering an ever-widening range of species (e.g. Hurley and Nudds, 2006). In this field, a wide range of evidence supporting the cognitive mediation of animal behaviour has emerged, such as, for example, evidence for foresight in pigs (Špinka *et al.*, 1998); yet scientific opinion on the extent to which such mediation reflects 'true understanding' remains deeply divided. There is much talk about 'lower-order' and 'higher-order' levels of intentionality, with recent studies proposing that it is not basic cognition, but 'metacognition', an ability apparently shown by few animal species, that reflects true intentionality (Smith, 2009). At present, it remains extremely difficult to find criteria of any kind that unambiguously distinguish 'true' from 'apparently' intentional behaviour, and discussions of what such a distinction might imply continue with unabated vigour.

But there is also a second way in which traditional views of agency are shifting. In this, the notion of agency is used to emphasize the integrity of the whole animal, and to discuss critically mechanistic/informational models that typically separate ('lower') bodily behaviour and ('higher') mental processing of information into different conceptual realms. Cognitive approaches have, for example, been criticized for overintellectualizing agency, making application to animals less likely (e.g. Hurley, 2006; Steward, 2009). A more holistic approach would regard animals generally as integrated sentient beings and, in this context, the notion of agency would not primarily refer to the direction of action by thought, but to the centrality of the whole animal in directing action (Hornsby 2004; Hurley, 2006). Such a notion does not encounter the need to distinguish between 'blind' and 'knowing' behaviour, because it regards all behaviour as sentient, and all animal sentience as embodied, and thus views sentience and intelligence as fundamental and gradually evolving properties of behavioural organization. Whether or not this is a reasonable, scientifically acceptable proposition is a question that goes to the heart of what it means to do science; clearly, there is not the space here to discuss that question at any length. One advantage of this approach for animal welfare, however, is that an animal's experience is conceived as an integrative aspect of its behavioural expression, and hence becomes more directly observable, and more amenable to description and interpretation, than it would have been if it were purely regarded as an 'internal mental state' (Wemelsfelder, 1997, 2007).

The material discussed in this chapter is not meant to support any of these approaches in particular – we have drawn equally on physiological, health-related, behavioural and cognitive studies wherever these were relevant to the theme of the chapter. That theme was to discuss and provide scientific support for the propensity and ability of animals to engage proactively with their environment, and to learn to deal skilfully and flexibly with novel and existing challenges. We think this ability is real, and that in addition to more specific abilities, plays a vital role in ensuring an animal's health and quality of life. Taking agency seriously as a topic for ethology and for animal welfare science and practice is bound to lead to more incisive observation of how animals behave, and therefore to inform the philosophical debate in scientifically relevant ways.

3.9 Conclusions

- Natural environments expose animals to many varied and novel challenges. We argue that animals possess an integrated yet multifaceted ability, which we call competence, to deal with such challenges.
- Competence is reinforced by an animal's agency, i.e. its intrinsic propensity to engage with the environment beyond the degree dictated by momentary needs, with the main purpose of gathering knowledge and enhancing the animal's skills for future use.
- As agency/competence concerns the integration of different levels of organization, it provides animal welfare scientists with an opportunity to address the wholeness of animals, an aspect of welfare that tends to be overshadowed by the focus on specific modules of animal welfare.
- The agency/competence complex is relevant for animal welfare for at least three reasons. First, performance of agency is directly rewarding for the animal. Secondly, when allowed its full

course, agency makes the animal competent to meet high challenges with high skills, a state that has been described as fulfilling in humans, and that presumably would also be in other animals. Thirdly, highly competent animals deal with challenges more efficiently and successfully than less competent ones, and thus end up healthier and less fearful.

- It is, therefore, likely that when captive environments deny animals the opportunity to unfold their agency, they prevent those animals from achieving better welfare in all three aspects: immediate reward value, long-term build-up of positive psychological constitution and the ability to maintain health and psychological balance in the face of challenges.

Acknowledgements

The authors would like to thank Lynda Birke for her contribution to and co-authorship of the precursor of this chapter, which was published in the first edition of the book. In addition, we thank the editors of this second edition of *Animal Welfare* for their patience in awaiting completion of our chapter, and for their helpful suggestions. During writing of this chapter, Marek Špinka was supported by grants No. GAČR P505/10/1411 and MZe MZE0002701404.

References

Appleby, M.C. (1999) *What Should We Do about Animal Welfare?* Blackwell Science, Oxford, UK.

Arakawa, H. (2007) Age-dependent change in exploratory behavior of male rats following exposure to threat stimulus: effect of juvenile experience. *Developmental Psychobiology* 49, 522–530.

Archer, J. and Birke, L.I.A. (eds) (1983) *Exploration in Animals and Humans*. Van Nostrand Reinhold, London.

Baldwin, J.D. and Baldwin, J.I. (1976) Effects of food ecology on social play: a laboratory simulation. *Zeitschrift für Tierpsychologie* 40, 1–14.

Bauer, E.B. and Smuts, B.B. (2007) Cooperation and competition during dyadic play in domestic dogs, *Canis familiaris. Animal Behaviour* 73, 489–499.

Beattie, V.E., O'Connell, N.E., Kilpatrick, D.J. and Moss B.W. (2000) Influence of environmental enrichment on welfare-related behavioural and physiological parameters in growing pigs. *Animal Science* 70, 443–450.

Bell, J.A., Livesey, P.J. and Meyer, J.F. (2009) Environmental enrichment influences survival rate and enhances exploration and learning but produces variable responses to the radial maze in old rats. *Developmental Psychobiology* 51, 564–578.

Berlyne, D.E. (1960) *Conflict, Arousal, and Curiosity.* McGraw-Hill, New York.

Blackshaw, J.K., Swain, A.J., Blackshaw, B.W., Thomas, F.J.M. and Gillies, K.J. (1997) The development of playful behaviour in piglets from birth to weaning in three farrowing environments. *Applied Animal Behaviour Science* 55, 37–49.

Boissy, A., Manteuffel, G., Jensen, M.B., Moe, R.O., Spruijt, B., Keeling, L.J., Winckler, C., Forkman, B., Dimitrov, I., Langbein, J., Bakken, M. and Aubert, A. (2007) Assessment of positive emotions in animals to improve their welfare. *Physiology and Behavior* 92, 375–397.

Broom, D.M. (1986) Stereotypes and responsiveness as welfare indicators in stall-housed sows. *Animal Production* 42, 438–439.

Burghardt, G.M. (2005) *The Genesis of Animal Play: Testing the Limits.* MIT Press, Cambridge, Massachusetts.

Byers, J.A. (1977) Terrain preferences in the play behaviour of Siberian ibex kids. *Zeitschrift für Tierpsychologie*, 45, 199–209.

Byers, J.A. (1984) Play in ungulates. In: Smith, P.K. (ed.) *Play in Animals and Humans*. Basil Blackwell, Oxford, UK, pp. 43–65.

Byers, J.A. and Walker, C. (1995) Refining the motor training hypothesis for the evolution of play. *American Naturalist* 146, 25–40.

Caro, T.M. (1980) The effects of experience on the predatory patterns of cats. *Behavioral and Neural Biology* 29, 1–28.

Chaloupková, H., Illmann, G., Neuhauserová, K., Tománek, M. and Vališ, L. (2007) Pre-weaning housing effects on the behavior and physiological measures of pigs during the sucking and fattening periods. *Journal of Animal Science* 85, 1741–1749.

Chau, M.J., Stone, A.I., Mendoza, S.P. and Bales, K.L. (2008) Is play behavior sexually dimorphic in monogamous species? *Ethology* 114, 989–998.

Clubb, R. and Mason, G. (2004) Pacing polar bears and stoical sheep: testing ecological and evolutionary hypotheses about animal welfare. *Animal Welfare* 13 (Supplement 1), 33–40.

Cohen, J.D., McClure, S.M. and Yu, A.J. (2007) Should I stay or should I go? How the human brain manages the trade-off between exploitation and exploration. *Philosophical Transactions of the Royal Society B – Biological Sciences* 362, 933–942.

Croft, D.P., Krause, J., Darden, S.K., Ramnarine, I.W., Faria, J.J. and James, R. (2009) Behavioural trait assortment in a social network: patterns and implications. *Behavioral Ecology and Sociobiology* 63, 1495–1503.

Csikszentmihalyi, M. (1975) *Beyond Boredom and Anxiety*. Jossey-Bass, San Francisco, California.

Csikszentmihalyi, M. (1992) *Flow. The Classic Work on How to Achieve Happiness*. Rider, London.

Dall, S.R.X., Giraldeau, L.A., Olsson, O., McNamara, J.M. and Stephens, D.W. (2005) Information and its use by animals in evolutionary ecology. *Trends in Ecology and Evolution* 20, 187–193.

Dawkins, M.S. (2008) The Science of Animal Suffering. *Ethology* 114, 937–945.

Deiss, V., Temple, D., Ligout, S., Racine, C., Bouix, J., Terlouw, C. and Boissy, A. (2009) Can emotional reactivity predict stress responses at slaughter in sheep? *Applied Animal Behaviour Science* 119, 193–202.

de Jonge, F.H., Tilly, S.L., Baars, A.M. and Spruijt B.M. (2008) On the rewarding nature of appetitive feeding behaviour in pigs (*Sus scrofa*): Do domesticated pigs contrafreeload? *Applied Animal Behaviour Science* 114, 359–372.

Diamond, J. and Bond, A.B. (2004) Social play in kaka (*Nestor meridionalis*) with comparisons to kea (*Nestor notabilis*). *Behaviour* 141, 777–798.

Duncan, I.J.H. (1996) Animal welfare defined in terms of feelings. *Acta Agriculturae Scandinavica, Section A – Animal Science* 27, 29–35.

Ernst, K., Puppe, B., Schon, P.C. and Manteuffel, G.A. (2005) Complex automatic feeding system for pigs aimed to induce successful behavioural coping by cognitive adaptation. *Applied Animal Behaviour Science* 91, 205–218.

Ernst, K., Tuchscherer, M., Kanitz, E., Puppe, B. and Manteuffel, G. (2006) Effects of attention and rewarded activity on immune parameters and wound healing in pigs. *Physiology and Behavior* 89, 448–456.

Fagen, R. (1992) Play, fun, and communication of well-being. *Play and Culture* 5, 40–58.

Fraser, D. and Weary, D.M. (2005) Applied animal behavior and animal welfare. In: Bolhuis, J.J. and Giraldeau, L.A. (eds) *The Behavior of Animals. Mechanisms, Function and Evolution*. Blackwell Publishing, Malden, Massachusetts, pp. 345–366.

Glanzer, M. (1958) Curiosity, exploratory drive, and stimulus satiation. *Psychological Bulletin* 55, 302–315.

Gunn, D. and Morton, D.B. (1995) Inventory of the behaviour of New Zealand White rabbits in laboratory cages. *Applied Animal Behaviour Science* 45, 277–292.

Hagen, K. and Broom, D.M. (2004) Emotional reactions to learning in cattle. *Applied Animal Behaviour Science* 85, 203–213.

Harlow, H.F. (1950) Learning and satiation of response in intrinsically motivated complex puzzle performance by monkeys. *Journal of Comparative Physiology and Psychology* 43, 289–294.

Harris, M.B. (2000) Correlates and characteristics of boredom proneness and boredom. *Journal of Applied Social Psychology* 30, 576–598.

Haskell, M., Wemelsfelder, F., Mendl, M., Calvert, S. and Lawrence, A.B. (1996) The effect of substrate-enriched and substrate-impoverished housing environments on the diversity of behaviour in pigs. *Behaviour* 133, 741–761.

Hebb, D.O. (1955) Drives and the C.N.S. (Conceptual nervous system). *Psychological Review* 62, 243–254.

Heinrich, B. and Smolker, R. (1998) Play in common ravens. In: Bekoff, M. and Byers, J.A. (eds) *Animal Play: Evolutionary, Comparative, and Ecological Perspectives*. Cambridge University Press, Cambridge, UK, pp. 27–45.

Held, S.D.E. and Špinka, M. (2011) Animal play and animal welfare. *Animal Behaviour* (in press) doi:10.1016/j.anbehav.2011.01.007.

Hessle, A., Rutter, M. and Wallin, K. (2008) Effect of breed, season and pasture moisture gradient on foraging behaviour in cattle on semi-natural grasslands. *Applied Animal Behaviour Science* 111, 108–119.

Hornsby, J. (2004) Agency and actions. In: Hyman, J. and Steward, H. (eds) *Agency and Action*. Cambridge University Press, Cambridge, UK, pp. 1–23.

Hurley, S. (2006) Making sense of animals. In: Hurley, S. and Nudds, M. (eds) *Rational Animals?* Oxford University Press, Oxford, UK, pp. 139–171.

Hurley, S. and Nudds, M. (eds) (2006) *Rational Animals?* Oxford University Press, Oxford, UK.

Hyman, J. and Steward, H. (eds) (2004) *Agency and Action*. Cambridge University Press, Cambridge, UK.

Inglis, I.R. (1983) Towards a cognitive theory of exploratory behaviour. In: Archer J. and Birke L.I.A. (eds) *Exploration in Animals and Humans*. Van Nostrand Reinhold, London, pp. 72–117.

Inglis, I.R. (2000) The central role of uncertainty reduction in determining behaviour. *Behaviour* 137, 1567–1599.

Inglis, I.R. and Langton, S. (2006) How an animal's behavioural repertoire changes in response to a changing environment: a stochastic model. *Behaviour* 143, 1563–1596.

Inglis, I.R., Forkman, B. and Lazarus, J. (1997) Free food or earned food? A review and fuzzy model of contrafreeloading. *Animal Behaviour* 53, 1171–1191.

Inglis, I.R., Langton, S., Forkman, B. and Lazarus, J. (2001) An information primacy model of exploratory and foraging behaviour. *Animal Behaviour* 62, 543–557.

Johnson, S.R., Patterson-Kane, E.G. and Niel, L. (2004) Foraging enrichment for laboratory rats. *Animal Welfare* 13, 305–312.

Jones, K.A. and Godin, J.G.J. (2010) Are fast explorers slow reactors? Linking personality type and anti-predator behaviour. *Proceedings of the Royal Society B – Biological Sciences* 277, 625–632.

Kalbe, C. and Puppe, B. (2010) Long-term cognitive enrichment affects opioid receptor expression in the amygdala of domestic pigs. *Genes, Brain and Behavior* 9, 75–83.

M. Špinka and F. Wemelsfelder

Kilbride, A.L., Gillman, C.E., Ossent, P. and Green, L.E. (2009a) A cross sectional study of prevalence, risk factors, population attributable fractions and pathology for foot and limb lesions in preweaning piglets on commercial farms in England. *BMC Veterinary Research* 5: 30.

Kilbride, A.L., Gillman, C.E. and Green, L.E. (2009b) A cross sectional study of the prevalence, risk factors and population attributable fractions for limb and body lesions in lactating sows on commercial farms in England. *BMC Veterinary Research* 5: 31.

Kleim, J.A. and Jones, J.A. (2008) Principles of experience-dependent neural plasticity: implications for rehabilitation after brain damage. *Journal of Speech Language and Hearing Research* 51, S225–S239.

Langbein, J., Nurnberg, G. and Manteuffel, G. (2004) Visual discrimination learning in dwarf goats and associated changes in heart rate and heart rate variability. *Physiology and Behavior* 82, 601–609.

Langbein, J., Siebert, K. and Nurnberg, G. (2009) On the use of an automated learning device by group-housed dwarf goats: do goats seek cognitive challenges? *Applied Animal Behaviour Science* 120, 150–158.

Laven, R.A. and Holmes, C.W. (2008a) A review of the potential impact of increased use of housing on the health and welfare of dairy cattle in New Zealand. *New Zealand Veterinary Journal* 56, 151–157.

Laven, R.A., Lawrence, K.E., Weston, J.F., Dowson, K.R. and Stafford, K.J. (2008b) Assessment of the duration of the pain response associated with lameness in dairy cows, and the influence of treatment. *New Zealand Veterinary Journal* 56, 210–217.

Lawrence, A. (1987) Consumer demand theory and the assessment of animal welfare. *Animal Behaviour* 35, 293–295.

Lewis, K.P. and Barton, R.A. (2006) Amygdala size and hypothalamus size predict social play frequency in nonhuman primates: a comparative analysis using independent contrasts. *Journal of Comparative Psychology* 120, 31–37.

Lindqvist, C. and Jensen, P. (2008) Effects of age, sex and social isolation on contrafreeloading in red junglefowl (*Gallus gallus*) and White Leghorn fowl. *Applied Animal Behaviour Science* 114, 419–428.

Manrod, J.D., Hartdegen, R. and Burghardt, G.M. (2008) Rapid solving of a problem apparatus by juvenile black-throated monitor lizards (*Varanus albigularis albigularis*). *Animal Cognition* 11, 267–273.

Manteuffel, G., Langbein, J. and Puppe, B. (2009) From operant teaming to cognitive enrichment in farm animal housing: bases and applicability. *Animal Welfare* 18, 87–95.

Martin, J.E. (2002) Early life experiences: activity levels and abnormal behaviours in resocialised chimpanzees. *Animal Welfare* 11, 419–436.

Mason, G. and Mendl, M. (1997) Do the stereotypies of pigs, chickens and mink reflect adaptive species differences in the control of foraging? *Applied Animal Behaviour Science* 53, 45–58.

Matzel, L.D. and Kolata, S. (2010) Selective attention, working memory, and animal intelligence. *Neuroscience and Biobehavioral Reviews* 34, 23–30.

McFarland, D.J. (1989) *Problems of Animal Behaviour*. Longman, Harlow, UK.

McFarland, S.E. and Hediger, R. (eds) (2009) *Animals and Agency*. Brill, Leiden, The Netherlands.

McLinn, C.M. and Stephens, D.W. (2006) What makes information valuable: signal reliability and environmental uncertainty. *Animal Behaviour* 71, 1119–1129.

McMillan, F. (ed.) (2005) *Mental Health and Well-Being in Animals*. Blackwell Publishing, Oxford, UK.

Meehan, C.L. and Mench, J.A. (2002) Environmental enrichment affects the fear and exploratory responses to novelty of young Amazon parrots. *Applied Animal Behaviour Science* 79, 75–88.

Meehan, C.L. and Mench, J.A. (2007) The challenge of challenge: can problem solving opportunities enhance animal welfare? *Applied Animal Behaviour Science* 102, 246–261.

Mettke-Hofmann, C. and Gwinner, E. (2004) Differential assessment of environmental information in a migratory and a nonmigratory passerine. *Animal Behaviour* 68, 1079–1086.

Moncek, F., Duncko, R., Johansson, B.B. and Ježová, D. (2004) Effect of environmental enrichment on stress related systems in rats. *Journal of Neuroendocrinology* 16, 423–431.

Moneta, G.B. and Csikszentmihalyi, M. (1996) The effect of perceived challenges and skills on the quality of subjective experience. *Journal of Personality* 64, 275–310.

Muller-Schwarze, D., Stagge, B. and Muller-Schwarze, C. (1982) Play-behavior – persistence, decrease, and energetic compensation during food shortage in deer fawns. *Science* 215, 85–87.

Napolitano, F., Knierim, U., Grasso, F. and De Rosa, G. (2009) Positive indicators of cattle welfare and their applicability to on-farm protocols. *Italian Journal of Animal Science* 8 (1s – [Special Issue] 'Criteria and Methods for the Assessment of Animal Welfare'], 355–365.

Newberry, R.C. (1999) Exploratory behaviour of young domestic fowl. *Applied Animal Behaviour Science* 63, 311–321.

Newberry, R.C., Špinka, M. and Cloutier, S. (2000) Early social experience of piglets affects rate of conflict resolution with strangers after weaning. In: *Proceedings of the 34th International Congress of the ISAE [International Society for Applied Ethology]*, 14–18 October 2000, Florianopolis, Brazil, p.67.

Palagi, E., Paoli, T. and Borgognini Tarli, S. (2006) Short-term benefits of play behavior and conflict prevention in *Pan paniscus*. *International Journal of Primatology* 27, 1257–1270.

Pellis, S. and Pellis, V. (2009) *The Playful Brain. Venturing to the Limits of Neuroscience.* Oneworld Publications, Oxford, UK.

Petrů, M., Špinka, M., Lhota, S. and Šípek, P. (2008) Head rotations in the play of hanuman langurs (*Semnopithecus entellus*): a description and an analysis of their function. *Journal of Comparative Psychology* 122, 9–18.

Petrů, M., Špinka, M., Charvátová, V. and Lhota, S. (2009) Revisiting play elements and self-handicapping in play: a comparative ethogram of five old world monkey species. *Journal of Comparative Psychology* 123, 250–263.

Puppe, B., Ernst, K., Schon, P.C. and Manteuffel, G. (2007) Cognitive enrichment affects behavioural reactivity in domestic pigs. *Applied Animal Behaviour Science* 105, 75–86.

Rennie, A.E. and Buchanan-Smith, H.M. (2006) Refinement of the use of non-human primates in scientific research. Part III: refinement of procedures. *Animal Welfare* 15, 239–261.

Rogers, L.J. (2008) Development and function of lateralization in the avian brain. *Brain Research Bulletin* 76, 235–244.

Schenck, E.L., McMunn, K.A., Rosenstein, D.S., Stroshine, R.L., Nielsen, B.D., Richert, B.T., Marchant-Forde, J.N., and Lay, D.C. (2008) Exercising stall-housed gestating gilts: effects on lameness, the musculo-skeletal system, production, and behavior. *Journal of Animal Science* 86, 3166–3180.

Seppanen, J.T., Forsman, J.T., Monkkonen, M. and Thomson, R.L. (2007) Social information use is a process across time, space, and ecology, reaching heterospecifics. *Ecology* 88, 1622–1633.

Sharpe, L.L. (2005) Play fighting does not affect subsequent fighting success in wild meerkats. *Animal Behaviour* 69, 1023–1029.

Smith, B.R. and Blumstein, D.T. (2008) Fitness consequences of personality: a meta-analysis. *Behavioral Ecology* 19, 448–455.

Smith, J.D. (2009) The study of animal metacognition. *Trends in Cognitive Sciences* 13, 389–396.

Spangenberg, E.M.F., Augustsson, H., Dahlborn, K., Essén-Gustavsson, B. and Cvek, K. (2005) Housing-related activity in rats: effects on body weight, urinary corticosterone levels, muscle properties and performance. *Laboratory Animals* 39, 45–57.

Spangenberg, E., Dahlborn, K., Essén-Gustavsson, B. and Cvek, K. (2009) Effects of physical activity and group size on animal welfare in laboratory rats. *Animal Welfare* 18, 159–169.

Špinka, M., Duncan, I. and Widowski, T. (1998) Do domestic pigs prefer short-term to medium-term confinement? *Applied Animal Behaviour Science* 58, 221–232.

Špinka, M., Newberry, R.C. and Bekoff, M. (2001) Mammalian play: training for the unexpected. *The Quarterly Review of Biology* 76, 141–168.

Stamps, J. (1995) Motor learning and the value of familiar space. *American Naturalist* 146, 41–58.

Sterelny, K. (2001) *The Evolution of Agency and Other Essays.* Cambridge University Press, Cambridge, UK.

Steward, H. (2009) Animal agency. *Inquiry* 52, 217–231.

Stolba, A. and Wood-Gush, D.G.M. (1981) The assessment of behavioral needs of pigs under free-range and confined conditions. *Applied Animal Ethology* 7, 388–389.

Stone, A.I. (2008) Seasonal effects on play behavior in immature *Saimiri sciureus* in eastern Amazonia. *International Journal of Primatology* 29, 195–205.

Svartberg, K., Tapper, I., Temrin, H., Radesater, T. and Thorman, S. (2005) Consistency of personality traits in dogs. *Animal Behaviour* 69, 283–291.

Tacconi, G. and Palagi, E. (2009) Play behavioural tactics under space reduction: social challenges in bonobos, *Pan paniscus. Animal Behaviour* 78, 469–476.

Toates, F. (2004) Cognition, motivation, emotion and action: a dynamic and vulnerable interdependence. *Applied Animal Behaviour Science* 86, 173–204.

Tosh, C.R., Krause, J. and Ruxton, G.D. (2009) Theoretical predictions strongly support decision accuracy as a major driver of ecological specialization. *Proceedings of the National Academy of Sciences of the United States of America* 106, 5698–5702.

Van der Harst, J.E. and Spruijt, B.M. (2007) Tools to measure and improve animal welfare: reward-related behaviour. *Animal Welfare* 16 (Supplement 1), 67–73.

Vanderschuren, L.J., Stein, E.A., Wiegant, V.M. and Van Ree, J.M. (1995) Social play alters regional brain opioid receptor binding in juvenile rats. *Brain Research* 680, 148–56.

Vanderschuren, L.J., Niesink, R.J.M. and Van Ree, J.M. (1997) The neurobiology of social play-behavior in rats. *Neuroscience and Biobehavioral Reviews* 21, 309–326.

Von Frijtag, J.C., Schot, M., van den Bos, R. and Spruijt, B.M. (2002) Individual housing during the play period results in changed responses to and consequences of a psychosocial stress situation in rats. *Developmental Psychobiology* 41, 58–69.

Watters, J.V. (2009) Toward a predictive theory for environmental enrichment. *Zoo Biology* 28, 609–622.

Weiss, A., King, J.E. and Perkins, L. (2006) Personality and subjective well-being in orangutans (*Pongo pygmaeus* and *Pongo abelii*). *Journal of Personality and Social Psychology* 90, 501–511.

Wemelsfelder, F. (1997) The scientific validity of subjective concepts in models of animal welfare. *Applied Animal Behaviour Science* 53, 75–88.

Wemelsfelder, F. (2005) Animal boredom: understanding the tedium of confined lives. In: Macmillan, F. (ed.) *Mental Health and Well-Being in Animals*. Blackwell Publishing, Oxford, UK, pp 79–93.

Wemelsfelder, F. (2007) How animals communicate quality of life: the qualitative assessment of animal behaviour. *Animal Welfare* 16 (Supplement 1), 25–31.

Wemelsfelder, F. and Birke, L.I.A. (1997) Environmental challenge. In: Appleby, M.C. and Hughes, B.O. (eds) *Animal Welfare*. CAB International, Wallingford, UK, pp. 35–47.

Wemelsfelder, F., Haskell, M., Mendl, M., Calvert, S. and Lawrence, A.B. (2000) Diversity of behaviour during novel object tests is reduced in pigs housed in substrate-impoverished conditions. *Animal Behaviour* 60, 385–394.

Wemelsfelder, F., Hunter, E.A., Mendl, M.T. and Lawrence, A.B. (2001) Assessing the 'whole animal': a free-choice-profiling approach. *Animal Behaviour* 62, 209–220.

Wemelsfelder, F., Nevison, I. and Lawrence, A.B. (2009) The effect of perceived environmental background on qualitative assessments of pig behaviour. *Animal Behaviour* 78, 477–484.

White, R.W. (1959) Motivation reconsidered: the concept of competence. *Psychological Review* 66, 297–333.

Wood-Gush, D.G.M. and Vestergaard, K. (1991) The seeking of novelty and its relation to play. *Animal Behaviour* 42, 599–606.

Wood-Gush, D.G.M. and Vestergaard, K. (1993) Inquisitive exploration in pigs. *Animal Behaviour* 45, 185–187.

Zanella, A.J., Broom, D.M., Hunter, J.C. and Mendl, M.T. (1996) Brain opioid receptors in relation to stereotypies, inactivity, and housing in sows. *Physiology and Behaviour* 59, 769–775.

4 Hunger and Thirst

ILIAS KYRIAZAKIS AND BERT TOLKAMP

Abstract

It has been argued that the recommendation of 'freedom from hunger and thirst' leads to confusion because it is generally assumed that hunger and thirst are required 'triggers' for an animal to start feeding or drinking. Here we consider what constitutes 'normal' feeding and drinking behaviour. All animals have such preferred or 'normal' behaviours, which are structured in bouts or meals. Such behaviours are very flexible and there is little evidence to suggest that deviations from a preferred feeding behaviour have significant welfare consequences. Because free and continuous access to a high-quality food leads to overweight, obesity and other problems, animal keepers impose a degree of quantitative or qualitative feeding restriction on animals, which may be associated with effects on animal behaviour. There is agreement between studies about behavioural changes that can be observed when food is qualitatively, compared with quantitatively restricted, but less so when physiological indicators of hunger are considered. These behavioural changes are interpreted by some as signs of improved welfare when animals are qualitatively restricted. Severe food restriction is considered here as undernutrition, and its effects on animal ill thrift, poor performance, severe deterioration of health and eventual death are well established. Malnutrition arises when an animal is given access to food that is inappropriately balanced. Severe malnutrition leads to problems similar to nutrient restriction and, in addition, may give rise to obesity. However, whether subtle cases of malnutrition lead to any behavioural changes and deteriorations in welfare is subject to debate. Finally, we suggest substituting the term thirst with that of 'water restriction', which may arise intentionally or unintentionally. Because of the many functions that water performs, even short-term restriction will lead to deterioration in health and welfare. There are many instances where water restriction can arise unintentionally, despite the fact that animal keepers frequently assume free and continuous availability of and access to water, and these are considered here.

4.1 Introduction

Feeding and drinking are the most natural of all animal behaviours, as they have a major effect on the survival of the individual and its reproductive success, i.e. its evolutionary fitness. Hunger and thirst can be considered as the motivational states that lead to the manifestation of these behaviours and, as a consequence, 'are the two most basic, primitive and unremitting of all motivating forces' (Webster, 1995). For this reason 'freedom from hunger and thirst' features as the first of the 'Five Freedoms' that form part of most codes of recommendation for animal welfare (e.g. FAWC, 1992; Animal Welfare Advisory Committee, 1994).

It has been argued, however, that the recommendation of 'freedom from hunger and thirst' leads to confusion because it is generally assumed (Le Magnen, 1985) that hunger and thirst are required 'triggers' for an animal to start feeding or drinking. What was, perhaps, in the minds of those making this recommendation is freedom from severe deprivation of food and water, because, *in extremis*, this will lead to ill health, inability of the animal to perform its physiological functions, poor welfare and, eventually, death. This, however, represents only one end of the spectrum. At the other end there are the many observations where unrestricted access to good-quality food can lead to rapid weight gains and obesity, with all its negative health, reproductive and welfare consequences (D'Eath *et al.*, 2009). Frequently, the dilemma with respect to the first of the Five Freedoms is, therefore, to avoid the negative health and welfare consequences of 'overfeeding' in animals that have free access to high-quality food and to minimize the hunger that can be associated with recommended restrictions of food intake. The apparent conflict between such recommendations and the 'freedom from hunger and thirst' concept has even led to a court case, in which the final judgment highlighted 'the lack of certainty as to meaning which can be attached to expressions such as the "hunger of chickens"' (Compassion In World Farming Ltd./ Defra, 2003).

Restriction of food (especially energy) intake can be achieved by feeding lower quality foods (*qualitative restriction*), which is common practice in ruminant production systems, or by restriction of the amount of high-quality food that animals receive (*quantitative restriction*). Under conditions of severe quantitative food restrictions (*undernutrition*), animals usually consume immediately any food made available and are left without food for the rest of the time. This means that they cannot express their normal feeding behaviour either (which conflicts with another of the five freedoms). This raises issues of what constitutes normal feeding behaviour and what the concept of hunger entails. For this reason, we start this chapter with a brief description of the structure of normal feeding behaviour of animals with free access to food. This allows us to judge the consequences of quantitative and qualitative restriction of food intake on animal behaviour and welfare, which are discussed next. Subsequently, we address the consequences of undernutrition and malnutrition for long-term animal health and welfare. Both these nutritional states are viewed as extremes in the feeding of animals. Finally, issues pertaining to thirst and drinking are discussed.

4.2 Normal Feeding Behaviour of Animals with Free Access to Food

Animals of most species do not feed continuously or randomly in time but do so in bouts. This means that, within bouts, actual feeding may be interrupted by short non-feeding intervals, while bouts, which can be called meals if they are properly identified, are separated by long non-feeding intervals. Animals of species as diverse as locusts, rats and cows terminate a meal as a result of satiety, and satiety will result in a low feeding motivation immediately after finishing a meal (Metz, 1975; Le Magnen, 1985; Simpson, 1995). Gradually, 'hunger' will increase and, therefore, also the probability that an animal will start another meal (Fig. 4.1(a)). This leads to a very typical frequency distribution of between-feeding interval lengths that may be observed in many species (see examples in Fig. 4.1).

To group such feeding behaviour into meals, novel methods to replace the conventional quantitative log-survivorship or log-frequency analyses (Slater and Lester, 1982; Langton *et al.*, 1995), which are clearly not appropriate under such conditions, have recently been developed (Tolkamp *et al.*, 1998; Yeates *et al.*, 2001; Howie *et al.*, 2009a). These methods rely on fitting mixed functions to the frequency distribution of log-transformed interval lengths. When animals do not drink during meals, these distributions (i.e. of intervals within and intervals between meals) are usually log normal (see Fig. 4.2(a) and 4.2(b) for examples). Drinking is, however, commonly associated with meals in many animal species (Bigelow and Houpt, 1988; Forbes *et al.*, 1991; Zorilla *et al.*, 2005) and normal feeding behaviour can only be characterized properly if this is taken into account (e.g. for cows, Yeates *et al.*, 2001; for rats, Zorrilla *et al.*, 2005). Figure 4.2(c) shows that animals that drink during meals show three populations of between-feeding intervals (Tolkamp and Kyriazakis, 1999; Yeates *et al.*, 2001). Finally, animals that only drink during meals (such as neonate mammals) seem to have a similar feeding behaviour structure to that of animals that do not drink at all during meals (compare Fig. 4.2(d) with Fig. 4.2(b)).

Patterns of normal feeding behaviour as characterized in Fig. 4.2 seem to be virtually unaffected by genetic selection for productive traits in intensively selected farm animals, such as broiler chickens (Howie *et al.*, 2009b). These findings contradict the view that animals intensively selected for productive traits are in a constant state of hunger (Burkhart *et al.*, 1983; Bokkers and Koene, 2003). In addition, the same meal patterns continue to be found when animals are subjected to qualitative restriction, such as dairy cows with free access to high-forage compared with high-concentrate diets (Tolkamp *et al.*, 2002).

Changes in the normal patterns of short-term feeding behaviour may be used for the early detection of health and welfare problems. Gonzalez *et al.* (2008) showed, for instance, that lame cows change their short-term feeding behaviour long before the condition is recognized by experienced farm staff. Presumably because standing is painful, lame cows may double or even triple their feeding rate and still maintain their intake while substantially reducing the duration of their meals. Similarly, automatic monitoring of milk consumption patterns can detect changes in normal drinking behaviour

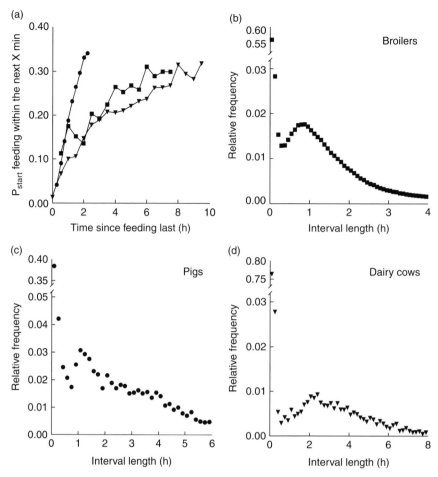

Fig. 4.1. The systematic increase in the probability of animals starting to feed (P_{start}) within the next X min (i.e. 15 min for broilers and 30 min for cows and pigs) in relation to time since feeding last of animals with continuous free access to food (a) leads to typical frequency distributions of between feeding intervals for (b) broilers(■), (c) pigs (•) and (d) dairy cows (▼). Graphs were calculated from data sets as described by Howie *et al.* (2009a) for broilers, Wilkinson (2007) for pigs and Yeates *et al.* (2001) for cows.

associated with the occurrence of disease in calves (Borderas *et al.*, 2009).

4.3 Constraining Meal Patterns Without Necessarily Reducing Daily Food Intake

It is evident that the normal patterns of feeding behaviour that have been described cannot be expressed unless animals have free and continuous access to food. There are, however, feeding regimes or conditions under which animals may be prevented from expressing such normal behaviour and still achieve their 'desired' or 'required' feed intake. Typical examples of such conditions are *ad libitum*

feed supply but relatively high animal pressure per feeder, or feed that is supplied once or twice daily only.

When Elizalde and Mayne (2009) increased the number of cows that had access to a feeder supplying silage they observed that the daily time spent feeding decreased by more than two-thirds but that eating rate more than tripled, with the result that there were no pronounced effects of feeder competition on average daily intake. Increasing the number of pigs per feeder from 5 to 10, 15 and 20 also resulted in a decrease in daily feeding time and an increase in feeding rate, with no significant effect on average daily intake or animal performance

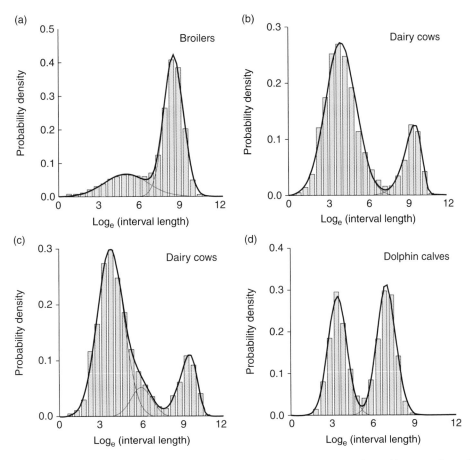

Fig. 4.2. Probability density functions (PDFs) fitted to the observed frequency distributions of $\log_e$-transformed between-feeding intervals (divided by bin width, i.e. 0.5 $\log_e$-units, shown by bars) recorded for: (a) broilers (after Howie *et al.*, 2010); dairy cows that do not (b), or do (c) drink during meals (both after Yeates *et al.*, 2001); and (d) neonate sucking bottlenose dolphin calves that only drink during meals (calculated from data kindly supplied by Jacobsen and Amundin; see Jacobsen *et al.*, 2003). Meal criteria are estimated as the interval length where the PDF that describes the long (i.e. between-meal) intervals crosses the PDF that describes the short (i.e. within-meal) intervals. Thin lines = individual functions; thick lines = the model as a whole.

(Nielsen *et al.*, 1995). When pigs are group housed and have access to a single feeder only, preferred diurnal patterns may disappear (Laitat *et al.*, 1999), and feeding behaviour may appear to be entirely random (Nielsen *et al.*, 1996; Wilkinson 2007). The same picture has been seen when high environmental temperatures prevented ruminants from grazing during the day, and they switched most of their feeding activity to the night, when the temperature dropped, with an increase in grazing activity near dawn (Brown and Lynch, 1972).

There are systems of production where food is presented to animals only once or twice daily and

the animals are allowed to eat as much food as they can, either over specified periods of time or while the food is available. An example of such systems is the 'to appetite' feeding system for pigs, which is practised with decreasing frequency nowadays. Providing that the food is easy to consume (e.g. with absence of competition) and highly digestible, then it is possible that animals achieve their required feed intake over very short periods of time but, of course, their preferred diurnal patterns entirely disappear. The question then is whether the disappearance of the preferred diurnal patterns – while daily food intake is maintained – matters to the animal, both in the short and long term. Some systemic fluctuations in the animal's physiological state would be expected when animals are fed thus (for a further discussion on these issues see also Section 4.4.4). The position taken here is that short-term feeding behaviour of animals is a device by which they exploit effectively their feeding environment or husbandry conditions. This behaviour, however, appears to relate very little to short-term fluctuation in their (internal) state. For this reason perhaps, the feeding pattern of animals is highly individualistic and differs from one day to the other, even for the same animal (e.g. chickens, Forbes and Shariatmadari, 1996).

4.4 Effects of Quantitative and Qualitative Restriction of Food Intake on Animals

4.4.1 Requirements and strategies

Animals with unrestricted access to high-quality food that can be obtained and ingested with little effort frequently consume large amounts of energy that is stored in the form of lipid. This applies not only to humans but also to companion animals (Gough, 2005; Czirjak and Chereji, 2008; Laflamme *et al.*, 2008; German *et al.*, 2009), zoo animals (Schwitzer and Kaumanns, 2001), laboratory rodents (Mela and Rogers, 1998; Home Office, 2003) and certain classes of farm animals (Meunier-Salaün *et al.*, 2001; Mench 2002; Tolkamp *et al.*, 2007). High-energy intakes, leading to obesity in the longer term, are associated with health problems such as diabetes, cardiovascular disease, reproductive problems, cancers and a reduction in longevity (Berg and Simms, 1960, 1961; Kealy *et al.*, 2002; Lawler *et al.*, 2005; German, 2006; Colman *et al.*, 2009). To avoid such problems, or correct cases of existing overweight, a restriction in energy intake (but not of essential nutrients) is generally recommended. In principle, three strategies to achieve this are currently used. In ruminant production systems, avoidance of excessive weight gains (or moderate weight loss in over-fat animals) is generally achieved by giving animals *ad libitum* access to lower quality foods (usually a forage), of which the voluntary energy intake is reduced. A typical example is the provision of dry cows with a limited amount of concentrate and *ad libitum* access to silage, hay or straw. The same aim is, however, frequently achieved by restricting access to high-quality food

in non-ruminant farm animals (such as broiler breeders and sows) and companion animals (e.g. dogs and cats). A third option is a combination of the two: provision of a restricted amount of lower quality (frequently bulkier) food in an attempt to minimize the negative welfare consequences of food restriction. The consequences of these strategies for animal health and welfare are discussed below.

4.4.2 Quantitative restriction and (feeding) behaviour

Under conditions of quantitative food restriction, the feeding motivation of animals is usually very high and they consume immediately any food that is made available. After feeding, various behaviours occur that are thought to be a result of unsatisfied feeding motivation (D'Eath *et al.*, 2009). Usually, quantitative restriction results in an increase in behaviours relating to foraging (Epling and Pierce, 1988; Dewasmes *et al.*, 1989; Koubi *et al.*, 1991; Weed *et al.*, 1997; Hebebrand *et al.* 2003), and in overall activity (Savory and Lariviere, 2000; Hocking, 2004). These are generally redirected oral behaviours, such as spot pecking by broiler breeders (Sandilands *et al.*, 2005, 2006), chain/bar chewing and rooting in any available substrate by sows (Appleby and Lawrence, 1987), tongue rolling in small cats (Shepherdson *et al.*, 1993), pacing, running in and out of the nest box and head nodding in mink (Mason, 1993) and wool stripping, wool chewing and slat biting in sheep (Cooper *et al.*, 1994). The drinker can also be a target and overdrinking or water spillage frequently occurs (Rushen, 1985a,b; Sandilands *et al.*, 2005). These redirected oral behaviours can become stereotypic (i.e. repetitive and unvarying while not serving an obvious function; Mason and Latham, 2004), and stereotypic behaviour generally increases with the level of food restriction in chickens (Savory *et al.*, 1996) and pigs (Brouns *et al.*, 1994; Bergeron *et al.*, 2000).

Some of the first clear and conclusive observations that related stereotypies and other 'abnormal' behaviours to food restriction were those of Rushen (1985b, p. 1064) on tethered sows. He observed that:

Rubbing the snout against bars, manipulating the drinker, bar-biting and head-weaving were clearly adjunctive behaviours, associated with the feeding period. These could be classified as either terminal responses (bar-biting, rubbing, and head weaving)

I. Kyriazakis and B. Tolkamp

that occurred immediately prior to the delivery of food, or interim responses that occurred immediately after food delivery. Interim responses involved either long duration drinking combined with rooting, rapid short duration drinking, or rapid rubbing. The last two behaviours tended to occur in stereotyped sequences. Vacuum chewing and playing with the chain however, appeared to occur with equal frequency before and after feeding.

The observations were subsequently confirmed (by comparing sows on different degrees of food restriction) by Appleby and Lawrence (1987). It is now widely believed that such stereotypies represent the expression of foraging behaviour, modified by the physical constraints of the environment (Lawrence et al., 1993). In some animals (e.g. mink and small cats) these behaviours are mostly performed before feeding and are more or less abolished once food is delivered (Mason, 1993), whereas in others (e.g. pigs and poultry) they occur at low levels before feeding and actually increase following food consumption (Terlouw et al., 1991). In addition, some of these behaviours are locomotory (e.g. pacing), whereas others are clearly oral (e.g. pecking and tongue rolling).

Terlouw et al. (1991, p. 988) were the first to argue that the above differences in food-related stereotypies could be explained by species-specific differences in the temporal sequences of foraging behaviour:

> In carnivores, in which [foraging] behaviour tends to precede the meal, stereotypies are largely found before feeding … In omnivores, such as the pig, in which [foraging behaviour] sequences may be interspersed over long periods of time, one might expect [foraging] behaviour to be stimulated by the ingestion of food, and result in stereotypy following a meal.

This idea has obvious attractions, but would have to be thoroughly tested before it could be used to predict the food-related stereotypic behaviour of animals in various husbandry conditions (Mason and Mendl, 1993).

These types of behaviours are substantially reduced if suitable substrates are provided to which more 'natural' foraging behaviour of food-restricted animals can be directed (Shepherdson et al., 1993; Spoolder et al., 1995). A similarly reduced incidence is observed when the bulkiness of the food increases but the amount of nutrients remains unaltered (Robert et al., 1993; Savory et al., 1996). Nevertheless, general activity levels continue to reflect the degree of food restriction and remain high.

4.4.3 Qualitative restriction and (feeding) behaviour

When food intake is quantitatively restricted, animals show (mainly behavioural) signs that they are stressed and hungry, which has led to considerable welfare concerns, as described above. In response to these concerns, a reduction in food quality has been proposed as an alternative to quantitative restriction (reviewed in Meunier-Salaün et al., 2001; Mench 2002; Hocking, 2004; Ru and Bao, 2004). When such lower quality food is offered ad libitum, this is called qualitative restriction (Sandilands et al., 2006). This can be an effective means of energy intake restriction because ad libitum-fed animals consume less energy from low-quality foods (Brouns et al. 1995; Savory et al. 1996; West and York, 1998; Whittemore et al., 2002; Tolkamp et al., 2005; Johnston et al., 2006), and is already routinely used in ruminant farm animals. To avoid the build-up of excessive lipid stores (or even to lose some in the condition when these have been built up), ruminants are usually fed at least one lower quality food source (a forage) ad libitum, even if they may get restricted amounts of high quality (e.g. concentrate) food. This allows ruminants to consume food in usual or 'preferred' patterns, and no obvious signs of hunger-related stress are then observed (Tolkamp et al., 2002). Ruminants will not deposit large amounts of lipid when they have access to a lower quality (e.g. grass hay) only and will not defend (but instead mobilize) lipid stores that have previously been built up on high-quality diets (Tolkamp et al., 2006, 2007).

For companion animals (Weber et al., 2007; Roudebush et al., 2008a,b) as well as non-ruminant farm animals (Meunier-Salaün et al., 2001; Hocking, 2004), attempts have been made to design alternative diets of which animals will limit voluntarily their energy (but not essential nutrient) intake, i.e. diets aimed at qualitative restriction of food intake. These alternative diets are usually based on 'diluting' a high-quality food with feedstuffs containing bulky ingredients or dietary fibre such as sugar beet pulp (Whittaker et al., 2000; Danielsen and Vestergaard, 2001; Weber et al., 2007; Bosch et al., 2009), wheat bran and cobs (Robert et al., 1993) or oat hulls (Sandilands et al., 2005), although appetite suppressants have also been used in companion animals (e.g. L-carnitine and dehydroepiandrosterone, Roudebush et al., 2008a) and in poultry (phenylpropanolamine, Oyawoye and Krueger, 1990; monensin sodium,

Savory *et al.*, 1996; calcium proprionate, Savory and Lariviere, 2000), sometimes in combination with fibre (Sandilands *et al.*, 2005, 2006).

The question of whether alternative diets aimed at qualitative restriction reduce hunger and enhance animal welfare in comparison with quantitative restriction is, however, controversial (D'Eath *et al.*, 2009). Qualitative, compared with quantitative, restriction of food intake generally leads to a number of behavioural changes. Feeding and foraging behaviour appears more 'natural', i.e. the animal is in control of its feeding behaviour, meals are ended while food is still available and more normal meal patterns can be observed over the day (Savory *et al.*, 1996; Meunier-Salaün *et al.*, 2001; Hocking, 2004). However, 'naturalness of behaviour' is considered more important for the assessment of animal welfare by some (Kiley-Worthington, 1989; FAWC, 1998 – with its 'Freedom to express normal behaviour') than by others (Dawkins, 1990; Broom and Johnson, 1993; Duncan, 1993).

In addition, there is generally a strong reduction in redirected oral behaviours and stereotypies when animals have access to qualitatively restrictive diets (D'Eath *et al.*, 2009). It has been argued that quantitative food restriction plays a central role in the development of post-feeding oral stereotypies (Dantzer, 1986; Appleby and Lawrence, 1987; Lawrence and Terlouw, 1993; Mason and Latham, 2004), that these appear to reflect frustrated motivation (Mason *et al.*, 2007) and, as they develop under suboptimal environmental conditions, are therefore a sign of reduced welfare (Dantzer, 1986; Mason and Latham, 2004). A reduction in redirected oral behaviours and stereotypies has, therefore, been interpreted as a sign of improved animal welfare (Bergeron *et al.*, 2000; Danielsen and Vestergaard, 2001; Zonderland *et al.*, 2004; Sandilands *et al.*, 2005).

Then again, the duration of feeding activity is usually much longer under qualitative than quantitative restriction and, as a result, the total duration of foraging-related oral behaviours is frequently similar under quantitative and qualitative restriction regimes (D'Eath *et al.*, 2009). It has been argued that oral behaviours are substitutable and are thus functionally equivalent outlets for unsatisfied feeding motivation and that, for this reason, a decrease of redirected oral behaviours and stereotypies under qualitative restriction should not be taken as a sign of improved welfare (Savory and Maros, 1993; Dailey and McGlone,

1997; McGlone and Fullwood, 2001). A reduction in stereotypic oral behaviour in animals subjected to qualitative as opposed to quantitative restriction is, therefore, not always seen as a sign of reduced hunger (D'Eath *et al.*, 2009).

4.4.4 Physiological indices of hunger

A number of physiological indices have been used, especially in studies with broiler breeders and sows, to measure hunger, to assess animal welfare or to improve understanding of the physiological processes underlying behavioural changes. These include blood plasma concentrations of energy substrates, such as glucose and non-esterified fatty acids (NEFA), and hormones, such as insulin and glucagon. These concentrations undergo characteristic changes during the day in relation to meal patterns, making point sampling difficult to interpret (de Jong *et al.*, 2003, 2005), especially when quantitative restriction is compared with other feeding regimes that usually result in different meal patterns and durations (Rushen *et al.*, 1999) and in differences in body weight and level of lipid reserves. For example, in broiler breeders, de Jong *et al.* (2003) observed that increasing quantitative restriction did not change blood glucose levels, but led to a decrease in blood NEFA; they therefore proposed to use the blood glucose/NEFA ratio as a measure of hunger. However, the differences in blood NEFA between treatments in this study may have simply been the result of variation in body lipid reserves, and de Jong *et al.* (2005) concluded that the lack of a 24 h profile makes it difficult to use this measure to compare quantitative restriction with other feeding regimes that lead to different meal patterns (and, therefore, different post-meal glucose peaks). Energy substrate concentrations may well vary with an animal's state of nutrition without necessarily corresponding to the animal's experience of hunger (D'Eath *et al.*, 2009).

Glucocorticoids (i.e. plasma cortisol or corticosterone concentration; PCC) have been used widely as a measure of stress and, therefore, of animal welfare (Mormède *et al.*, 2007). PCC is increased in response to a range of aversive situations, and the psychological component of the stressor is the main determinant of the magnitude of response (Mason *et al.*, 1968; Mason, 1971; Weiss, 1971; Veissier and Boissy, 2007). However, a number of problems exist with the use of PCC as a measure of hunger. Among these are that PCC may also rise in

I. Kyriazakis and B. Tolkamp

response to exciting stimuli (e.g. in anticipation of feeding), that it has a role in metabolic regulation (Mormède *et al.*, 2007) and that it is perhaps a better indicator of short-term stress than the longer term stress of hunger associated with prolonged restriction of food intake. In addition, there seems to be a confounding effect of age and weight on PCC in animals (at least in broiler breeders) subjected to different feeding regimes. PCC is frequently higher in birds that are quantitatively or qualitatively restricted compared with birds that have *ad libitum* access to high-quality food, if compared at the same age – when there will also be considerable differences between treatments in body weight. Fig. 4.3 shows that in *ad libitum* as well as in restricted fed birds, PCC concentrations

decrease with weight, and the graphs suggest that the decrease is similar under the different feeding strategies. Any differences between feeding treatments observed at a given age may, therefore, be (largely) the effect of differences in body weight between treatments rather than a good indicator of stress as a result of hunger associated with quantitative feed restriction.

In a study of stress in chickens, Gross and Siegel (1983) found that blood heterophil levels rose and lymphocytes fell in a more consistent way in response to different feeding treatments than did PCC, which suggests that these could be more useful as integrated measures of stress. The heterophil:lymphocyte ratio has subsequently been used frequently as a method of animal welfare

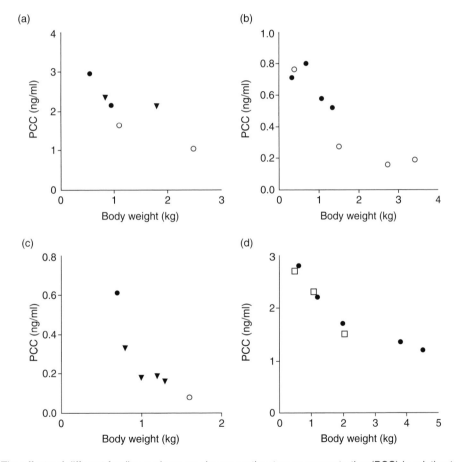

Fig. 4.3. The effects of different feeding regimes on plasma corticosterone concentration (PCC) in relation to broiler breeder body weight. Data shown are treatment means from (a) Savory *et al.* (1996), (b) de Jong *et al.* (2002), (c) de Jong *et al.* (2003) and (d) Sandilands *et al.* (2005). Feeding regimes were: quantitatively restricted according to breeders manual (•), *ad libitum* (○) and restricted between *ad libitum* and recommended restrictions (▼), all for birds fed high quality food, or qualitatively restricted, i.e. birds with *ad libitum* access to a lower quality food (□).

assessment, especially in poultry (Maxwell et al., 1990, 1992; Savory et al., 1992, 1993, 1996; Hocking et al., 1993, 2001, 2002). However, when all studies reporting measured heterophil: lymphocyte ratios in poultry are compared, the results are very confusing. Some studies find that in response to more severe levels of quantitative or qualitative restriction the ratio rises, but in other studies it falls or remains the same (D'Eath et al., 2009).

In summary, there is currently no strong evidence to suggest that qualititative restriction improves animal welfare if this is assessed by physiological indicators of stress.

4.4.5   Is qualitative restriction better for animal welfare than quantitative restriction?

In contrast to the results obtained when physiological indicators of hunger are considered, there is broad agreement between studies about the behavioural changes that can be observed when food is qualitatively, as opposed to quantitatively, restricted (D'Eath et al., 2009). These behavioural changes are interpreted by some authors as signs of improved animal welfare, but others dispute this interpretation. These disputes appear to arise – apart from different views on the importance of normal versus abnormal behaviour – from different (sometimes implicit) assumptions about the apparent food intake 'goal' of the animal.

Some authors (e.g. Savory et al., 1996; Savory and Lariviere, 2000) assume that the level of chronic hunger, at least in juveniles, is directly related to growth rate, which implies that any depression of an animal's growth will result in chronic hunger. A widely held position in livestock science is that an animal of a given physiological stage has a relatively fixed target for growth (or milk yield or mature size and level of body reserves). This target is frequently assumed to coincide with a (genetically determined) maximum performance level, which is associated with given energy and nutrient requirements (Kyriazakis, 1997). This would result in a feeding strategy for the animal that could be characterized as 'eating for requirements subject to constraints' (Owen, 1992; Kyriazakis, 1994; Day et al., 1997, 1998; Weston, 1996; Emmans, 1997). In this view, qualitative restriction prevents animals from achieving their desired target, for instance because of constraints on food processing by the gastrointestinal tract (Day et al., 1997; Emmans, 1997; Tsaras

et al., 1998; Whittemore et al. 2002). Such animals are sometimes thought to be experiencing 'metabolic' hunger (Day et al., 1997; Savory et al., 1996) and, for that reason, qualitative restriction is not considered to have real animal welfare benefits over quantitative restriction.

In some sections of the literature, there is a strong emphasis on the primacy in mature animals of physiological mechanisms to restore energy balance in the longer term, in particular, mechanisms to restore body weight following deprivation (Schwartz et al., 2000; Wynne et al., 2005). Other research allows for the possibility that there is no 'set point' of weight that is tightly regulated; instead, weight can depend on a number of other factors, including food quality (Mela and Rogers, 1998; Howarth et al., 2001; Levitsky, 2005; Rolls et al., 2005; Tolkamp et al., 2007).

An alternative view of what determines food intake is that, instead of fixed targets or 'set points', animals have more flexible goals for food intake and growth, depending on a balance of benefits and costs (see review by Illius et al., 2002). In this view, the balance of benefits and costs might vary with the available food quality. Food is beneficial, by providing energy and other nutrients essential for maintenance, growth and reproduction, but also carries extrinsic and intrinsic costs. Extrinsic costs include exposure to predation and other environmental costs during foraging, and the opportunity cost of spending time foraging which could be usefully spent on other activities. Intrinsic costs could include: (i) the cost of tissue damage from oxidative metabolism, resulting in reduced vitality and life span (Tolkamp and Ketelaars, 1992); (ii) exposure to toxins and parasitic organisms (Hutchings et al., 2000, 2003); and (iii) increased dental wear resulting in reduced foraging lifespan (Owen-Smith, 1994).

In support of this view are various observations that do not seem compatible with the idea of fixed constraints, for example the capacity of animals to increase intake of the same food when energy demands are high, e.g. during lactation or cold weather (Ketelaars and Tolkamp, 1992, 1996; Illius et al. 2002). Indeed, the size or function of some parts of the gastrointestinal tract (e.g. stomach or gizzard size) might appear to be a constraint in an individual animal under certain conditions, but in most cases this is not an absolute constraint (van Gils et al., 2003), and could change under natural selection if this increased Darwinian fitness (Barboza

I. Kyriazakis and B. Tolkamp

and Hume, 2006). In addition, it has been shown that for many animal species, restriction of energy (but not of essential nutrient) intake to below voluntary intake levels (i.e. caloric restriction, CR) has highly beneficial effects. These species range from nematodes (*Caenorhabditis elegans*) and fruit flies (*Drosophila melanogaster*) to rodents and primates (Berg and Simms, 1960, 1961; Finkel and Holbrook, 2000; Ramsey *et al.*, 2000; Hunt *et al.*, 2006; Terman and Brunk, 2006; Colman *et al.*, 2009). CR generally increases vitality, delays the onset of diseases and prolongs lifespan.

Moreover, it seems extremely counter-intuitive that animals would try to achieve a long-term target when they have unlimited access to food (i.e. a high energy intake and the resulting high level of lipid reserves) that has such obvious negative consequences for their health and longevity. Such behaviour is, perhaps, better understood from the point of view of animals using cues (internal cues, as well as cues from their environment) to optimize food intake. Such mechanisms could have evolved, and be very beneficial, if they resulted in higher energy intakes than required for maintenance during the (short) 'windows of opportunity' (Collier and Johnson, 1990) when high-quality foods could be obtained at low costs. This would allow the deposition of reserves that could be utilized in periods when (high-quality) food could not be obtained or could only be obtained at high costs. These same mechanisms would, however, result in uninterrupted periods of lipid deposition when animals are taken out of their natural environment and given continuous access to good-quality food that can be obtained at very low costs (Stubbs and Tolkamp, 2006). In this view, the provision of lower quality food would allow the animal to decide how much energy to consume, which mimics a natural environment in which high-quality foods are not readily available at low costs at all times. From this point of view, qualitative restriction would result in less hunger than quantitative restriction.

In summary, some scientists have concluded that feeding alternative diets improves animal welfare because it allows normal feeding behaviour, and the characteristics of such diets are thought to enhance satiety (Robert *et al.*, 1993; Zuidhof *et al.*, 1995; Zonderland *et al.*, 2004). This view is reflected in, for example, European Union (EU) policy, where there has been a requirement that sows receive 'a sufficient quantity of bulky or high-fibre food as well as high-energy food' (Council of Europe, 2001); this is implemented in some Member States through legislation (e.g. in the Netherlands, De Leeuw, 2004) or codes of practice (e.g. in the UK, Defra, 2003). Others have discussed the conflicting evidence and interpretations of the welfare benefits of alternative diets (Day *et al.*, 1997; Meunier-Salaün *et al.*, 2001; Mench, 2002; Hocking, 2004; Ru and Bao, 2004; de Jong *et al.*, 2005). A third group has argued that there is no evidence of any welfare benefits. Their position is that the hunger indicators of total oral behaviours and activity remain similar between quantitative restriction and alternative diets, and that if nutrient requirements and energy needs remain unsatisfied, 'metabolic hunger' will occur (Lawrence *et al.*, 1989; Owen, 1992; Savory *et al.*, 1996; Dailey and McGlone, 1997; Savory and Lariviere, 2000; McGlone and Fullwood, 2001).

4.5 Health and Welfare Consequences of Malnutrition and Undernutrition

The fact that energy, nutrient and water deprivation leads to ill thrift, poor growth and reproductive output, severe deterioration of health and eventual death is very well established, and constituted the early focus of nutritional science. If an animal's failure to obtain nutrients and water is severe then body reserves (fat, protein and major minerals in the bones) may be utilized for some of its physiological functions. Growing pigs are known to lose body fat and gain small amounts of protein under severe conditions of energy restriction (Kyriazakis and Emmans, 1992), while lactating mammals given restricted amounts of food utilize body reserves to produce milk. If prolonged, such a 'negative energy or nutrient balance' will lead to deterioration, illness and eventual death. Such deprivation is normally encompassed by the term undernutrition and can arise either directly or intentionally, most frequently as the outcome of certain (farming) practices, or indirectly, where food is freely available, but the animal is unable to consume a sufficient amount. The latter can be the outcome of extreme competition at the feeder, or when the environment is too hot and the animal is unable to consume sufficient food or modify its feeding behaviour, as discussed above. Although we will not deal any further in this chapter with the consequences of undernutrition for animal health, because they are relatively well established, we would like to point towards recent evidence that even moderate undernutrition can have detrimental effects on the functioning of the immune system and that maternal undernutrition can have detrimental effects on the long-term health of the offspring. In the former instance the animal may prioritize productive functions, such as pregnancy and lactation, over the functioning of the immune response. Such a prioritization makes it vulnerable to infection and may have long-term health consequences (Coop and Kyriazakis, 1999). In the latter case, there is increasing evidence that maternal nutrition during the different stages of pregnancy has long-term consequences for the metabolic health of the resulting offspring (Harding, 2001).

A more interesting case to consider is that of the malnutrition that arises when an animal is given access to a food which is inappropriately balanced (deficient or excessive in one or more nutrients in relation to energy content). It should be emphasized that malnutrition is considered to be different from the qualitative restriction detailed above. In the latter situation, it is usually the energy content of the food that is diluted with an inert material, so that the food still supplies adequate amounts of nutrients, whereas in the former situation the converse is usually the case. There are cases where farm animals are intentionally exposed to malnutrition to meet specific husbandry or production requirements. Veal calves are sometimes fed on a liquid diet deficient in iron to produce 'pale meat' (van Putten, 1982) and newly weaned pigs are given a low protein food to reduce the diarrhoea which frequently accompanies the abrupt change from milk to a solid diet (Wellock et al., 2009). However, the most common cases of malnutrition are probably those arising from the failure to meet the specific requirements of the individual animal given access to a single food, as food composition frequently targets the average rather than the individual animal in a group (Wellock et al., 2004).

Animals could attempt to meet requirements for the deficient nutrient(s) by increasing intake of the imbalanced food. Pigs on a low-protein food increase their intake to compensate for the protein deficiency (Kyriazakis et al., 1991), as do chickens on a vitamin A-deficient food (Ogunmodede, 1981). This situation is not usually observed in ruminants, and may reflect the effects of protein contents on digestible energy supply in this type of animal (Leng, 1990; Kyriazakis, 2003). Sometimes, however, the magnitude of the (specific) nutrient deficit is large or there are environmental or physical constraints and the

animal fails to compensate fully, perhaps because of adverse effects caused by increasing the amounts of other non-limiting nutrients. There is little evidence to suggest that, within limits, malnutrition has an effect upon animal health. However, the extra food consumed in attempts to meet the requirements for the deficient nutrient may lead to excessive fatness. Although excessive fatness does not appear to be a significant issue in farm animals grown for meat, owing to their short lifespan, it may be an issue in mature farm animals and in companion and zoo animals because it is associated with a reduced lifespan, as discussed above.

In the wild or under free-ranging conditions, an undernourished and/or malnourished animal will ordinarily direct most of its behaviour towards finding and consuming food items (foraging behaviour) until either it satisfies its nutrient requirements, or some constraint is imposed (e.g. time available for foraging). Sheep under extensive conditions are known to walk up to 20–25 km per day while foraging between nutrient-poor food patches (Squires and Wilson, 1971), while sows under free-ranging conditions (the 'Edinburgh Pig Park') and offered only a small amount of concentrate food, spent over 50% of their time foraging (Stolba and Wood-Gush, 1989). For animals in conditions where their nutrient requirements are not met and the expression of their foraging behaviour is restricted (e.g. intensively kept farm animals, laboratory animals), it is accepted that this foraging activity will be directed towards alternative available stimuli (Lawrence et al., 1993) in the manner discussed already. Jensen et al. (1993) observed that pigs offered access to a low-protein food in relation to energy spend more time standing and walking and were observed to root straw and engage in orally directed behaviours for longer than animals on a relatively balanced food. The question of whether even mild malnutrition, in the manner described above, has any significant effects on animal behaviour and, by extension, on welfare is the subject of debate.

4.6 Thirst and Drinking Behaviour

4.6.1 Water restriction

In their monograph entitled *Thirst*, Rolls and Rolls (1982) suggested that:

> [It] is a subjective sensation aroused by lack of water. … It can strictly only be studied directly in man, according to this definition. However, animals

I. Kyriazakis and B. Tolkamp

including man, when deprived of water, are in a state of drive in which they will search for and ingest water, and 'thirst' can be used in a different way to that described above as a name for this state of drive.

However, substituting the term thirst with 'drive for ingesting water' or perhaps with 'drinking tendency', when it is applied to animals, does not appear any more helpful. This is because the decision of an animal to perform, and the extent of a 'drive' or a 'tendency', will be the outcome of the interactions between internal factors (such as physiological state and diet) and external factors that arise from the environment the animal is kept in (such as environmental temperature, ease of access to a water source and water quality).

If, though, water is treated as a single nutrient – perhaps after oxygen the most important, because its lack leads to rapid death (Forbes, 1995) – then animals can be seen, by analogy with hunger, as being 'water restricted'. This definition, which we shall use in this chapter, implies that the only function of water intake is to meet physiological requirements. These physiological functions have been described by Brooks and Carpenter (1990, p. 116), as:

(i) the adjustment of body temperature, (ii) the maintenance of mineral-homeostasis, (iii) the excretion of the end products of digestion, (iv) the excretion of antinutritional factors ingested in the diets, (v) the excretion of drugs and drug residues, (vi) the achievement of satiety (gut fill) and (vii) the satisfaction of behavioural drives.

To those seven functions, we can add that of 'lubrication' for wetting food in the mouth and stomach(s) (Forbes, 1995). Functions (vi) and (vii) are not considered by many as physiological functions and this is the view that we shall take. Because of the many physiological functions that water performs, any serious water restriction would be expected to have more severe and immediate consequences than restriction of energy or any other nutrient.

The above view excludes instances where water consumption exceeds physiological requirements. This type of consumption has been termed 'secondary drinking' (Rolls and Rolls, 1982), 'adjunctive drinking' (Rushen, 1985b) or 'luxury drinking' (Fraser et al., 1990). Examples include hunger-induced drinking (Yang et al., 1981) and 'schedule-induced polydipsia' (Falk, 1961), where animals (most often rats), reduced to about 80% of normal body weight and given small amounts of foods at regular intervals, develop excessive drinking. These instances however, constitute extreme types of behaviour, induced only under certain experimental or husbandry conditions (Rolls and Rolls, 1982), and will not be considered further here.

Although short-term drinking behaviour has received significantly less attention than its equivalent feeding behaviour there are reasons to expect that it will also show similar 'bouting'. The evidence from animals that receive their food in liquid form (as in the example of dolphin calves detailed in Fig. 4.2(d)) strongly supports this view. In addition, there is anecdotal evidence that normal drinking behaviour may be affected by challenges to the health and welfare of animals in the manner described for feeding behaviour. It is possible, therefore, that changes in the normal patterns of short-term drinking behaviour of animals with free access to food may be used for the early detection of health and welfare problems (Borderas et al., 2009).

4.6.2 Sources of water restriction

Water restrictions fall into two categories: (i) intentionally imposed limitations on how much and when the animal may drink; and (ii) unintentional restriction as a result of physical, environmental and social factors. The first arises from systems or practices for intensively kept animals (mainly pigs or poultry), whereas the second is common to both intensively and extensively kept animals. Extensively kept animals (especially ruminants) in semi-arid or arid environments and high environmental temperatures are clearly water restricted, but will not be considered any further here.

Intentional water restriction

There are systems of production, mainly for pigs and poultry, where the water supply is either restricted to certain periods of the day or mixed with the food (liquid or wet feeding systems). Pigs may also be allowed to opt for a preferred water-to-food ratio by operating a water valve in their feeding trough (Brooks and Carpenter, 1990). The advantages of these systems are that delivery of both water and food can be automated; some researchers have also reported increases in productivity with this practice (Danish National Committee for Pig Breeding and Production, 1986). Water-restriction programmes are often employed for broiler breeders and sows as a means of reducing the overdrinking and wet droppings or water spillage that result from the severe food restriction programmes often imposed on these kinds of animals.

These systems of production and practices fail to recognize many of the physiological functions that water can serve and also their additive nature, which can lead to an increase in requirements (Brooks and Carpenter, 1990). Such systems can prove particularly risky when employed in conjunction with high environmental temperatures, where the animal has to regulate its body temperature, or where foods high in protein and/or minerals are offered (see below). On the grounds of both productivity and welfare, the justification for these systems seems limited.

Unintentional water restriction

The most common unintentional water restriction in intensively kept animals results from the failure of the water supply (e.g. blocked or burst pipes). There are however, more subtle features of the delivery system that could impose a variable degree of restriction. For example, water delivery rate can significantly restrict intake and performance of 3–6 week old weaned (Barber *et al.*, 1989) and 10–14 week old growing pigs (Nienaber and Hahn, 1984). Although pigs on a lower water delivery rate spent longer drinking, their intake was lower than for those on high water delivery rates. Inappropriate design of drinkers can similarly restrict the water intake of goats kept at high temperatures (Brooks and Carpenter, 1990). Social competition or ease of access to drinkers can result in water restriction, in particular for ruminants which take only a few large drinks per day, mostly associated with meals (Forbes *et al.*, 1991).

Water quality (e.g. cleanliness and presence of mineral and toxic substances) also plays a role in the water intake of animals, especially in pigs and poultry (NAC, 1974), which rely heavily on water supply (being fed on dry foods). Pigs prefer uncontaminated water (Brooks and Carpenter, 1990), while chickens reduced their water intake to about 75% of normal when it was adulterated with quinine (Yeomans and Savory, 1989). It is well established that water with a high mineral content or containing toxic substances reduces the water intakes of farm animals, and there is now an increased interest in reassessing how this affects the performance of, for example, high-yielding dairy cattle.

As with intentional water restriction, the importance of the above factors is exaggerated in conditions of high temperature or when animals are given access to certain foods. High dietary protein content (where animals have to excrete through urine the products of deamination), high mineral content (particularly sodium and potassium; Wahlstrom *et al.*, 1970), and high water-holding capacity (which traps water in the gastrointestinal tract; Kyriazakis and Emmans, 1995) also aggravate the effects of water restriction.

Extensively kept ruminants under temperate conditions are generally assumed to meet their water requirements through the water content of their food (Lynch *et al.*, 1972). This is probably true when their diet consists of high-water-content grass and root crops. However, if the availability of green grass declines and/or the physiological state of the animal changes (e.g. during lactation) then these animals may become (severely) water restricted. Water is increasingly being considered by many as 'the forgotten nutrient', mainly because animal keepers take its availability for granted. We have discussed above some subtle situations where water restriction can arise, and this will be expected to have adverse consequences on animal welfare and productivity. We trust that the approach taken in this chapter is an advance on the simple concepts of 'hunger and thirst' and the severe limitations that arise from them.

4.7 Conclusions

- All animals have a preferred or 'normal' feeding behaviour, which is structured in bouts or meals. Such feeding behaviour is flexible and reflects the ability of animals to exploit their feeding environment effectively. There is little evidence to suggest that deviations from a preferred feeding behaviour have significant welfare consequences.
- Because free and continuous access to a high-quality feed can lead to overweight and obesity, animal keepers may impose a degree of feed restriction on animals. Such restriction is either quantitative or qualitative and both of these are associated with effects on animal behaviour and, by extension, on welfare.
- There is broad agreement between studies about the behavioural changes that can be observed when food is qualitatively as opposed to quantitatively restricted, but less so when

I. Kyriazakis and B. Tolkamp

physiological indicators of hunger are considered. These behavioural changes are interpreted by some authors as signs of improved animal welfare when animals are qualitatively restricted, but others dispute this interpretation.

- Severe food restriction is considered in this chapter as undernutrition, and its effects on animal ill thrift, poor growth and reproductive output, severe deterioration of health and eventual death are well established.

- Malnutrition arises when an animal is given access to food that is inappropriately balanced (deficient or excessive in one or more nutrients, in relation, for example, to energy content). Severe malnutrition leads to problems similar to nutrient restriction and, in addition, may give rise to obesity. However, whether subtle cases of malnutrition lead to any significant behavioural changes and deteriorations in animal welfare is subject to debate.

- We suggest substituting the term 'thirst' with that of 'water restriction', which may arise intentionally or unintentionally. Because of the many physiological functions that water performs, even short-term restriction will lead to deterioration in health and welfare. There are many instances where water restriction can arise unintentionally, despite the fact that animal keepers frequently assume free and continuous availability of water.

References

Animal Welfare Advisory Committee (1994) *Annual Report, 1994*. Ministry of Agriculture and Forestry, Wellington, New Zealand.

Appleby, M.C. and Lawrence, A.B. (1987) Food restriction as a cause of stereotypic behavior in tethered gilts. *Animal Production* 45, 103–110.

Barber, J., Brooks, P.H. and Carpenter, J.L. (1989) The effects of water delivery rate on the voluntary food intake, water use and performance of early weaned pigs from 3 to 6 weeks of age. In: Forbes, J.M., Varley, M.A. and Lawrence, T.J.L. (eds) *The Voluntary Food Intake of Pigs*. Occasional Publication of British Society of Animal Production No. 13, Edinburgh, pp. 103–104.

Barboza, P.S. and Hume, I.D. (2006) Physiology of intermittent feeding: integrating responses of vertebrates to nutritional deficit and excess. *Physiological and Biochemical Zoology* 79, 250–264.

Berg, B.N. and Simms, H.S. (1960) Nutrition and longevity in the rat. II Longevity and the onset of disease with different level of food intake. *Journal of Nutrition* 71, 255–263.

Berg B.N. and Simms, H.S. (1961) Nutrition and longevity in the rat. II Food restriction beyond 800 days. *Journal of Nutrition* 74, 23–32.

Bergeron, R., Bolduc, J., Ramonet, Y., Meunier-Salaün, M.C. and Robert, S. (2000) Feeding motivation and stereotypies in pregnant sows fed increasing levels of fibre and/or food. *Applied Animal Behaviour Science* 70, 27–40.

Bigelow, J.A. and Houpt, T.R. (1988) Feeding and drinking patterns in young pigs. *Physiology and Behavior* 43, 99–109.

Bokkers, E.A.M. and Koene. P. (2003) Eating behaviour and prandial and postprandial correlations in male broiler and layer chickens. *British Poultry Science*, 44, 538–544.

Bokkers, E.A.M., Koene, P., Rodenburg, T.B., Zimmerman, P.H. and Spruijt, B.M. (2004) Working for food under conditions of varying motivation in broilers. *Animal Behaviour* 68, 105–113.

Borderas, T.F., Rushen, J., von Keyserlingk, M.A.G. and de Passille, A.M.B. (2009) Automated measurement of changes in feeding behavior of milk-fed calves associated with illness. *Journal of Dairy Science* 92, 4549–4554.

Bosch, G., Verbrugge, A., Hesta, M., Holst, J.J., van der Poel, A.F.B., Janssens, G.P.J. and Hendriks, W.H. (2009) The effects of dietary fibre type on satiety-related hormones and voluntary food intake in dogs. *British Journal of Nutrition* 102, 318–325.

Brooks, P.H. and Carpenter, J.L. (1990) The water requirement of growing-finishing pigs – theoretical and practical considerations. In: Haresign, W. and Cole, D.J.A. (eds) *Recent Advances in Animal Nutrition*. Butterworths, London, pp 115–136.

Broom, D.M. and Johnson, K.G. (1993) *Stress and Animal Welfare*. Chapman and Hall, London.

Brouns, F., Edwards, S.A. and English, P.R. (1994) Effect of dietary fiber and feeding system on activity and oral behavior of group-housed gilts. *Applied Animal Behaviour Science* 39, 215–223.

Brouns, F., Edwards, S.A. and English, P.R. (1995) Influence of fibrous feed ingredients on voluntary food intake of dry sows. *Animal Feed Science and Technology* 55, 301–313.

Brown, G.D. and Lynch, J.J. (1972) Some aspects of water-balance of sheep when deprived of drinking water. *Australian Journal of Agricultural Research* 23, 669–684.

Burkhart, C.A., Cherry, J.A., Vankrey, H.P. and Siegel, P.B. (1983) Genetic selection for growth-rate alters hypothalamic satiety mechanisms in chickens. *Behavioural Genetics* 13, 295–300.

Collier, G. and Johnson, D.F. (1990) The time window of feeding. *Physiology and Behavior* 48, 771–777.

Colman, R.J., Anderson, R.M., Johnson, S.C., Kastman, E.K., Kosmatka, K.J., Beasley, T.M., Allison, D.B., Cruzen, C., Simmons, H.A., Kemnitz, J.W. and

Weindruch, R. (2009) Caloric restriction delays disease onset and mortality in rhesus monkeys. *Science* 325, 201–204.

Compassion In World Farming Ltd. v Secretary of State for the Environment, Food and Rural Affairs (Defra) (2003) Court of Appeal – Administrative Court, November 27, 2003, EWHC 2850 (Admin) Judgment of the Honourable Mr Justice Newman, Case No: CO/1779/2003. High Court of Justice, Queen's Bench Division, Administrative Court, London.

Coop, R.L. and Kyriazakis, I. (1999) Nutrition–parasite interaction. *Veterinary Parasitology* 84, 187–204.

Cooper, J.J., Emmans, G.C. and Friggens, N.C. (1994) Effect of diet on behaviour of individually penned lambs. *Animal Production* 58, 441 (abstract).

Council of Europe (2001) Council Directive 2001/88/EC of 23rd October 2001 amending Directive 91/630/EEC laying down minimum standards for the protection of pigs. *Official Journal of the European Communities*, 316, 1–4. htttp://ec.europa.eu/food/animal/welfare/farm/pigs-undu-scoic-en.htm.

Czirjak, Z.T. and Chereji, A. (2008) Canine obesity – a major problem of pet dogs. *Fascicula: Ecotoxicologie, Zootehnie si Tehnologii de Industrie Alimentara* 7, 361–366.

Dailey, J.W. and McGlone, J.J. (1997) Oral/nasal/facial and other behaviors of sows kept individually outdoors on pasture, soil or indoors in gestation crates. *Applied Animal Behaviour Science* 52, 25–43.

Danielsen, V. and Vestergaard, E.M. (2001) Dietary fibre for pregnant sows: effect on performance and behaviour. *Animal Feed Science and Technology* 90, 71–80.

Danish National Committee for Pig Breeding and Production (1986) *Svinearl og-Production*. The National Committee for Pig Breeding, Health and Production, Copenhagen.

Dantzer, R. (1986) Behavioral, physiological and functional aspects of stereotyped behavior: a review and a re-interpretation. *Journal of Animal Science* 62, 1776–1786.

Dawkins, M.S. (1990) From an animal's point of view: motivation, fitness and animal welfare. *Behavioral and Brain Sciences* 13, 1–61.

Day, J.E.L., Kyriazakis, I. and Rogers, P.J. (1997) Feeding motivation in animals and humans: a comparative review of its measurements and uses. *Nutrition Abstracts and Reviews Series B*, 67, 69–79 and *Nutrition Abstracts and Reviews Series A*, 67, 107–117.

Day, J.E.L., Kyriazakis, I. and Rogers, P.J. (1998) Food choice and intake: towards a unifying framework of learning and feeding motivation. *Nutrition Research Reviews* 11, 25–43.

D'Eath, R.B., Tolkamp, B.J., Kyriazakis, I., Lawrence, A.B. (2009) 'Freedom from hunger' and preventing obesity: the animal welfare implications of reducing food quantity or quality. *Animal Behaviour* 77, 275–288.

Defra (Department for Environment, Food and Rural Affairs) (2003) *Code of Recommendations for the Welfare of Livestock: Pigs*. Defra Publications, London.

de Jong, I.C., van Voorst, S., Ehlhardt, D.A. and Blokhuis, H.J. (2002) Effects of restricted feeding on physiological stress parameters in growing broiler breeders. *British Poultry Science* 43, 157–168.

de Jong, I.C., van Voorst, A S. and Blokhuis, H.J. (2003) Parameters for quantification of hunger in broiler breeders. *Physiology and Behavior* 78, 773–783.

de Jong, I.C., Enting, H., van Voorst, A. and Blokhuis, H.J. (2005) Do low-density diets improve broiler breeder welfare during rearing and laying? *Poultry Science* 84, 194–203.

De Leeuw, J.A., Jongbloed, A.W. and Verstegen, M.W.A. (2004) Dietary fiber stabilizes blood glucose and insulin levels and reduces physical activity in sows (*Sus scrofa*). *Journal of Nutrition* 134, 1481–1486.

Dewasmes, G., Duchamp, C. and Minaire, Y. (1989) Sleep changes in fasting rats. *Physiology and Behavior* 46, 179–184.

Duncan, I.J.H. (1993) Welfare is to do with what animals feel. *Journal of Agricultural and Environmental Ethics* 6 (Supplement), 8–14.

Elizalde, H.F. and Mayne, C.S. (2009) The effect of degree of competition for feeding space on the silage dry matter intake and feeding behaviour of dairy cows. *Archivos de Medicina Veterinaria* 41, 27–34.

Emmans G.C. (1997) A method to predict the food intake of domestic animals from birth to maturity as a function of time. *Journal of Theoretical Biology* 186, 189–199.

Epling, W.F. and Pierce, W.D. (1988) Activity-based anorexia: a biobehavioral perspective. *International Journal of Eating Disorders* 7, 475–485.

Falk, J.L. (1961) Production of polydipsia in normal rats by intermittent food schedule. *Science* 133, 195–196.

FAWC (Farm Animal Welfare Council) (1992) FAWC updates on the five freedoms. *Veterinary Record* 131, 357.

FAWC (1998) Report on the Welfare of Broiler Breeders. FAWC, London.

Finkel, T. and Holbrook, N.J. (2000) Oxidants, oxidative stress and the biology of ageing. *Nature* 408, 239–247.

Forbes, J.M. (1995) *Voluntary Food Intake and Diet Selection in Farm Animals*. CAB International, Wallingford, UK.

Forbes, J.M. and Shariatmadari, F. (1996) Short-term effects of food protein content on subsequent diet selection by chickens and the consequences of alternate feeding of high- and low-protein foods. *British Poultry Science* 37, 597–607.

Forbes, J.M., Johnson, C.L. and Jackson, D.A. (1991) The drinking behaviour of lactating cows offered silage ad lib. *Proceedings of the Nutrition Society* 50, 97A.

Fraser, D., Patience, J.F., Phillips, P.A. and McLeese, J.M. (1990) Water for piglets and lactating sows,

---

I. Kyriazakis and B. Tolkamp

quantity, quality and quandaries. In: Haresign, W. and Cole, D.J.A. (eds) *Recent Advances in Animal Nutrition.* Butterworths, London, pp. 137–160.

German, A.J. (2006) The growing problem of obesity in dogs and cats. *Journal of Nutrition* 136, 1940S–1946S.

German, A.J., Holden, S.L., Bissot, T., Morris, P.J. and Biourge, V. (2009) A high protein high fibre diet improves weight loss in obese dogs. *The Veterinary Journal* 183, 294–297.

Gonzalez, L.A., Tolkamp, B.J., Coffey, M.P., Ferret, A. and Kyriazakis, I. (2008) Changes in feeding behavior as possible indicators for the automatic monitoring of health disorders in dairy cows. *Journal of Dairy Science* 91, 1017–1028.

Gough, A. (2005) Obesity in small animals and longevity in dogs. *Veterinary Times* 35, 10–11.

Gross, W.B. and Siegel, H.S. (1983) Evaluation of the heterophil lymphocyte ratio as a measure of stress in chickens. *Avian Diseases* 27, 972–979.

Harding, J.E. (2001) The nutritional basis of the fetal origins of adult disease. *International Journal of Epidemiology* 30, 15–23.

Hebebrand, J., Exner, C., Hebebrand, K., Holtkamp, C., Casper, R.C., Remschmidt, H., Herpertz-Dahlmann, B. and Klingenspor, M. (2003) Hyperactivity in patients with anorexia nervosa and in semistarved rats: evidence for a pivotal role of hypoleptinemia. *Physiology and Behavior* 79, 25–37.

Hocking, P.M. (2004) Measuring and auditing the welfare of broiler breeders. In: Weeks, C.A. and Butterworth, A. (eds) *Measuring and Auditing Broiler Welfare.* CAB International, Wallingford, UK, pp. 19–35.

Hocking, P.M., Maxwell, M.H. and Mitchell, M.A. (1993) Welfare assessment of broiler breeder and layer females subjected to food restriction and limited access to water during rearing. *British Poultry Science* 34, 443–458.

Hocking, P.M., Maxwell, M.H., Robertson, G.W. and Mitchell, M.A. (2001) Welfare assessment of modified rearing programmes for broiler breeders. *British Poultry Science* 42, 424–432.

Hocking, P.M., Maxwell, M.H., Robertson, G.W. and Mitchell, M.A. (2002) Welfare assessment of broiler breeders that are food restricted after peak rate of lay. *British Poultry Science* 43, 5–15.

Home Office (2003) Home Office Guidance Note: Water and Food Restriction for Scientific Purposes. Her Majesty's Stationery Office, London.

Howarth, N.C., Saltzman, E. and Roberts, S.B. (2001) Dietary fiber and weight regulation. *Nutrition Reviews* 59, 129–139.

Howie, J.A., Tolkamp, B.J., Avendaño, S. and Kyriazakis, I. (2009a) A novel flexible method to split feeding behaviour into bouts. *Applied Animal Behaviour Science* 116, 101–109.

Howie, J.A., Tolkamp, B.J., Avendaño, S. and Kyriazakis, I. (2009b) The structure of feeding behavior in commercial broiler lines selected for different growth rates. *Poultry Science* 88, 1143–1150.

Howie, J.A., Tolkamp, B.J., Bley, T.A.G. and Kyriazakis, I. (2010) Short-term feeding behaviour has a similar structure in broilers, ducks and turkeys. *British Poultry Science* 51, 714–724.

Hunt, N.D., Hyun, D.H., Allard, J.S., Minor, R.K., Mattson, M.P., Ingram, D.K. and de Cabo, R. (2006) Bioenergetics of aging and calorie restriction. *Ageing Research Reviews* 5, 125–143.

Hutchings, M.R., Kyriazakis, I., Papachristou, T.G., Gordon, I.J. and Jackson, F. (2000) The herbivores' dilemma: trade-offs between nutrition and parasitism in foraging decisions. *Oecologia* 124, 242–251.

Hutchings, M.R., Athanasiadou, S., Kyriazakis, I. and Gordon, I.J. (2003) Can animals use foraging behaviour to combat parasites? *Proceedings of the Nutrition Society* 62, 361–370.

Illius, A.W., Tolkamp, B.J. and Yearsley, J. (2002) The evolution of the control of food intake. *Proceedings of the Nutrition Society* 61, 465–472.

Jacobsen, T.B., Mayntz, M. and Amundin, M. (2003) Splitting suckling data of bottlenose dolphin (*Tursiops truncatus*) neonates in human care into suckling bouts. *Zoo Biology* 5, 477–488.

Jensen, M.B., Kyriazakis, I. and Lawrence, A.B. (1993) The activity and straw directed behaviour of pigs offered foods with different crude protein content. *Applied Animal Behavioural Science* 37, 211–221.

Johnston, S.L., Grune, T., Bell, L.M., Murray, S.J., Souter, D.M., Erwin, S.S., Yearsley, J.M., Gordon, I.J., Illius, A.W., Kyriazakis, I. and Speakman, J.R. (2006) Having it all: historical energy intakes do not generate the anticipated trade-offs in fecundity. *Proceedings of the Royal Society of London, Series B* 273, 1369–1374.

Kealy, R.D., Lawler, D.F., Ballam, J.M., Mantz, S.L., Biery, D.N., Greely, E.H., Lust, G., Segre, M., Smith, G.K. and Stowe, H.D. (2002) Effects of diet restriction on life span and age-related changes in dogs. *Journal of the American Veterinary Medical Association* 220, 1315–1320.

Ketelaars, J.J.M.H. and Tolkamp, B.J. (1992) Toward a new theory of feed intake regulation in ruminants 1. Causes of differences in voluntary feed intake: critique of current views. *Livestock Production Science* 30, 269–296.

Ketelaars, J.J.M.H. and Tolkamp, B.J. (1996) Oxygen efficiency and the control of energy flow in animals and humans. *Journal of Animal Science* 74, 3036–3051.

Kiley-Worthington, M. (1989) Ecological, ethological, and ethically sound environments for animals: toward symbiosis. *Journal of Agricultural Ethics* 2, 323–347.

Koubi, H.E., Robin, J.P., Dewasmes, G., Lemaho, Y., Frutoso, J. and Minaire, Y. (1991) Fasting-induced rise in locomotor activity in rats coincides with increased protein utilization. *Physiology and Behavior* 50, 337–343.

Kyriazakis, I. (1994) The voluntary food intake and diet selection of pigs. In: Wiseman, J., Cole, D.J.A. and Varley, M.A. (eds) *Principles of Pig Science*. Nottingham University Press, Nottingham, UK, pp. 85–105.

Kyriazakis, I. (1997) The nutritional choices of farm animals. In: Forbes, J.M., Lawrence, T.L.J., Rodway, R.G. and Varley, M.A. (eds) *Animal Choice*. Occasional Publication of the British Society of Animal Science No. 20, Edinburgh, pp. 55–65.

Kyriazakis, I. (2003) What are ruminant herbivores trying to achieve through their feeding behaviour and food intake? In: 't Mannetie, L., Ramirez-Aviles, L., Sandoval-Castro, C.A., Ku-Vera, J.C. (eds) *Matching Herbivore Nutrition to Ecosystems Biodiversity*. Universidad Autonóma de Yucatán, Mérida, Mexico, pp 154–173.

Kyriazakis, I. and Emmans, G.C. (1992) The effects of varying protein and energy intakes on the growth and body composition of pigs. 1. The effects of energy intake at constant, high protein intake. *British Journal of Nutrition* 68, 603–613.

Kyriazakis, I. and Emmans, G.C. (1995) The voluntary food intake of pigs given foods based on wheat bran, dried citrus pulp and grass meal, in relation to measurements of food bulk. *British Journal of Nutrition* 73, 191–207.

Kyriazakis, I., Emmans, G.C. and Whittemore, C.T. (1991) The ability of pigs to control their protein intake when fed in three different ways. *Physiology and Behavior* 50, 1197–1203.

Laflamme, D.P., Abood, S.K., Fascetti, A.J., Fleeman, L.M., Freeman, L.M., Michel, K.E., Bauer, C., Kemp, B.L.E., van Doren, J.R. and Willoughby, K.N. (2008) Pet feeding practices of dog and cat owners in the United States and Australia. *Journal of the American Veterinary Medical Association* 232, 687–694.

Laitat, M., Vandenheede, M., Desiron, A., Canart, B. and Nicks, B. (1999) Comparison of feeding behaviour and performance of weaned pigs given food in two types of dry feeders with integrated drinkers. *Animal Science* 68, 35–42.

Langton, S.D., Collett, D. and Sibly, R.M. (1995) Splitting behavior into bouts – a maximum-likelihood approach. *Behaviour* 132, 781–799.

Lawler, D.F., Evans, R.H., Larson, B.T., Spitznagel, E.L., Ellersieck, M.R. and Kealy, R.D. (2005) Influence of lifetime food restriction on causes, time, and predictors of death in dogs. *Journal of the American Veterinary Medical Association* 226, 225–231.

Lawrence, A.B. and Terlouw, E.M.C. (1993) A review of the behavioural factors involved in the development and continued performance of stereotypic behaviors in pigs. *Journal of Animal Science* 71, 2815–2825.

Lawrence, A.B., Appleby, M.C., Illius, A.W. and Macleod, H.A. (1989) Measuring hunger in the pig using operant conditioning: the effect of dietary bulk. *Animal Production* 48, 213–220.

Lawrence, A.B., Terlouw, E.M.C. and Kyriazakis, I. (1993) The behavioral effects of undernutrition in confined farm animals. *Proceedings of the Nutrition Society* 52, 219–229.

Le Magnen, J. (1985) *Hunger*. Cambridge University Press, Cambridge, UK.

Leng, R.A. (1990) Factors affecting the utilisation of 'poor quality' foragers by ruminants particularly under tropical conditions. *Nutrition Research Reviews* 3, 277–303.

Levitsky, D.A. (2005) The non-regulation of food intake in humans: hope for reversing the epidemic of obesity. *Physiology and Behavior* 86, 623–632.

Lynch, J.J., Brown, G.D., May, P.F. and Donnelly, J.B. (1972) The effect of withholding drinking water on wool growth and lamb production of grazing Merino sheep in a temperate climate. *Australian Journal of Agricultural Research* 23, 659–668.

Mason, G.J. (1993) Age and context affect the stereotypies of caged mink. *Behaviour* 127, 191–229.

Mason, G. and Latham, N. (2004) Can't stop, won't stop: is stereotypy a reliable animal welfare indicator. *Animal Welfare* 13 (Supplement 1), 57–69.

Mason, G. and Mendl, M. (1993) Why is there no simple way of measuring animal welfare? *Animal Welfare* 2, 310–319.

Mason, G., Clubb, R., Latham, N. and Vickery, S. (2007) Why and how should we use environmental enrichment to tackle stereotypic behaviour? *Applied Animal Behaviour Science* 102, 163–188.

Mason, J.W. (1971) Re-evaluation of concept of non-specificity in stress theory. *Journal of Psychiatric Research* 8, 323–333.

Mason, J.W., Wool, M.S., Mougey, E.H., Wherry, F.E., Collins, D.R. and Taylor, E.D. (1968) Psychological vs. nutritional factors in the effects of 'fasting' on hormonal balance. *Psychosomatic Medicine* 30, 554–555.

Maxwell, M.H., Robertson, G.W., Spence, S. and McCorquodale, C.C. (1990) Comparison of hematological values in restricted-fed and ad-libitum-fed domestic fowls: white blood cells and thrombocytes. *British Poultry Science* 31, 399–405.

Maxwell, M.H., Hocking, P. and Robertson, G. (1992) Differential leukocyte responses to various degrees of food restriction in broilers, turkeys and ducks. *British Poultry Science* 33, 177–187.

McGlone, J.J. and Fullwood, S.D. (2001) Behavior, reproduction, and immunity of crated pregnant gilts: effects of high dietary fiber and rearing environment. *Journal of Animal Science* 79, 1466–1474.

Mela, D.J. and Rogers, P.J. (1998) Food, Eating and Obesity. The Psychobiological Basis of Appetite and Weight Control. Chapman and Hall, London.

Mench, J.A. (2002) Broiler breeders: feed restriction and welfare. *World's Poultry Science Journal* 58, 23–29.

Metz, J.H.M. (1975) *Time Patterns of Feeding and Rumination in Domestic Cattle.* Communications of the Agricultural University, Wageningen, The Netherlands, Bull. 75-12.

Meunier-Salaün, M.C., Edwards, S.A. and Robert, S. (2001) Effect of dietary fibre on the behaviour and health of the restricted fed sow. *Animal Feed Science and Technology* 90, 53–69.

Mormède, P., Andanson, S., Auperin, B., Beerda, B., Guemene, D., Malnikvist, J., Manteca, X., Manteuffel, G., Prunet, P., van Reenen, C.G., Richard, S. and Veissier, I. (2007) Exploration of the hypothalamic–pituitary–adrenal function as a tool to evaluate animal welfare. *Physiology and Behavior* 92, 317–339.

NAC (National Academy of Sciences) (1974) *Nutrients and Toxin Substances in Water for Livestock and Poultry.* Washington, DC.

Nielsen, B.L., Lawrence, A.B. and Whittemore, C.T. (1995) Effect of group size on feeding behaviour, social behaviour, and performance of growing pigs using single-space feeders. *Livestock Production Science* 44, 73–85.

Nielsen, B.L., Lawrence, A.B. and Whittemore, C.T. (1996) Feeding behaviour of growing pigs using single or multi-space feeders. *Applied Animal Behaviour Science* 47, 235–246.

Nienaber, J.A. and Hahn, G.L. (1984) Effects of water flow restriction and environmental factors on performance of nursery-age pigs. *Journal of Animal Science* 59, 1423–1429.

Ogunmodede, B.K. (1981) Vitamin A requirement of broiler chicks in Nigeria. *Poultry Science* 60, 2622–2627.

Owen, J.B. (1992) Genetic aspects of appetite and feed choice in animals. *Journal of Agricultural Science* 119, 151–155.

Owen-Smith, N. (1994) Foraging responses of kudus to seasonal changes in food resources: elasticity in constraints. *Ecology* 75, 1050–1062.

Oyawoye, E.O. and Krueger, W.F. (1990) Potential of chemical regulation of food intake and body weight of broiler breeder chicks. *British Poultry Science* 31, 735–742.

Ramsey, J.J., Harper, M.E. and Weindruch, R. (2000) Restriction of energy intake, energy expenditure and aging. *Free Radical Biology and Medicine* 29, 946–968.

Robert, S., Matte, J.J., Farmer, C., Girard, C.L. and Martineau, G.P. (1993) High-fiber diets for sows: effects on stereotypies and adjunctive drinking. *Applied Animal Behaviour Science* 37, 297–309.

Rolls, B.J. and Rolls, E.T. (1982) *Thirst.* Cambridge University Press, Cambridge, UK.

Rolls, B.J., Drewnowski, A. and Ledikwe, J.H. (2005) Changing the energy density of the diet as a strategy for weight management. *Journal of the American Dietetic Association* 105 (5, Supplement), 98–103.

Roudebush, P., Schoenherr, W.D. and Delaney, S.J. (2008a) An evidence-based review of the use of nutraceuticals and dietary supplementation for the management of obese and overweight pets. *Journal of the American Veterinary Medicine Association* 232, 1646–1655.

Roudebush, P., Schoenherr, W.D. and Delaney, S.J. (2008b) An evidence-based review of the use of therapeutic foods, owner education, exercise, and drugs for the management of obese and overweight pets. *Journal of the American Veterinary Medicine Association* 233, 717–725.

Ru, Y.J. and Bao, Y.M. (2004) Feeding dry sows ad libitum with high fibre diets. *Asian-Australasian Journal of Animal Sciences* 17, 283–300.

Rushen, J. (1985a) Stereotypies, aggression and the feeding schedules of tethered sows. *Applied Animal Behaviour Science* 14, 137–147.

Rushen, J. (1985b) Stereotyped behaviour, adjunctive drinking and the feeding periods of tethered sows. *Animal Behaviour* 32, 1059–1067.

Rushen, J., Robert, S. and Farmer, C. (1999) Effects of an oat-based high-fibre diet on insulin, glucose, cortisol and free fatty acid concentrations in gilts. *Animal Science* 69, 395–401.

Sandilands, V., Tolkamp, B.J. and Kyriazakis, I. (2005) Behaviour of food restricted broilers during rearing and lay – effects of an alternative feeding method. *Physiology and Behavior* 85, 115–123.

Sandilands, V., Tolkamp, B.J., Savory C.J. and Kyriazakis, I. (2006) Behaviour and welfare of broiler breeders fed qualitatively restrictive diets during rearing: are there viable alternatives to quantitative restriction? *Applied Animal Behaviour Science* 96, 53–67.

Savory, C.J. and Lariviere, J.M. (2000) Effects of qualitative and quantitative food restriction treatments on feeding motivational state and general activity level of growing broiler breeders. *Applied Animal Behaviour Science* 69, 135–147.

Savory, C.J. and Maros, K. (1993) Influence of degree of food restriction, age and time of day on behavior of broiler breeder chickens. *Behavioural Processes* 9, 179–190.

Savory, C.J., Seawright, E. and Watson, A. (1992) Stereotyped behavior in broiler breeders in relation to husbandry and opioid receptor blockade. *Applied Animal Behaviour Science* 32, 349–360.

Savory, C.J., Carlisle, A., Maxwell, M.H., Mitchell, M.A. and Robertson, G.W. (1993) Stress, arousal and opioid peptide-like immunoreactivity in restricted-fed and ad-lib-fed broiler breeder fowls. *Comparative Biochemistry and Physiology Part A: Molecular and Integrative Physiology* 106, 587–594.

Savory, C.J., Hocking, P.M., Mann, J.S. and Maxwell, M.H. (1996) Is broiler breeder welfare improved by using qualitative rather than quantitative food restriction to limit growth rate? *Animal Welfare* 5, 105–127.

Schwartz, M.W., Woods, S.C., Porte, D., Seeley, R.J. and Baskin, D.G. (2000) Central nervous system control of food intake. *Nature* 404, 661–671.

Schwitzer, C. and Kaumanns, W. (2001) Body weights of ruffed lemurs (*Varecia variegata*) in European zoos with reference to the problem of obesity. *Zoo Biology* 20, 261–269.

Shepherdson, D.J., Carlstead, K., Mellen, J.D. and Seidensticker, J. (1993) The influence of food presentation on the behaviour of small cats in confined environment. *Zoo Biology* 12, 203–216.

Simpson, S.J. (1995) Regulation of a meal: chewing insects. In: Chapman, R.F. and de Boer, G. (eds) *Regulatory Mechanisms in Insect Feeding*. Chapman and Hall, New York, pp. 26–45.

Slater, P.J.B. and Lester, N.P. (1982) Minimising errors in splitting behaviour into bouts. *Behaviour* 79, 153–161.

Spoolder, H.A.M., Burbridge, J.A., Edwards, S.A., Simmins, P.H. and Lawrence, A.B. (1995) Provision of straw as a foraging substitute reduces the development of excessive chain and bar manipulation in food restricted sows. *Applied Animal Behaviour Science* 43, 249–262.

Squires, V.R. and Wilson, A.D. (1971) Distance between food and water supply and its effect on drinking frequency, and food and water intake of Merino and Border Leicester sheep. *Australian Journal of Agricultural Research* 22, 283–290.

Stolba, A. and Wood-Gush, D.G.M. (1989) The behaviour of pigs in a semi-natural environment. *Animal Production* 48, 419–425.

Stubbs, R.J. and Tolkamp B.J. (2006) Control of energy balance in relation to energy intake and energy expenditure in animals and humans: an ecological perspective. *British Journal of Nutrition* 95, 657–675.

Terlouw, E.M.C., Lawrence, A.B. and Ilius, A.W. (1991) Influences of feeding level and physical restriction on development of stereotypies in sows. *Animal Behaviour* 42, 981–991.

Terman, A. and Brunk, U.T. (2006) Oxidative stress, accumulation of 'garbage', and aging. *Antioxid Redox Signal* 8, 197–204.

Tolkamp, B.J. and Ketelaars, J.J.M.H. (1992) Toward a new theory of feed-intake regulation in ruminants 2. Costs and benefits of feed consumption: an optimization approach. *Livestock Production Science* 30, 297–317.

Tolkamp, B.J. and Kyriazakis, I. (1999) To split behaviour into bouts, log-transform the intervals. *Animal Behaviour* 57, 807–817.

Tolkamp, B.J., Allcroft, D.J., Austin, E.J., Nielsen, B.L. and Kyriazakis, I. (1998) Satiety splits feeding behaviour into bouts. *Journal of Theoretical Biology* 194, 235–250.

Tolkamp, B.J., Friggens, N.C., Emmans, G.C., Kyriazakis, I. and Oldham J.D. (2002) Meal patterns of dairy cows consuming diets with a high or a low ratio of concentrate to grass silage. *Animal Science* 74, 369–382.

Tolkamp, B.J., Sandilands, V. and Kyriazakis, I. (2005) Effects of qualitative food restriction during rearing on the performance of broiler breeders during rearing and lay. *Poultry Science* 84, 1286–1293.

Tolkamp, B.J., Emmans, G.C. and Kyriazakis, I. (2006) Body fatness affects feed intake of sheep at a given body weight. *Journal of Animal Science* 84, 1778–1789.

Tolkamp, B.J., Yearsley, J.M., Gordon, I.J., Illius, A.W., Speakman, J.R. and Kyriazakis, I. (2007) Predicting the effects of body fatness on the intake and performance of sheep. *British Journal of Nutrition* 97, 1206–1215.

Tsaras, L.N., Kyriazakis, I. and Emmans, G.C. (1998) The prediction of the voluntary food intake of pigs on poor quality foods. *Animal Science* 66, 713–723.

van Gils, J.A., Piersma, T., Dekinga, A. and Dietz, M.W. (2003) Cost–benefit analysis of mollusc-eating in a shorebird II. Optimizing gizzard size in the face of seasonal demands. *Journal of Experimental Biology* 206, 3369–3380.

van Putten, G. (1982) Welfare in veal calf units. *Veterinary Record* 111, 437.

Veissier, I. and Boissy, A. (2007) Stress and welfare: two complementary concepts that are intrinsically related to the animal's point of view. *Physiology and Behavior* 92, 429–433.

Wahlstrom, R.C., Taylor, A.R. and Seerley, R.W. (1970) Effects of lysine in the drinking water of growing swine. *Journal of Animal Science* 30, 368–373.

Weber, M., Bissot, T., Servet, E., Sergheraert, R., Biourge, V. and German, A.J. (2007) A high protein, high fiber diet designed for weight loss improves satiety in dogs. *Journal of Veterinary Internal Medicine* 21, 1203–1208.

Webster, A.J.F. (1995) *Animal Welfare – A Cool Eye Towards Eden*. Blackwell Science, Oxford, UK.

Weed, J.L., Lane, M.A., Roth, G.S., Speer, D.L. and Ingram, D.K. (1997) Activity measures in rhesus monkeys on long-term calorie restriction. *Physiology and Behavior* 62, 97–103.

Weiss, J.M. (1971) Effects of coping behavior in different warning signal conditions on stress pathology in rats. *Journal of Comparative and Physiological Psychology* 77, 1–13.

Wellock, I.J., Emmans, G.C. and Kyriazakis, I. (2004) Modelling the effects of stressors on the performance of populations of pigs. *Journal of Animal Science* 82, 2442–2450.

Wellock, I.J., Houdijk, J.G.M., Miller, A.C., Gill, B.P. and Kyriazakis, I. (2009) The effect of weaner diet protein content and diet quality on the long-term performance of pigs to slaughter. *Journal of Animal Science* 8, 1261–1269.

West, D.B. and York, B. (1998) Dietary fat, genetic predisposition, and obesity: lessons from animal models. *American Journal of Clinical Nutrition* 67, 505S–512S.

Weston, R.H. (1996) Some aspects of constraints to forage consumption by ruminants. *Australian Journal of Agricultural Research* 47, 175–197.

Whittaker, X., Edwards, S.A., Spoolder, H.A.M., Corning, S. and Lawrence, A.B. (2000) The performance of group-housed sows offered a high fibre diet *ad libitum. Animal Science* 70, 85–93.

Whittemore, E.C., Kyriazakis, I., Tolkamp, B.J. and Emmans, G.C. (2002) The short-term feeding behavior of growing pigs fed foods differing in bulk content. *Physiology and Behavior* 76, 131–141.

Wilkinson, S. (2007) The short-term feeding behaviour of commercially reared group-housed growing-finishing pigs. MRes thesis, University of Edinburgh, Edinburgh, UK.

Wynne, K., Stanley, S., McGowan, B. and Bloom, S. (2005) Appetite control. *Journal of Endocrinology* 184, 291–318.

Yang, T.S., Howard, B. and MacFarlane, W.V. (1981) Effects of food on drinking behaviour of growing pigs. *Applied Animal Ethology* 7, 259–270.

Yeates, M.P., Tolkamp, B.J., Allcroft, D.J. and Kyriazakis, I (2001) The use of mixed distribution models to determine bout criteria for analysis of animal behaviour. *Journal of Theoretical Biology* 213, 413–425.

Yeomans, M.R. and Savory, C.J. (1989) Altered spontaneous and osmotically induced drinking for fowls with permanent access to quinine. *Physiology and Behavior* 46, 917–922.

Zonderland, J.J., De Leeuw, J.A., Nolten, C. and Spoolder, H.A.M. (2004) Assessing long-term behavioural effects of feeding motivation in group-housed pregnant sows; what, when and how to observe. *Applied Animal Behaviour Science* 87, 15–30.

Zorrilla, E.P., Inoue, K., Fekete, E.M., Tabarin, A., Valdez, G.R. and Koob, G.F. (2005) Measuring meals: structure of prandial food and water intake of rats. *American Journal of Physiology – Regulatory Integrative and Comparative Physiology* 288, R1450–R1467.

Zuidhof, M.J., Robinson, F.E., Feddes, J.J.R., Hardin, R.T., Wilson, J.L., McKay, R.I. and Newcombe, M. (1995) The effects of nutrient dilution on the wellbeing and performance of female broiler breeders. *Poultry Science* 74, 441–456.

5 Pain

IGNACIO VIÑUELA-FERNÁNDEZ, DANIEL M. WEARY
AND PAUL FLECKNELL

Abstract

Pain in animals is a significant welfare concern that occurs under a variety of circumstances – for example as a result of methods of housing or husbandry, following surgical procedures or as a result of disease processes. The problems associated with pain are exacerbated by our limited ability to assess the magnitude of pain experienced by individual animals. Both physiological and behavioural responses are used to assess pain, and increased recognition of pain should lead to changes in our methods of care and use of animals and a reduction in the magnitude of this problem.

5.1 Introduction

Pain has been defined by the International Association for the Study of Pain (1979, p. 250) as 'an unpleasant sensory and emotional experience associated with actual or potential tissue damage, or described in terms of such damage'. This definition relies heavily on language and self-report, as highlighted by Anand and Craig (1996), and is therefore of limited use for understanding animal pain. Anand and Craig argue that self-report is just one of the 'efferent' responses to pain and that the inability to communicate does not presuppose inability to feel pain. Prompted by this critique, Molony (1997, p. 293) provided a working definition of animal pain that focuses on efferent responses other than verbal self-report:

> Animal pain is an aversive sensory and emotional experience representing an awareness by the animal of damage or threat to the integrity of its tissues; it changes the animal's physiology and behaviour to reduce or avoid the damage, to reduce the likelihood of recurrence and to promote recovery.

Such pain-related changes in physiology and behaviour form the basis of the pain assessment strategies that will be discussed below. Additionally, Molony's definition acknowledges the multidimensional nature of the experience of pain, which comprises not only a sensory-discriminative component (relating to stimulus location, intensity and type) but also an emotional component, thereby involving some form of awareness by the animal. Although it is generally accepted that animal species that are closely related to us, and similar in anatomy, physiology and behaviour, probably experience pain in a similar manner to ourselves, it appears more difficult to infer pain in species that are dissimilar from us.

Paramount to the discussion on the ability of a particular species to experience pain is differentiation between the concepts of nociception and pain. Nociception relates to the detection of and ability to respond to noxious stimuli, but pain also includes an unpleasant emotional experience that depends upon central nervous system (CNS) processing. Both nociceptive systems and the emotional aspects of pain probably evolved to serve a protective function and thus might be expected to be present in many species. The distinction between nociception and pain was first made to allow a description of the responses seen in decorticate animals, or ones in which the spinal cord had been severed, under anaesthesia, at a level below the brainstem (Sherrington, 1906). If it is accepted that conscious awareness of pain requires a functioning cortex – or equivalent brain structure – then responses of decorticate or anaesthetized animals represent nociceptive responses, but not pain. Data from people support the view that a fully conscious perception of the emotional component of pain involves cortical activity (e.g. Baliki *et al.*, 2006), but it is possible that some level of consciousness may not require a cortex. Recent studies of pain in fish have stirred the debate on criteria to be assessed when trying to ascertain if a particular species might experience pain. Sneddon (2004) proposes a framework for investigation and scientific discussion, based on the criteria discussed by Bateson (1992), which is summarized in Box 5.1. It should be pointed out that the criteria presented in Box 5.1 do not provide an all-or-nothing test for the experience of pain, but rather evidence that should be considered as an integrated whole in order to make an informed assessment of the animal's experience.

Pain itself is a major animal welfare concern. Additionally, it may have incapacitating effects that can further impair welfare and compromise survival by interfering with an animal's ability to find food, water or shelter. Pain can also have detrimental effects on physical health, such as by causing metabolic changes that retard and impair wound healing or result in immunosuppression. The issue of hypersensitive and often persistent (chronic) pain states which develop following nerve damage or inflammation is particularly important. Such pain can occur not only in association with disease states such as lameness, but also when adequate analgesia for surgery or routine procedures such as castration and tail docking is not provided.

Despite the considerable progress made in recent years, pain management in animals remains suboptimal. This failure to alleviate pain may be due to difficulties associated with its recognition, assessment and the narrow range of treatment strategies used in animals; these will be discussed further in this chapter.

5.2 The Neurobiology of Pain: a Brief Overview

Pain perception involves the detection of actual or potential tissue damage by populations of anatomically and physiologically specialized peripheral sensory neurons, the activity of which is conducted by thin myelinated Aδ fibres or unmyelinated C fibres. These so-called nociceptors are characterized by sensor proteins localized in their free nerve endings, which transduce noxious (mechanical, thermal and chemical) stimuli, generating action potentials that travel along their afferent fibres to the dorsal horn of the spinal cord. Synaptic release of neurotransmitters by nociceptors activates nociceptive dorsal horn neurons, which mediate withdrawal reflexes and relay nociceptive activity to the brain. This ascending activity is transmitted through and modulated by multiple relay stations (the thalamus, reticular formation, pons, amygdala) on its way to the cerebral cortex, where it can be registered and interpreted as a pain experience (Fig. 5.1). Transmission through these nociceptive pathways is subject to ascending and descending modulation (which can be inhibitory or facilitatory) at every level of the neuroaxis; for a more extensive review of the neurophysiology of pain see Basbaum and Jessell, 2000.

Long-lasting changes in these nociceptive pathways (neuronal plasticity) are crucial for transition from normal acute 'nociceptive' pain (which has a protective role) to 'pathological' pain states that are generally not useful. These can develop following tissue damage, inflammation and nerve injury (for a review, see Woolf and Salter, 2000). Pathological pain states are characterized by hypersensitivity to potentially painful stimulation and are clinically manifested as spontaneous pain (pain in the absence

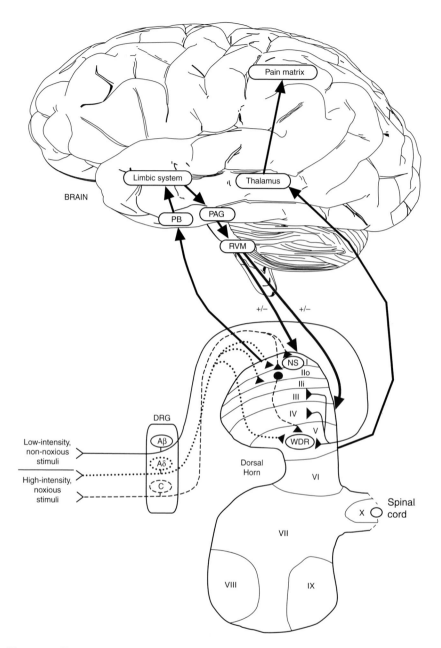

Fig. 5.1. Diagram to illustrate the ascending pain pathways conveying noxious information from the periphery to the brain. Nociceptive inputs enter the dorsal horn of the spinal cord through Aδ (dotted blue) and C fibres (dashed blue) which terminate in the superficial dorsal horn at laminae I and IIo, and in the deep dorsal horn at lamina V. Nociceptive-specific (NS) neurons are contacted by C fibres in the superficial dorsal horn. Noxious information is relayed, via the spinoparabrachial tract (left thick black arrow), to the parabrachial area (PB) before reaching the limbic system. Information from neurons in the deep dorsal horn ascends via the spinothalamic tract (right thick black arrow) to reach the thalamus and from there to the cortical areas comprising the 'pain matrix'. Descending control from noradrenergic and serotonergic pathways originates in the periaqueductal grey (PAG) region and the rostral ventromedial medulla (RVM) (thick blue arrows), and plays a role in the modulation of spinal processing. DRG = dorsal root ganglion; WDR = wide-dynamic range neurons. (From D'Mello and Dickenson, 2008.)

I. Viñuela-Fernández *et al.*

of any stimulation), as well as hyperalgesia (an exaggerated response to a noxious stimulus) and allodynia (the presence of a pain response to a non-noxious stimulus such as gentle touch). Hypersensitivity to pain is generated both at the level of the peripheral sensory neurons and centrally, in the spinal cord. Following inflammation, nociceptors are sensitized by chemicals released from damaged cells, inflammatory cells and peripheral sensory nerve terminals. Peripheral sensitization generally resolves after treatment and healing; however, sustained nociceptive input into the spinal cord caused by peripheral nerve injury and inflammation can lead to sensitization of central spinal relays, where enhanced excitability of nociceptive relay neurons results in prolonged (chronic) hyperalgesia and allodynia. Chronic pain states are difficult to treat and clearly reduce the animal's welfare, highlighting the importance of adequate analgesia perioperatively for surgery and for painful husbandry procedures such as castration and tail docking in order to prevent this central sensitization.

5.3 The Occurrence of Pain

5.3.1 Farm animals

Some of the best-known examples of animal pain come from procedures performed on farm animals. Examples include mulesing of sheep, dehorning and branding of cattle, beak or bill trimming of poultry, and castration and tail docking of pigs. All of these procedures cause pain, but they are often performed without anaesthetics or analgesics. In some cases, producers may choose not to use pain-relieving drugs because they feel that the animals do not experience pain, or that this pain is unimportant. In other cases, they may believe that the pain is difficult to treat, or that treatment protocols are impractical or economically prohibitive.

One approach to avoid this type of procedural pain is to rethink whether the procedure is really needed. For example, tail docking of dairy cattle has become popular in some parts of the world in the belief that it improves cow hygiene and ultimately udder health. If tail docking really improved udder cleanliness and health one might argue that the trade-off is worthwhile, but a series of experiments involving thousands of animals with docked and intact tails has failed to show these benefits (Schreiner and Ruegg, 2002). This type of evidence is now leading many dairy producers to leave tails intact.

In some cases a procedure that is considered necessary can be avoided through selective breeding. The best example is the development of polled (i.e. genetically hornless) cattle. This approach has been especially successful for some breeds of beef cattle, eliminating the need for dehorning in these herds.

Sometimes the pain associated with a procedure can be reduced by means of a simple refinement. For example, piglets often have their ears notched to provide a permanent mark. A refinement to this procedure is to insert an ear tag. Marchant-Forde *et al.* (2009) compared the two approaches, and found that although both procedures caused pain the response was less to tagging than to notching. Alternatives are not always successful. The study by Marchant-Forde *et al.* also compared two methods of removing the tips of piglets' needle teeth, which can otherwise lacerate the faces of litter mates and the sow's udder. Grinding the teeth was expected to cause less pain than the more conventional approach of clipping the tips, but grinding took more time to perform and was associated with a stronger vocal response by the piglets.

5.3.2 Laboratory animals

Unlike farm animals, national legislation in many countries requires that painful procedures performed on laboratory animals be performed with the benefit of anaesthesia and analgesia unless there is a sound experimental reason not to do so. However, provision of effective pain relief is not uniform, especially for small rodents (Coulter *et al.*, 2009; Stokes *et al.*, 2009). One reason for this may be a failure to recognize the occurrence of pain (Richardson and Flecknell, 2005), and much of the recent work on pain in laboratory animals has focused on how best to recognize post-procedural pain, with the aim of developing better treatment protocols.

One challenge in improving the skills of caregivers is that individuals experienced with the animals are sometimes incorrectly assumed to be adept at pain assessment. This is perhaps especially true for the highly trained veterinarians, researchers and animal technicians responsible for the care of laboratory animals. Roughan and Flecknell (2006) tested the performance of hundreds of animal care professionals, blind to treatment, in evaluating pain from a video of post-surgical rats treated with various doses of analgesic. Participants first assessed the video using only their professional experience and a visual analogue scale to rate the pain. Participants then received 10 min of training in recognizing behaviours associated with this type of post-operative pain (e.g. arching the back, staggering, writhing and twitching), and rescored the video. Before training only 54% of observers were able to differentiate among treatments, but performance improved to 75% after training. These results illustrate the value of clear criteria and training in pain assessment, even for seasoned professionals.

One troubling topic is the use of animals as models in pain research, as this can involve deliberately exposing conscious animals to pain. Despite efforts to reduce the use of animals in research, and to refine procedures such that research animals experience less suffering, animal use in pain research appears to be on the rise. For example, a recent bibliographic analysis shows a steady increase in the use of animals in pain research over the past two decades (Mogil *et al.*, 2009).

5.3.3 Companion animals

Painful surgeries on cats and dogs are normally performed under anaesthesia, taking us beyond the more obvious welfare issues associated with completely untreated pain. The long history of pain treatment for companion animals has also allowed for the development of some sophistication in pain treatment protocols.

Of particular interest is combining different classes of drugs to treat different aspects of pain, such as local anaesthetics to prevent the immediate pain resulting from surgery and non-steroidal anti-inflammatory drugs to control the post-operative pain. Even when the immediate perception of pain is prevented using general anaesthesia, trauma caused by surgery can sensitize the surrounding nerves. Combining general anaesthesia with a local block can prevent this sensitization, thereby lowering post-operative pain and aiding recovery. For example, one recent study showed that providing topical local anaesthetic around the area of incision reduced the amount of general anaesthetic required to maintain anaesthesia in cats undergoing ovariectomy (Zilberstein *et al.*, 2008).

Cultural factors play an important role in determining when pain is considered important and worth treating. A survey of Canadian veterinarians showed that approximately half of the dogs and cats undergoing ovariohysterectomy received analgesics post-

operatively (Hewson *et al.*, 2006). The distribution of this treatment was bimodal, with some practitioners rarely or never prescribing the analgesics and others consistently treating the post-operative pain. Much of the variation in treatment could be explained by the veterinarian's perception of the pain that the animals were experiencing. Younger veterinarians (who had received their training more recently) were more likely to identify animals as being in pain.

In addition to the research on procedural pain, work on companion animals has also addressed chronic pain associated with ailments like osteoarthritis. One limitation in treating this pain has been that ongoing use of some drugs can result in complications such as gastric ulcers. Now though newer classes of non-steroidal anti-inflammatory drugs are available that can provide effective treatment for chronic pain caused by osteoarthritis in cats, with little or no side effects (Gunew *et al.*, 2008).

5.3.4 Wild animals

Relatively little research has focused on pain in wildlife, perhaps because in this case human action or inaction is typically not the cause of pain experienced by these animals. There are, however, some notable exceptions, including painful procedures performed on captive wildlife and pain associated with hunting or angling.

One example is the harvesting of velvet antler from farmed deer. The antler is popular in traditional Chinese medicine, but antler removal causes pain. Although this pain can be controlled using a ring block with local anaesthetic, marketers of the velvet have expressed concern that the drugs may contaminate the product. Compression, using a stretched rubber ring under high tension around the pedicle of the antler, has been suggested as an alternative, drug-free, form of local analgesia. As with painful procedures performed on farm animals, the animals are physically restrained. The restraint itself can cause a pronounced physiological stress response, and constrains many of the behaviours that might otherwise be useful in pain assessment. To avoid these problems, Johnson *et al.* (2005) compared electroencephalographic (EEG) responses of red deer undergoing both procedures. Their results showed that the local anaesthetic was effective at controlling the EEG responses to antler removal. In contrast, the compression showed no effect in controlling the pain due to antler removal. Indeed, the compression itself caused a pain response (Fig. 5.2).

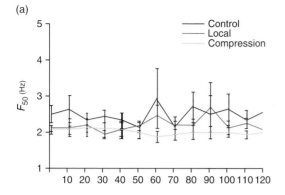

(a)

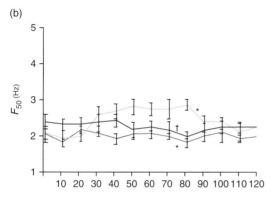

(b)

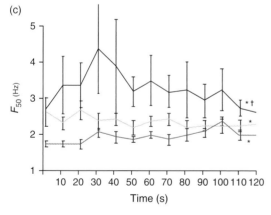

(c)

Fig. 5.2. Electroencephalographic (EEG) frequencies (F_{50} (Hz); mean ± SEM) recorded during (a) baseline, (b) analgesic treatment and (c) antler removal. Experimental groups comprised red deer stags allocated to one of the following treatments: no analgesic (control), a 2% lidocaine 'ring block' (local) and a compression rubber band (compression). (From Johnson *et al.* (2005) Comparison of analgesic techniques for antler removal in halothane-anaesthetized red deer (*Cervus elaphus*): electroencephalographic responses. *Veterinary Anaesthesia and Analgesia* 32, 61–71, Wiley-Blackwell.)

Pain perception in fish has generated much debate in the recent scientific literature. The prefrontal cortex of the brain is believed to be an important structure in allowing conscious experiences in humans. Some authors have suggested that animals (such as fish) that lack a well-developed prefrontal cortex may therefore be unable to experience pain consciously although, of course, similar functions may be served by different structures in different species. As explained by Braithwaite and Huntingford (2004, p. S88):

> Comparisons of avian and mammalian visual systems clearly illustrate how different taxonomic groups can perceive and process the same type of information but through different pathways and neural structures.

Thus, painful stimuli might be perceived by fish as unpleasant through other brain structures. Indeed, fish are taxonomically and neuroanatomically diverse, so how and what fish experience might vary considerably from species to species. Recent work has shown that different species of fish respond differently to the same type of stimulus: zebra fish and rainbow trout responded to acetic acid injections in their lips by increasing their ventilation and swim rates, but common carp exposed to the same or more severe stimuli did not show this response (Reilly *et al.*, 2008). These results should not be used to infer that carp feel less pain, as it is also possible that these fish simply respond to pain in other ways.

5.4 The Recognition of Animal Pain

5.4.1 Physiological responses to pain

Pain and nociceptive stimuli are stressors, and so provoke a biological stress response. Consequently, the production of such stress responses has been used to infer the presence of pain in animals. As discussed above, an immediate problem that arises is that the stress responses observed do not require conscious perception of pain, but only activation of nociceptive pathways. Clear evidence for this is provided by studies in anaesthetized animals, which cannot have conscious experiences, and yet noxious stimuli produce stress responses such as increases in heart rate and blood pressure. The interpretation of these physiological responses to pain is further confounded by the stress response to tissue damage. For example, surgical stimulation occurring during surgery, or tissue damage during procedures such as application of rubber rings to the tail or scrotum as part of husbandry procedures, cause a stress response (Bailey and Child, 1987; Thornton and Waterman-Pearson, 1999). Some of the response is due to nociceptor activation (Haga and Ranheim, 2005), some (in conscious animals) to the perception of pain and some to the tissue damage caused by the procedure. In addition, it has been shown that anaesthesia itself causes a stress response (Derbyshire and Smith, 1984), and, in conscious animals, restraint and other handling procedures may also provoke a stress response.

Despite these limitations, the physiological responses to presumed painful procedures have been used to assess the degree of pain that these procedures may cause. For such studies to provide persuasive evidence that the occurrence of pain and its intensity is being assessed, the following criteria should be met:

1. The responses should vary in magnitude in parallel with the likely degree of pain (e.g. based on the degree of tissue trauma).
2. Appropriate control groups should be included in which the stress response to other variables can be quantified.

The validity of these physiological indicators of pain is strengthened if they can be shown to correlate with other potential markers (e.g. behavioural responses).

Cardiovascular and respiratory responses

Acute pain in humans results in a major stress response, with catecholamine release and marked elevations in heart rate and blood pressure. These responses have been assessed in animals in an attempt to incorporate them into practically applicable pain scoring schemes. Results to date have been disappointing (Cambridge et al., 2000; Price et al., 2003), probably because of the numerous other factors that can influence these responses. Changes in heart rate and blood pressure are used routinely to monitor the depth of anaesthesia in animals and humans. They are often recommended as means of monitoring anaesthesia when neuromuscular blocking agents are used, as animals are then unable to make movement responses. However, the degree of elevation of these cardiovascular parameters that indicates a return of consciousness has not been established, and in humans,

measurement of awareness during anaesthesia has shifted to the use of EEG responses (e.g. bi-spectral index monitors), a technique that can be applied to animals in some circumstances (Murrell and Johnson, 2006; Otto, 2008).

Although simple measurement of heart rate changes has not proven particularly useful as a marker of pain, more sophisticated analysis, for example by assessing heart rate variability (HRV), may prove to be of value (Rietmann et al., 2004). Once again, it must be emphasized that this is not a specific measure of pain or nociception, and that other factors may also influence HRV.

Endocrine responses

The endocrine responses to presumed painful procedures, for example the dehorning of cattle (McMeekan et al., 1998), and castration and tail docking in lambs (Molony and Kent, 1997), have been extensively investigated. Although, as mentioned above, these are indirect markers of pain, the graded responses shown to a range of procedures in lambs, and the correlation of these changes with behavioural markers, provide valuable supportive measures for the assessment of the relative pain intensity caused by such procedures (Lester et al., 1996).

Neurophysiological and electrophysiological responses

Stimulation of the production of biomarkers, such as the early gene product c-*fos*, has been used to determine the occurrence and intensity of nociceptor stimulation (Coggeshall, 2005). In pigs, c-*fos* activation has been demonstrated after surgical procedures, and this activation has been shown to be blocked by administration of local anaesthetic (Lykkegaard et al., 2005). Similarly, in rats, intraperitoneal administration of pentobarbital (with pH >10) has been shown to produce c-*fos* activation, which can be blocked by injecting the pentobarbital along with a local anaesthetic (Svendsen et al., 2007). These and other biomarkers can only be assessed post-mortem, and only under carefully controlled conditions, so are useful primarily as research tools, rather than as more broadly applicable methods of assessing pain.

Measurement of changes in the peripheral and central nervous systems produced by activation of nociceptors and by processes that can increase nociception (for example inflammation) have also been used to infer the presence of pain. The development of hyperalgesia and allodynia have been used as indicators that procedures may cause pain as assessed by quantitative sensory testing (see below). Central changes in the EEG of lightly anaesthetized animals have been used to demonstrate activation by presumed painful stimuli in a number of different species (Johnson *et al.*, 2005). This approach has been proposed as a more humane way of investigating the degree of pain associated with procedures such as the dehorning of cattle, and, in particular, for demonstrating the efficacy of measures designed to reduce the pain associated with these procedures (Murrell and Johnson, 2006).

It is important to appreciate that many of these markers indicate activation of nociceptor pathways, but not conscious perception of pain. However, they are important tools, because demonstration of such activation is good evidence that pain may be experienced. Recent advances in imaging techniques have enabled studies to assess the activation of brain areas associated with pain processing. Demonstration of the activation of brain structures in response to noxious stimuli can be used to provide evidence of central processing of this information. Demonstration that the brain areas involved are not simply primary somatosensory regions, but areas that are associated (in humans) with the emotional component of pain, implies that similar processing is occurring in animals. In other words, these animals may be capable of experiencing pain, not simply nociception (Hess *et al.*, 2007). It is important to note that, to date, such studies have only been conducted in anaesthetized animals, so they demonstrate only the capacity for this type of processing of noxious stimuli, and not conscious awareness of the emotional component of pain.

5.4.2 Quantitative sensory testing

Quantitative sensory testing (QST) techniques are used to quantify particular aspects of sensory function in the assessment of human neurological conditions. The methods are psychophysical, non-invasive tests based on the application of standard physical stimuli (e.g. heat) and the end point of the test is determined by the patient's report of stimulus detection or perception of a change in its intensity.

In animals, QST techniques have been adapted to detect nociceptive responses as a means of assessing changes in the sensitivity of tissues to potentially painful stimulation. In this context, QST involves the controlled application of standard noxious stimuli (mechanical or thermal) at particular anatomical locations in order to elicit a withdrawal or avoidance reaction, which determines the end point of the test. Common parameters of QST are reaction time (time elapsed from the onset of the stimulus to the withdrawal response) and the nociceptive threshold (the minimum magnitude of the stimulus required to elicit an avoidance reaction).

For a particular QST technique or protocol to be reliable it should meet standards of repeatability and sensitivity. Repeatability can be defined as the level of agreement between measurements from a particular animal obtained by tests separated by short periods of time. Repeatability can be adversely affected by the animal learning to react to the stimulus before it becomes noxious (Chambers *et al.*, 1990), and can depend on the precise location of the site that is stimulated (Haussler *et al.*, 2007). Sensitivity refers to the ability of the method to detect changes in the quantity measured. Several studies have shown decreased nociceptive thresholds in animals after surgery or in animals with chronic pain, while other studies have shown increased nociceptive thresholds after analgesic treatment (Welsh and Nolan, 1995; Ley *et al.*, 1996; Lascelles *et al.*, 1997, 1998; Slingsby *et al.*, 2001; KuKanich *et al.*, 2005).

However, it remains to be determined whether any QST techniques are sensitive and robust enough to be clinically useful as a means of monitoring disease progression or treatment efficacy. A number of factors can influence QST measurements. Mechanical nociceptive thresholds differ between tissues and body regions, e.g. in horses, thresholds were lower when measured on soft tissue than when measured on bony landmarks, and lower thresholds have been measured over spines of the cervical vertebrae than over those of the lumbosacral region (Haussler and Erb, 2006). Additional factors potentially affecting QST measurements are sex, height, weight and age, which is particularly relevant when QST measurements from clinical cases are compared with reference measurements made from control animals (i.e. those not in pain).

A further limitation on QST is that it requires some cooperation from the animal and therefore the animal's temperament may be a limiting factor.

5.4.3 Behavioural assessment

Types of responses

Changes in behaviour have long been used in both scientific and clinical assessments of pain in animals. Most intuitive are the simple avoidance responses of animals, such as the struggling and escape responses typical in farm animals subjected to routine procedures such as dehorning and branding. For example, the horn buds of goat kids are sometimes removed using a hot iron, and kids show more than twice the amount of struggling when this procedure is performed compared with when subjected to a sham procedure with identical handling (Alvarez et al., 2009). Some studies have used very sophisticated methods of quantifying the amount of struggling. Schwartzkopf-Genswein et al. (1998) quantified the struggling of beef cattle using strain gauges and load cells to assess the forces that the animals exerted against the restraining chute during hot-iron branding.

Other behavioural responses during painful procedures are not necessarily associated with escape. For example, young animals still dependent upon their parents for care and protection often vocalize in response to pain. Among the best-studied examples are the vocal responses of piglets to painful procedures. For example, Leidig et al. (2009) used automated, spectrographic analysis of piglet vocalizations during castration to show that this response is much reduced when animals are given a local anaesthetic. Indeed a lowered vocal response has been used to evaluate alternative methods of castration, tail docking, teeth clipping and other procedures (Marchant-Forde et al., 2009). Older animals have less to gain by calling attention to their pain, but vocal responses can still be useful in some situations. For example, Grandin (1998) monitored commercial slaughterhouses and found that about 10% of the animals vocalized during handling and stunning, almost always in association with painful events such as prodding or unsuccessful stunning. On this basis, vocalizations are now included as an audit point in assessing welfare in many slaughterhouses.

Behavioural responses to pain are easiest to interpret when they are directed towards the site of

the injury, such as calves licking the tail area after tail docking, and rubbing their heads after disbudding. These responses may be accentuated if the area of injury is stimulated. For example, lambs that have been castrated show a bucking response if the scrotum is palpated (Thornton and Waterman-Pearson, 1999).

In other instances, the pain responses after the injury are less specific. Animals experiencing pain may show reduced levels of activity, including decreased locomotion and feeding. Behavioural changes that come at a clear cost to the animal, such as reduced feed intake leading to weight loss or a check in growth, are arguably the easiest to interpret from a welfare perspective. In some instances, changes in behaviour may also provide insights into the nature of the pain experienced. Moya et al. (2008) followed the behaviour of piglets in the hours after castration; castrated piglets spent less time walking and dog sitting compared with sham-castrated controls, perhaps because both behaviours aggravated the area of incision.

All the responses described above would be expected to be reduced if animals were provided with the appropriate analgesics. Indeed, testing animals with and without analgesics is recognized as a key method of validating pain responses. Some studies have taken this logic to the next level, by assessing pain using the amount of analgesics that the animal chooses to consume. In one study, Danbury et al. (2000) provided chickens with two coloured feeds, one that contained an analgesic. In comparison with sound birds, lame chickens ate more of the feed containing the analgesic (Fig. 5.3), with the most severely lame birds showing the highest intakes.

An especially interesting scientific challenge is assessing the emotional component of animal pain. One approach has been to assess pain responses in animals experiencing some other form of emotional upset. For example, Boccalon et al. (2006) showed that mice that were exposed to chronic stress showed stronger pain responses than did non-stressed controls. Other work has focused on the changes in cognition associated with emotional states. For example, a number of studies have now shown that depressed and anxious humans and other animals are more likely to show a 'pessimistic' response when faced with a novel opportunity, suggesting that this type of cognitive bias could be used to assess the emotional state of the animals. To our knowledge, no work to date has tested cognitive bias in animals

I. Viñuela-Fernández et al.

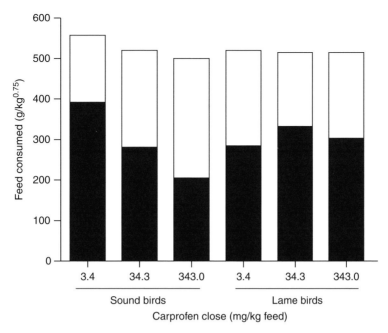

Fig. 5.3. Mean consumption of drugged (■) versus undrugged (❑) feed (in g/kg$^{0.75}$) containing three different doses of carprofen (3.4, 34.3 and 343.0 mg/kg feed) in both sound and lame broiler chickens. Sound chickens avoided the drugged feed, especially at higher doses, but lame birds showed a preference for the drugged feed (from Danbury *et al.*, 2000).

exposed to pain, but this type of study could be helpful in assessing the emotional component of animal pain (Mendl *et al.*, 2009).

Clinical scores

The behavioural approaches described above, such as strain gauge measures of struggling, spectrographic analysis of vocalizations, and self-medication trials, are useful in the scientific assessment of pain and these studies can provide some general guidance for how pain may best be prevented or controlled. However, veterinarians, farmers and others working with animals also require practical methods of assessing pain, for example so that they can recognize which animals require treatment and for how long.

Lameness in both dairy cows and broiler chickens is an important welfare problem. Lameness is a common ailment, and when left untreated animals may experience pain for weeks or longer. Contributing to the problem is the fact that many cases remain unrecognized by caregivers. For example, one study in the USA showed that only a third of lame dairy cows were recognized as such by the

farm staff responsible for their care (Espejo *et al.*, 2006). A variety of clinical scoring systems has been developed for assessing lameness. One of the best known is the 5-point numerical score developed by Sprecher *et al.* (1997); dairy cows are scored as '1' if they walk and stand with a flat back, but higher scores are assigned if cows walk or stand with an arched back. These clinical assessments require human judgement, such that different assessors can come to different conclusions. Training using clear criteria can much improve assessments and agreement among independent observers. For example, early work on lameness in broiler chickens showed poor agreement among observers, but when observers were trained with clear criteria, independent assessments were nearly identical (Garner *et al.*, 2002). Even more important than inter- and intra-observer reliability is the validity of the assessment. Lameness scores for dairy cows have been validated in many ways, including comparing the gait of animals before and after treatment with analgesics and local anaesthetics, comparing animals with and without injuries, and comparing animals walking on concrete versus a resilient, high-traction rubber surface (Flower and Weary, 2009).

With the growing size of many farms, and the increased use of automation, such as the use of robotic milkers, it is becoming more feasible to use automated measures of animal behaviour in detecting painful ailments such as lameness. In one elegant example, Pastell and Kujala (2007) were able to identify lame cows successfully using load cells in the floor of a robotic milker. The load cells provided data on how the cows distributed their weight, and these data were processed using a probabilistic neural network to identify cases of lameness.

5.5 Pain Management

Although our methods of pain assessment remain relatively limited, it is now generally assumed that many animals can experience pain and that whenever possible this pain should be alleviated. However, application of this general principle remains variable across different animal groups. For example painful husbandry procedures are still common in farm animals, either without any analgesia or anaesthesia (e.g. for castration and tail docking of lambs), or with a level of local anaesthesia that is unlikely to be of sufficient duration. Pain management in companion animals following acutely painful procedures, such as surgery, is thought to have improved since surveys conducted some 5–10 years ago. This assumption is based largely on the increased number of analgesics available for veterinary use, and anecdotal evidence from the companies marketing these products. There is also increased veterinary interest and concern with pain management generally, as shown by the formation of the International Academy of Veterinary Pain Management, and the change in name of the *Journal of Veterinary Anaesthesia* to the *Journal of Veterinary Anaesthesia and Analgesia*. How far this interest translates into effective pain management remains uncertain.

Treatment of chronic pain in companion animals, such as lameness, is increasing, based on the increased sales of anti-arthritic and analgesic products. However, management of pain associated with other conditions, such as otitis, ocular conditions and cancer pain appears to be relatively neglected. In laboratory animals, data on reported analgesic use suggest that this remains limited in small rodents (Stokes *et al.*, 2009), but is more extensive in large species (Coulter *et al.*, 2009). The

reasons for this apparent underuse are believed to include both a concern related to possible interactions of analgesic administration with the aims of particular research protocols and an inability to recognize pain (Richardson and Flecknell, 2005). The factors limiting analgesic use in laboratory animals have parallels in all groups of animals. It is difficult to manage pain effectively if its nature, intensity and duration cannot be assessed. Failure to recognize pain can lead to a complete failure to administer analgesics or introduce other measures to reduce or eliminate pain. A further major factor is the misconception regarding the potential side effects of analgesics. This may be a factor that accounted for the lower reported use of analgesics in cats compared with dogs (Capner *et al.*, 1999), as there is a long-held misconception that opioids cannot be safely used in cats and that non-steroidal anti-inflammatory drugs are toxic to them.

It is important to emphasize that these concerns are not clinically relevant for the vast majority of animals. It is critically important that veterinary graduates are educated appropriately in the use of analgesics, but also that all concerned give an appropriate weight to the distressing nature of animal pain and its adverse effects on animal welfare. It is encouraging to note that many of these issues have parallels in the development of pain management in people, when a failure to assess pain, a concern regarding side effects, and a general lack of appreciation of the longer term effects of acute pain, led to a systemic failure to control pain in people. This situation has been transformed over the past few decades, and it provides hope that a similar transformation can be achieved in relation to the control of animal pain. One limiting factor for some species will be the lack of availability of analgesics with a licensed indication for animal use. The limit this places on a veterinarian's ability to manage pain effectively varies greatly in different countries, and also varies depending upon the type of animal involved. If the animal is considered likely to enter the food chain, then this may restrict the products that can be used. In companion animals, the options are usually less restrictive. Although national legislation may permit the administration of a range of analgesic agents, if appropriate clinical trials data in the target species are not available, then estimating an appropriate dose rate may be difficult. In addition, limitations in pain assessment may preclude assessment of efficacy in the individual animal that is being treated.

Improvements in pain management require more than changes in the availability of analgesics, and changes in the attitudes of the veterinary profession and of others with responsibilities for animal welfare. Managing pain in farm animals, for example, raises economic issues. For example, the pain after hot-iron disbudding can be readily controlled using analgesics, but farmers may see little direct economic benefit from providing the pain relief. Painful ailments such as lameness in dairy cattle might be largely avoided with changes in breeding, housing and management, but these changes can be costly for producers. Future research can help to resolve these issues by identifying more economically feasible methods of preventing or controlling pain, and documenting the economic advantages associated with these methods. In some cases, changes in policy may also be required to provide a level playing field for those responsible for animal care, thus allowing caregivers to put more emphasis on pain prevention, assessment and treatment.

5.6 Conclusions

- Animal pain remains a major welfare problem and there are challenges in the appropriate assessment and treatment of this pain.
- Although advances in pharmacology have provided opportunities to treat animal pain effectively, numerous other factors act to reduce analgesic use. These include economic considerations, a failure to recognize pain, an inability to assess its severity in many circumstances and a failure to appreciate the importance and welfare significance of pain.
- We suggest that much progress can be made through improved training in pain assessment, and through developing treatment protocols that are both effective for the animal and practical for the caregiver.
- We also suggest that post-operative pain, and ongoing pain associated with injury and disease, are special priorities for future work, as this pain is often undertreated.

Acknowledgment

Professor Vince Molony was a co-author of the version of this chapter in the first edition of this book. The authors would also like to thank him for his valuable input in the drafting of this edition.

References

Alvarez, L., Nava, R.A., Ramírez, A., Ramírez, E. and Gutiérrez, J. (2009) Physiological and behavioural alterations in disbudded goat kids with and without local anaesthesia. *Applied Animal Behaviour Science* 117, 190–196.

Anand, K.J.S. and Craig, K.D. (1996) New perspectives on the definition of pain. *Pain* 67, 3–6.

Bailey, P.M. and Child, C.S. (1987) Endocrine response to surgery. In: Kaufman, L. (ed.) *Anaesthesia: Review 4*. Churchill Livingstone, Edinburgh, pp. 100–116.

Baliki, M.N., Chialvo, D.R., Geha, P.Y., Levy, R.M., Harden R.N., Parrish, T.B. and Apkarian, A.V. (2006) Chronic pain and the emotional brain: Specific brain activity associated with spontaneous fluctuations of intensity of chronic back pain. *Journal of Neuroscience* 26, 12165–12173.

Basbaum, A.I. and Jessell, T.M. (2000) The perception of pain. In: Kandel, E.R., Schwartz, J.H. and Jessell, T.M. (eds) *Principles of Neural Science*, 4th edn. McGraw-Hill, New York, pp. 472–491.

Bateson, P. (1992) Assessment of pain in animals. *Animal Behaviour* 42, 827–839.

Boccalon, S., Scaggiante, B. and Perissin, L. (2006) Anxiety stress and nociceptive responses in mice. *Life Sciences* 78, 1225–1230.

Braithwaite, V.A. and Huntingford, F.A. (2004) Fish and welfare: do fish have the capacity for pain perception and suffering? *Animal Welfare* 13 (Supplement 1), 87–92.

Cambridge, A.J., Tobias, K.M., Newberry, R.C. and Sarkar, D.K. (2000) Subjective and objective measurements of postoperative pain in cats. *Journal of the American Veterinary Medical Association* 217, 685–690.

Capner, C.A., Lascelles, B.D. and Waterman-Pearson, A.E. (1999) Current British veterinary attitudes to perioperative analgesia for dogs. *Veterinary Record* 145, 95–99.

Chambers, J.P., Livingston, A. and Waterman, A.E. (1990) A device for testing nociceptive thresholds in horses. *Journal of the Association of Veterinary Anaesthetists* 17, 42–44.

Coggeshall, R.E. (2005) Fos, nociception and the dorsal horn. *Progress in Neurobiology* 77, 299–352.

Coulter, C.A., Flecknell, P.A. and Richardson, C.A. (2009) Reported analgesic administration to rabbits, pigs, sheep, dogs and non-human primates undergoing experimental surgical procedures. *Laboratory Animals* 43, 232–238.

Danbury, T.C., Weeks, C.A., Chambers, J.P., Waterman-Pearson, A.E. and Kestin, S.C. (2000) Self-selection of the analgesic drug carprofen by lame broiler chickens. *Veterinary Record* 146, 307–311.

Derbyshire, D.R. and Smith, G. (1984) Sympathoadrenal responses to anaesthesia and surgery. *British Journal of Anaesthesia* 56, 725–739.

D'Mello, R. and Dickenson, A.H. (2008) Spinal cord mechanisms of pain. *British Journal of Anaesthesia* 101, 8–16.

Espejo, L.A., Endres, M.I. and Salfer, J.A. (2006) Prevalence of lameness in high-producing Holstein cows housed in freestall barns in Minnesota. *Journal of Dairy Science* 89, 3052–3058.

Flower, F.C. and Weary, D.M. (2009) Gait assessment in dairy cattle. *Animal* 3, 87–95.

Garner, J., Falcone, C., Wakenell, P., Martin, M. and Mench, J. (2002) Reliability and validity of a modified gait scoring system and its use in assessing tibial dyschondroplasia in broilers. *British Poultry Science* 43, 355–363.

Grandin, T. (1998) The feasibility of using vocalization scoring as an indicator of poor welfare during cattle slaughter. *Applied Animal Behaviour Science* 56, 121–128.

Gunew, M.N., Menrath, V.H. and Marshall, R.D. (2008) Long-term safety, efficacy and palatability of oral meloxicam at 0.01–0.03 mg/kg for treatment of osteoarthritic pain in cats. *Journal of Feline Medicine and Surgery* 10, 235–241.

Haga, H.A. and Ranheim, B. (2005) Castration of piglets: the analgesic effects of intratesticular and intrafunicular lidocaine injection. *Veterinary Anaesthesia and Analgesia* 32, 1–9.

Haussler, K.K. and Erb, H.N. (2006) Mechanical nociceptive thresholds in the axial skeleton of horses. *Equine Veterinary Journal* 38, 70–75.

Haussler, K.K., Hill, A.E., Frisbie, D.D. and McIlwraith, C.W. (2007) Determination and use of mechanical nociceptive thresholds of the thoracic limb to assess pain associated with induced osteoarthritis of the middle carpal joint in horses. *American Journal of Veterinary Research* 68, 1167–1176.

Hess, A., Sergejeva, M., Budinsky, L., Zeilhofer, H.U. and Brune, K. (2007) Imaging of hyperalgesia in rats by functional MRI. *European Journal of Pain* 11, 109–119.

Hewson, C.J., Dohoo, I.R. and Lemke, K.A. (2006) Factors affecting the use of postincisional analgesics in dogs and cats by Canadian veterinarians in 2001. *Canadian Veterinary Journal* 47, 453–459.

International Association for the Study of Pain (1979) Pain terms: a list with definitions and notes on usage. *Pain* 6, 247–252.

Johnson, C.B., Wilson, P.R., Woodbury, M.R. and Caulkett, N.A. (2005) Comparison of analgesic techniques for antler removal in halothane-anaesthetized red deer (*Cervus elaphus*): electroencephalographic responses. *Veterinary Anaesthesia and Analgesia* 32, 61–71.

KuKanich, B., Lascelles, B.D.X. and Papich, M.G. (2005) Assessment of a von Frey device for evaluation of the antinociceptive effects of morphine and its application in pharmacodynamic modelling of morphine in dogs. *American Journal of Veterinary Research* 66, 1616–1622.

Lascelles, B.D.X., Cripps, P.J., Jones, A. and Waterman, A.E. (1997) Postoperative central hypersensitivity and pain: the pre-emptive value of pethidine for ovariohysterectomy. *Pain* 73, 461–471.

Lascelles, B.D.X., Cripps, P.J., Jones, A. and Waterman, A.E. (1998) Efficacy and kinetics of carprofen, administered preoperatively or postoperatively, for the prevention of pain in dogs undergoing ovariohysterectomy. *Veterinary Surgery* 27, 568–582.

Leidig, M.S., Hertrampf, B., Failing, K., Schumann, A. and Reiner, G. (2009) Pain and discomfort in male piglets during surgical castration with and without local anaesthesia as determined by vocalization and defence behaviour. *Applied Animal Behaviour Science* 116, 174–178.

Lester, S.J., Mellor, D.J., Holmes, R.J., Ward, R.N. and Stafford, K.J. (1996) Behavioural and cortisol responses of lambs to castration and tailing using different methods. *New Zealand Veterinary Journal* 44, 45–54.

Ley, S.J., Waterman, A.E. and Livingston, A. (1996) Measurement of mechanical thresholds, plasma cortisol and catecholamines in control and lame cattle: a preliminary study. *Research in Veterinary Science* 61, 172–173.

Lykkegaard, K., Lauritzen, B., Tessem, L., Weikop, P. and Svendsen, O. (2005) Local anaesthetics attenuates spinal nociception and HPA-axis activation during experimental laparotomy in pigs. *Research in Veterinary Science* 79, 245–251.

Marchant-Forde, J.N., Lay, D.C. Jr, McMunn, K.A., Cheng, H.W., Pajor, E.A. and Marchant-Forde, R.M. (2009) Postnatal piglet husbandry practices and well-being: the effects of alternative techniques delivered separately. *Journal of Animal Science* 87, 1479–1492.

McMeekan, C.M., Stafford, K.J., Mellor, D.J., Bruce, R.A., Ward, R.N. and Gregory, N.G. (1998) Effects of regional analgesia and/or non-steroidal anti-inflammatory analgesic on the acute cortisol response to dehorning calves. *Research in Veterinary Science* 64, 147–150.

Mendl, M., Burman, O.H.P., Parker, R.M.A. and Paul, E.S. (2009) Cognitive bias as an indicator of animal emotion and welfare: emerging evidence and underlying mechanisms. *Applied Animal Behaviour Science* 118, 161–181.

Mogil, J.S., Simmonds, K. and Simmonds, M.J. (2009) Pain research from 1975 to 2007: a categorical and bibliometric meta-trend analysis of every research paper published in the journal Pain. *Pain* 142, 48–58.

Molony, V. (1997) Comments on Anand and Craig (Letters to the Editor). *Pain* 70, 293.

Molony, V. and Kent, J.E. (1997) Assessment of acute pain in farm animals using behavioral and physiological measurements. *Journal of Animal Science* 75, 266–272.

I. Viñuela-Fernández *et al.*

Moya, S.L, Boyle, L.A., Lynch, P.B. and Arkins, S. (2008) Effect of surgical castration on the behavioural and acute phase responses of 5-day-old piglets. *Applied Animal Behaviour Science* 111, 133–145.

Murrell, J.C. and Johnson, C.B. (2006) Neurophysiological techniques to assess pain in animals. *Journal of Veterinary Pharmacology and Therapeutics* 29, 325–335.

Otto, K.A. (2008) EEG power spectrum analysis for monitoring depth of anaesthesia during experimental surgery. *Laboratory Animals* 42, 45–61.

Pastell, M.E. and Kujala, M. (2007) A probabilistic neural network model for lameness detection. *Journal of Dairy Science* 90, 2283–2292.

Price, J., Catriona, S., Welsh, E.M. and Waran, N.K. (2003) Preliminary evaluation of a behaviour-based system for assessment of post-operative pain in horses following arthroscopic surgery. *Veterinary Anaesthesia and Analgesia* 30, 124–137.

Reilly, S.C., Quinn, J.P., Cossins, A.R. and Sneddon, L.U. (2008) Behavioural analysis of a nociceptive event in fish: comparisons between three species demonstrate specific responses. *Applied Animal Behaviour Science* 114, 248–259.

Richardson, C. and Flecknell, P.A. (2005) Anaesthesia and post-operative analgesia following experimental surgery in laboratory rodents – are we making progress? *ATLA – Alternatives to Laboratory Animals* 33, 119–127.

Rietmann, T.R., Stauffacher, M., Bernasconi, P., Auer, J.A. and Weishaupt, M.A. (2004) The association between heart rate, heart rate variability, endocrine and behavioural pain measures in horses suffering from laminitis. *Journal of Veterinary Medicine A* 51, 218–225.

Roughan, J.V. and Flecknell, P.A. (2006) Training in behaviour-based post-operative pain scoring in rats–An evaluation based on improved recognition of analgesic requirements. *Applied Animal Behaviour Science* 96, 327–342.

Schreiner, D.A. and Ruegg, P.L. (2002) Responses to tail docking in calves and heifers. *Journal of Dairy Science* 85, 3287–3296.

Schwartzkopf-Genswein, K.S., Stookey, J.M., Crowe, T.G. and Genswein, B.M.A. (1998) Comparison of image analysis, exertion force, and behavior measurements for use in the assessment of beef cattle responses to hot-iron and freeze branding. *Journal of Animal Science* 76, 972–979.

Sherrington C.S. (1906) *The Integrative Action of the Nervous System*. Charles Scribner's Sons, New York.

Slingsby, L.S., Jones, A. and Waterman-Pearson, A.E. (2001) Use of a fingermounted device to compare mechanical nociceptive thresholds in cats given pethidine or no medication after castration. *Research in Veterinary Science* 70, 243–246.

Sneddon, L.U. (2004) Evolution of nociception in vertebrates: comparative analysis of lower vertebrates. *Brain Research Reviews* 46, 123–130.

Sprecher, D.J., Hostetler, D.E. and Kaneene, J.B. (1997) A lameness scoring system that uses posture and gait to predict dairy cattle reproductive performance. *Theriogenology* 47, 1179–1187.

Stokes, E.L., Flecknell, P.A. and Richardson, C.A. (2009) Reported analgesic and anaesthetic administration to rodents undergoing experimental surgical procedures. *Laboratory Animals* 43, 149–154.

Svendsen, P., Kok, L. and Lauritzen, B. (2007) Nociception after intraperitoneal injection of a sodium pentobarbitone formulation with and without lidocaine in rats quantified by expression of neuronal c-fos in the spinal cord – a preliminary study. *Laboratory Animals* 41, 197–203.

Thornton, P.D. and Waterman-Pearson, A.E. (1999) Quantification of the pain and distress responses to castration in young lambs. *Research in Veterinary Science* 66, 107–118.

Welsh, E.M., and Nolan, A.M. (1995) Effect of flunixin meglumine on the thresholds to mechanical stimulation in healthy and lame sheep. *Research in Veterinary Science* 58, 61–66.

Woolf, C.J. and Salter, M.W. (2000) Neuronal plasticity: increasing the gain in pain. *Science* 288, 1765–1768.

Zilberstein, L.F., Moens, Y.P. and Leterrier, E. (2008) The effect of local anaesthesia on anaesthetic requirements for feline ovariectomy. *The Veterinary Journal* 178, 214–218.

6 Fear and Other Negative Emotions

BRYAN JONES AND ALAIN BOISSY

Abstract

Concern for animal welfare stems from the recognition that animals are sentient beings capable of experiencing negative emotions, such as fear, pain and frustration. Fear is a major emotion in determining how animals perceive and respond to their social and physical environment. We tackle three main areas. Our first section on the assessment of fear embraces the diversity of frightening events and fear responses, various ways of assessing the fear state, the concept of fearfulness as related to temperament, and the need for stringent validation of tests and measures. Emotions are generally viewed as the result of how an individual evaluates a triggering situation according to its relevance, the likely consequences and the potential for control. Secondly, we review reported interactions between cognition and emotion and explore a possible role for cognitive ability in the assessment of emotional states: we ask whether fear-induced cognitive bias can be used to probe long-lasting affective states, and we examine the role of specific aspects, such as predictability and controllability. Thirdly, we revisit the damaging effects of fear and other negative emotions. Harmful consequences of intense and/or prolonged fear include health and welfare problems (withdrawal, injury, reduced immunocompetence, pathological anxiety), management problems (e.g. difficulty in handling and moving animals), as well as damaging effects on production, product quality and profitability. Finally, before concluding, we briefly mention the main ways of alleviating fear and the need for such strategies to be practical.

6.1 Introduction

The Brambell Committee (1965) stated that animal welfare embraces physical and mental well-being. Consideration of mental well-being in animals reflects the now widespread acceptance that they are 'sentient beings' (Dawkins, 2001). Indeed, the Amsterdam treaty of the European Union (Council of Europe, 1997) stipulates that measures shall be applied 'to ensure improved protection and respect for the welfare of animals as sentient beings'. This requires recognition that animals have emotional capacities and will attempt to minimize exposure to stimuli eliciting negative emotions such as fear, frustration, distress and anxiety (Dawkins, 1990; Duncan, 1996; Jones, 1998). Recent results demonstrate the existence of sentience and emotions in farm and laboratory animals and also a role for cognitive ability (Boissy et al., 2007; Mendl et al., 2009). Improving our understanding of the range and depth of emotions that animals can experience is essential for the design of effective measures to safeguard/improve their welfare. This chapter focuses primarily on fear because this is one of the better understood and the potentially most damaging of the negative emotions, as well as the

precursor of many others (Jones, 1997; Boissy, 1998; Boissy et al., 2007). Although it is often adaptive in ideal circumstances the sudden, unpredictable, intense, prolonged or inescapable elicitation of fear can severely harm the mental and physical well-being, growth and reproductive performance of laboratory, zoo and farm animals (Jones and Waddington, 1992; Jones, 1997; Faure et al., 2003).

Fear is generally defined as a response to the perception of actual danger, whereas anxiety is regarded as the reaction to a potential (as yet unreal) threat (Boissy, 1998). Fear-related reactions are characterized by physiological and behavioural responses that prepare the animal to deal with the danger. From an evolutionary standpoint, such defensive reactions promote fitness: an animal's life expectancy is obviously increased if it can avoid sources of danger such as predators. Although captive animals have few natural predators, the mechanisms, emotions and behavioural responses persist (Dwyer, 2004). Moreover, domestic animals reared in a range environment may still experience severe predation by wild animals or dogs (Asheim and Mysterud, 2005).

Farm animals also show predator-avoidance reactions to human contact, even though reduced fear of human beings is a major component of domestication (e.g. in herbivores, Price, 1984; and in poultry, Jones, 1997). Routine management procedures such as shearing, castration, tail docking, beak trimming, dehorning, vaccination, harvesting, herding and transportation also cause fear and distress in cattle, sheep and poultry (Gentle *et al.*, 1990; Hargreaves and Hutson, 1990; Wohlt *et al.*, 1994; Manteca *et al.*, 2009). In addition, intense fear can cause chronic stress, which may compromise the expression of fundamental behaviours (social, sexual, parental, etc.) and reduce productivity and product quality in cattle, sheep, pigs and poultry (Hemsworth and Coleman, 1998; Jones, 1998; Bouissou *et al.*, 2001; Fisher and Matthews, 2001; Faure and Jones, 2004).

This chapter follows three main threads. First, we discuss the assessment of negative emotions, some of the difficulties involved and the need for better validation of various experimental approaches, particularly in the larger farm animal species. Secondly, we review the interactions between emotions and cognition and explore the possibilities of using cognitive ability and cognitive bias as new approaches to the assessment of fear and longer lasting negative mood. Thirdly, we describe some implications of high levels of negative emotions for the welfare, management and performance of farm animals.

6.2 Assessment of Fear and Temperament

6.2.1 What is fear?

Although there is no single general definition, we believe that an emotion can be best defined as an intense but short-lived affective response to an event that is associated with specific bodily changes (Boissy *et al.*, 2007). An emotion is classically described in terms of three components (Dantzer, 1988): a subjective component, i.e. the emotional experience (what one feels), and two expressive components, behavioural (what an animal shows to others, e.g. facial expressions, movements) and neurophysiological (how the body responds, e.g. stress reactions). Emotions differ from sensations, which are simply physical consequences of exposure to particular stimuli (e.g. heat, pressure), and from feelings, which only

designate internal states with no specific reference to external reactions.

The complexity of the fear concept is illustrated by its numerous definitions. Examples include: a state of alarm or dread, an unpleasant or painful emotion elicited by danger, an expectation of pain, a behaviour system that has evolved to ensure survival, a hypothetical state of the brain and neuroendocrine system, a defensive motivational state and an emotional response to perceived danger (Gray, 1987; Jones, 1987a,b,c, 1996; Boissy, 1998). Nevertheless, there is considerable overlap between these definitions, particularly in regard to the protective aspects of fear. Indeed, fear is a major negative emotion determining how humans and other animals respond to physical and social challenge. Ideally, it is an adaptive state with fear behaviour serving to protect the animal from injury (Jones, 1987a; 1996).

The classical approach to studying fear (and all emotions) embraces three stages. The first necessarily involves exposure to frightening stimulation. The second stage involves observing and measuring the animal's responses, for example cautious investigation, fight, flight or immobility. These responses may be altered and integrated according to changes in the perceived potency of the threatening stimulus and in the consequent intensity of the emotional state. The third stage embraces the continued development and validation of appropriate and robust experimental tests and measures.

6.2.2 Diversity of frightening events

According to Gray (1987), the fear-eliciting properties of an event reflect its general characteristics, for example, novelty, movement, intensity, duration, suddenness or proximity. Fear may also be elicited by specific stimuli, such as height and darkness, in relation to the evolutionary history of the species (ancestral/innate fears). Additionally, a stimulus can elicit fear through association with previous experience of another frightening event (conditioned fear).

Gregariousness is common to most domestic animals (Keeling and Gonyou, 2001), and the variety of social stimuli that accompany social cohesion and structure may elicit or modulate fear reactions. Social signals may represent particular cases of the types of fear-eliciting stimulation mentioned above. Some social signals are characterized by their unfamiliarity, for example the novelty of the neonate that affects maternal behaviour in primiparous females (Poindron *et al.*, 1984), alarm calls that can spontaneously elicit fear (Boissy *et al.*, 1998), and certain social odours – such as the mother hen's uropygial secretion – can reduce fear and distress in the recipients (Madec, 2008). Some fear-inducing social signals, for example threatening behaviours, may also be acquired (Bouissou *et al.*, 2001). Additionally, social isolation is one of the most stressful components of many fear tests when members of a social species are tested individually. Most domestic species are strongly motivated to rejoin conspecifics when isolated, and may consequently suffer more from separation anxiety than from the presumed fear-eliciting event itself.

6.2.3 Diversity in fear-related responses

Fear-related behaviours vary greatly depending on the nature of the threat. They can sometimes seem contradictory, as both active and passive strategies may be observed in a challenging situation: these strategies include active defence (attack, threat), active avoidance (flight, hiding, escape) or passive avoidance (immobility) (Jones, 1987a; Erhard and Mendl, 1999). Other responses can also be regarded as fear indicators, for example expressive movements such as head postures and facial expressions, alarm calls and the release of alarm odours or pheromones. These behaviours play important social roles by serving as alerting signals for conspecifics. Particularly noxious odours may even deter predators (Jones and Roper, 1997). Fear-eliciting stimuli can affect the activity in which the animal is engaged: low levels of fear can enhance activity (e.g. attention, investigation), whereas intense fear can disturb or terminate an ongoing activity. Indeed, fear inhibits all other behaviour systems, including feeding, exploration and sexual and social interactions (Jones, 1987a). Finally, conflict between a negative emotional state and a positive motivation may generate a compulsive behaviour, for example a disturbed or hungry pig bites a chain or bar. Activation of the sympathetic nervous system, including the adrenal medulla, and the hypothalamo–pituitary–adrenal system are major neuroendocrine responses associated with negative emotions (von Borell *et al.*, 2007; Mormède *et al.*, 2007), but they also accompany positive rewards such as food delivery and sexual interaction. Frightening stimulation also elicits a range of complex changes in central nervous mechanisms such as neural pathways and neurotransmitters (Gray, 1987; Rosen and Schulkin, 1998; Phillips *et al.*, 2003; Rosen, 2004).

6.2.4 The various ways of assessing fear

Many and varied experimental situations have been designed to study fear in domestic animals and several are variations of tests originally developed for laboratory species. Since Hall's classic work (1936), the open-field or novel arena test has been extensively used in rodents (Archer, 1973). Generally, a single animal is placed in a large novel arena and the amount of defecation and activity is interpreted as reflecting the emotional response to novelty. Later work has shown that this test also incorporates many other threatening stimuli, such as absence of shelter and landmarks, human contact, social isolation and bright light. So one must take factors such as social reinstatement motivation and exploration into account as well as fear (Jones, 1987b, 1997). Many other paradigms devised to assess fear in rodents, for example exposure to a predator or a novel object, confinement, handling, inescapable noxious stimulation and passive or active avoidance conditioning (Ramos *et al.*, 1997) have also been used in farm animals, from the 1970s onwards (Forkman *et al.*, 2007), although the designs were generally modified to suit the circumstances. Fear of novelty is evaluated using the open-field test or exposure to a novel object in ruminants, pigs, horses and poultry, while specific tests have been developed to assess an animal's fear of humans: it is either approached by a human (the 'Forced Approach test') or is free to approach the experimenter (the 'Voluntary Approach test'). A forced approach test is more likely to elicit an active response, whereas the likelihood of observing either no response or a passive one is greater in the voluntary approach test (Waiblinger *et al.*, 2006). Restraint tests are also commonly used, for example by restraining the animal in a chute or, more commonly, by inducing a tonic immobility reaction (Jones, 1986; Forkman *et al.*, 2007). Finally, fear can be induced and evaluated in terms of the reactions of the animals when exposed either to a natural predator (the 'Predator test'), to a sudden sound or visual stimulation (the 'Startle test'), or to a signal that had been previously associated with a nociceptive or otherwise disturbing event such as electric shock (the 'Conditioned Fear test').

B. Jones and A. Boissy

6.2.5 Temperament

Terms and concepts used for defining long-term emotional states differ in the degree of structure they provide, ranging from 'individual differences' to the concepts of 'personality' and 'temperament'. Distinction between these concepts, according to species and age, has not been consistently maintained (Jones and Gossling, 2005). For present purposes we concur with the definition of temperament as the characteristics of individuals that describe and account for consistent patterns of feeling and behaving (Pervin and John, 1997). Whereas dimensions of personality are generally discussed in terms of coping style or fearfulness, it has been proposed that there is currently no single concrete method for assessing a propensity to feel and express emotions (Gossling, 2001). However, by identifying long-term emotional states, we can examine whether characteristic behavioural patterns are linked to each other and to particular neuroendocrine states and if they are correlated within individuals and over time. This may enable us to predict more firmly an animal's responses in one situation based upon its reactions in another. Underlying fearfulness (inherent and/or acquired propensity to be easily frightened; Jones, 1987a) is one such case. Several reports indicate that fearful animals are more likely to show exaggerated fear responses than their less fearful counterparts, regardless of the nature of the threatening stimulus (Jones, 1987a, 1998; Faure *et al.*, 2003; Van Reenen *et al.*, 2005). Indeed, Boissy (1995, p. 183) concluded that 'fearfulness has to be considered as a component of personality' and (p. 165) that:

> [Fearfulness is] a basic feature of the temperament of each individual, one that predisposes it to respond similarly to a variety of potentially alarming challenges, but is nevertheless continually modulated during development by the interaction of genetic traits of reactivity with environmental factors, particularly in the juvenile period.

Such interaction(s) may explain much of the inter-individual variability observed in adaptive responses.

There is no universally accepted definition of 'individual variation' or 'individual differences'. Erhard and Schouten (2001) structured individual variation in animal behaviour using the Eysenck (1967) description of personality. This approach organizes the different aspects of personality into three levels – 'state', 'trait' and 'type'. First, state or mood reflects the behaviour that an individual performs at a specific moment in time and in a specific situation. Secondly, if an individual repeatedly shows similar states in similar situations, we can make assumptions about the underlying personality trait (e.g. fearfulness or happiness). Thirdly, when several trait dimensions are linked in such a way that an individual's position on one dimension predicts its position in another, the individual can be categorized by its position on a particular trait dimension. This approach brings the study of individual differences in non-human animals closer to that of humans (Mendl and Deag, 1995; Erhard and Schouten, 2001; Paul *et al.*, 2005).

Several studies have attempted to evaluate emotional states in farm animals. For instance, from a collective examination of their responses in fear and learning tests, dairy ewes were categorized as non-emotive (calm, adequate learning), emotive (nervous, fear susceptible, inadequate learning) and intermediate (Dimitrov and Djorbineva, 2001). The emotional state also influenced various adaptive behaviours; non-emotive ewes showed stable maternal care whereas postpartum anxiety was increased in emotive ewes. Furthermore, non-emotive ewes earned more rewards and showed better reproduction and production (Dimitrov *et al.*, 2005). The notion that personality has different 'dimensions' was supported by a report of different components (general agitation and avoidance) of personality in cattle and sheep (Kilgour *et al.*, 2006). Reports that high fearfulness and a poor milking temperament were related to low milk yield in cattle (Jones and Manteca, 2009; Van Reenen *et al.*, 2009), that calm ewes produce better quality milk with higher protein concentration than nervous ones (Sart *et al.*, 2004) and that they were better mothers (i.e. they spent longer with their lambs, had a shorter flight distance when disturbed and returned to their lambs faster, Murphy *et al.*, 1994), suggest that temperament should be incorporated into breeding programmes for dairy animals. The value of using fear tests as selection criteria was also indicated when fearful 'broiler' quail showed more pronounced adrenocortical stress responses, greater fluctuating asymmetry, slower growth and poorer egg production (Satterlee *et al.*, 2000; Jones and Satterlee, 2002). Strategies to maximize offspring survival through a combination of management, nutrition and genetic selection for temperament have also been proposed (Martin *et al.*, 2004). Temperament may also influence other aspects of the reproductive process, all of which might be improved by genetic selection for calmness. These include the duration of the oestrous cycle (Przekop *et al.*, 1984), ovulation rate (Doney *et al.*, 1976), the proportion of ewes mated and sexual behaviour (Gelez *et al.*, 2003). Finally, temperament can exert major effects on other aspects of production in many farm animal species; these include growth rate (Voisinet *et al.*, 1997; Burrow, 1998; Fel *et al.*, 2003), feed conversion efficiency (Jones *et al.*, 1993; Hemsworth and Coleman, 1998), immune function (Fel *et al.*, 2003), milk yield (Lawstuen *et al.*, 1988, Van Reenen *et al.*, 2005) and meat quality (Jones and Hocking, 1999; Reverter *et al.*, 2003).

Much debate has centred on whether fear varies over a unitary scale. According to Archer (1979, p. 57), 'this corresponds to asking "How frightened are you? Just a little, or scared out of your wits?"' Like Hinde (1985), but unlike Gray (1987), Archer (1979, p. 57) regarded 'such a unitary representation as too great an oversimplification to be useful for precise analysis' and proposed that fear should not be regarded as a unitary variable because correlations between different measures of fear were weak, at least in laboratory rodents and human beings. He also argued that putative fear responses must be considered as specific to the test stimulus, species, age and sex of the subject. If these suggestions were valid, this would severely constrain attempts to reduce fear in practical situations because the manipulation of narrow stimulus-specific responses would be of little value in modifying general adaptability and responsiveness to challenge. However, the limited numbers of tests and measures and the small sample sizes examined in the studies on which Archer's (1979) criticism was based were unlikely to have yielded significant correlations, and one should not expect correlations close to unity when measuring a complex, behaviourally disruptive phenomenon such as fear (Tachibana, 1982; Gray, 1987). Furthermore, Duncan (1993) proposed that although fear is really a hypothetical intervening variable, it can still be defined operationally and measured in the same way as hunger or thirst. Thus, fear can be assessed functionally in terms of the animal's attempts to avoid danger. Because fear competes with and inhibits all other behaviour systems, we can infer how frightened an animal is by monitoring its responses in test situations intuitively regarded as more or less frightening (Jones, 1987b, 1996).

Behavioural methods commonly used to measure fear in farm animals, for example open-field, emergence, approach/avoidance and tonic immobility tests, are described elsewhere (Jones, 1987b, 1996; Forkman *et al.*, 2007). Many researchers also record physiological measures of alarm or stress, for example, heart rate and concentrations of plasma catecholamines, corticosterone or cortisol. These indices are most useful when measured in conjunction with behavioural observation; indeed, behavioural and physiological information should be regarded as complementary rather than alternative measures of fear.

Encouragingly, strong intra-individual associations were found across scores in several tests of fear in chickens and Japanese quail (Jones *et al.*, 1991; Jones and Waddington, 1992), and in goats, sheep, cattle and service dogs (Goddard and Beilharz, 1984; Lyons *et al.*, 1988; Romeyer and Bouissou, 1992; Fordyce *et al.*, 1996). These findings strongly suggest that the tests were all measuring the same intervening variable, perhaps underlying fearfulness, and not just stimulus-specific responses. If so, the ability to measure this underlying characteristic or personality trait offers exciting opportunities for manipulating non-specific responsiveness to threat. Indeed, the selective breeding of animals for reduced fear and stress responses in a variety of challenging situations could not only result in improved productivity, product quality and profitability, it could also help to safeguard animal welfare, improve public perception of farming and raise the ethical status of the industry (Blokhuis *et al.*, 2003; Martin *et al.*, 2004).

6.2.6 A need for greater validation of tests and measures

Many tests of fear were originally designed for laboratory rodents and, unfortunately, several of these have been used for farm animals with insufficient regard for their biological relevance. For example, while laboratory rodents are nocturnal and thigmotaxic (wall seeking), most farm animals are diurnal and, apart from poultry, their ancestors generally occupied open areas/ranges. Cattle may actually perceive the open field test as an enclosed area. Furthermore, most farm animals are highly gregarious, many have exclusive mother–young relationships and the young are generally precocial. In view of such species differences in ecological

B. Jones and A. Boissy

experience and motivation, we must avoid testing animals in inappropriate environments that may elicit motivational states unrelated to the one under study, thereby leading to inaccurate estimations of fear and inconsistencies between studies. Reconsideration of the ecological contexts of domestic species may help us to develop more reliable and valid tests and measures of fear.

Despite these cautionary notes, studies of laboratory animals may still guide the interpretation of a range of emotional reactions observed when farm animals are exposed to aversive situations. From an ecological perspective, suddenness and novelty, which underpin many fear tests (Boissy, 1998), are key features of predatory attack, and we must remember that domestic ungulates and poultry kept on range may still experience predation by wild animals and dogs (Shelton and Wade, 1979). The ability to cope with an aversive event can also strongly influence the animal's emotional experience. For instance, cows with the strongest tendency to approach a novel object voluntarily are also the most reactive to humans, but the opposite is true if they are forced to move towards that novel object (Murphey et al., 1981).

The expression of fear, like that of all the emotions, is a brief affective reaction that is often difficult to measure, particularly in farm animals kept in commercial conditions. Research in this area would benefit substantially if more effort was devoted to overcoming three particular weaknesses. First, many tests used to measure fear, particularly in large farm animals, need more rigorous examination of validity and repeatability (see below). Secondly, there is a pressing need for standardized protocols to ensure harmonization of methods across research teams and countries; this would reduce unnecessary duplication of effort while greatly facilitating meaningful interpretation of the results and their general acceptance. This was a major feature of the recent Welfare Quality® project (Keeling, 2009). Thirdly, the assessment protocols have to be robust and practical: it must be possible to carry them out in an acceptable time frame and without causing undue disturbance to the herd/flock, to animal productivity or to farmers' routines – otherwise they will not be accepted by farmers, breeders or abattoir managers. In other words, it is not enough simply to transfer a laboratory test to the field without making appropriate modifications to fit the biology and perceptions of the animals concerned, the environmental

conditions and the farmers' requirements. We may need to 'think out of the box' and design whole new methodologies.

Martin and Bateson (1993) defined validity as the combination of accuracy (the degree of freedom from systematic errors that might over- or underestimate the measured variable), specificity (the extent to which a measured variable reflects what it is supposed to and nothing else) and scientific validity (the extent to which the method gives relevant answers to the hypothesis). Martin and Bateson (1993) defined reliability as the degree to which the measures are free from random errors. This is partly determined by the repeatability or the consistency of the measures, i.e. repeated measures of the same construct should yield the same result. Assessing the repeatability of fear tests is difficult as it involves repeating the same test several times and, while some animals may become used to it and reduce their responses (habituation), others may become more fearful (sensitization). This is especially true for tests based on novelty, as the unfamiliarity of the test situation would decrease with repeated exposure. In validation studies, it might be advisable to focus only on the animal's immediate responses in the first test and then determine the correlations between these reactions and those shown by the same animal to other biologically related situations (i.e. inter-test consistency), for instance by using different novel stimuli. Recording micro-behavioural expressions, for example, posture of the head and tail and specific alarm calls (Bouissou et al., 2001), may also illuminate the animal's fear state and its perception of that test situation.

Many studies in laboratory (and some farm) animals provide convincing evidence for consistency in the individuals' responses across different aversive situations. For example, rats exhibiting low performance during active avoidance conditioning show higher defecation rates in an open-field test (Brush et al., 1985), and defecation rates in rats subjected to a conditioned fear stimulus and then to an open-field test are positively correlated (Ley, 1975). Strong correlations between responses to various aversive stimuli were also reported in dogs (Goddard and Beilharz, 1984). Likewise, numerous correlations are found between behavioural reactions to open-field, novel object, emergence and tonic immobility tests in chicks (Jones and Mills, 1983), hens (Jones, 1987a, 1988) and Japanese quail (Mills and Faure, 1986). Furthermore, Japanese quail from two genetic lines selected for contrasting stress-induced adrenocortical activation also show divergence in related behavioural, haematological, reproductive and morphological characteristics (Jones and Satterlee, 1996, 2002; Jones *et al.*, 2002; Satterlee *et al.*, 2002). Similarly, several correlations were found between the fear responses of domestic ungulates in different aversive situations: for heifers, Boissy and Bouissou, 1995; for cattle, Van Reenen *et al.*, 2005; for sheep, Romeyer and Bouissou, 1992; Vandenheede and Bouissou, 1993). The fear behaviour of calves is also sensitive to neurochemical manipulation (Van Reenen *et al.*, 2009). Collectively, these findings indicate a strong tendency for individual characteristics to manifest themselves across a variety of aversive situations. Validation through the evaluation of inter-test consistency should thus be given more attention in future studies (Forkman *et al.*, 2007).

A recent comprehensive review of the methodologies used to assess general fear (Forkman *et al.*, 2007) in farm animals was a very useful first step in bringing together some of the above-mentioned issues; it describes the various methods used for cattle, pigs, horses, sheep, goats, chickens and quail, and gives indications of their repeatability and validity. Similarly, fear tests involving exposure to or physical contact with humans were discussed in a review of human–animal interactions (Waiblinger *et al.*, 2006). The Welfare Quality® project (Keeling, 2009), a large EU-funded, multidisciplinary venture, also made substantial progress in developing, refining, validating and standardizing numerous measures covering many aspects of animal welfare, including negative and positive emotions, and fear of humans. Protocols for the assessment of welfare in cattle, pigs and poultry are now available (Welfare Quality®, 2009). However, there is still a need for continued development and refinement of measures, especially for fear.

6.3 Cognitive Abilities and the Assessment of Emotions and Moods in Animals

6.3.1 A need for an explanatory framework

Despite the substantial literature on neurobiological (cardiac acceleration, neurotransmitter release, adrenocortical activation) and behavioural (startle, escape, attack, immobility, etc.) responses to frightening events, the exact nature of animal emotions is still poorly understood. Indeed, the question of whether animals feel emotions (the psychological component) remains controversial (Duncan, 2006). The emotional experience (the subjective component) is generally accessed in humans by means of verbal self-reports, but such conceptual–psychological scales cannot be used in animals; hence the emotional experience can only be inferred from indicators such as the above behavioural and physiological reactions.

Because of the complexity of the mechanisms underlying fear-related responses, it is very difficult to attribute a given behaviour clearly to any single emotion (Boissy, 1998). First, for instance, the response to a novel object is strongly influenced by the features of the test environment. For example, the open field behaviour of domestic chicks was sensitive to whether or not the floor was delineated into distinct areas with ink lines (Jones and Carmichael, 1997), the colour of the walls and floor (Jones, 1989) and whether or not the observer was visible (Jones, 1987c). Secondly, it is becoming increasingly accepted that a seemingly straightforward measure such as open-field activity may reflect fear, coping style, exploration and/or social reinstatement (Jones, 1989; Van Reenen *et al.*, 2005). Therefore, a measure used as an emotional indicator in one situation cannot necessarily be extrapolated directly to other situations. Indeed, there is no such thing as a single 'objective and perfect' measurement of fear.

The attribution of emotions to animals is often considered anthropomorphic, naive and unhelpful because anthropomorphism introduces the risk of misinterpreting the animals' responses. For instance, we may think that a pig wallowing in mud is happy, whereas it is actually just trying to cool itself down. Conversely, it has been argued that the danger lies not only in anthropomorphism but also in 'anthropodenial': animals and humans do share common characteristics, and cautious anthropomorphism may help us investigate the human-like characteristics of animals as well as the animal-like characteristics of humans (de Waal, 1999). This led Wynne (2004) to conclude that: 'progress will surely be most rapid when we adopt explanatory frameworks that are concrete and unambiguous'.

In addition to emotions, which are by definition fleeting, welfare involves more persistent affective states that influence how the individual perceives and reacts to its environment (Lazarus, 1991). Despite a clear need, there are relatively few

B. Jones and A. Boissy

long-term studies of the effects of repeated short-lasting emotional experiences on farm animal welfare. We now review reported interactions between emotions and cognition and explore the potential value of using such interactions as a new way of assessing negative emotions and mood in animals.

6.3.2 Cognition as a possible means of accessing the emotional experience of animals

Explanatory frameworks are central to our understanding of animal emotions, especially those that take into account the cognitive processes underlying the emotional states. Appraisal theories developed in cognitive psychology to probe human emotions may provide a strong candidate framework because appraisal processes can be simple, rapid and automatic (Désiré et al., 2002). Human emotions are the result of how an individual evaluates a triggering situation per se followed by the responses to that situation (Arnold, 1945; Lazarus et al., 1970), and Scherer (2001) proposed that such evaluation involves a sequence of checks grouped into four classes:

- relevance of the situation, including checks for novelty, suddenness, predictability, intrinsic pleasantness, and relevance for the individual's goals;
- implications of the situation for the individual, including checks of the probability of experiencing the expected consequences as well as the resultant match with expectations;
- coping potential, including checks for controllability (offered by the environment) and abilities (within the individual); and
- normative significance, including checks for internal and social standards.

These checks do not necessarily require high cognitive processing. Some are fairly automatic and subconscious, especially in the first check of relevance, while others are more complex (Kappas, 2006). Each check operates at several different levels according to the intensity of cognitive processing required: a sensorimotor level involving automatic processes; a schematic level requiring memorizing of emotional experiences and involving conditioned responses; and a conceptual level that is voluntarily and consciously activated (e.g. comparisons between the real-world situation and conscious plans or self-representation) (Leventhal and Scherer, 1987). Checks for suddenness seem to require only sensorimotor processes; the assessment of familiarity, predictability and controllability require schematic processing; and checks for normative significance are likely to require conceptual processing.

The outcomes of these checks determine the psychological component of an emotion which, in turn, affects physiological and behavioural responses. Typical (human) emotions, such as fear, anger or happiness, are linked to the outcome of the evaluation (Table 6.1). For instance, fear is elicited by exposure to an unpleasant event that is sudden, unfamiliar, unpredictable and inconsistent with expectations; rage ('hot anger') is experienced in similar situations except that the individual's evaluation is that he/she can control this situation; happiness is triggered by an event evaluated as slightly sudden, quite predictable, very pleasant and consistent with expectations; and so on. The internal component, i.e. the feeling, may be essential to emotions but it does not imply that the individual is conscious of his/her own emotions. A gradient of emotional responses can be assumed, from the mere expression of automatic reactions to the experience of emotional feelings and conscious emotional experiences. Simple checks are likely to

Table 6.1. Human emotions in relation to the outcome of their evaluation of a triggering situation (from Sander et al., 2005).

	Fear	Anger	Despair	Rage	Boredom	Happiness	Pride	Shame	Disgust
Suddenness	High	Low	High	High	Very low	Low	–	Low	–
Familiarity	Low	–	Very low	Low	High	–	–	–	Low
Predictability	Low	Medium	Low	Low	Very high	Medium	–	–	Low
Pleasantness	Low	–	–	–	–	High	–	–	Very low
Consistency with expectations	Low	–	Very low	Low	High	High	–	–	–
Control	Open	High	Very low	High	Medium	–	–	–	–
Social norms	–	Low	–	Low	–	–	High	Low	–
Emotion	**Fear**	**Anger**	**Despair**	**Rage**	**Boredom**	**Happiness**	**Pride**	**Shame**	**Disgust**

generate automatic responses (e.g. startle responses to a sudden event), checks requiring schematic processes may induce proper emotional experiences (i.e. feelings, as reported verbally by people), while those requiring conceptual processing lead to more conscious emotions.

Ongoing studies based on similar human paradigms have improved our understanding of how an animal evaluates its environment and responds to it emotionally by simply transposing the framework proposed by appraisal theories (Désiré et al., 2004, 2006; Greiveldinger et al., 2007, 2009; Veissier et al., 2009). Experimental situations designed to activate one or several evaluation checks were developed in order to ascertain which ones are relevant to animals. Checks of suddenness or unfamiliarity would only require automatic processes, while those of familiarity, predictability, consistency with expectations and controllability would imply the existence of emotional feelings. Cardiac and behavioural reactions were also recorded to probe the links between presumed appraisal and measurable emotional outcomes. Not unexpectedly, the sudden presentation

of an object induced a startle response and brief cardiac acceleration (tachycardia), while presentation of an unfamiliar object elicited behavioural orientation and a transitory increase in heart rate variability in sheep, goats and quail (Désiré et al., 2004; Roussel et al., 2005; Valance et al., 2007). Furthermore, as would be assumed from the dynamic sequential model of appraisal (Scherer, 1987), the combination of different checks enhanced behavioural and cardiac responsiveness. In sheep, the combination of suddenness with unfamiliarity or unpredictability had a synergistic effect; the heart rate acceleration specific to suddenness was accentuated if the sudden event was also unfamiliar (Fig. 6.1), while both startle and tachycardia responses to a sudden event were less marked when the animal could predict its occurrence (Fig. 6.2). Sheep can develop expectations, and discrepancy between these and the actual situation induced behavioural agitation and cardiac acceleration; the controllability of a disturbing situation also affected the emotional response (Greiveldinger et al., 2007). Pigs were sensitive to the predictability of aversive

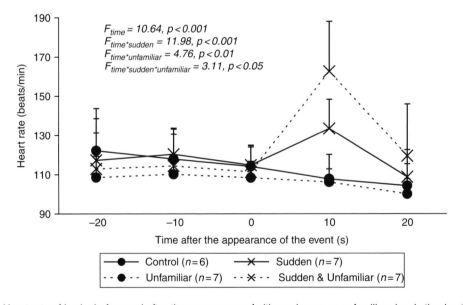

Fig. 6.1. Heart rate of lambs before and after the appearance of either a known or unfamiliar visual stimulus that was presented either slowly or suddenly. Statistical analyses were performed with a mixed-mode analysis of variance for repeated measures with a random effect for lambs. Cardiac activity recorded before and after the appearance of the object during the 10s time windows were taken as the repetition factor. Thus, the fixed effects of time (five levels), suddenness (two levels) and unfamiliarity (two levels) and their interactions were assessed. Heart rate increased after the sudden appearance of the object ($F = 11.98$, $p < 0.001$), and the increase was greater if the object was also unfamiliar ($F = 3.11$, $p < 0.05$). (After Désiré et al., 2006.)

B. Jones and A. Boissy

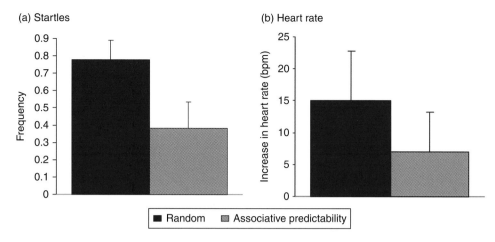

Fig. 6.2. Lambs' (a) startle responses and (b) heart rates (mean beats per min (bpm) ± SE) during food delivery when a white-and-blue panel appeared suddenly and randomly behind the trough, or when its appearance was preceded by a light signal (associative predictability). Specific emotional responses to suddenness were significantly reduced when the lambs could predict the sudden event. (After Greiveldinger *et al.*, 2007.)

events (Duepjan *et al.*, 2008) and the variability of food rewards (de Jonge *et al.*, 2008). Finally, the emotional responses of a sheep to a disturbing situation are respectively more internalised or more overt if it is subordinate or dominant to a companion sheep (Greiveldinger *et al.*, unpublished).

Collectively, the results described above suggest that sheep may experience a wide range of emotions: (i) fear and anger, as they are sensitive to suddenness, unpredictability, controllability and social norms; (ii) rage, as they respond to suddenness, unfamiliarity, unpredictability, discrepancy from expectations, controllability and social norms; (iii) despair, as they react to suddenness, unfamiliarity, unpredictability, discrepancy from expectations and controllability; and (iv) boredom, as they are sensitive to suddenness, unfamiliarity, unpredictability, discrepancy from expectations and controllability.

Based on the framework of appraisal theory, the results that have been described support the hypothesis that non-human animals can experience emotions. They further suggest that the evaluative checks validated in humans are also valid in other animal species. Therefore, emotional processing in animals should be assumed to comprise a cognitive component as well as behavioural, physiological and subjective components. Using a similar rationale, species comparisons across phyla could provide critical insights into the kind of emotions they can feel and, thereby, help to refine regulations designed to safeguard animal welfare.

6.3.3 Fear-induced cognitive bias as a means of understanding long-lasting affective states

While the above results begin to elucidate the emotional consequences of various acute situations in farm animals, less attention has been focused on the effects of repeated exposure to fear-eliciting events. In modern farming systems, animals may be exposed to a large variety of potentially stressful events from hatch or weaning through to puberty and adulthood. These include social mixing, transport, transfer from a rearing or nursing environment to a growing environment, dietary change, new social partners, exposure to different stockpersons, harvesting, etc. Emotional responses can bias human attention, memory and judgement of a situation, for example anxiety induces attentional shift towards potential threats (Bradley *et al.*, 1995), and emotionally charged events are more readily remembered than neutral ones (Reisberg and Heurer, 1995). Likewise, moderately strong emotions improve memory while extreme emotions damage it (Mendl *et al.*, 2001). Finally, the prevailing emotional state may affect judgement: people exposed to strongly negative events tend to interpret *all* subsequent ambiguous events negatively (Wright and Bower, 1992). Such emotional modulations of cognitive processes may have adaptive value by helping fearful or anxious individuals to pay attention to, memorize and make judgements in threatening circumstances. Such effects are not necessarily restricted to humans (Paul *et al.*, 2005). In rodents, the startle response induced by exposure to a sudden event is faster and larger under negative emotional states (Lang *et al.*, 1998). Heifers subjected to a potent stressor were unable to abandon a previously learned behaviour that was no longer rewarded, this prevented them acquiring a new, more appropriate behaviour (Lensink *et al.*, 2006). In contrast, exposure to a moderate stressor can facilitate learning, for example rats given a catecholamine injection mimicking the physiological component of a moderate emotion are more attentive and display improved memory (Sandi *et al.*, 1997).

We need to establish if and how emotions might cumulatively modify an animal's cognitive functions in the long term and consequently damage or improve well-being. Laboratory studies already imply considerable and long-lasting modulation of evaluation capacities by emotional experiences. For example, rats or mice subjected to repeated unpredictable (and hence frightening) events respond less readily to an ambiguous stimulus signalling the delivery of a positive event, thus indicating reduced capacities for judgement and decision making (Pardon *et al.*, 2000; Harding *et al.*, 2004). Likewise, rats housed under social stress or in isolation show less behavioural agitation between the presentation of a conditioned stimulus predicting sucrose reward and its arrival, suggesting reduced anticipation owing to impaired judgment (von Frijtag *et al.*, 2000). Such research was notably lacking in domestic animals until a recent pioneer study in sheep. After training sheep in a spatial location task that required a go/no-go response according to the location of a bucket (one location associated with food reward and the other with a negative reinforcer), half were subjected to restraint and isolation stress. Sheep that had experienced isolation and restraint approached three buckets at new locations sited between the two learned ones more than did control sheep, suggesting that stress may have induced a more optimistic-like judgement bias (Doyle *et al.*, 2010). The expression of such biases (optimistic versus pessimistic) as indicators of a persistent state of comfort or, by contrast, distress may merit further investigation. Establishing the nature and frequency of cognitive processes used to evaluate the environment may help us to understand why chronic stress sometimes results in apathy or blunted emotion, while in other cases it

leads to heightened emotional reactivity. Apathy would be likely to develop when the animal has no way of altering negative events, whereas hyperreactivity would occur when it thinks it can control such events. Developing this approach in farm animals might guide the development of new indicators of good or bad welfare; a possible candidate might be an enhanced expectation of positive events (Spruijt *et al.*, 2001).

As well as determining whether the methods and outcomes of studies on cognitive abilities are useful for assessing animal emotions, investigating the effects of emotions on cognition might provide deeper insight into the relationships between emotions and more persistent affective states defining welfare. Taken together, these approaches show how cognitive approaches might be used to probe animals' emotions as short-term affective experiences, and to probe their welfare as persistent affective experiences.

6.4 Damaging Effects of Negative Emotions

Negative emotions, particularly fear, can seriously damage the welfare of all animals. Here, we focus on the broad recognition that they represent a major problem facing the animal farming industry. For example, the many harmful effects of fear on the welfare, management, performance and profitability of poultry are outlined in Box 6.1.

6.4.1 Welfare problems

The activities of advisory committees (e.g. the Brambell Committee, 1965) and consumer groups, together with the terrible consequences of BSE (bovine spongiform encephalopathy), foot-and-mouth disease, avian influenza, etc., have resulted in a growing public awareness that animal production should be more than just an industry. Indeed, animal welfare is now a perceived attribute of food quality (Blokhuis and Jones, 2010). The UK's Farm Animal Welfare Council (FAWC, 1993) had already recommended that animals should have 'freedom from fear and distress' and there was mounting evidence that fear is both a state of suffering and a powerful, potentially damaging stressor in its own right (Jones, 1996; Boissy *et al.*, 2005).

Although fear and stress are adaptive in ideal circumstances, the questions of degree and context are crucial because restrictions imposed by many farming systems can interfere with an animal's ability to respond appropriately to frightening events; for example, caged hens or tethered sows cannot run away. Both acute and chronic fear states can seriously harm an animal's welfare (Jones, 1996). First, for example, panic-induced injuries are seen in cattle, pigs, horses, rabbits, poultry, etc., particularly during transportation, and these are a major welfare (and production) problem because they can lead to chronic pain, infection, physical debilitation and death. Secondly, if the animal cannot escape from a frightening stimulus, if it fails to find an alternative coping strategy, or if it learns that no such strategies are available, it may enter the dangerous states of hopelessness, learned helplessness and behavioural depression (Job, 1987). These states can cause psychosomatic symptoms, or even death (Jones, 1997). Furthermore, repeated or prolonged elicitation of fear can result in a damaging state of pathological anxiety in both humans and other animals (Rosen and Schulkin, 1998). Thirdly, fear inhibits behaviour patterns generated by all other motivational systems (Jones, 1996), thereby reducing an animal's ability to adapt to environmental change, to learn new tasks, to utilize new resources and to interact successfully with its companions. For instance, constant stress compromises a sheep's ability to attend selectively and learn appropriate responses to environmental changes (Kendrick et al., 2001). Fourthly, fear and stress compromise immunocompetence, disease status and welfare in many farm animal species (Zulkifli and Siegel, 1995; Hemsworth and Coleman, 1998).

Generic knowledge of cognition can often improve our understanding of potentially negative aspects of animal living conditions. For instance, sudden noises or movements occur in all farm environments, and animals that are offered new foods, mixed with strangers, or moved to a new barn can all experience fear. Cows, for instance, are very sensitive to shouting and find novel situations aversive (Rushen et al., 1999; Herskin et al., 2004). Predictability is also influential. For example, it is common practice to feed or milk animals at the same time(s) every day and the animals may be distressed if this regime is changed (Johannesson and Ladewig, 2000); pigs can develop gastric ulcers if feeding is delayed or when the usual signals of feeding (such as noise in pipes) are not followed by food delivery. Non-controllability of even positive events such as food delivery can also damage welfare, for example by causing hypalgesia in mice (Tazi et al., 1987) and learned helplessness in hens (Haskell et al., 2004). Tethering large farm animals when they are housed indoors (e.g. cows in winter) seriously reduces their control over the environment, whereas some modern systems, like loose housing and automatic feeding systems in which the animals operate a device for food, offer far better control. Similarly, some robotic milking facilities enable the cows to be milked when they wish. Increased control by animals also appears beneficial; for example, pigs that could control food delivery showed enhanced healing abilities (Ernst et al., 2006).

Unpredictability and/or uncontrollability of aversive events exert long-term deleterious effects and heighten negative emotions such as fear in humans and animals (Adkin et al., 2006; Armfield, 2006; Carlsson et al., 2006). Unpredictability can also induce negative cognitive bias whereby neutral situations are more likely to be perceived as negative (Harding et al., 2004); this may cause anxiety (Zvolensky et al., 2000), depression (Anisman and Matheson, 2005) or neurosis (Mineka and

Kihlstrom, 1978). Similarly, repeated exposure to uncontrollable events can induce chronic stress: rats that cannot control the termination of an electric shock develop more gastric ulcers than yoked counterparts (Weiss, 1972; Milde et al., 2005), and restricted movement promotes stereotypies and apathy in sows (Broom, 1987; Terlouw et al., 1991). We propose that the mechanism underpinning such long-term effects may be the repeated elicitation of (short-term/acute) negative emotions. This is consistent with earlier suggestions that a series of acute frightening events can develop into a chronic stress syndrome (Jones, 1997; Hemsworth and Coleman, 1998).

6.4.2 Management problems

Despite centuries of domestication, the predominant reaction of many farm animals to people is one of fear (Jones, 1997). For example, naive chickens not only find human contact alarming (Duncan, 1993), they are also thought to perceive people as predators rather than as benevolent caretakers (Suarez and Gallup, 1983). Indeed, many human–animal interactions in current farm practice are inherently frightening, for example restraint, beak trimming, disbudding, veterinary treatment, while only a few, other than feeding, are positively reinforcing. Both acute and chronic fear states can lead to pronounced handling difficulties in most farm animals (Jones, 1996; Boivin et al., 2003; Hemsworth, 2003). Animals that are predisposed, (genetically or through past experience), to be easily frightened are also very difficult to handle and manage. This could greatly exacerbate the problems encountered during routine examination, translocation, artificial insemination, veterinary treatment, etc. As well as the damaging effects on the animals, animal intractability could place a greater demand on stockpersons' time, increase the risk of their injury and disaffection, and harm their attitude and behaviour towards the animals, thereby instigating a vicious cycle (Jones, 1996; Boivin et al., 2003; Faure et al., 2003; Hemsworth, 2003; Waiblinger and Spoolder, 2007). It has been argued that such findings present a case for the replacement of stockpersons by automated systems. However, as it is neither possible nor desirable to eliminate the human–animal relationship, human contact could then become even more traumatic, as increasing automation reduces the opportunities for the animals to become accustomed to people. Instead, we should identify ways of improving the human–animal relationship to maximize mutual benefits (see Chapter 15, Human Contact).

Fear of novelty can cause management problems if animals are translocated from one environment to another, encounter an unfamiliar stockperson, or are fed a new diet (Jones, 1997; Jones and Roper, 1997; Hemsworth and Coleman, 1998). Furthermore, feelings of vulnerability and susceptibility to attack by predators might explain why many modern-day chickens seem reluctant to leave a poultry house and move on to free range (Grigor and Hughes, 1993). Remedial measures might include the incorporation of 'reassuring' features, such as shelter, feeding sites and familiar stimuli, as well as selective breeding for increased willingness to use free range.

6.4.3 Fear and its production and economic consequences

There is growing evidence that animals subjected to acute or chronic fear and stress show impaired growth, poorer food conversion efficiency, delayed maturation, reduced reproductive performance and poorer product quality (Jones, 1997; Hemsworth and Coleman, 1998). For example, a series of studies reported negative (and probably causal) correlations between fear of humans and productivity in the dairy, pork, egg and broiler chicken industries (reviewed by Hemsworth, 2003). More specifically, fear of humans was associated with decreased egg production in laying hens (Hemsworth et al., 1994), poorer food conversion in broilers (Hemsworth et al., 1994) and reduced growth and reproductive performance in pigs (Hemsworth, 2003). Furthermore, stress-induced reduction in meat quality was more pronounced in a genetic line of Japanese quail showing high fear (Faure et al., 2003), the incidence of pale and soft exudative meat was greater in nervous pigs (Grandin and Deesing, 1998), and fearful cattle had more bruised as well as tougher meat (Fordyce et al., 1988).

The production losses, reduced product quality and related economic penalties of exposure to other frightening stimuli, for example low-flying aircraft, thunder, and sudden and unfamiliar noises, are also increasingly apparent in a range of farm animals. In poultry, for example, such fear is associated with reduced growth rates, downgrading of carcasses at slaughter, eggshell abnormalities and poorer reproductive efficiency (Mills and Faure, 1991; Jones, 1997; Faure et al., 2003).

B. Jones and A. Boissy

Clearly, there is a strong argument for alleviating fear from an economic as well as from an ethical viewpoint.

6.5 Alleviation of Fear

The main remedial measures for the alleviation of fear include environmental enrichment, improved stockmanship/management and targeted breeding programmes. These are described elsewhere in this book. In reality, an integrated approach involving all three approaches is likely to be the most effective strategy (Faure and Jones, 2004; Jones, 2004; Boissy et al., 2005). However, practicality is an overriding requirement: whatever strategy is adopted must be safe, relatively easy to implement and affordable; otherwise, it will not be implemented.

6.6 Conclusions

- Fear is an extremely complex concept. It is generally accepted as one of the primary emotions governing the way that animals respond to their social and physical environment.
- Prolonged or intense fear is an undesirable state of suffering and a powerful stressor which seriously damages an animal's welfare, management, performance and profitability.
- Fear, distress and other undesirable emotional states must be reduced (from the animal's, the public's and the owner's viewpoints), but manipulation of narrow stimulus-specific responses would have little practical value. The ability to measure and modify underlying fearfulness is crucial, and we need to develop, refine and validate tests further in all farm animal species, and particularly in the larger ones. The use of cognitive ability and cognitive bias may also prove useful.
- Fear and stress can be reduced by environmental enrichment, positive human stimulation and genetic selection; but we must ensure that these strategies are practical and free of deleterious side effects.
- Satisfying these requirements may help us to identify the most appropriate approach to achieving optimal levels of sensory stimulation, fear and stress.
- Increased collaboration between researchers, industry and policy groups can strengthen the relevance, timeliness and practicality of welfare research, while increasing the likelihood of knowledge transfer and the implementation of results.

References

Adkin, A.L., Quant, S., Maki, B.E. and McIlroy, W.E. (2006) Cortical responses associated with predictable and unpredictable compensatory balance reactions. *Experimental Brain Research* 172, 85–93.

Anisman H. and Matheson K. (2005) Stress, depression, and anhedonia: caveats concerning animal models. *Neuroscience and Biobehavioral Reviews* 29, 525–546.

Archer, J. (1973) Tests for emotionality in rats and mice: a review. *Animal Behaviour* 21, 205–235.

Archer, J. (1979) Behavioural aspects of fear. In: Sluckin, W. (ed.) *Fear in Animals and Man.* Van Nostrand Reinhold, New York, pp. 56–85.

Armfield, J.M. (2006) Cognitive vulnerability: a model of the etiology of fear. *Clinical Psychology Review* 26, 746–768.

Arnold, M.B. (1945) Psychological differentiation of emotional states. *Psychological Reviews* 52, 35–48.

Asheim, L.J. and Mysterud, I. (2005) External effects of mitigating measures to reduce large carnivore predation on sheep. *Journal of Farm Management* 12, 206–213.

Blokhuis, H. and Jones, B. (2010) Welfare quality. In: Beilage. E. and Blaha, T. (eds) *Proceedings of the Second European Symposium on Porcine Health Management*, 26–28 May 2010, Hanover, Germany, pp. 17–20.

Blokhuis, H.J., Jones, R.B., Geers, R., Miele, M. and Veissier, I. (2003) Measuring and monitoring animal welfare: transparency in the food product quality chain. *Animal Welfare* 12, 445–455.

Boissy, A. (1995) Fear and fearfulness in animals. *Quarterly Review of Biology* 70, 165–191.

Boissy, A. (1998) Fear and fearfulness in determining behavior. In: Grandin, T. (ed.) *Genetics and the Behaviour of Domestic Animals.* Academic Press, San Diego, California, pp. 67–111.

Boissy, A. and Bouissou, M.F. (1995) Assessment of individual differences in behavioural reactions of heifers exposed to various fear-eliciting situations. *Applied Animal Behaviour Science* 46, 17–31.

Boissy, A., Terlouw, C. and Le Neindre, P. (1998) Presence of cues from stressed conspecifics increases reactivity to aversive events in cattle: evidence for the existence of alarm substances in urine. *Physiology and Behavior* 63, 489–495.

Boissy, A., Fisher, A.D., Bouix, J., Hinch, G.N. and Le Neindre, P. (2005) Genetics of fear in ruminant livestock. *Livestock Production Science* 93, 23–32.

Boissy, A., Arnould, C., Chaillou, E., Désiré, L., Duvaux-Ponter, C., Greiveldinger, L., Leterrier, C., Richard, S., Roussel, S., Saint-Dizier, H., Meunier-Salaün, M.C., Valance, D. and Veissier, I. (2007) Emotions and cognition: a new approach to animal welfare. *Animal Welfare* 16, 37–43.

Boivin, X., Lensink, B.J., Tallet, C. and Veissier, I. (2003) Stockmanship and farm animal welfare. *Animal Welfare* 12, 479–492.

Bouissou, M.F., Boissy, A., Le Neindre, P. and Veissier, I. (2001) The social behaviour of cattle. In: Keeling, L. and Gonyou, H. (eds) *Social Behaviour in Farm Animals*. CAB International, Wallingford, UK, pp. 113–146.

Bradley, B.P., Mogg, K. and Williams, R. (1995) Implicit and explicit memory for emotion congruent information in depression and anxiety. *Behaviour Research and Therapy* 33, 755–770.

Brambell Committee (1965) *Report of the Technical Committee to Enquire into the Welfare of Animals Kept under Intensive Husbandry Systems*. Command Paper 2836, Her Majesty's Stationery Office, London.

Broom, D.M. (1987) Applications of neurobiological studies to farm animal welfare. In: Wiepkema, P.R. and Van Adrichem, P.W.M. (eds) *Biology of Stress in Farm Animals: An Integrative Approach*. Martinus Nijhoff Publishers: Dordrecht, The Netherlands/Boston, Massachusetts/Lancaster, UK, pp. 101–110.

Brush, F.R.; Baron, S., Froehlich, J.C., Ison, J.R., Pellegrino, L.J., Phillips, D.S., Sakellaris, P.C. and Williams, V.N. (1985) Genetic differences in avoidance learning by *Rattus norvegicus*: escape/avoidance responding, sensitivity to electric shock, discrimination learning and open-field behavior. *Journal of Comparative Psychology* 99, 60–73.

Burrow, H.M. (1998) The effects of inbreeding on productive and adaptive traits and temperament of tropical beef cattle. *Livestock Production Science* 55, 227–243.

Carlsson, K., Andersson, J., Petrovic, P., Petersson, K.M., Hman, A. and Ingvar, M. (2006) Predictability modulates the affective and sensory-discriminative neural processing of pain. *NeuroImage* 32, 1804–1814.

Council of Europe (1997) The Amsterdam treaty modifying the treaty on European Union, the treaties establishing the European communities, and certain related facts. *Official Journal of the European Communities* C 340, 1–144.

Dantzer, R. (1988) *Les Émotions*. PUF (Presses Universitaires de France), Paris.

Dawkins, M.S. (1990) From an animal's point of view: motivation, fitness, and animal welfare. *Behavioral and Brain Sciences* 13, 1–61.

Dawkins, M.S. (2001) How can we recognise and assess good welfare? In: Broom, D.M. (ed.) *Coping with Challenge: Welfare in Animals Including Humans*. Dahlem University Press, Berlin, pp. 63–76.

de Jonge, F.H., Ooms, M., Kuurman, W.W., Maes, J.H.R. and Spruijt, B.M. (2008) Are pigs sensitive to variability in food rewards? *Applied Animal Behaviour Science* 114, 93–104.

de Waal, F.B.M. (1999) Anthropomorphism and anthropodenial: consistency in our thinking about humans and other animals. *Philosophical Topics* 27, 255–280.

Désiré, L., Boissy, A. and Veissier, I. (2002) Emotions in farm animals: a new approach to animal welfare in applied ethology. *Behavioral Processes* 60, 165–180.

Désiré, L., Veissier, I., Després, G. and Boissy, A. (2004) On the way to assess emotions in animals: do lambs evaluate an event through its suddenness, novelty or unpredictability? *Journal of Comparative Psychology* 118, 363–374.

Désiré, L., Veissier, I, Després, G., Delval, E., Toporenko, G. and Boissy, A. (2006) Appraisal process in sheep: interactive effect of suddenness and unfamiliarity on cardiac and behavioural responses. *Journal of Comparative Psychology* 120, 280–287.

Dimitrov, I.D. and Djorbineva, M.K. (2001) The influence of emotional reactivity over maternal behavior and lactating in dairy ewes. In: Schäfer, D. and Borell, E. V. (eds) *Tierschutz und Nutztierhalung. Proceedings, 15th International Symposium on Applied Ethology*, 4–6 October, Halle, Germany (ISBN 3-86010-634-1), pp. 115–118.

Dimitrov, I., Djorbineva, M., Sotirov, L. and Tanchev S. (2005) Influence of fearfulness on lysozyme and complement concentrations in dairy sheep. *Revue de Médecine Vétérinaire* 156, 445–448.

Doney, J.M., Gunn, R.G. and Smith, W.F. (1976) Effects of premating environmental stress, ACTH, cortisone acetate or metyrapone on oestrus and ovulation in sheep. *Journal of Agricultural Science* 87, 127–132.

Doyle, R.E., Fisher, A.D., Hinch, G.N., Boissy, A. and Lee, C. (2010) Release from restraint generates a positive judgement bias in sheep. *Applied Animal Behaviour Science* 122, 28–34.

Duepjan, S., Schoen, P.C., Puppe, B., Tuchscherer, A. and Manteuffel, G. (2008) Differential vocal responses to physical and mental stressors in domestic pigs (*Sus scrofa*). *Applied Animal Behaviour Science* 114, 105–115.

Duncan, I.J.H. (1993) Welfare is to do with what animals feel. *Journal of Agricultural and Environmental Ethics* 6, 8–14.

Duncan, I.J.H. (1996) Animal welfare in terms of feelings. *Acta Agriculturae Scandinavica, Section A – Animal Science* 27, 29–35.

Duncan, I.J.H. (2006) The changing concept of animal sentience. *Applied Animal Behaviour Science* 100, 11–19.

Dwyer, C.M. (2004) How has the risk of predation shaped the behavioural responses of sheep to fear and distress? *Animal Welfare* 13, 269–281.

Erhard, H.W. and Mendl, M. (1999) Tonic immobility and emergence time in pigs: more evidence for behavioural strategies. *Applied Animal Behaviour Science* 61, 227–237.

B. Jones and A. Boissy

Erhard, H.W. and Schouten, W.G.P. (2001) Individual differences and personality. In: Keeling, L.J. and Gonyou, H.W. (eds) *Social Behaviour in Farm Animals*. CAB International, Wallingford, UK, pp. 333–352.

Ernst, K., Tuchscherer, M., Kanitz, E., Puppe, B. and Manteuffel, G. (2006) Effects of attention and rewarded activity on immune parameters and wound healing in pigs. *Physiology and Behavior* 89, 448–456.

Eysenck, H.J. (1967) *The Biological Basis of Personality*. Charles C. Thomas, Springfield, Illinois.

Faure, J.M. and Jones, R.B. (2004) Genetic influences on resource use, fear and sociality. In: Perry, G.C. (ed.) *Welfare of the Laying Hen, 27th Poultry Science Symposium of the World's Poultry Science Association (UK Branch)*, held in Bristol in July 2003. Poultry Science Symposium Series, CAB International, Wallingford, UK, pp. 99–108.

Faure, J.M., Bessei, W. and Jones, R.B. (2003) Direct selection for improvement of animal well-being. In: Muir, W. and Aggrey, S. (eds) *Poultry Breeding and Biotechnology*. CAB International, Wallingford, UK, pp. 221–245.

FAWC (Farm Animal Welfare Council) (1993) *Report on Priorities for Animal Welfare Research and Development*. FAWC, Tolworth Tower, Surbiton, UK.

Fel, L.R., Colditz, I.G., Walker, K.H. and Watson, D.L. (2003) Associations between temperament, performance and immune function in cattle entering a commercial feedlot. *Australian Journal of Experimental Agriculture* 39, 795–802.

Fisher, A. and Matthews, L. (2001) The social behaviour of sheep. In: Keeling, L. and Gonyou, H. (eds) *Social Behaviour in Farm Animals*. CAB International, Wallingford, UK, pp. 211–245.

Fordyce, G., Dodt, R.M. and Wythes, J.R. (1988) Cattle temperaments in extensive beef herds in northern Queensland. 1. Factors affecting temperament. *Australian Journal of Experimental Agriculture* 28, 683–687.

Fordyce, G., Howitt, C.J., Holroyd, R.G., O'Rourke, P.K. and Entwistle, K.W. (1996) The performance of Brahman-Shorthorn and Sahiwal-Shorthorn beef cattle in the dry tropics of northern Queensland. 5. Scrotal circumference, temperament, ectoparasite resistance, and the genetics of growth and other traits in bulls. *Australian Journal of Experimental Agriculture* 36, 9–17.

Forkman, B., Boissy A., Meunier-Salaün, M.C., Canali, E. and Jones, R.B. (2007) A critical review of fear tests used on cattle, pigs, sheep, poultry and horses. *Physiology and Behavior* 92, 340–374.

Gelez, H., Lindsay, D.R., Blache, D., Martin, G.B. and Fabre-Nys, C. (2003) Temperament and sexual experience affect female sexual behaviour in sheep. *Applied Animal Behaviour Science* 84, 81–87.

Gentle, M.J., Waddington, D., Hunter, L.N. and Jones, R.B. (1990) Behavioural evidence for persistent pain following partial beak amputation in the chicken. *Applied Animal Behaviour Science* 27, 149–157.

Goddard, M.E. and Beilharz, R.G. (1984) A factor analysis of fearfulness in potential guide dogs. *Applied Animal Behaviour Science* 12, 253–265.

Gossling, S.D. (2001) From mice to men: what can we learn about personality from animal research? *Psychological Bulletin* 127, 45–86.

Grandin, T. and Deesing, M.J. (1998) Genetics and behavior during handling, restraint, and herding. In: Grandin, T. (ed.) *Genetics and the Behavior of Domestic Animals*. Academic Press, San Diego, California, pp. 113–144.

Gray, J. (1987) *The Psychology of Fear and Stress*, 2nd edn. Problems in Behavioural Sciences, Cambridge University Press, Cambridge, UK.

Greiveldinger, L., Veissier, I. and Boissy A. (2007) Emotional experiences in sheep: predictability of a sudden event lowers subsequent emotional responses. *Physiology and Behavior* 92, 675–683.

Greiveldinger, L., Veissier, I. and Boissy, A. (2009) Behavioural and physiological responses of lambs to controllable versus uncontrollable aversive events. *Psychoneuroendocrinology* 34, 805–814.

Grigor, P.N. and Hughes, B.O. (1993) Does cover affect dispersal and vigilance in free-range domestic fowls? In: Savory, C.J. and Hughes, B.O. (eds) *Proceedings of Fourth European Symposium on Poultry Welfare*, 18–21 September, Edinburgh, UK. World's Poultry Science Association, Beekbergen, The Netherlands/Universities Federation for Animal Welfare, Wheathampstead, UK, pp. 246–247.

Hall, C.S. (1936) Emotional behaviour in the rat: III The relationship between emotionality and ambulatory activity. *Journal of Comparative Psychology* 22, 345–352.

Harding, E.A., Elizabeth, E.S. and Mendl, M. (2004) Animal behaviour: cognitive bias and affective state. *Nature* 427, 312.

Hargreaves, A.L. and Hutson, G.D. (1990) The effect of gentling on heart rate, flight distance and aversion of sheep to a handling procedure. *Applied Animal Behaviour Science* 26, 243–252.

Haskell, M.J., Coerse, N.C.A., Taylor, P.A.E. and Mccorquodale, C. (2004) The effect of previous experience over control of access to food and light on the level of frustration-induced aggression in the domestic hen. *Ethology* 110, 501–513.

Hemsworth, P.H. (2003) Human–animal interactions in livestock production. *Applied Animal Behaviour Science* 81, 185–198.

Hemsworth, P.H. and Coleman, G.J. (eds) (1998) *Human–Livestock Interactions: The Stockperson and the Productivity and Welfare of Intensively-farmed Animals*. CAB International, Wallingford, UK.

Hemsworth, P.H., Coleman, G.J., Barnett J.L. and Jones, R.B. (1994) Behavioural responses to humans and the productivity of commercial broiler chickens. *Applied Animal Behaviour Science* 41, 101–114.

Herskin, M.S., Kristensen, A.M. and Munksgaard, L. (2004) Behavioral responses of dairy cows toward novel stimuli presented in the home environment. *Applied Animal Behaviour Science* 89, 27–40.

Hinde, R.A. (1985) Was the 'expression of the emotions' a misleading phrase? *Animal Behaviour* 33, 985–992.

Job, R.F.S. (1987) Learned helplessness in chickens. *Animal Learning and Behavior* 15, 347–350.

Johannesson, T. and Ladewig J. (2000) The effect of irregular feeding times on the behaviour and growth of dairy calves. *Applied Animal Behaviour Science* 69, 103–111.

Jones, A.C. and Gossling, S.D. (2005) Temperament and personality in dogs (*Canis familiaris*): a review and evaluation of past research. *Applied Animal Behaviour Science* 95, 1–53.

Jones, R.B. (1986) The tonic immobility reaction of the domestic fowl: a review. *World's Poultry Science Journal* 42, 82–96.

Jones, R.B. (1987a) The assessment of fear in the domestic fowl. In: Zayan, R. and Duncan, I.J.H. (eds) *Cognitive Aspects of Social Behaviour in the Domestic Fowl*. Elsevier, Amsterdam, pp. 40–81.

Jones, R.B. (1987b) Social and environmental aspects of fear in the domestic fowl. In: Zayan, R. and Duncan, I.J.H. (eds) *Cognitive Aspects of Social Behaviour in the Domestic Fowl*. Elsevier, Amsterdam, pp. 82–149.

Jones, R.B. (1987c) Open field behaviour in domestic chicks (*Gallus domesticus*): the influence of the experimenter. *Biology of Behaviour* 12, 100–115.

Jones R.B. (1988) Repeatability of fear ranks among adult laying hens. *Applied Animal Behaviour Science* 19, 297–304.

Jones, R.B. (1989) Avian open-field research and related effects of environmental novelty: an annotated bibliography, 1960–1988. *The Psychological Record* 39, 397–420.

Jones R.B. (1996) Fear and adaptability in poultry: insights, implications and imperatives. *World's Poultry Science Journal* 52, 131–174.

Jones, R.B. (1997) Fear and distress. In: Appleby, M.C. and Hughes, B.O. (eds) *Animal Welfare*. CAB International, Wallingford, UK, pp. 75–87.

Jones, R.B. (1998) Alleviating fear in poultry. In: Greenberg, G. and Haraway, M. (eds) *Comparative Psychology: A Handbook*. Garland Press, New York, pp. 339–347.

Jones, R.B. (2004) Environmental enrichment: the need for bird-based practical strategies to improve poultry welfare. In: Perry, G.C. (ed.) *Welfare of the Laying Hen, 27th Poultry Science Symposium*. CAB International, Wallingford, UK, pp. 215–225.

Jones, R.B. and Carmichael, N.L. (1997) Open-field behavior in domestic chicks tested individually or in pairs: differential effects of painted lines delineating subdivisions of the floor. *Behavior Research Methods, Instruments and Computers* 29, 396–400.

Jones, R.B. and Hocking, P.M. (1999) Genetic selection for poultry behaviour: big bad wolf or friend in need? *Animal Welfare* 8, 343–359.

Jones, R.B. and Manteca, X. (2009) Best of breed. *Public Science Review* 18, 562–563.

Jones, R.B. and Mills, A.D. (1983) Estimation of fear in two lines of the domestic chick: correlations between various methods. *Behavioral Processes* 8, 243–253.

Jones, R.B. and Roper, T.J. (1997) Olfaction in the domestic fowl: a critical review. *Physiology and Behavior* 62, 1009–1018.

Jones, R.B. and Satterlee, D.G. (1996) Threat-induced behavioral inhibition in Japanese quail genetically selected for contrasting adrenocortical response to mechanical restraint. *British Poultry Science* 37, 465–470.

Jones, R.B. and Satterlee, D.G. (2002) Divergent selection for adrenocortical responsiveness affects fear, sociality and sexual maturation in quail. In: *Proceedings of 11th European Poultry Conference*, 6–10 September 2002, Bremen, Germany. European Federation World's Poultry Science Association, Beekbergen, The Netherlands, pp. 1–9.

Jones, R.B. and Waddington, D. (1992) Modification of fear in domestic chicks, *Gallus gallus domesticus*, via regular handling and early environmental enrichment. *Animal Behaviour* 43, 1021–1033.

Jones, R.B., Mills, A.D. and Faure, J.M. (1991) Genetic and experiential manipulation of fear-related behavior in Japanese quail chicks (*Coturnix coturnix japonica*). *Journal of Comparative Psychology* 105, 15–24.

Jones, R.B., Hemsworth, P.H. and Barnett, J.L. (1993) Fear of humans and performance in commercial broiler flocks. In: Savory, C.J. and Hughes, B.O. (eds) *Proceedings of Fourth European Symposium on Poultry Welfare*, Edinburgh, World's Poultry Science Association, Beekbergen, The Netherlands/Universities Federation for Animal Welfare, Wheathampstead, UK, pp. 292–294.

Jones, R.B., Marin, R.H., Satterlee, D.G. and Cadd, G.G. (2002) Sociality in Japanese quail (*Coturnix japonica*) genetically selected for contrasting adrenocortical responsiveness. *Applied Animal Behaviour Science* 75, 337–346.

Kappas, A. (2006) Appraisals are direct, immediate, intuitive, and unwitting … and some are reflective … *Cognition and Emotion* 20, 952–975.

Keeling, L. (ed.) (2009) *Welfare Quality. An Overview of the Development of the Welfare Quality® Project Assessment Systems*. Welfare Quality Reports Series No. 12, Welfare Quality® (Science and Society Improving Animal Welfare in the Food Quality Chain, EU-funded project FOOD-CT-2004-506508), Cardiff, UK.

B. Jones and A. Boissy

Keeling, L. and Gonyou, H. (eds) (2001) *Social Behaviour in Farm Animals*. CAB International, Wallingford, UK.

Kendrick, K.M., da Costa, A.P., Leigh, A.E., Hinton, M.R. and Peirce, J.W. (2001) Sheep don't forget a face. *Nature* 414, 165–166.

Kilgour, R.J., Melville, G.J. and Greenwood, P.L. (2006) Individual differences in the reaction of beef cattle to situations involving social isolation, close proximity of humans, restraint and novelty. *Applied Animal Behaviour Science* 99, 21–44.

Lang, F.R., Staudinger, U.M. and Carstensen, L.L. (1998) Perspectives on socio-emotional selectivity in late life: how personality and social context do (and do not) make a difference. *Journal of Gerontology* 53, 21–30.

Lawstuen, D.A., Hansen, L.B., Steuernagel, G.R. and Johnson, L.P. (1988) Management traits scored linearly by dairy producers. *Journal of Dairy Science* 71, 788–799.

Lazarus, R.S. (1991) Progress on a cognitive-motivational-relational theory of emotion. *American Psychologist* 46, 819–834.

Lazarus, R.S., Averill, J.R. and Opton, E.M. Jr (1970) Towards a cognitive theory of emotion. In: Arnold, M. (ed.) *Feelings and Emotions: The Loyola Symposium*. Academic Press, New York, pp. 207–232.

Lensink, B.J., Veissier, I. and Boissy, A. (2006) Enhancement of performances in a learning task in suckler calves after weaning and relocation: motivational versus cognitive control? A pilot study. *Applied Animal Behaviour Science* 100, 171–181.

Leventhal, H. and Scherer, K. (1987) The relationship of emotion to cognition: a functional approach to a semantic controversy. *Cognition and Emotion* 1, 3–28.

Ley, R. (1975) Open-field behavior, emotionality during fear conditioning and fear-motivated instrumental performance. *Bulletin of the Psychonomic Society* 6, 598–600.

Lyons, D.M., Price, E.O. and Moberg, G.P. (1988) Individual differences in temperament of domestic dairy goats: constancy and change. *Animal Behaviour* 36, 1323–1333.

Madec, I. (2008) Effets du semiochemique MHUSA (Mother Hens' Uropygial Secretion Analogue) sur le stress des poulets de chair. Approches zootechnique, physiologique et comportementale. PhD thesis, University of Toulouse, France.

Manteca, X., Velarde, A. and Jones, R.B. (2009) Animal welfare components. In: Smulders, F.J.M. and Algers, B. (eds) *Welfare of Production Animals: Assessment and Management of Risks*. Wageningen Academic Publishers, Wageningen, The Netherlands, pp. 61–77.

Martin, G.L., Milton, J.T.B., Davidson, R.H., Banchero Hunzicker, G.E., Lindsay, D.R. and Blache, D. (2004) Natural methods for increasing reproductive efficiency in small ruminants. *Animal Reproduction Science* 82, 231–246.

Martin, P. and Bateson, P. (1993) *Measuring Behaviour: An Introductory Guide*, 2nd edn. Cambridge University Press, Cambridge, UK.

Mendl, M. and Deag, M.J. (1995) How useful are the concepts of alternative strategy and coping strategy in applied studies of social behavior. *Applied Animal Behaviour Science* 44, 119–137.

Mendl, M., Burman, O., Laughlin, K. and Paul, E. (2001) Animal memory and animal welfare. *Animal Welfare* 10 (Supplement 1), 17–25.

Mendl, M., Burman, O.H.P., Parker, R.M.A. and Paul, E.S. (2009) Cognitive bias as an indicator of animal emotion and welfare: emerging evidence and underlying mechanisms. *Applied Animal Behaviour Science* 18, 161–181.

Milde, A.M., Arslan, G., Overmier, J.B., Berstad, A. and Murison, R. (2005) An acute stressor enhances sensitivity to a chemical irritant and increases ^{51}CrEDTA permeability of the colon in adult rats. *Integrative Physiological and Behavioral Science* 40, 35–44.

Mills, A.D. and Faure, J.M. (1986) The estimation of fear in domestic quail: correlations between various methods and measures. *Biology of Behaviour* 11, 235–243.

Mills, A.D. and Faure, J.M. (1991) Divergent selection for duration of tonic immobility and social reinstatement behavior in Japanese quail (*Coturnix coturnix japonica*) chicks. *Journal of Comparative Psychology* 105, 25–38.

Mineka, S. and Kihlstrom, J.F. (1978) Unpredictable and uncontrollable events: a new perspective on experimental neurosis. *Journal of Abnormal Psychology* 87, 256–271.

Mormède, P., Andanson, S., Aupérin, B., Beerda, B., Guémené, D., Malmkvist, J., Manteca, X., Manteuffel, G., Prunet, P., Van Reenen, C.G., Richard, S. and Veissier, I. (2007) Exploration of the hypothalamic-pituitary-adrenal function as a tool to evaluate animal welfare. *Physiology and Behavior* 92, 317–339.

Murphey, R.M., Duarte, F.A.M. and Penedo, M.C.T. (1981) Responses of cattle to humans in open spaces: breed comparisons and approach. avoidance relationships. *Behavior Genetics* 11, 37–48.

Murphy, P.M., Purvis, I.W., Lindsay, D.R., Le Neindre, P., Orgeur, P. and Poindron, P. (1994) Measures of temperament are highly repeatable in Merino sheep and some are related to maternal behaviour. *Animal Production in Australia: Proceedings of the Australian Society of Animal Production* 20, 247–250.

Pardon, M.C., Perez-Diaz, F., Joubert, C. and Cohen-Salmon, C. (2000) Influence of a chronic ultramild stress procedure on decision-making in mice. *Journal of Psychiatry and Neuroscience* 25, 167–177.

Paul, E.S., Harding, E.J. and Mendl, M. (2005) Measuring emotional processes in animals: the utility of a cognitive approach. *Neuroscience and Biobehavioral Reviews* 29, 469–491.

Pervin, L.A. and John, O.P. (1997) *Personality and Research*. Wiley, New York.

Phillips, M.L., Drevets, W.C., Rauch, S.L. and Lane, R. (2003) Neurobiology of emotion perception 1: the neural basis of normal emotion perception. *Biological Psychiatry* 54, 505–514.

Poindron, P., Raksanyi, I., Orgeur, P. and Le Neindre, P. (1984) Comparaison du comportement maternel en bergerie à la parturition chez des brebis primipares ou multipares de race Romanov, Préalpes du Sud et Ile de France. *Genetics, Selection, Evolution* 16, 503–522.

Price, E.O. (1984) Behavioural aspects of animal domestication. *Quarterly Review of Biology* 59, 1–32.

Przekop, P., Wolinska-Witord, E., Mateusiak, K., Sadowski, B. and Domanski, E. (1984) The effect of prolonged stress on the oestrus cycles and prolactin secretion in sheep. *Animal Reproduction Science* 7, 333–342.

Ramos, A., Berton, O., Mormède, P. and Chaouloff, F. (1997) A multiple test study of anxiety related behaviours in six inbred rat strains. *Behavioural Brain Research* 85, 57–69.

Reisberg, D. and Heuer, F. (1995) Emotion's multiple effects on memory. In: McGaugh, J.L., Weiberger, N.M. and Lynch, G. (eds) *Brain and Memory: Modulation and Mediation of Neuroplasticity*. Oxford University Press, Oxford, UK, pp. 84–92.

Reverter, A., Johnson, D.J., Ferguson, D.M., Perry, D., Goddard, M.E., Burrow, H.M., Oddy, V.H., Thompson, J.M. and Bindon, B.M. (2003) Genetic and phenotypic characterisation of animal, carcass and meat quality traits from temperate and tropically adapted beef breeds. 4. Correlations among animal, carcass and meat quality traits. *Australian Journal of Agricultural Research* 54, 149–158.

Romeyer, A. and Bouissou, M.F. (1992) Assessment of fear reactions in domestic sheep, and influence of breed and rearing conditions. *Applied Animal Behaviour Science* 34, 93–119.

Rosen, J.B. (2004) The neurobiology of conditioned and unconditioned fear: a neurobehavioral analysis of the amygdala. *Behavioral and Cognitive Neuroscience Reviews* 3, 23–41.

Rosen, J.B. and Schulkin, J. (1998) From normal fear to pathological anxiety. *Psychology Review* 105, 325–350.

Roussel, S., Boissy, A., Montigny, D., Hemsworth, P.H. and Duvaux-Ponter, C. (2005) Gender-specific effects of prenatal stress on emotional reactivity and stress physiology of goat kids. *Hormones and Behavior* 47, 256–266.

Rushen, J., de Passillé, A.M. and Munskgaard, L. (1999) Fear of people by cows and effects on milk yield, behavior, and heart rate at milking. *Journal of Dairy Science* 82, 720–727.

Sander, D., Grandjean, D. and Scherer, K.R. (2005) Special Issue. A systems approach to appraisal mechanisms in emotion: emotion and brain. *Neural Networks* 18, 317–352.

Sandi, C., Loscertales, M. and Guaza, C. (1997) Experience-dependent facilitating effect of corticosterone on spatial memory formation in the water maze. *European Journal of Neurosciences* 9, 637–642.

Sart, S., Bencini, R., Blache, D. and Martin, G.B. (2004) Calm ewes produce milk with more protein than nervous ewes. In: *Animal Production in Australia: Proceedings of the Australian Society of Animal Production* 25, 307.

Satterlee, D.G., Cadd, G.G. and Jones, R.B. (2000) Developmental instability in Japanese quail genetically selected for contrasting adrenocortical responsiveness. *Poultry Science* 79, 1710–1714.

Satterlee, D.G., Marin, R.H. and Jones, R.B. (2002) Selection of Japanese quail for reduced adrenocortical responsiveness accelerates puberty in males. *Poultry Science* 81, 1071–1076.

Scherer, K.R. (1987) Toward a dynamic theory of emotion: the component process model of affective states. *Geneva Studies in Emotion and Communication* 1, 1–98.

Scherer, K.R. (2001) Appraisal considered as a process of multi-level sequential checking. In: Scherer, K.R., Schorr, A. and Johnstone, T. (eds) *Appraisal Processes in Emotion: Theory, Methods, Research*. Oxford University Press, New York/Oxford, UK, pp. 92–120.

Shelton, M. and Wade, D. (1979) Predatory losses: a serious livestock problem. *Annals of Industry Today* 2, 4–9.

Spruijt, B.M., van den Bos, R. and Pijlman, F.T. (2001) A concept of welfare based on reward evaluating mechanisms in the brain: anticipatory behavior as an indicator for the state of reward systems. *Applied Animal Behaviour Science* 72, 145–171.

Suarez, S.D. and Gallup, G.G. Jr (1983) Social reinstatement and open-field testing in chickens. *Animal Learning and Behavior* 11, 119–126.

Tachibana, T. (1982) Open-field test for rats: correlational analysis. *Psychological Reports* 50, 899–910.

Tazi, A., Dantzer, R. and Le Moal, M. (1987) Prediction and control of food rewards modulate endogenous pain inhibitory systems. *Behavioural Brain Research* 23, 197–204.

Terlouw, E.M.C., Lawrence, A.B. and Illius, A.W. (1991) Influences of feeding level and physical restriction on development of stereotypies in sows. *Animal Behaviour* 42, 981–991.

Valance, D., Boissy, A., Després, G., Constantin, P. and Leterrier, C. (2007) Emotional reactivity modulates autonomic responses to an acoustic challenge in quail. *Physiology and Behavior* 90, 165–171.

Van Reenen, C.G., O'Connell, N.E., Van der Werf, J.T.N., Korte, S.M., Hopster, H., Jones, R.B. and Blokhuis,

H.J. (2005) Responses of calves to acute stress: individual consistency and relations between behavioral and physiological measures. *Physiology and Behavior* 85, 557–570.

Van Reenen, C.G., Hopster, H., van der Werf, J.T.N., Engel, B., Buist, W.G., Jones, R.B. and Blokhuis, H.J. (2009) The benzodiazpeine brotizolam reduces fear in calves exposed to a novel object test. *Physiology and Behavior* 96, 307–314.

Vandenheede, M. and Bouissou, M.F. (1993) Sex differences in fear reactions in sheep. *Applied Animal Behaviour Science* 37, 39–55.

Veissier, I., Boissy, A., Désiré, L. and Greiveldinger, L. (2009) Animals' emotions: studies in sheep using appraisal theories. *Animal Welfare* 18, 347–354.

Voisinet, B.D., Grandin, T., Tatum, J.D., O'Connor, S.F. and Struthers J.J. (1997) Feedlot cattle with calm temperaments have higher average daily weight gains than cattle with excitable temperament. *Journal of Animal Science* 75, 892–896.

von Borell, E., Langbein, J., Després, G., Hansen, S., Leterrier, C., Marchant-Forde J., Marchant-Forde R., Minero, M., Mohr, E., Prunier, A., Valance, D. and Veisseir, I. (2007) Heart rate variability as a measure of autonomic regulation of cardiac activity for assessing stress and welfare in farm animals a review. *Physiology and Behavior* 92, 293–316.

von Frijtag, J.C., Reijmers, L.G., van der Harst, J.E., Leus, I.E., van den Bos, R. and Spruijt, B.M. (2000) Defeat followed by individual housing results in long-term impaired reward- and cognition-related behaviors in rats. *Behavioural Brain Research* 117, 137–146.

Waiblinger, S. and Spoolder, H. (2007) Quality of stockpersonship. In: Velarde, A. and Geers, R. (eds) *On Farm Monitoring of Pig Welfare*. Wageningen Academic Publishers, Wageningen, The Netherlands, pp. 159–166.

Waiblinger, S., Boivin, S., Pedersen, V., Tosi, M., Janczak, A.M., Visser, E.K. and Jones, R.B. (2006) Assessing the human–animal relationship in farmed species: a critical review. *Applied Animal Behaviour Science* 101,185–242.

Weiss, J.M. (1972) Psychological factors in stress and disease. *Scientific American* 226, 104–113.

Welfare Quality® (Science and Society Improving Animal Welfare in the Food Quality Chain) (2009) First European Protocols Assessing Farm Animal Welfare. Available at: http://www.welfarequality.net/everyone/43148/9/0/22 (accessed 14 December 2010).

Wohlt, J.E., Allyn, M.E., Zajac, P.K. and Katz, L.S. (1994) Cortisol increases in plasma of Holstein heifer calves from handling and method of electrical dehorning. *Journal of Dairy Science* 77, 3725–3729.

Wright, W.F. and Bower, G.H. (1992) Mood effects on subjective probability assessment. *Organizational Behavior and Human Decision Processes* 52, 276–291.

Wynne, C.D. (2004) The perils of anthropomorphism. *Nature* 428, 606.

Zulkifli, I. and Siegel, P.B. (1995) Is there a positive side to stress? *World's Poultry Science Journal* 51, 63–76.

Zvolensky, M.J., Eifert, G.H., Lejuez, C.W., Hopko, D.R. and Forsyth J.P. (2000) Assessing the perceived predictability of anxiety-related events: a report on the perceived predictability index. *Journal of Behavior Therapy and Experimental Psychiatry* 31, 201–218.

7 Behavioural Restriction

GEORGIA J. MASON AND CHARLOTTE C. BURN

Abstract

Captivity often restricts the abilities of animals to perform natural behaviour. Here, we review how this constraint affects psychological welfare by preventing motivations from being satisfied. One means by which this happens is through frustrating specific motivations pertaining to particular behavioural systems. This can occur when constrained behaviours are 'behavioural needs': activities that animals have instincts to perform even in environments where they are not biologically necessary for fitness (e.g. non-nutritive sucking by calves). It can also occur when deficits or external cues in the environment elicit strong motivations to behave a certain way (e.g. the lack of burrow-like structures triggering digging attempts in gerbils). Furthermore, given that humans suffer boredom in monotonous conditions that resemble the environments of many captive animals, and that many animals actively seek stimulation, it seems likely that, at least for some individuals in some species, behavioural restriction also harms welfare by thwarting general motivations to seek variety and/or to avoid monotony, thus causing boredom.

7.1 Introduction

If captive animals are well fed and physically healthy, is that enough to ensure good welfare? This has long been doubted: as long ago as 1863, Blake wrote the famous line 'A robin redbreast in a cage puts all heaven in a rage'. Captive conditions are often too 'impoverished' (i.e. barren – without appropriate stimuli or substrates), or too small, to allow animals to perform behaviour patterns that they would display in more natural settings, a constraint known as behavioural restriction. This holds for billions of animals worldwide: intensively farmed poultry, pigs, cattle, sheep, mink and foxes; laboratory rats, mice, primates and others; stabled horses; many zoo animals; fish in tanks and aquaria; caged pet parrots, canaries and other birds; and cats, dogs and wild animals kept in breeding facilities (Fig. 7.1). Behavioural restriction often elicits concern, perhaps because it is so negative for humans to experience; indeed, from sending a naughty child to its room to locking a prisoner into solitary confinement, behavioural restriction plays a major part in human punishment. The issue of whether behavioural restriction truly matters to captive animals was first taken up by biologists following the Brambell Committee's 1965 Report to the UK government (Brambell Committee, 1965) – a report focusing on intensive

farming, but addressing issues relevant far beyond farms. This report concluded that:

> The degree to which the behavioural urges of the animals are frustrated under the particular conditions of the confinement, must be a major consideration in determining its acceptability or otherwise.

William Thorpe's influential ethological appendix to the Brambell Report argued that animals cannot simply abandon natural behaviours when captivity prevents them or makes them redundant:

> A very large part of animal behaviour is basically determined by instinctive or innate abilities, proclivities and dispositions.

The UK's Farm Animal Welfare Council (FAWC) followed this in 1979 by including 'the freedom to express normal behaviour by providing sufficient space, proper facilities, and the company of the animal's own kind' in its 'Five Freedoms' prerequisites for good welfare (FAWC, 2009).

This chapter covers what biologists have since discovered about the psychological effects of behavioural restriction, and how it can impair welfare even when physical health is fine. Several psychological aspects of impoverished environments are relevant here, including that sensory environments may be inappropriate (e.g. over- or under-stimulating), and that captive animals' lack of control over their

©CAB International 2011. *Animal Welfare*, 2nd Edition (eds M.C. Appleby *et al.*)

lives may be inherently stressful. Our emphasis, however, is on the motivational effects of such environments – on consequences such as specific frustrations and more generalized boredom. First, we briefly review the techniques used to identify behaviour patterns that animals need to perform if welfare is not to be compromised (Section 7.2). Next, we give examples, focusing on the best understood and emphasizing the specific natural behaviour patterns that concerned Thorpe and Brambell (Section 7.3). We end by discussing whether impoverished environments also cause boredom (Section 7.4).

(a)

Fig. 7.1. Animals in behaviourally restrictive conditions. The sight of animals housed in small, barren enclosures often raises welfare questions. For example, Fig 7.1a shows canaries in a Greek pet shop (Photo: iStock Photo): do they miss social contact or being able to fly? Fig. 7.1b shows intensively farmed pigs (Photo: iStock Photo): does it matter that they were taken from their mothers while still suckling? Do they miss natural behaviours like being able to root?

Continued

Fig. 7.1. Continued.

7.2 Identifying Natural Behaviour Patterns That Are Important for Welfare, and Understanding Why

7.2.1 Which natural behaviour patterns, in principle, are important for welfare?

Encouraging captive animals to perform all the behaviour patterns evident in the wild is neither sensible nor humane. Animals in nature often have lives that could be described as 'poor, nasty, brutish and short' (Hobbes, 1651). In such conditions, many behaviours are just responses to adversity: animals compete aggressively for resources such as food and mates; they eat non-preferred food items because preferred diets are unavailable; they defend large home ranges when resource availability is low; they migrate huge distances to find water; they eat their own offspring; they hide or flee from predators, screaming if caught; and so on. All these behaviours are perfectly natural, but their absence in captivity should not raise welfare concerns because they are elicited by external stimuli or changes in physiological state that should be absent in animals whose safety, health and nutritional needs are ensured.

Welfare researchers, therefore, do not catalogue all natural behaviours to try and ensure that each is performed in captivity. Instead, they try to identify the subset whose performance is likely to benefit welfare, because driven by high motivations that still occur in captivity: 'behavioural

deprivation implies not only that an animal is prevented from performing behaviour … but that adverse effects arise as a result' (Dawkins, 1988). These key behaviours comprise responses known or assumed to be elicited by external stimuli present, desirably or undesirably, in captivity (e.g. cues from food, potential mates, infants or potential predators); and responses elicited by internal physiological mechanisms that – again – are unavoidable or even desirable in captivity (e.g. hormonal changes preceding parturition/egg laying). Critical here is the idea that frustrating motivated behaviours compromises welfare. Next, we therefore discuss: links between motivation and emotion; how frustration is identified; and how so-called 'behavioural needs' both differ from and resemble other strong motivations.

7.2.2 Frustrated motivation as an animal welfare issue

Motivational states are states within the brain that determine the likelihood and intensity of a given behavioural pattern, and the efforts that animals will make in order to perform it. Motivated behaviours have four key characteristics (Hughes and Duncan, 1988; Jensen and Toates, 1993; Mason and Bateson, 2009). First, the degree to which they are motivated (i.e. the magnitude of the underlying motivational state) is typically determined by a

G.J. Mason and C.C. Burn

combination of physiological state (e.g. blood sugar levels) and external stimuli (e.g. food odours). Secondly, motivated behaviours broadly have two phases: an 'appetitive' phase, in which the animals flexibly search or prepare for the opportunity to perform a 'consummatory' phase (e.g. courtship paves the way for mating). This consummatory behaviour is often more stereotyped and species-typical, involves engagement with external stimuli in a way that reduces the underlying motivation (e.g. eating reduces motivations to forage), and typically seems elicited by these very stimuli, once successful appetitive behaviour has brought them into proximity. Thirdly, the actual performance of these activities, independent of physiological outcomes or effects on the animal's world, can sometimes help reduce the underlying motivation; thus, the very act of ingesting food, for example, plays a role in satiety. The fourth key characteristic of motivated behaviours is that emotions appear to be important in their control. In particular, satisfying strong motivations seems emotionally positive, while being unable to satisfy strong motivations – a situation known as 'frustration' – seems very negative. Evidence for this comes from several sources: from human self-report, when able or unable to consummate a strong motivation; from neuroscience studies showing that the forebrain regions central to the motivational control of behaviour are involved in feelings of reward (animals will lever press to have these brain regions electrically stimulated, and both animals and humans will work to stimulate them with recreational drugs); and from findings that animals will avoid frustrating situations if they can (for more information see Dawkins, 1990; Rolls, 1999, 2007). Further evidence comes from ethological studies showing that anxiety-reducing drugs reduce the behavioural consequences of frustration (as described in more detail in Section 7.2.3), including pacing in hens (Duncan and Wood-Gush, 1974) and displacement activities in primates (Maestripieri et al., 1992). These links with emotion are why satisfying motivations is crucial for animal welfare.

Concerns that inabilities to perform natural behaviour compromise welfare are therefore to a large extent concerns about unsatisfied motivations: that animals are frustrated – with the negative emotions that this involves – if captive conditions either lack the right stimuli or substrates, or are too cramped and physically constraining, to enable consummatory behaviours. As Dawkins (1988) summarizes:

> Suffering [is] caused by the absence of certain conditions, where the animal is motivated to perform a behaviour but unable to perform it because of physical restraint or lack of suitable stimuli.

Next, we consider how to identify frustration that is welfare-compromising.

7.2.3 Research methods for identifying motivational frustration

Experimental techniques for frustrating strong motivations include exposing hungry animals to situations in which they have learned to expect food that is unexpectedly withheld, or to food that they can detect but not obtain. By inducing frustration in a controlled way, such protocols allow its effects to be characterized and catalogued. As already mentioned, frustration is aversive, and so escape attempts often occur, along with distress signals (olfactory and/or vocal; see Table 7.1). Other typical behavioural reactions include repeated attempts to eat the absent or inaccessible food, aggression (if housed with a conspecific), and apparently irrelevant 'displacement activities' like briefly drinking, preening, grooming or self-touching. As Hinde (1970) summarizes for the frustrated animal:

> Its behaviour may take a number of forms. One possibility is the appearance of investigatory activity or trial-and-error ... Another is a response to the normally inadequate stimulus situation, making possible the completion of the sequence albeit possibly in a non-functional manner. In other cases, the animal may show ... displacement or aggressive behaviour.

Duncan and Wood-Gush (1972) further showed that the escape attempts of frustrated hens develop into sustained, repetitive, stereotyped pacing, with more recent studies confirming that sustained repetitions of frustration responses underlie several forms of stereotypic behaviour. For example, the motor patterns involved in feather-plucking by caged hens are morphologically identical to foraging pecks, while bar-chewing by laboratory mice develops from repeated escape attempts (for more on stereotypic behaviour see Latham and Mason, 2010, and Chapter 9). Frustration typically also induces physiological effects, especially sympathetic responses such as elevated blood pressure, and corticosteroid release by the adrenal cortex (see Table 7.1).

Table 7.1. Effects of motivational frustration, as modified from Papini (2003), with supplementary examples from other papers.

Effect of frustration as summarized by Papini (2003)	Species (from Papini, 2003)	Supplementary examples from other papers
Displacement activities: polydipsia (excess drinking); increased eating in sexually frustrated animals	Rats	Object manipulation and self-scratching increase in orang-utans playing a frustrating computer game (Elder and Menzel, 2001). Self-preening increases in food-frustrated hens (Duncan and Wood-Gush, 1974), as does oral manipulation of objects and pen mates in food-frustrated pigs (Lewis, 1999). Displacement drinking and self-scratching increase in male rats exposed to oestrous females that they cannot reach (Hansen and Drake af Hagelsrum, 1984)
Aggression	Rats, pigs and humans	Aggressive pecks and 'running attacks' increase in hens frustrated of access to food or water (Haskell *et al.*, 2000)
Escape/avoidance responses and increased activity levels	Rats and humans	Food- or water-frustrated hens pace (Duncan and Wood-Gush, 1974; Haskell *et al.*, 2000). Food-frustrated pigs become more active (Lewis, 1999), and food/water-frustrated hens spend more time away from the test arena, becoming reluctant to approach it (Haskell *et al.*, 2000)
Calling (crying/ultrasonic vocalizations)	Humans (crying) and rats (ultrasound)	'Gakel' calls increase in hens frustrated from obtaining food, water, a dust bath or nesting pre-lay (Zimmerman *et al.*, 2000)
Stress odour release	Rats, wood rats, gerbils	
Corticosteroid release	Rats, monkeys	Kawasaki and Iwasaki (1997) also found corticosterone increases in rats denied food that they were expecting in a runway task
Decreased heart rate (paradoxically, as this suggests parasympathetic response)	Rats	Garcia-Leon *et al.* (2003) report similar decreases in heart rate for frustrated humans
Increased blood pressure	Humans	

These studies suggest that displacement activities, and attempts to perform highly motivated behaviours that cannot be consummated, can be used to identify frustration. Highly repetitive stereotypic behaviours deriving from these activities could also be evidence of specific frustrations – although these may be exacerbated by a wide range of stressors, and are additionally caused by brain dysfunction (see Latham and Mason, 2010, and Chapter 9). Physiological 'stress responses' (see Chapter 10) could also indicate frustration – although, again, they are not specific to frustration – occurring in other negative states (e.g. fear) and in positive states involving elevated activity (e.g. copulation). Finally, chronic frustration (as just one of many potential causes of long-term stress) may be inferred when behaviourally restrictive environments cause poor reproduction, poor or aberrant growth, and/or decreased resistance to disease, because the sustained elevation of corticosteroids and sympathetic responses, such as hypertension, compromise both health and reproduction (as reviewed by Blache *et al.* in Chapter 10).

Measuring preference is another way to identify strong motivations (see Chapter 11). This involves assessing whether animals will learn arbitrary operant responses such as lever pressing to perform natural behaviours; and/or making it difficult to do so (e.g. increasing the number of lever presses required to gain access, or imposing natural barriers such as gaps to squeeze though, or heavily weighted doors to push (Fig. 7.2)), in order to assess motivations for various goals.

Such experiments have yielded important findings. However, they are not ideal for investigating behavioural restriction, because the experiences of subjects being tested are likely to increase their motivations over those of naive animals. These 'priming effects' include the motivational consequences of: learning the reward

(a)

(b)

(c)

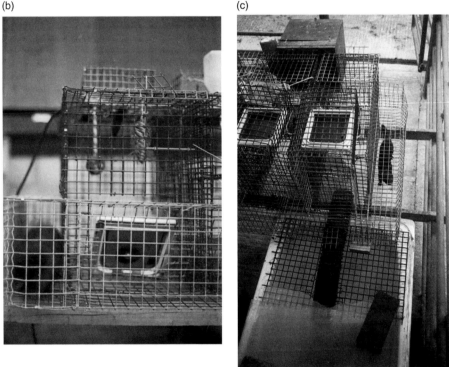

Fig. 7.2. Assessing the reward value of enrichments: an example with mink. Progressively weighting a one-way door (Fig. 7.2a) that leads to a resource can help to reveal the costs that mink are prepared to pay to perform a particular behaviour, and thus their motivations to do so. In one experiment, mink would push an average maximum of 1.06 kg to reach cat toys they could manipulate and chew (Fig. 7.2b); over the whole 6-week-long experiment they each pushed a total of 34 kg to reach these toys. In the same experiment, the mink pushed an average maximum of 1.25 kg to reach a bath of water in which they could head dip, paddle, swim and dive (Fig. 7.2c) – behaviours that are part of their natural foraging behaviour (see Box 7.1); over the 6 weeks of the experiment they each pushed 130 kg in total to reach this resource (Mason *et al.*, 2001). In another experiment, using differently designed weighted doors, mink pushed 2.5 kg to reach a water bath: not significantly different from the weight that they pushed for food in that same study (Cooper and Mason, 2000; Photos: Georgia Mason).

Box 7.1. Water-based foraging and captive American mink (*Mustela vison*).

Along the coasts, lakes and rivers of parts of northern Europe, feral American mink thrive. Despite being captive bred on fur farms for generations, they seem to retain an affinity for water; for example, Vorontsov (1985), describing mink released near a river, says: 'An overwhelming majority … set out towards water. Many of the tiny animals tried to bathe right there.' Mink can wreak havoc on indigenous aquatic species, sometimes swimming several kilometres to predate island seabird colonies (e.g. Nordström and Korpimäki, 2004). (Once, unfortunately for a study cited in our nest-building section, feral mink even decimated feral hens introduced by applied ethologists to a Scottish island; Duncan *et al.*, 1978). Baths of water in which to swim, paddle and 'head dip' also seem valued if added to caged mink pens, despite not allowing true hunting, which is the function of these behaviours in the wild. Thus, water baths increase juvenile play (Vinke *et al.*, 2005), and mink will lever-press (Hansen and Jensen, 2006) and push weighted doors to reach them (see Fig. 7.2c). Furthermore, mink given baths, but then denied them, can show behavioural and physiological signs of frustration (though not all studies report such effects; Vinke *et al.*, 2008). So could interacting with water be a behavioural need? Most data suggest not. In several Dutch and Danish studies across multiple farms (Vinke *et al.*, 2008), baths failed to reduce stereotypic behaviour, or improve reproduction or immune function, compared with control mink. However, work by a Finnish group differs, with water baths reducing stereotypic behaviour on one farm (Mononen *et al.*, 2008; Vinke *et al.*, 2008). Other enrichments, such as balls, rope to chew and wire tunnels, seemed instead to be far more effective (Hansen *et al.*, 2007). One possible explanation for this pattern of results is that water-based foraging is generally not a need; it is just rewarding when it occurs, an illustration of the 'priming' effects discussed in Section 7.2.3. Another explanation is that the populations studied by different researchers truly varied – an issue that crops up in several cases in Section 7.3.2. To illustrate, in the British project

Photo: Claudia Vinke

where mink worked hard for baths (e.g. Cooper and Mason, 2000; Mason *et al.*, 2001; and others), all subjects used baths when access was free (58/58 mink over five experiments); in contrast, only 26/40 did in a Danish study finding that baths did not reduce stereotypic behaviour (Skovgaard *et al.*, 1997). This is a significant difference ($\chi^2 = 23.78$, df = 1, $p < 0.0001$) that could reflect genetic differences between the populations, and/or subtle differences in either the control housing or the types of baths provided. In sum, there is little doubt that mink find water-based foraging activities rewarding and enriching. However, it is currently less certain that they miss them if they have never had them – and it is even possible that some mink do while others do not.

---

value of consummatory behaviour (as in the sexual behaviour of male rats) (Lopez and Ettenberg, 2002; Hosokawa and Chiba, 2005); and of exposure to eliciting stimuli in the test apparatus (cf. Section 7.2.1 on the role of external stimuli in motivation). So, being able to see toys, for example, increases the effort that mink are prepared to make to reach them (Fig. 7.3), and similar effects may even induce motivations

that would, otherwise, probably be absent (see Fig. 7.4)! Hence, because subjects performing preference tests do not model animals living in resource-poor environments (Mendl, 1990; Warburton and Mason, 2003), preference tests can identify natural behaviours that might benefit welfare, but not show us whether animals without experience of such behaviours suffer frustration.

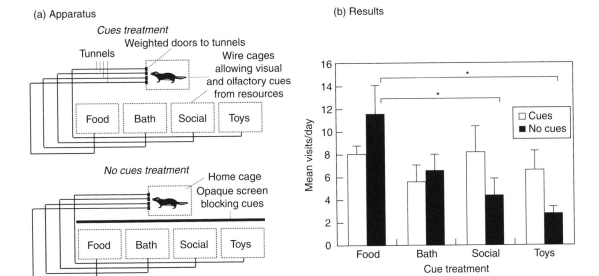

Fig. 7.3. Resource cues as a cause of the 'priming effects' that can affect animals' preferences. Mink were tested for their motivations to leave their home cages, push a weighted door and travel along a tunnel to reach either food, a water bath, social contact or a cat toy, in two different types of apparatus: one in which they could see and smell cues from these four resources at the point where they made their decisions to visit them (the 'Cues' condition), and the other in which these cues were instead screened off from them at this point ('No cues'): see Fig. 7.3a. (Redrawn by Melissa Bateson from Warburton and Mason, 2003.) Note that mink are naturally solitary, so inspecting a conspecific (as in the social contact cue) may reflect the type of defensive problem solving shown in Fig. 7.4. As shown in Fig. 7.3b, when mink could not see the toys or social contact, they visited them less often than they visited food. This was not true in the 'Cues' condition. Being able to see toys also increased the number of visits that mink made to them each day, compared with the 'No Cues' condition; furthermore, when the access doors were made increasingly heavy to push, the mink seemed slightly more willing to pay these costs to reach toys in the 'Cues' condition than in the 'No Cues' condition. (See Warburton and Mason, 2003.)

Overall, behaviour patterns important for good welfare are therefore identified by assessing which are strongly preferred and, more importantly, which reduce behavioural and physiological signs of frustration compared with control animals unable to perform these behaviours.

7.2.4 Types of strong behavioural motivation: from 'behavioural needs' to elicited motivations

The potential role of external stimuli has stimulated discussions about how behaviourally restrictive housing might induce strong motivations. At one end of a spectrum of highly motivated behaviours are so-called 'behavioural needs'. These are activities that animals have instinctive, intrinsic propensities to perform whatever the environment is like, thus even when the physiological needs that the behaviour serves are

fulfilled, and even when these behaviours are not necessary for fitness (Jensen and Toates, 1993). Hypothetical examples include: being motivated to perform natural foraging even when nutritionally sated; or prey species retaining strong motivations to hide under cover, despite being safe, not exposed to predator-like stimuli, and having been captive and not predated for generations. Other potential examples (returned to in Section 7.3.2) are motivations to nest build even when the environment provides a ready-made nest, as if performing this activity is satisfying per se. Behavioural needs are, therefore, largely driven by internal factors, and their motivations are best reduced via performance. At the other end of the spectrum of highly motivated behaviours are activities induced by the animal's external circumstances: thus, deficits or external cues in the environment elicit the motivations. Hypothetical cases include: being highly motivated to forage

(a)

(b)

Fig. 7.4. Working to reach rivals: motivations to perform defensive problem solving. Male zebra finches (a) were placed in an apparatus in which hopping from perch to perch acted as an operant, opening a door that allowed birds see other members of their own species. It was found that a male would work harder to see a female than a male, and harder to see his own mate than another female. However, he would work hardest of all when behind the door was a threatening scenario: his own mate with another male (McFarland, 1982; Photo: iStock Photo). Likewise, male silver foxes (a colour type of the red fox *Vulpes vulpes*) will perform an operant response, pulling a lever (b), to get close to other foxes. They will do this to reach vixens, who they wag their tails at – but also to reach males, who they then bare their teeth and growl at (Hovland *et al.*, 2006; Photo: Anne Lene Hovland).

because nutritionally deprived; the cover-seeking species mentioned above being motivated to escape from cages that offer no seclusion; or animals being motivated to hide because of external stimuli being perceived as threatening (e.g. human noise). Another case, returned to in Section 7.3, is when animals are motivated to create nests or shelters because their environments lack such structures.

This spectrum is conceptually useful for appreciating the different ways in which frustrations may be alleviated by improving housing. If strong motivations are largely internally generated, then the best way to reduce the frustration of captive animals is to allow them to perform the highly motivated behaviour (or to take the challenging approach of trying to alter the underlying physiological states of the animals). Conversely, if strong motivations are largely externally driven, then removing, providing or altering eliciting stimuli in the captive environment (or allowing animals to do so themselves) will alleviate frustration. In practice, however, it can be hard to categorize any behaviour as being strictly at one end or other of this spectrum. Indeed, as both have similarly negative implications for welfare if frustrated, and both can often be equally addressed by adding naturalistic stimuli, agonizing about such distinctions

may be unimportant from a practical viewpoint; Jensen and Toates' excellent 1993 paper 'Who needs "behavioural needs"?' argues this further.

7.3. Frustrated Natural Behaviour and Animal Welfare: Some Examples

Here, we first present examples of the benefits of general social/physical 'enrichment' of the captive environment that enables a wide range of natural behaviours. We then review in more detail experiments that manipulate just one or two aspects of that environment to pinpoint specific natural behaviours that are important for welfare. Finally, we present some natural behaviours that are known or suspected to be rewarding, but whose impact on the welfare of captive animals remains untested.

7.3.1 General welfare effects of environmental complexity and compatibility with natural lifestyles

Many studies have demonstrated the harms of captive environments that prevent or restrict natural behaviours. For example, being removed

G.J. Mason and C.C. Burn

from the mother too early has lasting negative effects on many mammals, including on their stress responses, stereotypic behaviour and lifespan (reviewed by Latham and Mason, 2008). For naturally social species, isolation is also detrimental. Compared with socially housed conspecifics, isolation-housed rats show more asymmetrical body plans, indicating disrupted development (Sorenson *et al.*, 2005), exaggerated corticosterone responses to stressors and impaired wound healing (Hermes *et al.*, 2006), and shorter lifespans (Shaw and Gallagher, 1984; though cf. Skalicky *et al.*, 2001); isolated horses show more stereotypic weaving (McAfee *et al.*, 2002) and greater reactivity to acute stressors like trailer transport (Visser *et al.*, 2008, Kay and Hall, 2009); while isolated monkeys are, again, more stereotypic (Lutz and Novak, 2003), and have elevated heart rates, increased risks of atherosclerosis, and signs of immunosuppression (Watson *et al.*, 1998; Lilly *et al.*, 1999, Schapiro *et al.*, 2000). Similarly, small, physically impoverished captive environments are also often detrimental compared with those that are larger and/or 'enriched' (more naturalistic or complex). To give just a few examples, impoverished environments decrease growth rates and increase aggression in goats (Flint and Murray, 2001), decrease reproductive success and increase stereotypic behaviour in pandas (Zhang *et al.*, 2004), and increase the fearfulness of laboratory rats, impair their abilities to heal brain injuries and reduce their longevity in old age (Passineau *et al.*, 2001; Balcome, 2006; Bell *et al.*, 2009). As one last illustration, from a slightly different perspective, across the Carnivora, species that are naturally wide ranging and travel large distances daily in the wild show the most stereotypic behaviour and highest infant mortality rates when caged (Fig. 7.5). Together, this large and diverse body of work dramatically demonstrates that unnatural restrictive captive conditions compromise animal welfare.

However, these studies cannot tell us whether any specific behaviour pattern is important: only that 'something about large home ranges' is important for carnivores, 'something about social contact' is important for primates, and so on. For more precision, we need studies involving fewer variables – specific environmental enrichments identifying particular natural behaviours whose performance reduces frustration. Examples of this type of finding include: that male rats live longer and are healthier into old age if they can mate

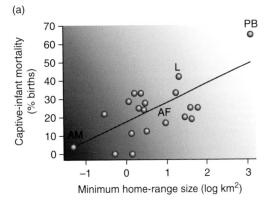

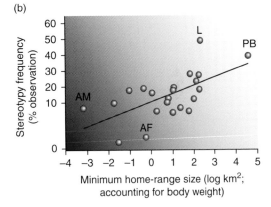

Fig. 7.5. Constraints on natural ranging predict welfare issues in carnivores. Welfare indices in captivity for species in the Order Carnivora are predicted by aspects of natural behavioural biology that are constrained by enclosure – natural home range size (shown here) and daily distance travelled. For example, the average smallest range sizes extracted from field studies of approximately 20 species predict (a) rates of infant mortality (deaths before 30 days) in captivity, and (b) the severity of stereotypic behaviour in animals developing that behaviour. PB = polar bear; L = lion, AF = Arctic fox and AM = mink. (Clubb and Mason, 2003; Figure reprinted with permission from *Nature*).

regularly (Salmon *et al.*, 1990); that devices attracting foraging or foraging pecks reduce feather-pecking in caged hens (Huber-Eicher and Wechsler, 1998, McAdie *et al.*, 2005, Dixon *et al.*, 2008); and that farmed rabbits given wooden sticks to gnaw spend less time gnawing the bars of their cage, are less aggressive and grow better (Princz *et al.*, 2007, 2009). As one final example, clouded leopards in zoos show lowered faecal corticoid output if provided with hiding places (Shepherdson *et al.*, 2004).

Together, these cases show both specific and general ways in which impoverished housing should be improved (with further examples given below), and to some extent, what they reveal about the harms of behavioural restriction is obvious. However, they often leave unanswered fundamental questions about exactly how the added opportunities exert their benefits – such as whether the actual performance of the behaviour is important, or instead whether its outcome is important. In the rabbit case above, for instance, the act of gnawing itself could be the key, or instead improved tooth wear (in which case tooth trimming might be just as effective); while in the clouded leopard case, stress might be reduced by the very ability to hide, or instead by the consequences of hiding (in which case removing stressful stimuli outside the enclosure would be equally effective). Furthermore, when experiments are replicated, or potential enrichments studied by multiple research groups, subtleties and complexities often become apparent, including differences between genetically distinct populations, and large impacts of slight variations in the enrichments being offered. We deal with some of these complexities next, using, as examples, sucking by calves, belly nosing in piglets, nest building by sows and hens, and burrowing in rodents. These cases seemingly represent a spectrum from unambiguous behavioural needs through to motivations driven by deficits in the captive environment. In Box 7.1, we also discuss motivations for swimming/paddling in mink, for which a corpus of recent work reveals great variation between studies, for all or some of the reasons illustrated in the examples below.

7.3.2 Well-studied examples of specific frustrations

Dairy calves are typically removed from their mothers soon after birth, and then fed milk or 'milk replacer', often from buckets. These maternally deprived calves often suck protruding features of the pen and other calves' ears or prepuces – behaviour which may be accompanied by head butting resembling the udder butting of mother-reared calves. Several experiments have identified the stimuli triggering this 'non-nutritive sucking' and its apparent consequences

(de Passillé *et al.*, 1993; Haley *et al.*, 1998; Veissier *et al.*, 2002; de Passillé and Rushen, 2006). Researchers used non-nutritive rubber teats offered alongside or after a bucket of milk, and also compared calves that obtained their milk by sucking rubber teats with those having to drink directly from buckets. This allowed them to disentangle the effects of milk ingestion from the act of teat sucking itself. Non-nutritive teat sucking proved to be elicited by milk ingestion (therefore peaking after drinking milk but not water), lactose being the key component. The quantity ingested – thus nutritional satiety – seems to play little role in the termination of the behaviour which, instead, is caused by two independent processes: a time-dependent decay in sucking motivation following the last intake of milk, and the very act of sucking itself. Furthermore, the behaviour seems functionally important: allowing early-weaned, bucket-fed calves to suck rubber teats increases the release of satiety hormones (Fig. 7.6); while allowing calves to suck milk from teats instead of buckets decreases their heart rates and increases heart-rate variability (signs of reduced sympathetic activation), and seems to induce calmer, more restful states (Veissier *et al.*, 2002). This treatment also reduced the abnormal oral behaviours directed at pen mates and at the pen's bars (though, perhaps surprisingly, providing hay to eat is effective too; Haley *et al.*, 1998). Overall then, the act of sucking seems to be a behavioural need, whose performance per se is welfare enhancing.

Other well-studied examples of behavioural frustrations come from farmed pigs. In the weeks after weaning, young piglets can similarly develop abnormal feeding-related behaviours, nosing repetitively at each others' bellies, sometimes even causing lesions. Leaving suckling piglets longer with their mothers greatly reduces their 'belly nosing' post-weaning (Widowski *et al.*, 2003; Main *et al.*, 2005). This, and the morphology of the abnormal behaviour, has led to hypotheses that it represents frustrated udder massage that would be directed at sows in more natural circumstances. Consistent with this, artificial nipple- or udder-like stimuli greatly reduce belly nosing in early-weaned piglets (Widowski *et al.*, 2005; Bench and Gonyou, 2006, 2007); these stimuli seem to do this by offering the animals alternative, preferred outlets for similar nosing activities.

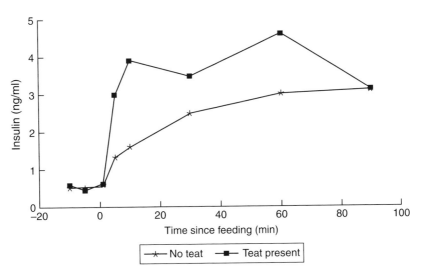

Fig. 7.6. The consequences of non-nutritive sucking by calves. Mean insulin concentrations in the hepatic portal vein of young calves after drinking milk. The calves either could or could not suck a rubber teat (which did not deliver milk) after drinking milk from a bucket. Similar effects are suggested to occur in human babies sucking pacifiers or dummies (Uvnäs-Moberg *et al.*, 1987). (Figure redrawn by Jeff Rushen and Carol Petherick, based on data in de Passillé *et al.*, 1993).

So enrichments that only allowed rooting or chewing did not seem effective, at least in one study (Bench and Gonyou, 2006). However, another study – albeit using piglets with rather high levels of belly nosing – suggests that rather less 'mother-like' stimuli (a suspended tyre and rope to chew) can also reduce this behaviour, as well as making piglets less fearful of humans (Rodarte *et al.*, 2004); and even more diverse enrichments (e.g. straw (Day *et al.*, 2002) or working for food (Puppe *et al.*, 2007)) can reduce belly nosing in older 'growing pigs'. Overall, it seems clear that something about premature loss of the sow and/or of suckling induces abnormal oral behaviour in newly weaned piglets, and that something about artificial stimuli that attract oro-nasal behaviour also reduces belly nosing (as well as probably improving welfare). However, it remains unknown exactly which enrichments might best compensate for this loss; or whether nosing smooth, soft materials benefits piglet physiology. The causes and welfare significance of breed differences in belly-nosing severity (Bench and Gonyou, 2007) also remain unknown.

Turning to adult pigs, females about to give birth undergo hormonal changes that both alter what they regard as suitable environments, and precipitate species-typical maternal activities such as nesting. In natural environments, pre-parturient sows seek a secluded site where they dig a depression in the earth. They then vigorously gather plant materials, which they carry to this nest site (see Fig. 2.5) and manipulate it with their snouts to construct a soft bed (Wischner *et al.*, 2009). Little work seems to have been conducted on the motivations of sows to have particular types of nest site (unlike hens, as discussed below), but much has examined motivations to obtain nesting material, and the effects of a lack of nesting material on frustration-related variables. Sows will perform operant responses to collect straw, especially as parturition approaches (Arey, 1992), although their interest seems to vary between studies; in some, all sows offered a straw dispenser use it (Arey *et al.*, 1991), while in others, only some do (Hutson, 1988; Widowski and Curtis, 1990). Furthermore, a lack of nesting material seems frustrating, for instance in inducing restless, stereotyped rooting and ground pawing (Wischner *et al.*, 2009), although once again studies vary. Some experiments compare smaller bare crates with larger strawed pens, so examining effects of substrate and enclosure size combined; the crates tend to elevate the cortisol levels of pre-parturient females and prolong the process of parturition (Jarvis et al., 1997, 2001; Oliviero et al., 2008). Thodberg and colleagues (1999) showed similarly beneficial effects on the speed of parturition from just providing sand and nest-making materials (although another similar study found quite the opposite effect; Jarvis et al., 2004). Such sows also suckled their piglets for longer, showed less piglet crushing, and more responsiveness to piglet distress cries, than conspecifics in a similar pen with a concrete floor and no straw (Herskin et al., 1998, 1999).

But is it the acquiring of a nest or actually creating a nest that is the key for good sow welfare? The presence of a satisfactory nest does seem important, quite independently of the behaviour involved. Sows provided with a ready-made depression in the sand and a straw bed greatly reduce their collection of straw from a dispenser, so the pre-built nests reduce this motivation (Arey et al., 1991). Furthermore, if sows are allowed to nest build but their nests are repeatedly removed, their heart rates elevate, and by parturition their cortisol is near double that of a control group allowed to keep their nests (Damm et al., 2003). This effect occurred even though the total amount of nest building performed was similar between the two groups, again suggesting that the nest itself, not the nesting behaviour, is most important. However, not all nesting behaviour is abolished if made redundant by a ready-made nest; moving of and pawing at nesting material is unchanged, but rooting at it actually increases (Arey et al. 1991). This could mean that sows simply 'have' to perform some aspects of nesting themselves (although equally, it might just show that humans are bad at making satisfactory sow nests!). Future tests of this hypothesis could include giving sows choices between pre-made sow-built nests (e.g. their own from a previous parturition), and piles of raw materials; or between nests that cannot be manipulated and similar ones that can (perhaps a foam replica of a pre-made nest versus foam pieces). Pending more detailed ethological studies, for sows, the current consensus seems to be that they probably do gain motivational satisfaction from some elements of the nest-building behaviour itself. Either way, however, the take-home message is the same: it generally benefits sows to have nesting material.

Hens about to lay similarly seek secluded, enclosed nest sites, and their motivations to do so have been well studied. Thus, as Weeks and Nicol (2006) summarize, collating several authors' research:

> Hens place a high value on access to discrete closed nest sites and ... their behavioural priority to access one increases the closer they get to the time of egg laying (oviposition). They are prepared to pay 'high prices' such as squeezing though narrow gaps or opening doors to gain access to nest boxes before egg laying. Moreover, hens have been found to work as hard for a nest site during the pre-laying period as they would for food following short periods of deprivation ... hens would squeeze though narrow gaps of up to 95 mm width (compared with an average hen body width of 120 mm) to access a nest box before oviposition, but would go without food for an average of 8 hours before passing through such a small gap. Using a load-recording push door, the work rate for the nest site at 40 minutes before the expected time of oviposition was comparable to the work rate for food at 4 hours' deprivation.

This motivation is not just induced by cues from or experience with the proffered nest sites (cf. Section 7.2.4). Hens without suitable nest sites at this time exhibit 'gakel' calls apparently typifying food frustration (see Table 7.1), along with stereotypic pacing seemingly derived from escape attempts, and much apparent searching behaviour. They may even delay oviposition (Walker and Hughes, 1998; Yue and Duncan, 2003). A suitable nest site is, therefore, important for hen welfare.

However, this behaviour by hens again raises the question: do hens need to nest or just a nest? Behavioural aspects of nesting per se have been studied in feral hens (Duncan et al., 1978) as well as in caged birds, and have often been argued to be important. Once in a suitable nest site, hens show rotation movements to make a cup-shaped depression (and if litter is available they may manipulate it), and the presence of a ready-made nest does not seem to 'turn off' this behaviour. So laying hens experimentally given either a nest box of wood shavings levelled out so as to have a flat surface, or a similar box containing a nest depression made by the same hen the day before, perform similar amounts of rotating, and produce eggs similarly readily – after about 45 minutes – following first entering the box (Hughes *et al.*, 1989). This has been widely held to indicate that the very activity of nest building is motivating in its own right: a behavioural need. However, while this interpretation may be true, the findings could perhaps equally be explained in terms of body-rotation pre-lay being simply a habit-like 'fixed action pattern', triggered by a suitable nest but not necessarily rewarding. Consistent with this view, hens offered an enclosed nest box with a ready-made, non-mouldable Astroturf surface accept this readily as a nesting location (Appleby *et al.*, 1993), and lay soon after entering the box (*c.* 40 minutes; Struelens *et al.*, 2008), suggesting that the simple non-functional nesting movements possible on Astroturf do not frustrate the hen. Furthermore, when hens are offered a choice, substrates such as peat, which allow the creation of a nest hollow and can be manipulated, seem no more preferred than artificial turf, which does not allow this. Thus, Struelens *et al.* (2005) found that both these nest types were greatly preferred over wire, but attracted the same number of eggs laid. In further work, peat was found to elicit longer sitting and more pecking than artificial turf but, aside from that, its behavioural effects seemed small (Struelens *et al.*, 2008). Overall, this suggests that good captive environments must supply a suitably enclosed nest area, but as long as this is there, hens can forgo true nest making itself.

Note that some variation between studies occurs on this topic too, with nest-box properties (e.g. degree of privacy) having quite strong effects on hen responses, and the types of artificial nest sites accepted as suitable also varying between strains and even between individual females (Yue and Duncan, 2003; Weeks and Nicol, 2006). The provision of unsuitable nest boxes might explain why some cages furnished with nests fail to have the benefits one might expect. Barnett *et al.* (2009), for example, found that providing a nest box did not reduce corticosteroid levels, nor enhance immunity, compared with control hens in barren cages – but only half to two-thirds of their experimental birds chose to lay their eggs in the provided nest boxes, compared with the 99–100% 'approval rating' seen in other experiments (Appleby *et al.*, 1993; Weeks and Nicol 2006).

Our last example concerns tunnel creation and digging in naturally burrowing rodents.

Some elegant experiments have investigated stereotypic digging by gerbils. This abnormal behaviour is unaffected by providing loose sand in which the animals can dig (Wiedenmayer, 1996, 1997a). Instead, it is nearly abolished by the provision of a ready-made naturalistic burrow (Wiedenmayer, 1997b), in which gerbils also prefer to hide, sleep and suckle their young compared with the shelters provided in standard laboratory cages. Non-naturalistic plastic 'burrows' have similar effects, as long as they consist of a tunnel leading to a chamber, and have opaque, rather than transparent, walls. Like a naturalistic burrow, gerbils rapidly adopt these as their preferred homes when given a choice, and stereotypic digging all but vanishes (Wiedenmayer, 1997a; Waiblinger and Konig, 2004). Such findings reveal that frustrated motivations to dig in captivity persist only when gerbils lack the natural goal of this behaviour: a burrow leading to a nest chamber. Digging per se, in contrast, does not seem to be a behavioural need.

These five well-studied cases illustrate that behavioural needs and environmental deficits can be hard to tell apart, but that both cause frustration in typical housing systems. They also generate several other 'take-home messages'. One is that enrichments that seem similar to us (e.g. different types of nest box for hens; plastic nest chambers with or without a short tunnel for gerbils) may be perceived very differently by the animals they are offered to. This may help to explain why studies do not always replicate each others' findings. Genetic differences between populations studied in different experiments (best documented for hens and piglets, but possibly widespread) may also help to explain the lack of replication. Finally, they also remind us that if specific frustrations cause welfare problems, we should expect equally specific enrichments to be effective. However, sometimes surprisingly, diverse enrichments can have similar effects on signs of frustration (as in both piglet belly nosing and calf sucking). Box 7.1 illustrates all these issues, using recent work on mink.

7.3.3. And finally . . .

There are many apparently rewarding behaviours whose potential welfare impact is still unknown, and many more that remain completely uninvestigated. Thus laboratory rats will bar press for live mice to kill (Van Hemel, 1972), but whether other predators find killing reinforcing is unknown, as are any benefits to welfare (carnivore welfare rather than that of their prey!). Hens also perform operant tasks to reach litter or straw to scratch and peck at (they may dust bathe in it too; Dawkins and Beardsley, 1986; Gunnarsson *et al.*, 2000), while pigs do likewise to gain access to peat and other materials in which they can root (Jensen and Pedersen, 2007). However, the reward value and welfare importance of many other natural foraging behaviours is still little understood (for example, whether browsers need to browse, and grazers to graze). Given the chance, laboratory mice and several species of deer mice will perform operant tasks in order to dig burrows in sand or peat (King and Weisman, 1964; Sherwin *et al.*, 2004), but we do not know if the absence of digging opportunities – or of completed burrows as for the gerbils described above – compromises mouse welfare in standard cages. The reward values of only a few specific social behaviours have been investigated so far: being groomed was reinforcing for a rhesus monkey (Taira and Rolls, 1996), play is a reinforcer for juvenile rats (Humphreys and Einon, 1981), and female mammals in oestrus will pay costs to mate with males (reviewed by Wallen, 1990). However, whether a lack of any of these is frustrating and stressful, we do not know. Opportunities to display aggressively or even to fight are also reinforcing for males, in species as diverse as chickens, Siamese fighting fish, rats and mice (Fish *et al.*, 2002; reviewed May and Kennedy, 2009; see also Fig. 7.4b); and male chaffinches will perform a perch-hopping operant to hear the songs of other males (Stevenson-Hinde and Roper, 1975). However, yet again it is unknown whether aggression (or perhaps just winning) improves the welfare of males, or whether these findings instead merely reflect defensive motivations elicited when rivals appear to be close by. To end, among the many further unanswered questions are whether captive horses need to gallop, captive cheetahs to sprint, captive beavers to make dams, caged birds need to fly or, as discussed below, whether any animals experience 'boredom' caused by the sheer lack of anything relevant to do at all.

7.4 Do Impoverished Environments Cause Boredom in Animals?

For humans, the frustration of specific natural behaviour patterns is not the only problem caused by impoverished environments. Human boredom arises when there is 'nothing to do' (i.e. there is an absence of behavioural opportunities) or when a monotonous environment offering little chance for behavioural diversification makes us 'tired of doing the same thing' (Larson and Richards, 1991). Humans find boredom aversive (Harris, 2000), and it can be seen as a form of frustration in that it is a thwarting of a motivation to 'do something else' or to explore something new.

If animals can experience boredom like humans, then unlike frustrations arising from specific thwarted motivations, boredom will arise when animals are not motivated to perform any of the behaviours that are possible in their environments because these are too limited or monotonous. Furthermore, this will cause poor welfare. The issue of whether boredom is truly a welfare problem for behaviourally restricted animals can be approached from three angles: is boredom a problem for humans who live in conditions similar to those of captive animals?; do captive animals seek novelty and diversity generally, rather than just seeking to satisfy specific motivations?; and do animals in environments likely to induce boredom show responses similar to those of bored humans?

To address the first question, there are ample reports that boredom is a profound problem for human prisoners. For example, a prisoner quoted in Wahidin (2006) says:

> The boredom and the isolation … every day is the same. It is just dreadful. Every single day is the same. It drives you mad! … every single day is just actually the same as the day before. Your meals are the same time. Everything. Oh it's horrible.

Bracke *et al.* (2006) attribute boredom in human prisoners to 'routinized activities lacking intrinsic meaning and lacking emphasis on task completion'. Experimentally induced sensory deprivation similarly induces aversive states. As Berlyne (1960) reports:

> Sensory deprivation becomes aversive when internal factors cause a rise in arousal and the lack of stimulation renders the cortex incapable of keeping arousal

G.J. Mason and C.C. Burn

within bounds; Lying motionless in a quiet dark room … is extremely trying … when one is healthy and has had enough sleep.

Normal everyday life can also induce boredom if monotonous; for example, in 'Bleak House', Dickens (1853) describes the importance of irrelevant repetition and lack of novelty:

My Lady Dedlock has been bored to death. Concert, assembly, opera, theatre, drive, nothing is new to my Lady under the worn-out heavens … They meet again at dinner – again, next day – again, for many days in succession. Lady Dedlock is … terribly liable to be bored to death.

Routine, irrelevant repetition and monotonous stimuli – the causes of boredom in humans – arguably characterize many captive animals' lives, which leads us to the second question. Boredom will only be problematic if animals, like humans, seek novelty and diversity generally, rather than just the satisfaction of specific motivations. In fact, many species will work to obtain sensory or exploratory experiences, at least when housed in research laboratories or zoos. Here, they will invest time and energy in manipulating novel objects without any obvious rewards; for example, primates perform operant responses to obtain stones, sand or other manipulable objects, and rats – which are usually averse to light – will lever press to access flashing lights in the absence of other stimulation (reviewed Berlyne, 1960). Animals may even forgo usually favoured food items in preference for less favoured, but more novel, items. For example, rats and hamsters fed chow containing a preferred non-nutritive flavour for several days will discard this in favour of a novel chow containing a less preferred flavour (Galef and Whiskin, 2005). Some animals in barren conditions also seem to 'sensation seek' (cf. Zuckerman, 1994), seeking out almost any stimuli that increase arousal; individually housed rats even self-administer corticosterone, delivered intraperitoneally via a catheter, and also to some extent the saline control (which slightly increases endogenous corticosterone production) (Piazza *et al.*, 1993). Such behaviours are most common in neophilic (novelty-seeking) species, almost by definition (Stevenson, 1983; Kirkden, 2000), and are sometimes assumed most evident in species with greater cognitive flexibility (Wood-Gush *et al.*, 1983). As well as species differences, individuals differ in motivations for

stimulation. For example, the rats showing the highest propensity to self-administer corticosterone were also the most exploratory (Piazza *et al.*, 1993), suggesting a common 'sensation-seeking' underlying mechanism. Overall, it is thus possible that at least some animals have behavioural needs for stimulation. For such animals, behavioural restriction may well frustrate general, non-specific motivations for sensory or cognitive feedback.

The third, most difficult, question is whether captive animals actually show responses consistent with boredom. Human boredom and proneness to boredom (reviewed by Berlyne, 1960; Fazzi *et al.*, 1999; Newberry and Duncan, 2001) are associated with signs of poor welfare, including depression and lethargy, social dysfunction, aggression and cognitive deficits. They are also associated with 'sensation-seeking' behaviours like dangerous driving, drug abuse and gambling. Finally, boring situations induce repetitive behaviours (e.g. knee joggling, hair twirling and other 'fidgets' as akin to displacement activities, along with more stereotypic pacing, rocking and similar), which could reflect frustration, be attempts to generate stimulation, or both. Turning to animals, neophilic species such as pigs spend longer interacting with novel objects if kept in more barren environments than do pigs in more complex environments, indicating that barren environments increase urges to seek stimulation (reviewed by Wood-Gush *et al.*, 1983). Also, rats kept individually in standard cages showed significantly greater consumption of self-administered morphine than those housed socially in a large, highly enriched pen (Alexander *et al.*, 1978). More studies are needed to see if such effects are widespread, and what exactly are their causes. For captive animals not offered novelty in this way, boredom is already often suggested as a cause of 'depression', 'apathy', or abnormal and destructive behaviours (e.g. reduced activity, Dawkins, 2001; tail-chasing in dogs, Hartigan, 2000; primate stereotypies, Tarou *et al.*, 2005) – although these hypotheses have not been directly tested. Stereotypic behaviours caused or exacerbated by insufficient stimulation generally, rather than by one specific frustration, should be alleviated by almost any 'environmental enrichment' that offers stimulation. Some indirect evidence for this comes from meta-analyses of the effects of enrichments in zoos (Shyne, 2006; Swaisgood and Shepherdson, 2006) showing that the stereotypies of zoo animals can be reduced by a diverse array of enrichments (e.g. polar bear pacing can be reduced by food puzzles, olfactory stimuli or physical toys alike). Swaisgood and Shepherdson (2006) suggest that:

> in the most extreme situations, one might argue that 'something, anything' added to a stimulus-poor environment may have equally meaningful effects on stereotypy (and perhaps other indices of welfare)

although they admit some equally plausible alternative explanations.

Together these reports are consistent with boredom being an animal welfare problem. However, more evidence is needed to demonstrate true parallels between human and animal responses to monotonous environments and, thus, true boredom, because to date most consistent findings are rather circumstantial, post hoc, or open to alternative explanations.

7.5 Conclusions

- A vast amount of research demonstrates that, compared with standard barren cages, socially naturalistic and/or physically complex environments yield behavioural and physiological benefits to animals that indicate reduced stress.
- Well-fed, physically healthy animals do not then automatically have good welfare, if – as is still the case for many species in many husbandry systems – their cages are small, impoverished or otherwise behaviourally restrictive.
- Frustrating specific strongly motivated behaviours is one potential cause of welfare problems in barren environments. Strong specific motivations may be induced by the presence of certain external stimuli, the lack of appropriate resources in the environment and/or intrinsic requirements to perform certain natural activities ('behavioural needs').
- Research has also identified a diverse array of natural behaviours that animals prefer to perform when they can, but whose absence has an uncertain or unknown effect on the welfare of animals who have never experienced these behaviours (mouse-killing by rats and swimming by mink being two examples).
- When the welfare implications of specific frustrations are investigated by different research groups

working in different ways and in different locations, findings sometimes vary. This could reflect subtle (to us) differences between the behavioural opportunities offered to the animals, subtle (to us) differences in the control environments where these are denied, genetic differences between populations or non-specific effects like boredom.

- Such findings potentially open up future possibilities of improving welfare by selective breeding and genetic selection or even genetic engineering (cf. Chapter 16), and suggest that offering captive animals a range of resources or substrates rather than one single type may best allow the flexible accommodation of their motivations.
- In addition to specific behavioural frustrations, more general, non-specific boredom is also a real possibility, especially for neophilic species and individuals. For example, sometimes rather dissimilar enrichments, which appear as outlets for quite different motivations, can have surprisingly similar effects on the abnormal behaviour and stress physiology of animals. This suggests that 'cognitive enrichments' (cf. Manteuffel et al., 2009) could be a valuable way to improve animal welfare, enhancing it via cognitive stimulation regardless of whether the tasks or the actions involved resemble anything natural.

Acknowledgements

With thanks to Joy Mench for her help and patience as an editor, and to the University of Guelph Animal Behaviour and Welfare Group, especially Ian Duncan, Kati Poczta, Kim Sheppard, Stephanie Yue-Cottee, Stephanie Torrey and Tina Widowski.

References

Alexander, B.K., Coambs, R.B. and Hadaway, P.F. (1978) The effect of housing and gender on morphine self-administration in rats. *Psychopharmacology* 58, 175–179.

Appleby, M.C., Smith, S.F. and Hughes, B.O. (1993) Nesting, dust bathing and perching by laying hens in cages: effects of design on behavior and welfare. *British Poultry Science* 34, 835–847.

Arey, D.S. (1992) Straw and food as reinforcers for pre-partal sows. *Applied Animal Behaviour Science* 33, 217–226.

Arey, D.S., Petchey, A.M. and Fowler, V.R. (1991) The preparturient behaviour of sows in enriched pens and effect of pre-formed nests. *Applied Animal Behaviour Science* 31, 61–68.

Balcome, J.P. (2006) Laboratory environments and rodents' behavioural needs: a review. *Laboratory Animals* 40, 217–235.

Barnett, J.L., Tauson, R., Downing, J.A., Janardhana, V., Lowenthal, J.W., Butler, K.L. and Cronin, G.M. (2009) The effect of a perch, dust bath and nest bow, either alone or in combination as used in furnished cages on the welfare of laying hens. *Poultry Science* 88, 456–470.

Bell, J.A., Livesey, P.F. and Meyer, J.F. (2009) Environmental enrichment influences survival rates and enhances exploration and learning but produces variable responses to the radical maze in old rats. *Developmental Psychobiology* 51, 564–578.

Bench, C.J. and Gonyou, H.W. (2006) Effect of environmental enrichment at two stages of development on belly nosing in piglets weaned at fourteen days of age. *Journal of Animal Science* 84, 3397–3403.

Bench, C.J. and Gonyou, H.W. (2007) Effect of environmental enrichment and breed line on the incidence of belly nosing in piglets weaned at 7 and 14 days-of-age. *Applied Animal Behaviour* 105, 26–41.

Berlyne, D.E. (1960) *Conflict, Arousal, and Curiosity*. McGraw-Hill, New York.

Blake, W. (1863) Auguries of innocence. In: Rossetti, D.G. (ed.) *Poems*.

Blom, H.J.M., Baumans, V., Van Vorstenbosch, C.J.A.H.V., Van Zutphen, L.F.M. and Beynen, A.C. (1993) A preference test with rodents to assess housing conditions. *Animal Welfare* 2, 81–87.

Bracke, P., Bruynooghe, K. and Verhaeghe, M. (2006) Boredom during day activity programs in rehabilitation centers. *Sociological Perspectives* 49, 191–215.

Brambell Committee (1965) *Report of the Technical Committee to Enquire into the Welfare of Animals Kept under Intensive Husbandry Systems.* Command Paper 2836, Her Majesty's Stationery Office, London.

Clubb, R. and Mason, G. (2003) Captivity effects on wide-ranging carnivores. *Nature* 425, 473–474.

Cooper, J. and Mason, G. (2000) Costs of switching cause behavioural rescheduling in mink: implications for the assessment of behavioural priorities. *Applied Animal Behaviour Science* 66, 135–151.

Damm, B.I., Pedersen, L.J., Marchant-Forde, J.N. and Gilbert, C.L. (2003) Does feed-back from a nest affect periparturient behaviour, heart rate and circulatory cortisol and oxytocin in gilts? *Applied Animal Behaviour Science* 83, 55–76.

Dawkins, M.S. (1988) Behavioural deprivation: a central problem in animal welfare. *Applied Animal Behaviour Science* 20, 209–225.

Dawkins, M.S. (1990) From an animal's point of view: motivation, fitness, and animal welfare. *Behavioral and Brain Sciences* 13, 1–61.

Dawkins, M.S. (2001) How can we recognize and assess good welfare? In: Broom, D.M. (ed.) *Coping with Challenge: Welfare in Animals including Humans*. Dahlem University Press, Berlin, pp. 63–78.

Dawkins, M.S. and Beardsley, T. (1986) Reinforcing properties of access to litter in hens. *Applied Animal Behaviour Science* 15, 351–364.

Day, J.E.L., Burfoot, A., Docking, C.M., Whitaker, X., Spoolder, H.A.M. and Edwards, S.A. (2002) The effects of prior experience of straw and the level of straw provision on the behaviour of growing pigs. *Applied Animal Behaviour Science* 76, 189–202.

de Passillé, A.M.B., Christopherson, R. and Rushen, J. (1993) Nonnutritive suckling by the calf and postprandial secretion of insulin, CCK, and gastrin. *Physiology and Behaviour* 54, 1069–1073.

de Passillé, A.M.B. and Rushen, J. (2006) What components of milk stimulate suckling in calves? *Applied Animal Behaviour Science* 101, 243–252.

Dickens, C. (1853) *Bleak House*. Bradbury and Evans, London.

Dixon, L., Duncan, I.J.H. and Mason, G.J. (2008) What's in a peck? Using fixed action patterns to identify the motivation behind feather-pecking. *Animal Behaviour* 76, 1035–1042.

Duncan, I.J.H. and Wood-Gush, D.G.M. (1972) Thwarting of feeding behaviour in the domestic fowl. *Animal Behaviour* 20, 444–451.

Duncan, I.J.H. and Wood-Gush, D.G.M. (1974) The effect of a rauwolfia tranquillizer on stereotyped movements in frustrated domestic fowl. *Applied Animal Ethology* 1, 67–76.

Duncan, I.J.H., Savory, C.J. and Wood-Gush, D.G.M. (1978) Observations on the reproductive behaviour of domestic fowl in the wild. *Applied Animal Ethology* 4, 29–42.

Elder, C.M. and Menzel, C.R. (2001) Dissociation of cortisol and behavioral indicators of stress in orang-utan (*Pongo pygmaeus*) during a computerized task. *Primates* 42, 345–357.

FAWC (Farm Animal Welfare Council) (2009) *Five Freedoms*. Available at: http://www.fawc.org.uk/freedoms.htm (accessed July 2010).

Fazzi, E., Lanners, J., Danova, S., Ferrarri-Ginevra, O., Gheza, C., Luparia, A., Balottin, U. and Lanzi, G. (1999) Stereotyped behaviours in blind children. *Brain Development* 21, 522–528.

Fish, E.W., de Bold, J.F. and Miczek, K.A. (2002) Aggressive behaviour as a reinforce in mice: activation by allopregnanolone. *Psychopharmacology* 163, 459–466.

Flint, M. and Murray, P.J. (2001) Lot-fed goats – the advantages of using an enriched environment. *Australian Journal of Experimental Agriculture* 41, 473–476.

Galef, B.G. Jr and Whiskin, E.E. (2005) Differences between golden hamsters (*Mesocricetus auratus*) and Norway rats (*Rattus norvegicus*) in preference for the sole diet that they are eating. *Journal of Comparative Psychology* 119, 8–13.

García-León, A., del Paso, G.A.R., Robles, H. and Vila, J. (2003) Relative effects of harassment, frustration, and task characteristics on cardiovascular reactivity. *International Journal of Psychophysiology* 47, 159–173.

Gunnarsson, S., Matthews, L.R., Forste, T.M. and Temple, W. (2000) The demand for straw and feathers as litter substrates by laying hens. *Applied Animal Behaviour Science* 65, 321–330.

Haley, D.B., Rushen, J., Duncan, I.J.H., Widowski, T.M. and de Passillé, A.M. (1998) Effects of resistance to milk flow and the provision of hay on nonnutritive sucking by dairy calves. *Journal of Dairy Science* 81, 2165–2127.

Hansen, S. and Drake af Hagelsrum, L.J. (1984) Emergence of displacement activities in the male-rat following thwarting of sexual-behavior. *Behavioral Neuroscience* 98, 868–883.

Hansen, S.W. and Jensen, M.B. (2006) Quantitative evaluation of the motivation to access a running wheel or a water bath in farm mink. *Applied Animal Behaviour Science* 98, 127–144.

Hansen, S.W., Malmkvist, J., Palme, R. and Damgaard, B.M. (2007) Do double cages and access to occupational materials improve the welfare of farmed mink? *Animal Welfare* 16, 63–76.

Harris, M.B. (2000) Correlates and characteristics of boredom proneness and boredom. *Journal of Applied Social Psychology* 30, 576–598.

Hartigan, P.J. (2000) Compulsive tail chasing in the dog: a mini-review. *Irish Veterinary Journal* 53, 261–264.

Haskell, M., Coerse, N.C.A. and Forkman, B. (2000) Frustration-induced aggression in the domestic hen: the effect of thwarting access to food and water on aggressive responses and subsequent approach tendencies. *Behaviour* 137, 531–546.

Hermes, G.L., Rosenthal, L., Montag, A. and McClintock, M.K. (2006) Social isolation and the inflammatory response: sex differences in the enduring effects of a prior stressor. *American Journal of Physiology – Regulatory, Integrative and Comparative Physiology* 290, 273–282.

Herskin, M.S., Jensen, K.H. and Thodberg, K. (1998) Influence of environmental stimuli on maternal behaviour related to bonding, reactivity and crushing of piglets in domestic sows. *Applied Animal Behaviour Science* 58, 241–254.

Herskin, M.S., Jensen, K.H. and Thodberg, K. (1999) Influence of environmental stimuli on nursing and suckling behaviour in domestic sows and piglets. *Animal Science* 68, 27–34.

Hinde, R.A. (1970) *Animal Behaviour: A Synthesis of Ethology and Comparative Psychology*, 2nd edn. McGraw-Hill Kogakusha, Tokyo.

Hobbes, T. (1651) *Leviathan or The Matter, Forme and Power of a Common Wealth Ecclesiasticall and Civil.* Andrew Crooke and William Cooke, London.

Hosokawa, N. and Chiba, A. (2005) Effects of sexual experience on the conspecific odor preference and estrous odor-induced activation of the vomeronasal projection pathway and the nucleus accumbens in male rats. *Brain Research* 1066, 101–108.

Hovland, A.L., Mason, G., Bøe, K.E., Steinham, G. and Bakken, M. (2006). Evaluation of 'maximum price paid' as an index of motivational strength for farmed silver foxes (*Vulpes vulpes*). *Applied Animal Behaviour Science* 100, 258–279.

Huber-Eicher, B. and Wechsler, B. (1998) The effect of quality and availability of foraging materials on feather pecking in laying hen chicks. *Animal Behaviour* 55, 861–873.

Hughes, B.O. and Duncan, I.J.H. (1988) The notion of ethological 'need', models of motivation and animal welfare. *Animal Behaviour* 36, 1696–1707.

Hughes, B.O., Duncan, I.J.H., and Brown, M.F. (1989) The performance of nest building by domestic hens: is it more important than the construction of a nest? *Animal Behaviour* 37, 210–214.

Humphreys, A.P. and Einon, D.F. (1981) Play as a reinforcer for maze-learning in juvenile rats. *Animal Behaviour* 29, 259–270.

Hutson, G.D. (1988) Do sows need straw for nest building? *Australian Journal of Experimental Agriculture* 28, 187–194.

Jarvis, S., Lawrence, A.B., McLean, K.A., Deans, L.A., Chirnside, J. and Calvert, S.K. (1997) The effect of environment on behavioural activity, ACTH, beta-endorphin and cortisol in pre-farrowing gilts. *Animal Science* 65, 465–472.

Jarvis, S., Van der Vegt, B.J., Lawrence, A.B., McLean, K.A., Deans, L.A., Chirnside, J. and Calvert, S.K. (2001) The effect of parity and environmental restriction on behavioural and physiological responses of pre-parturient pigs. *Applied Animal Behaviour Science* 71, 203–216.

Jarvis, S., Reed, B.T., Lawrence, A.B., Calvert, S.K. and Stevenson, J. (2004) Peri-natal environmental effects on maternal behaviour, pituitary and adrenal activation and the progress of parturition in the primiparous sow. *Animal Welfare* 13, 171–181.

Jensen, M.B. and Pedersen, L.J. (2007) The value assigned to six different rooting materials by growing pigs. *Applied Animal Behaviour Science* 108, 31–44.

Jensen, P. and Toates, F.M. (1993) Who needs 'behavioural needs'? Motivational aspects of the needs of animals. *Applied Animal Behaviour Science* 37, 161–181.

Kawasaki, K. and Iwasaki, T. (1997) Corticosterone levels during extinction of runway response in rats. *Life Sciences* 61, 1721–1728.

Kay, R. and Hall, C. (2009) The use of a mirror reduces isolation stress in horses being transported by trailer. *Applied Animal Behaviour Science* 116, 237–243.

King, J.A. and Weisman, R.G. (1964) Sand digging contingent upon bar pressing in deermice. *Animal Behaviour* 12, 446–450.

Kirkden, R.D. (2000) Assessing motivational strength and studies of boredom and enrichment in pigs. PhD thesis. University of Cambridge, Cambridge, UK.

Larson, R.W. and Richards, M.H. (1991) Boredom in the middle school years: blaming schools versus blaming students. *American Journal of Education* 99, 418–443.

Latham, N. and Mason, G.J. (2008) Maternal separation and the development of stereotypies: a review. *Applied Animal Behaviour Science* 110, 84–108.

Latham, N. and Mason, G.J. (2010) Frustration and perseveration in stereotypic captive animals: is a taste of enrichment worse than none at all? *Behaviour and Brain Research* 211, 96–104.

Lewis, N.J. (1999) Frustration of goal-directed behaviour in swine. *Applied Animal Behaviour Science* 64, 19–29.

Lilly, A.A., Mehlman, P.T. and Higley, J.D. (1999) Trait-like immunological and haematological measures in female rhesus across varied environmental conditions. *American Journal of Primatology* 48, 197–223.

Lopez, H.H. and Ettenberg, A. (2002) Exposure to female rats produces differences in c-fos inductions between sexually-naïve and experienced male rats. *Brain Research* 947, 57–66.

Lutz, C., Well, A. and Novak, M. (2003) Stereotypic and self-injurious behavior in rhesus macaques: a survey and retrospective analysis of environment and early experience. *American Journal of Primatology* 60, 1–15.

Maestripieri, D., Schino, G., Aureli, F. and Troisi, A. (1992) A modest proposal: displacement activities as an indicator of emotions in primates. *Applied Animal Behaviour Science* 44, 967–979.

Main, R.G., Dritz, S.S., Tokach, M.D., Goodband, R.D., Nelssen, J.L. and Loughlin, T.M. (2005) Effects of weaning age on postweaning belly-nosing behaviour and umbilical lesions in a multi-site production system. *Journal of Swine Health and Production* 13, 259–264.

Manteuffel, G., Langbein, J. and Puppe, B. (2009) From operant learning to cognitive enrichment in farm animal housing: bases and applicability. *Animal Welfare Science* 18, 87–95.

Mason, G. and Bateson, M. (2009) Motivation and the organization of behaviour. In: Jensen, P. (ed.) *The Ethology of Domesticated Animals*, 2nd edn. CAB International, Wallingford, UK, pp. 38–56.

Mason, G., Cooper, J. and Clarebrough, C. (2001) The welfare of fur-farmed mink. *Nature* 410, 35–36.

May, M.E. and Kennedy, C.H. (2009) Aggression as positive reinforcement in mice under various ratio- and time-based reinforcement schedules. *Journal of Experimental Analysis of Behavior* 91, 185–196.

McAdie, T.M., Keeling, L.J., Blokhuis, H.J. and Jones, R.B. (2005) Reduction in feather pecking and improvement of feather condition with the presentation of a string device to chickens. *Applied Animal Behaviour Science* 93, 67–80.

McAfee, L.M., Mills, D.S. and Cooper, J.J. (2002) The use of mirrors for the control of stereotypic weaving behaviour in the stabled horse. *Applied Animal Behaviour Science* 78, 159–173.

McFarland, D. (1982) *The Oxford Companion to Animal Behaviour,* 1st edn. Oxford University Press, Oxford, UK.

Mendl, M. (1990) Developmental experience and the potential for suffering: does "out of experience" mean "out of mind?" *Behavioral and Brain Science* 13, 28–29.

Mononen, J., Mohaibes, M., Savolainen, S. and Ahola, L. (2008) Water baths for farmed mink: intra-individual consistency and inter-individual variation in swimming behavior, and effects on stereotyped behavior. *Agricultural and Food Science* 17, 41–52.

Newberry, A.L. and Duncan, R.D. (2001) Roles of boredom and life goals in juvenile delinquency. *Journal of Applied Social Psychology* 31, 527–541.

Nordström, M. and Korpimäki, E. (2004) Effects of island isolation and feral mink removal on bird communities on small islands in the Baltic Sea. *Journal of Animal Ecology* 73, 424–433.

Oliviero, C., Heinonen, M., Valros, A., Häilli, O. and Peltoniemi, O.A.T. (2008) Effect of the environment on the physiology of the sow during late pregnancy, farrowing and early lactation. *Animal Reproduction Science* 105, 365–377.

Papini, M.R. (2003) Comparative psychology of surprising non-reward. *Brain, Behavior and Evolution* 62, 83–95.

Passineau, M.J., Green E.J. and Dietrich, W.D. (2001) Therapeutic effects of environmental enrichment on cognition functions and tissue integrity following severe traumatic brain injury in rats. *Experimental Neurology* 168, 373–384.

Piazza, P.V., Deroche, V., Deminiere, J.M., Maccari, S., Le Moal, M. and Simon, H. (1993) Corticosterone in the range of stress-induced levels possesses reinforcing properties: implications for sensation-seeking behaviors. *Proceedings of the National Academy of Sciences of the United States of America* 90, 11738–11742.

Princz, Z., Orova, Z., Nagy, I., Jordan, D., Stuhec, I., Luzi, F., Verga, M. and Szendrö, Zs. (2007) Application of gnawing sticks in rabbit housing. *World Rabbit Science* 15, 29–36.

Princz, Z., Zotte, A.D., Metzger, Sz., Radnai, I., Biró-Németh, E., Orova, Z. and Szendrö, Zs. (2009) Response of fattening rabbits reared under different housing conditions: live performance and health status. *Livestock Science* 121, 86–91.

Puppe, B., Ernst, K., Schön, P. and Manteuffel, G.. (2007) Cognitive enrichment affects behavioural reactivity in domestic pigs. *Applied Animal Behaviour Science* 105, 75–86.

Rodarte, L.F., Docing, A., Galindo, F., Romano, M.C. and Valdez, R.A. (2004) The effect of environmental manipulation on behaviour, salivary cortisol, and growth of piglets weaned at 14 days of age. *Journal of Applied Animal Welfare Science* 7, 171–179.

Rolls, E.T. (1999) *The Brain and Emotion*. Oxford University Press, New York.

Rolls, E.T. (2007) *Emotion Explained*. Oxford University Press, Oxford, UK.

Salmon, G.K., Leslie, G., Roe, F.J.C. and Lee, P.N. (1990) Influence of food intake and sexual segregation on longevity organ weights and the incidence of non-neoplastic and neoplastic diseases in rats. *Food and Chemical Toxicology* 28, 39–48.

Schapiro, S.J., Nehete, P.N., Perlman, J.E., and Sastry, K.J. (2000) A comparison of cell-mediated immune responses in rhesus macaques houses singly, in groups or in pairs. *Applied Animal Behaviour Science* 68, 67–84.

Shaw, D.C. and Gallagher, R.H. (1984) Group or singly housed rats? In: *Standards in Laboratory Animal Management, Proceedings of a LASA/UFAW Symposium*. The Universities Federation for Animal Welfare, Potters Bar, UK, pp. 65–70.

Shepherdson, D.J., Carlstead, K.C. and Wielebnowski, N. (2004) Cross-institutional assessment of stress responses in zoo animals using longitudinal monitoring of faecal corticoids and behaviour. *Animal Welfare* 13, 105–113.

Sherwin, C.M., Haug, E., Terkelsen, N. and Vadgama, M. (2004) Studies on the motivation for burrowing by laboratory mice. *Applied Animal Behaviour Science* 88, 343–358.

Shyne, A. (2006) Meta-analytic review of the effects of enrichment on stereotypic behavior in zoo mammals. *Zoo Biology* 25, 317–337.

Skalicky, M., Narath, E. and Viidik, A. (2001) Housing conditions influence the survival and body composition of ageing rats. *Experimental Gerontology* 36, 159–170.

Skovgaard, K., Jeppesen, L.L. and Hansen, C.P.B. (1997) The effect of swimming water and cage size on the behaviour of ranch mink (*Mustela vison*). *Scientifur* 21, 253–259.

Sorensen, D.B., Stub, C., Jegstrup, I.M., Ritskes-Hoitinga, M. and Hansen, A.K. (2005) Fluctuating asymmetry in relation to stress and social status in inbred male Lewis rats. *Scandinavian Journal of Laboratory Animal Science* 32, 117–123.

Stevenson, M.F. (1983) The captive environment: its effect on exploratory and related behavioural responses in wild animals. In: Archer, J. and Birke, L.I.A. (eds) *Exploration in Animals and Humans*. Van Nostrand Reinhold, London, pp. 198–208.

Stevenson-Hinde, J. and Roper, R. (1975) Individual differences in reinforcing effects of song. *Animal Behaviour* 23, 729–734.

Struelens, E., Tuyttens, F.A.M., Janssen, A., Leroy, T., Audoorn, L., Vranken, E., de Baere, K., Odberg, F., Berckmans, D., Zoons, J. and Sonck, B. (2005) Design of laying nests in furnished cages: influence of nesting material, nest box position and seclusion. *British Poultry Science* 46, 9–15.

Struelens, E., Van Nuffel, A., Tuyttens, F.A.M., Audoorn, L., Vraken, E., Zoons, J., Berckmans, D., Ödberg, F., Van Dongen, S. and Sonck, B. (2008) Influence of nest seclusion and nesting material on pre-laying behaviour of laying hens. *Applied Animal Behaviour Science* 112, 106–119.

Swaisgood, R. and Shepherdson, D. (2006) Environmental enrichment as a strategy for mitigating stereotypies in zoo animals: a literature review and meta-analysis. In: Mason, G. and Rushen, J. (eds) *Stereotypic Behaviour in Captive Animals: Fundamentals and Applications to Welfare*. CAB International, Wallingford, United Kingdom pp. 256–285.

Taira, K. and Rolls, E.T. (1996) Receiving grooming as a reinforcer for the monkey. *Physiology and Behaviour* 59, 1189–1192.

Tarou, L.R., Bloomsmith, M.A. and Maple, T.L. (2005) Survey of stereotypic behavior in prosimians. *American Journal of Primatology* 65, 181–196.

Thodberg, K., Jensen, H.K., Herskin, M.S. and Jorgensen, E. (1999) Influence of environmental stimuli on nest building and farrowing behaviour in domestic sows. *Applied Animal Behaviour Science* 63, 131–144.

Uvnäs-Moberg, K., Widström, A.M., Marchini, G. and Winberg, J. (1987) Release of GI hormones in mother and infant by sensory stimulation. *Acta Paediatrica Scandinavica* 76, 851–860.

Van Hemel, P.E. (1972) Aggression as a reinforcer: operant behaviour in the mouse-killing rat. *Journal of the Experimental Analysis of Behavior* 17, 237–245.

Veissier, I., de Passillé, A.M.B., Després, G., Rushen J., Charpentier, I., Ramirez de la Fe, A.R. and Pradel, P. (2002) Does nutritive and non-nutritive sucking reduce other oral behaviours and stimulate rest in calves? *Journal of Animal Science* 80, 2574–2587.

Vinke, C.M., van Leeuwen, J. and Spruijt, B.M. (2005) Juvenile farmed mink (*Mustela vison*) with additional access to swimming water play more frequently than animals housed with a cylinder and platform, but without swimming water. *Animal Welfare* 14, 53–60.

Vinke, C.M., Hansen, S.W., Mononen, J., Korhonen, H., Cooper, J.J., Mohaibes, M., Bakken, M., and Spruijt,

B.M. (2008) To swim or not to swim: an interpretation of farmed mink's motivation for a water bath. *Applied Animal Behaviour Science* 111, 1–27.

Visser, E.K., Ellis, A.D. and Van Reenen, C.G. (2008) The effect of two different housing conditions on the welfare of young horses stabled for the first time. *Applied Animal Behaviour Science* 114, 521–533.

Vorontsov, Y.N. (1985) Some observations on captive minks released for the purpose of introducing them into the local biocoenosis. In: Safanov, V.G. (ed.) *Biology and Pathology of Farm-bred Fur-bearing Animals* – Abstracts of Papers Presented at the Second All Union Scientific Conference, Kirov, 26–29 July 1977. Amerind Publishing, New Delhi, pp. 74–75.

Wahidin, A. (2006) *Time and the Prison Experience.* Available at: www.socresonline.org.uk/11/1/wahidin.html (accessed 11 October 2009).

Waiblinger, E. and Konig, B. (2004) Refinement of gerbil housing and husbandry in the laboratory. *Alternatives to Laboratory Animals* 32, 163–169.

Walker, A.W. and Hughes, B.O. (1998) Egg shell colour is affected by laying cage design. *British Poultry Science* 39, 696–699.

Walker, K. (1990) Desire and ability: hormones and the regulation of female sexual behaviour. *Neuroscience and Biobehavioural Reviews* 14, 233–241.

Wallen K. (1990) Desire and ability: hormones and the regulation of female sexual behavior. *Neuroscience Biobehavioural Reviews* 14, 233–241.

Warburton, H. and Mason, G. (2003) Is out of sight, out of mind? The effects of resource cues on motivation in the mink (*Mustela vison*). *Animal Behaviour* 65, 755–762.

Watson, S.L., Shively, C.A., Kaplan, J.R. and Line, S.W. (1998) Effects of chronic social separation on cardiovascular disease in female cynomogus monkeys. *Atherosclerosis* 137, 259–266.

Weeks, C.A. and Nicol, C.J. (2006) Behavioural needs, priorities and preferences of laying hens. *World's Poultry Science Journal* 62, 296–307.

Widowski, T.M. and Curtis, S.E. (1990) The influence of straw, cloth tassel, or both on the prepartum behavior of sows. *Applied Animal Behaviour Science* 27, 53–71.

Widowski, T.M., Cottrell, T., Dewey, C.E. and Friendship, R.M. (2003) Observations of piglet-directed behaviour patterns and skin lesions in eleven commercial swine herds. *Journal of Swine Health and Production* 11, 181–185.

Widowski, T.M., Yuan, Y. and Gardner, J.M. (2005) Effect of accommodating sucking and nosing on the behavior of artificially reared piglets. *Laboratory Animals* 39, 240–250.

Wiedenmayer, C. (1996) Effect of cage size on the ontogeny of stereotyped behaviour in gerbils. *Applied Animal Behaviour Science* 47, 225–233.

Wiedenmayer, C. (1997a) Causation of the ontogenetic development of stereotypic digging in gerbils. *Animal Behaviour* 53, 461–470.

Wiedenmayer, C. (1997b) Stereotypies resulting from a deviation in the ontogenetic development of gerbils. *Behavioural Processes* 39, 215–221.

Wischner, D., Kemper, N. and Krieter, J. (2009) Nest-building behaviour in sows and consequences for pig husbandry. *Livestock Science* 124, 1–8.

Wood-Gush, D.G., Stolba, A. and Miller, C. (1983) Exploration in farm animals and animal husbandry. In: Archer, J. and Birke, L.I.A. (eds.) *Exploration in Animals and Humans.* Van Nostrand Reinhold, London, pp. 198–208.

Yue, S. and Duncan, I.J.H. (2003) Frustrated nesting behaviour: relation to extra-cuticular shell calcium and bone strength in White Leghorn hens. *British Poultry Science* 39, 696–699.

Zhang, G., Swaisgood, R.R. and Zhang, H. (2004) Evaluation of behavioral factors influencing reproductive success and failure in captive giant pandas. *Zoo Biology* 23, 15–31.

Zimmerman, P.H., Koene, P. and Van Hooff, J.A.R.A.M. (2000) Thwarting of behaviour in different contexts and the gakel-call in the laying hen. *Applied Animal Behaviour Science* 69, 255–264.

Zuckerman, M. (1994) *Behavioral Expressions and Biosocial Bases of Sensation Seeking.* Cambridge University Press, Cambridge, UK.

8 Health and Disease

MICHAEL S. COCKRAM AND BARRY O. HUGHES

Abstract

Health and disease are important components in the broader concept of welfare. Understanding the relationship between health and welfare requires inferences about subjective feelings such as pain and distress. Similar types of disease occur in animals and humans – our assessment of how animals suffer depends on behavioural, physiological and clinical observations and on our own experiences. In this chapter, the implications of disease for welfare, the underlying pathophysiological principles, the types of disease, their assessment and control, and care and treatment options are described and discussed. Suffering can occur both with acute conditions such as foot-and-mouth disease or canine distemper and with chronic disease – disorders of the joints such as arthritis, foot conditions such as laminitis and foot rot, and digestive disturbances, which may be especially painful. Skin conditions such as sheep scab or red mite infestation can cause intense discomfort. In modern agriculture, genetic selection for maximum productivity, such as milk output, broiler growth or egg number, can adversely affect health in terms of metabolic diseases, lameness or bone fractures. In both farm and companion animals, selection for conformation can affect normal functioning. The development of intensive systems has emphasized the importance of first-class stockmanship. This involves not only careful attention to the animals' daily needs but also a detailed programme of disease control with vaccination, treatment, environmental maintenance, hygiene, record keeping and biosecurity.

8.1 Introduction

8.1.1 Animal health, disease and welfare

There are many definitions of health and disease (Gunnarsson, 2006), but the following are helpful when considering the relationships between health and welfare. A disease is a physical or mental condition where a normal function of an animal is disturbed and harmed. Illness is the subjective sensation of experiencing a diseased state. Sickness is the state of being ill, whereas health is the absence of illness or injury. Good health also includes positive attributes, such as fitness, soundness and/or vigour.

Suffering as a result of disease or injury is an important aspect of animal welfare, but the importance of animal health in relation to animal welfare is sometimes underestimated. The Brambell Committee Report (1965, p. 11) stated:

A principal cause of suffering in animals, as it is in men, is disease. Many veterinary witnesses have drawn our attention to this and to the necessity of taking it into account fully in assessing the welfare of animals … Accordingly, we lay stress on the incidence of disease and on the guarantees that a sick animal will be quickly recognised and appropriately treated or slaughtered.

Health indicators such as disease, lameness, injuries and measures of immune function form part of many outcome-based welfare assessments. Freedom from pain, injury and disease – by prevention or rapid diagnosis and treatment – forms one of the UK Farm Animal Welfare Council's Five Freedoms (FAWC, 2009).

Although animal health has a vital input to animal welfare, it is not the sole determinant. There is more to good welfare than good health (Bayvel, 2004; Ladewig, 2008). If animals are healthy it does not mean that other aspects of their welfare are always satisfactory, and all animals with ill health are not necessarily suffering. However, animals that are cared for appropriately and in accordance with acceptable welfare standards are more likely to be healthy and, conversely, animals kept in poor welfare conditions are often at greater risk of disease. Improving health can improve growth, reproduction and productivity and this can, to some extent, indicate good biological functioning. However, excessively high productivity can adversely affect health and welfare, for example, high-yielding dairy cattle or rapidly growing broiler chickens.

The importance of animal health was recognised by Dawkins (2008) who considered that focusing exclusively on the current emotional state of an animal may not completely encompass their welfare because what animals choose or will work for may not be good for their long-term health. She considered (p. 942) that:

> Any assessment of animal welfare must similarly take into account what improves physical health, both what reduces disease, deformity and injury as well as what promotes positive health, good growth and longevity … We need to know both what the animals themselves want and what is good for their health.

This follows the concept of Webster (2001, pp. 229, 232) who considered that:

> The welfare of a farm animal depends on its ability to sustain fitness and avoid suffering … The phrase 'sustain fitness' implies physical welfare; e.g. freedom from disease, injury and incapacity, and this acquires particular importance when these problems can be directly attributed to the conditions in which the animals are reared.

The magnitude of a welfare problem has been defined by the incidence, severity and duration of the problem (Willeberg, 1991; Webster, 1998). However, it is difficult to judge the relative impact of different forms of disease or injury on animal welfare and it is difficult to get reliable and valid information on the occurrence of disease and injury (Rushen *et al.*, 2008). The economically important diseases of livestock are those that affect groups of animals and result in impaired productivity or mortality, or are zoonotic. However, a different emphasis is required when considering the welfare aspects of disease. As discussed by Kirkwood (2007), welfare concerns related to animal health are not focused on the disease itself, but on how the animal experiences the consequences of disease. Inferences about how sick animals feel are based on behavioural, clinical or other observations of the animal, knowledge of its biology and of our own experiences of pleasant and unpleasant feelings. This anthropomorphic approach is useful as many diseases are common to humans and animals, and there are shared clinical and pathological responses to many different types of disease. An alternative approach was proposed by Rushen *et al.* (2008) who compared different diseases in cattle by the severity of common clinical signs. For example, reduced feed intake and activity and increased resting occur in many disease states, and the relative severity of different diseases might be judged by comparing the relative magnitude of these changes. However, as Gregory (2004, p. 183) proposed:

> To understand the suffering in disease one has to appreciate its pathophysiology and the feelings experienced by humans in comparable situations.

The literature on animal health and disease is enormous, but relatively few studies have attempted to evaluate the implications of disease for animal welfare.

Clinical and pathological investigations are integral to cases of animal cruelty arising from abuse and neglect. They can quantify the state of the animal, provide information on the likely cause of the problem, and identify whether disease was present and whether the animal had received appropriate treatment (Munro, 2008). For example, poor body condition does not necessarily result from starvation due to neglect; it might also have been influenced by conditions such as neoplasia, chronic diseases, dental problems, parasites and lactation (Green and Tong, 2004).

8.1.2 Effects of disease on welfare

These can be summarized as follows:

- Diseases that are likely to cause pain and suffering are of particular concern. For example, viral diseases such as foot-and-mouth disease cause vesicular lesions in the oral mucosa that makes eating and drinking painful, and similar lesions on the feet cause lameness. The amount of tissue damage and the sensitivity of the tissue to pain can in some circumstances give an approximation of the associated pain (Rutherford, 2002), but damage is not always required to cause pain, for example, visceral distension.
- Diseased animals are likely to feel ill (for example, inappetence, thirst, fever and nausea). Figure 8.1 shows a feverish calf showing typical behavioural signs. Animals might also experience other negative emotional states, such as fear (because of disorientation or reduced ability to respond to perceived danger) and distress (for example, hypoxia from impaired oxygen supply).

Fig. 8.1. The calf on the right shows behavioural signs associated with fever – lying prone, head and neck down, ears floppy and limbs drawn in. The calf on the left is healthy – sitting upright, head and neck raised, with an alert appearance and ears extended. (Photographic image kindly provided by Dr Jose M. Peralta.)

- Some diseases can cause discomfort (for example irritation from skin diseases such as mite infestation), and this can also reduce rest and sleep.
- A disease might also weaken an animal, reducing its ability to compete for limited resources, such as food, which in turn leads to malnutrition, while prolonged expenditure of energy in the immune response to combat disease (Colditz, 2002) can result in fatigue.
- Weakness or impaired perception as a result of disease can reduce the ability of an animal to avoid attack from conspecifics or predators.
- Fever, or prolonged immobility, can increase heat loss and the animal might experience cold sensations, while reduced ability to move away from environmental threats such as draughts, precipitation or solar radiation, especially if combined with inappetence, dehydration or nutrient malabsorption, could affect thermal comfort (Balsbaugh *et al.*, 1986).
- Reduced mobility from lameness or injury can prevent an animal from moving easily to gain access to resources such as food, water or a comfortable lying area. For example, lame

broiler birds that are unable to reach food and water can die from starvation and dehydration (Butterworth *et al.*, 2002).
- Lameness, together with other factors, such as slippery floors or restriction of movement, can cause insecurity in response to perceived threats such as humans or aggressive conspecifics (Blowey, 1998).
- Prolonged lying as a result of lameness or disease can result in pressure injuries and possibly muscular and skin damage.
- Some diseases can cause emaciation (such as Johne's disease or bovine viral diarrhoea in cattle), reduced function or loss of function (blindness, deafness, paralysis and congestive heart failure), and this might decrease vigour and ability to perform normal biological functions, and reduce the opportunity to experience pleasurable activities (McMillan, 2003).

Using examples of diseases of newborn animals, Mellor and Stafford (2004) discuss how potentially aversive experiences such as breathlessness, hypothermia, hunger, sickness or pain might be

M.S. Cockram and B.O. Hughes

mitigated by any associated hypoxaemia, hypothermia, drowsiness, sleep and/or unconsciousness. These reduce the ability of the animal to be sufficiently aware or conscious to experience these sensory inputs and interpret them as noxious.

8.2 Inflammation, Immunity and Pathological Responses to Disease

Inflammatory diseases are a major cause of pain. The response to infection and tissue injury (acute phase response) consists of a local inflammatory and a systemic (whole body) response. The inflammatory response involves a series of vascular and interstitial tissue changes that increase blood flow to the affected area and result in redness, heat, pain, swelling and loss of function (Smith *et al.*, 1972). The pain arises from the stimulation of nerve endings by cytokines and other mediators of inflammation (including histamine, prostaglandins and growth factors) that are released from damaged cells, peripheral sensory neuron terminals and inflammatory cells (Viñuela-Fernández *et al.*, 2007). Some pain also comes from the swelling associated with the increased pressure of inflammatory exudates at the site of injury. Hyperalgesia (increased sensitivity to noxious stimulation) is also present in several inflammatory conditions, such as chronic mastitis and foot rot in sheep (Fitzpatrick *et al.*, 2006). The purpose of the inflammatory response is to deal with the injury and initiate tissue repair. Increased local vascular permeability causes release of plasma proteins and leucocytes. Cytokines are released from activated leucocytes at the site of tissue damage and also act on the brain (Dantzer *et al.*, 2008) to cause non-specific behavioural signs of sickness associated with fever: depression, lethargy, inappetence and thirst (Hart, 1988; Gregory, 1998). Other cytokines promote the production of leucocytes and acute-phase proteins. When an antigen gains entry to the body there is stimulation of lymphocytes that produce antibodies. This initial exposure results in a rapid response to subsequent exposure to that specific antigen. This is the basis of vaccines that sensitize the animal to an infectious organism so that during subsequent exposure to the pathogen, there is protection against disease caused by that organism. Significant achievements have been made in improving animal health and welfare through vaccination. Vaccines, unlike therapeutic treatments, are a means of avoiding animal suffering because they prevent disease (Pastoret, 1999). However, some vaccines can cause systemic effects, the animals require handling and restraint, and there might be local inflammation and bacterial contamination of the injection site.

The stress associated with physical and psychological stimuli has been associated with decreased immunocompetence and increased susceptibility of animals to infectious and other diseases (Kelly, 1980; Griffin, 1989; Peterson *et al.*, 1991; Gross and Siegel, 1997). If animals with infections that only produce subclinical disease are exposed to stressors, clinical disease can develop (Andrews, 1992). This sensitivity of the immune system to external stimuli has resulted in the use of immune function tests as a bioassay to provide information considered relevant to animal welfare. However, as discussed by Vedhara *et al.* (1999), it is necessary to consider the clinical significance of any altered immunity, and whether the magnitude of any stress-associated immune changes is sufficient to alter immunocompetence.

Chronic inflammation can follow persistent, unresolved acute inflammation or develop as a chronic process. It is characterized by simultaneous destruction and healing of the tissue. Specific types of inflammation occur: granulomatous (in response to some bacterial infections), fibrinous (on mucous and serous membranes such as the respiratory tract), purulent (pus that can form an abscess), serous (skin blisters) and ulcerative (where necrosis leads to a loss of tissue from the surface, exposing lower layers to form an ulcer). If microorganisms are not contained by the actions of acute inflammation they can spread to other parts of the body and cause inflammation (such as fasciitis and cellulites) in the surrounding tissue, and may spread via the circulatory or lymphatic systems to produce bacteraemia or viraemia. Marked and prolonged viraemia can cause multi-system disease. Bacteraemia can lead to septic shock and death (Smith *et al.*, 1972; Hardie and Rawlings, 1983).

Steroidal anti-inflammatory drugs such as corticosteroids can provide symptomatic relief in the treatment of both acute and chronic inflammatory conditions. Non-steroidal anti-inflammatory drugs reduce inflammation, pain and fever, and are used to treat conditions such as pneumonia, mastitis and osteoarthritis (Mathews, 2000a; Nolan, 2000). These are aspirin-like drugs and inhibit pro-inflammatory mediators such as prostaglandins (Livingston, 2000).

8.3 Infectious Diseases

Infectious diseases can be species-specific, shared with other species, or zoonotic (transmissible between animals and humans).

8.3.1 Simple pathogen–host diseases

These occur when a susceptible animal is exposed to a sufficient dose of a specific pathogen that is itself sufficient to cause disease. The disease can be controlled by removing infected animals (isolation or culling) or removal of the pathogen (by hygiene practices or by vaccination).

8.3.2 Viral diseases

Viruses produce disease by replicating inside cells to cause degeneration and cell death (this can cause inflammation and sickness, as described in Section 8.2). For example, infectious bovine rhinotracheitis virus can cause epithelial necrosis and ulceration. Viral diseases are frequently very contagious and can cause high mortality (for example canine distemper).

8.3.3 Bacterial diseases

Pathogenic bacteria cause disease by producing necrosis and pus (mastitis in cattle, strangles in horses and arthritis in pigs), by secreting toxins (*Escherichia coli* can cause acute enterotoxic colibacillosis in pigs, calves and lambs with symptoms of watery diarrhoea, dehydration and acidosis), or by replicating within macrophages and host cells (salmonellosis) (Cheville, 2006). These pathophysiological changes can cause inflammation and sickness as described in Section 8.2. Bacterial infections can be secondary to or combine with viruses to cause disease. Antibiotics are important for the treatment of clinical bacterial infections (therapy) and can be used for preventing clinical infections (prophylaxis) (Refsdal, 2000; McEwen, 2006). However, there are increasing restrictions on antibiotic use because of extensive antibiotic resistance, and there can be a tendency to use antibiotics to mask disease rather than to correct underlying defects.

8.3.4 Parasites

Some protozoa can kill host cells, for example cryptosporidia damage intestinal epithelial cells in poultry and young ruminants and cause diarrhoea, while babesia, transmitted by ticks, can destroy erythrocytes to cause anaemia in cattle. Adult parasitic worms can cause disease by mechanical obstruction of ducts (ascarides in pig intestines), by sucking blood and causing anaemia (hookworms in dog intestines) and by causing diarrhoea (parasitic gastroenteritis in sheep). Enteric parasites can debilitate animals by causing inappetence, protein loss (from leakage of plasma protein and damage to the lining of the gastrointestinal tract) and weight loss (Holmes, 1987). Larvae can pass through the body and cause damage (granulomatous lesions) to various organs. Culicoides (biting midges) can cause serious irritation and transmit pathogenic viruses such as bluetongue. Lice can cause skin irritation, while mites can cause intense irritation, pain, distress and blood loss in poultry. In sheep, the mites that cause sheep scab can cause scratching, inappetence and sometimes mortality from secondary infections, emaciation and dehydration (Milne *et al.*, 2008). Sea lice erode the skin of farmed fish causing tissue damage, and may also act as a vector of other diseases (Ashley, 2007). Parasites can cause varying degrees of inflammation and discomfort, and sometimes disease that is associated with sickness. Treatment and control of parasites is based on the life cycle of the particular parasite. Adult worms lay eggs that pass out of the host animal in the faeces. Some parasites have an intermediate host (snails for liver fluke and earthworms for lungworms in pigs). Because an animal can ingest the parasite while it grazes, control is based on separation of the animal from its faeces, or on reducing pasture contamination by avoiding overstocking and by grazing rotation. Medicinal products can reduce the numbers of eggs passed in the faeces or kill external parasites. Vaccines can stimulate immunity to some parasites (for example, lungworm in cattle).

8.3.5 Complex, pathogen–animal–environment diseases

These diseases can be produced by simultaneous infection with one or more pathogens and by interaction between the infectious agents and predisposing, enabling or reinforcing factors, such as the genetics or age of the animal; and their nutrition, environment and management system. The diseases can occur when the potential pathogen is present in the environment, or on or in the animal, but does not cause disease until the equilibrium between pathogen, animal and environment is disturbed (Webster, 1992; Thrusfield, 2005). Control of these types of diseases requires a whole herd/flock approach and consideration of the husbandry, management, nutrition and environment. Examples of poor husbandry that can predispose to disease include: overcrowding, mixing of different ages of animals, obtaining animals from several sources, poor air hygiene, poor drainage and bedding, unhygienic food and watering equipment, inappropriate nutrition, and inadequate cleaning and disinfection (Sainsbury, 1998). The same principles can apply to wild animals: UFAW's Garden Bird Initiative emphasizes the importance of clean bird feeders in reducing the incidence of *Trichomonas* in finches (UFAW, 2008).

8.4 Disease Control

8.4.1 Health risks

The underlining philosophy is that disease 'prevention is better than cure'. However, maintenance of the health status of animals is a constant challenge. Understanding the underlying causes and the mechanisms by which disease spreads is vital to controlling disease. Contagious diseases can be transmitted by direct physical contact with other animals, while infectious diseases can be transmitted not only by infected animals, but also via air, water, food and many other vectors, including wild mammals and birds, invertebrates, vehicles, humans and environmental contamination. When large numbers of animals are housed on one site, the risk of infectious disease (such as enzootic pneumonia and enteric disorders in pigs and calves) is high.

In the European Union (EU), welfare codes and legislation (Defra, 2003) require that management risk factors that have the potential to cause health problems in livestock are controlled. These include: the genotype and phenotype of the animal; the materials used in accommodation so that they are not harmful and are capable of being thoroughly cleaned and disinfected; the presence of no sharp edges or protrusions likely to cause injury; air hygiene and pollutants kept within limits which are not harmful to the animals; protection from adverse weather conditions and predators; inspection of all essential automated or mechanical equipment; and a wholesome diet which is appropriate to age and species which is fed in sufficient quantity.

Each system of husbandry has its own characteristic disease problems, but the health of the animals within each system is also dependent on the stockmanship and the disease prevention and control measures that are in place. For example, in laying hens there is an increased risk of mortality from bacterial diseases, parasites and cannibalism in litter-based and free-range systems compared with cages, but a reduction in viral conditions (Fossum *et al.*, 2009) (Table 8.1). However, in Switzerland, the change from conventional battery cage housing systems to alternative systems was not followed by increased mortality resulting from coccidia and other parasites. There was a reduction in mortality due to viral diseases, but mortality due to bacterial infection increased. Vaccination against viral diseases and coccidiosis, together with de-worming strategies, paddock rotation, biosecurity and other disease control measures accompanied the changes in housing system (Kaufmann-Bart and Hoop, 2009).

Table 8.1. Causes of mortality in 914 laying hens from different housing systems submitted for post-mortem examination to the National Veterinary Institute in Uppsala, Sweden between 2001 and 2004 (adapted from Fossum *et al.*, 2009).

Housing system[a]	% of hens in Sweden housed in each system	No. of flocks examined with increased mortality	% of flocks within each housing system diagnosed with:			
			Bacterial diseases	Viral diseases	Parasitic diseases	Cannibalism
Cages	56	20	65	30	10	5
Litter-based	39	129	73	12	18	19
Free range	5	23	74	4	22	26

[a]Cages include conventional and enriched. Litter-based includes single-tiered floor and multi-tiered aviary systems. Free range includes hens housed indoors on litter with access to outdoor pens and includes organic systems.

8.4.2 Health plans

A written health and welfare plan is a management tool matched to the individual needs of each livestock unit that is drawn up in consultation with the unit's veterinarian, to ensure that preventive and treatment regimes are planned, that health performance is recorded and reviewed, and that appropriate action plans are developed (Main *et al.*, 2003). Health plans form part of many quality assurance schemes and their presence can be used in welfare assessments as evidence of best practice. Such plans set out health and husbandry activities that cover the whole year's cycle of production, and include strategies to prevent, treat or limit existing disease problems. They provide for regular veterinary visits to advise on animal health, and include: biosecurity arrangements; procedures for purchased stock; vaccination policy and timing; isolation procedures; external and internal parasite control; the timing and dose of any necessary medical treatments; and any specific disease programmes. Important variables are recorded, including number of animals, age, breed, etc. and performance values such as production, water consumption, number of animals found dead and number of culls. A veterinary intervention point is set to decide when normal values have reached an unacceptable level. When the veterinarian visits, clinical notes, postmortem examinations or laboratory work that has been carried out are added to the health plan. The vaccination programme and use of medicines are also recorded. Useful health information can also be recorded using feedback from pathology found at the slaughterhouse during meat inspection (Green *et al.*, 1997).

8.4.3 Biosecurity

Biosecurity on animal units is required to reduce the risk of disease occurring or spreading to other animals. Best practice involves relatively prescriptive guidelines designed to prevent disease-causing agents from entering, spreading within, or leaving a property and spreading to other units. Animals newly brought on to the unit present the greatest risk of infectious disease spread. Information on their health status should be obtained and the animals isolated for a suitable period. Only essential visitors should be allowed on to certain units such as pig, poultry or laboratory animal sites. They should follow disinfection procedures, wear unit

clothing and footwear, avoid visits to other units for an agreed period before and afterwards, and record their visit(s). Loading facilities and feed bins should be sited at the unit perimeter. Vehicles that visit other units should be kept off the unit wherever possible. Domestic pets and wild animals (birds and rodents) should be discouraged. Free-range poultry flocks might be required to be housed during an outbreak of avian influenza to prevent disease transmission via wild birds. Animal units should be sited as far as is practicable from other units, as this will reduce the risk of spread of airborne infectious diseases. The animals should be kept in groups in all-in/all-out systems with a cleansing and disinfection programme that is documented, implemented and checked for effectiveness (Defra, 2008). Cleaning and disinfection of animal housing is facilitated by flat, featureless walls and floors, and an absence of internal structure. Enhancements to animal housing to facilitate behaviour can be constrained by concerns about disease transmission associated with social housing, increased contact with excreta, and difficulty cleaning and disinfecting enclosures containing soil and natural and porous materials. Although care should be taken, especially when housing young animals, some disease concerns have not proved to be major obstacles to the adoption of alternative housing (Newberry, 1995).

8.4.4 Culling to control infectious disease

In many countries, considerable resources have been allocated to control or eradicate important viral diseases successfully. However, globalization and increased international trade of animals and animal products have increased the risk of disease spread (Thiermann, 2004). Examples are provided by Zepeda *et al.* (2001) and Fèvre *et al.* (2006) of the consequences associated with animal movement and the spread of infectious disease. The World Organisation for Animal Health (OIE) web site (www.oie.int) provides information on current disease distribution throughout the world. The means of dealing with a disease outbreak such as foot-and-mouth disease or avian influenza include early detection of disease, rapid killing of all known infected animals, tracing of all high-risk contacts, application of herd quarantine, testing of populations at risk and, in some instances, the application of pre-emptive slaughter or strategic vaccination (Whiting, 2003). Culling large numbers of animals can be justified on welfare grounds to eliminate suffering in diseased animals, to prevent suffering in susceptible animals owing to the spread of disease, and to prevent welfare problems due to overcrowding or other deteriorating animal husbandry conditions because of movement restrictions (Whiting, 2003; Raj, 2008). The killing of large numbers of animals in a short period of time is difficult to achieve humanely owing to limited availability of skilled slaughtermen, handling problems, time constraints and, in some cases, a lack of a suitable humane method of killing (Crispin *et al.*, 2002; Whiting, 2003). A guide to the practice of humane culling is provided by the OIE (2009).

8.5 Production-related Diseases

Some diseases are considered to have particular welfare significance because they are likely to have occurred as a direct consequence of the management system used, but are tolerated because they do not reduce the economic profitability associated with the management system. Genetic selection has increased production, but in some cases this has been accompanied by an increased risk of health problems (Rauw *et al.*, 1998). Many metabolic diseases are associated with increased metabolism, rapid growth rate or high production which results in the failure of a body system because of the increased workload on that organ or system (Julian, 2005). Selection of dairy cows for increased milk yield leads in general to a higher risk of mastitis (a bacterial infection that can cause pain (Milne *et al.*, 2003), sometimes fever and sometimes death), metabolic diseases and lameness. Metabolic diseases in dairy cattle, such as hypocalcaemia, hypomagnesaemia and ketosis, are associated with imbalances in the input and output of metabolites required for milk production. In broilers, difficulty in providing sufficient oxygen to enlarged muscles can cause hypertrophy of the right ventricle of the heart and ascites (Julian, 2000). Osteoporosis is a progressive decrease in mineralized structural bone that leads to bone fragility and susceptibility to spontaneous bone fractures (Gregory and Wilkins, 1989; Whitehead, 2004).When it occurs later in the laying cycle it has been called cage-layer fatigue and can be a serious animal welfare problem resulting in acute and chronic pain and debility from bone fractures that is sufficient to cause mortality (Riddell *et al.*, 1968; Webster, 2004). Fractured sternums occur especially in aviaries when hens

collide with perches, while the wing and leg bones can break if handling is rough during depopulation. Genetic selection for laying hens that remain in reproductive condition over a prolonged period increases susceptibility to osteoporosis. During this time, medullary bone, which acts as a labile source of calcium for eggshells is produced in preference to structural bone. However, as both medullary and structural bone are resorbed over time, there is a progressive loss of structural bone throughout the skeleton (Whitehead, 2004). The strength of bone is dependent on its load-bearing activity, and birds kept in housing systems that encourage physical activity have stronger bones.

8.6 Pain and Welfare

8.6.1 Lameness

Lameness (impaired movement or deviation from normal gait) is a serious welfare problem in all species and pain is the most common cause. However, it is possible to have mechanical lameness that is not associated with pain. Foot and leg lesions are common reasons for culling animals. In dairy cattle, sheep, broiler chickens and horses there is a high incidence of lameness that can be associated with discomfort and pain over a long duration (Ley *et al.*, 1989; Whay *et al.*, 1998; Bradshaw *et al.*, 2002; Egenvall *et al.*, 2008; Knowles *et al.*, 2008; Laven *et al.*, 2008). Lameness can increase sensitivity to pain (hyperalgesia), reduce food intake and lead to a loss of body condition. Non-steroidal anti-inflammatory drugs can reduce lameness in cattle (Flower *et al.*, 2008), broilers (McGeown *et al.*, 1999), dogs and cats (Peterson and Keefe, 2004; Clarke and Bennett, 2006; Mansa *et al.*, 2007), horses (Owens *et al.*, 1995; Hu *et al.*, 2005) and reduce hyperalgesia in lame cows (Whay *et al.*, 2005) and sheep (Welsh and Nolan, 1995). Lame broilers have been shown to be capable of selecting a non-steroidal anti-inflammatory drug in their feed and the amount of non-steroidal anti-inflammatory drug consumed increases with the severity of lameness (Danbury *et al.*, 2000).

Most lameness in cattle originates from foot lesions (sole ulcer, white line disease, digital dermatitis and interdigital necrobacillosis) (Whay *et al.*, 1998; Dyer *et al.*, 2007). There are many environmental, genetic and nutritional predisposing factors for lameness in dairy cattle. Sheep are susceptible to bacterial lameness that varies from mild inflammation of the interdigital space to severe underrunning and separation of the sole and hoof wall, exposure of underlying sensitive tissues and abscess development (Winter, 2008). These lesions can be associated with raised plasma cortisol (Ley *et al.*, 1994) and catecholamine concentrations (Ley *et al.*, 1992), which suggests that the sheep are under stress. Pigs are also susceptible to hoof problems and, in housed pigs, rough concrete floors and slatted or slotted floors with sharp edges can damage feet and legs, resulting in pain and possibly secondary bacterial infection (Sainsbury, 1998). Secondary infection can spread up the leg from foot lesions to cause tenosynovitis and cellulitis. In horses, there are many causes of lameness associated with pain, including foot wounds, laminitis, navicular bone lesions, arthritis, fractures, and tendon and ligament strains (Blunden *et al.*, 2005; Dyson *et al.*, 2005).

Selection for rapid growth, greater body weight and increased breast muscle has affected broiler shape, walking ability and increased mechanical stresses on legs and hip joints. Reduced time spent exercising and increased time spent resting by lame broilers increases susceptibility to leg weakness and contact dermatitis (hock burn and breast blisters from poor litter caused by chemical burning from ammonia). Factors associated with difficulty in walking include age, genotype, type of feed, short dark periods, high stocking density and not using antibiotics (Knowles *et al.*, 2008). Lameness in poultry is often caused by infection in the bone and joints (Butterworth, 1999), so effective prevention and control of viral and bacterial disease is essential on the farm, in the parent flock and hatchery. Common infectious disorders in poultry are osteomyelitis, chondritis and suppurative arthritis (Thorp, 1994). Tibial dyschondroplasia is a growth plate disorder that causes lameness, reduces dust bathing and increases the duration of tonic immobility (Vestergaard and Sanotra, 1999). Angular deformities of the long bones are frequently accompanied by slippage of the gastrocnemius tendon. Cartilage abnormalities can also act as foci for bacterial infections, resulting in osteomyelitis that can lead to necrosis and degeneration of cartilage and adjacent bone tissue in the proximal part of the tibia or femur (femoral head necrosis or proximal femoral degeneration) (Waldenstedt, 2006).

Arthritis (inflammation of a joint) is a common cause of lameness that is associated with pain upon movement of the joint, swelling, excessive fluid and crepitus (Renberg, 2005). Repeated trauma can cause serous inflammation and an enlarged joint capsule. Infection from local wounds, or more often from septicaemic or pyaemic infections, can cause fibrinous or purulent inflammation – joint ill in calves, lambs and piglets. In older pigs, polyarthritis can become chronic and is often associated with pus (Smith, 1988). Chronic arthritis, such as osteoarthritis, and degenerative joint disease, such as osteochondrosis, can follow repeated trauma, infection or excessive wear from structural defects, such as hip dysplasia in dogs (Scott, 1999). In adult breeding turkeys, the pain associated with degenerative hip lesions was demonstrated by injections of anti-inflammatory steroid that increased behavioural activity (Duncan *et al.*, 1991). In chronic arthritis, irregular fibrous tissue and bone can develop in the joint between articular surfaces; this causes pain and the smooth articular surface can be destroyed (Smith *et al.*, 1972).

8.6.2 Neoplastic diseases

Neoplasia can be a problem in companion animals, especially as they grow old. As tumours grow they can exert pressure on surrounding tissues and cause pain, as in bone marrow tumours (Smith *et al.*, 1972; Cheville, 2006). Tumours on the skin or mucous membranes can ulcerate. Pain can also be caused by direct tumour involvement of pain-sensitive structures such as soft tissue, bone, nerves and viscera, or via bone metastases (Lester and Gaynor, 2000). Malignant tumours grow at the site of origin and also spread via the circulation and lymphatic system to other sites such as the lungs, liver, spleen and kidneys. They can cause ill health (emaciation and anaemia) and death.

8.6.3 Parturition and neonatal care

Dystocia (difficult birth) caused by fetal oversize or malposition and obstetric problems such as vaginal or uterine prolapse can be associated with considerable pain, discomfort and health risks (McGuirk *et al.*, 2007). Veterinary procedures, the use of analgesia and antibiotics, and hygienic practices can reduce the adverse welfare implications associated with obstetric problems (Scott, 2005). However, they cannot eliminate all of the associated suffering and if the incidence of dystocia is high, breeding policy should be reviewed – for example, in the

M.S. Cockram and B.O. Hughes

double-muscled Belgian Blue breed, calves often have to be delivered by Caesarean section, and there is now a case for selecting for smaller calves (Kolkman *et al.*, 2010a,b). Dystocia also causes problems in newborn animals from direct physical trauma, delayed passive antibody transfer and subsequent bacterial infections. Newborn animals are susceptible to mortality from hypothermia, infections, injuries and predation, and require additional care (Mellor and Stafford, 2004).

8.6.4 Genetic diseases

Many diseases have a genetic component. Breed standards for some pedigree dogs may have encouraged breeders to select for characteristics that result in health problems. Certain features of some breeds (such as coat, weight, skin, eyes, shortness of muzzle) have been exaggerated to the detriment of health, and some breeds are susceptible to hereditary diseases (Stafford, 2006). Disorders such as entropion and hip dysplasia are painful, while others require surgery or prolonged treatment, and in brachiocephalic breeds the head shape can cause problems during whelping and breathing and the bulging eyes (exophthalmoses) are susceptible to injury.

8.6.5 Other painful conditions

Mathews (2000b) and Hansen (2000) provide other examples of conditions that are likely to be associated with pain. They include:

- neural damage (neuropathic pain) and intervertebral disc herniation;
- extensive inflammation of body tissue (meningitis, peritonitis, fasciitis, cellulitis) and organs (nephritis);
- excessive stretching of tissues, such as capsular pain as a result of enlarged organs (pyelonephritis, hepatitis, splenitis, splenic torsion) or hollow organ distension (accumulation of gas in bloat or gastric torsion, intestinal obstruction and colic in horses; Thoefner *et al.*, 2003);
- torsions – mesenteric, gastric and testicular; thrombosis and ischaemia; and
- obstruction to ducts – ureteral, urethral, or biliary.

Sheep veterinarians asked to score pain intensity associated with some common diseases of sheep ranked foot rot, fly strike and chronic mastitis in that order of severity (Fitzpatrick *et al.*, 2006). Sheep fly strike is a painful condition associated with irritation and, sometimes, death. Blowflies lay their eggs on the sheep and the maggots burrow into the flesh and poison the sheep with the ammonia that they secrete. Cattle veterinarians identified the following cattle conditions as painful: dystocia, pelvic and limb fracture, acute toxic mastitis, white line disease with an abscess, digital dermatitis, uveitis, joint ill and pneumonia (Huxley and Whay, 2006).

8.7 Measurement of Disease

Measurements of disease should form part of a welfare assessment of a unit or system. When assessing the welfare relevance of disease and injury it is important to consider the epidemiology of the condition, the length of time that the animal has been suffering and any treatment or prevention that has been undertaken. Thrusfield (2005, p. 22) defines epidemiology as 'the study of disease in populations and of the factors that affect its occurrence', and outlines methods for quantifying disease. Surveys of disease prevalence are subject to various types of bias and care is required in their conduct and interpretation. Morbidity (amount of disease) and mortality rates are useful to assess poor welfare associated with disease and lack of care. However, they cannot be used exclusively as an indication of welfare (Ortiz-Pelaez *et al.*, 2008). If death is quick and without suffering it is not a welfare issue, however, when it is prolonged and associated with feelings such as sickness, pain and fear it is a welfare concern (Broom, 1988). In livestock farming, mortality can be confounded with culling, so different criteria for culling influence the incidence of mortality. Culling is the selection (often on the basis of inferior quality or performance) and subsequent removal or killing of surplus animals from an animal population. The decision to cull an animal depends on many factors:

- animal factors, such as age, production, health status, and reproductive performance; and/or
- economic factors, such as product price, the price of culled animals, and the price and availability of replacement animals (Bascom and Young, 1998).

Ideally, mortality and culling rate should be low, as this would indicate that the animals were healthy and productive. However, in situations where the animals experience health problems associated with suffering, humane on-site euthanasia is an option consistent with good welfare.

Many diseases with a known cause can be diagnosed precisely by clinical signs, laboratory tests and other clinical procedures. The reliability of measurements of disease is dependent on factors such as the clinical skill of the observer, the validity of the diagnostic procedures undertaken, and the accuracy and consistency of records. As part of a brief welfare assessment, it might be possible to make some general observations of the animals for signs of ill health, and more detailed studies can describe the severity of clinical signs or lesions by defining categories according to specified criteria. For example, in dairy cattle: coughing, coat condition (hair loss, dullness), skin condition, swellings, ulceration, claw condition, body condition and locomotion scores have been used (Whay *et al.*, 2003). In addition, inspection of records for treatment of clinical conditions, drugs used (bottles and containers, sale receipts and medicine records; Scott *et al.*, 2007), dystocia, sudden death, casualty slaughter, and culling and pathology identified post-mortem at the slaughterhouse are beneficial in obtaining an impression of current and past health problems on a unit (Main *et al.*, 2001; Whay, 2007). For example, in culled sows, common lesions found at the slaughterhouse were abscesses at various sites and skin surface injuries from bite wounds or trauma. Some sows were identified with chronic arthritis, decubital ulcers, healed fractures and osteomyelitis (Cleveland-Nielsen *et al.*, 2004). Although there are many other factors that can affect production and fertility, examination of these records might also provide evidence of the potential effects of disease on productivity (Edwards, 2007).

8.8 Stockmanship

Within the EU, livestock farmers have to manage their animals by conforming to detailed legal requirements (European Council, 1998). Animals must be cared for by a sufficient number of staff who possess the appropriate ability, knowledge and professional competence.

8.8.1 Inspection

In the EU, all animals kept in husbandry systems in which their welfare depends on frequent human attention must be inspected at least once a day. Animals in other systems, such as sheep on the open hill must be inspected at intervals sufficient to avoid any suffering. Adequate lighting is necessary so that the animals can be thoroughly inspected at any time. All stockkeepers should be familiar with normal behaviour and should watch for any signs of distress or disease. To do this, it is important that stockkeepers have enough time to inspect the stock, to check equipment and to take action to deal with any problem. The stockkeeper should be aware of the signs of ill health in the relevant species. Examples of signs of ill health are shown in Box 8.1.

8.8.2 Treatment

Although animals possess immunological and other mechanisms for responding to disease and, after a period of illness, some animals might fully or partially recover without intervention, this recovery is likely to take longer, and be associated with more

---

Box 8.1. Examples of signs of ill health.

- Listlessness
- Separation from the group
- Unusual behaviour
- Lack of coordination
- Loss of body condition
- Loss of appetite
- Change in water consumption
- Sudden fall in production – such as milk yield or egg production or quality
- Constipation or diarrhoea
- Discharge from the nostrils or eyes
- Excessive saliva
- Lack of rumination

- Vomiting
- Persistent coughing or sneezing
- Rapid or irregular breathing
- Abnormal resting behaviour
- Difficulty in moving (assessed using a locomotion score) or lameness
- Swollen joints or navel
- Mastitis
- Visible wounds, abscesses or injuries
- Scratching or rubbing
- Shivering
- Discoloration or blistering of the skin

M.S. Cockram and B.O. Hughes

pathological changes and more suffering than if prompt and appropriate care and treatment are provided. However, there are many factors affecting the choice of treatment. Veterinarians and owners often have to consider whether to treat or not to treat an animal (Webster, 1995; Main, 2006) and what action to take, for example, euthanasia (the killing of an animal for its own benefit to prevent further suffering, Broom, 2007).

Some of the following factors can affect treatment decisions and the type of action taken:

- **Severity of suffering:** this should affect the promptness of any action, the use of analgesia and whether euthanasia is considered.
- **Prognosis:** the likelihood of the animal recovering and having a good quality or a productive life. Veterinarians use quality of life to help guide decisions about the treatment or euthanasia options for their patients (McMillan, 2003). This is particularly the case in companion animal practice, where the animal is not kept for financial gain and the owner usually wants to do what is best for their animal.
- **Attitude of the owner or carer:** this is likely to be influenced by many factors such as the owner's relationship with the animal and their ethical views; for example, organic farming can restrict treatment options. In organic farming, there is a potential contradiction between the principles of organic farming that are supposed to promote good welfare and the increased risk to animal health. Many veterinarians have a theoretical concern about the lack of optimal treatment (von Borell and Sørensen, 2004) and lack of use of optimal preventive measures (Lund and Algers, 2003) in organic farming. If the owner has sufficient money or insurance, many advances in human medicine and surgery, such as therapy for cancer and transplantation, are now available for companion animals. Whether the procedures used and the final outcomes are always compatible with optimal animal welfare requires careful consideration in each case (Christiansen and Forkman, 2007; Soulsby, 2007).
- **Cost of treatment:** this can be expensive and can include factors such as the costs of veterinary services, diagnostic tests (blood tests, radiography, ultrasound, endoscopy and histopathology), medicines and treatment procedures (surgical procedures can include premedication, anaesthesia, analgesia, intravenous

fluids, surgical consumables, operating theatre procedures, monitoring of physiological variables, medication and hospitalization). Some owners can be reluctant to pay for the services of a veterinarian or may delay too long before requesting a veterinarian.

- **Economics:** treatment options can be affected by the cost of treatment in relation to the animal's replacement value, by loss of productivity associated with the disease (decreased growth, reduced milk yield), by losses resulting from drug withdrawal times before products are safe to eat, by slaughter value for human consumption, by cost of euthanasia and carcass disposal, and by potential loss of premium associated with lack of conformity with any quality assurance scheme and the biosecurity implications for the control of disease. The cost of preventing or controlling certain livestock production diseases, such as leg problems in broiler chickens, may be higher than the economic gain and a certain amount of disease is tolerated.
- **Ability to provide appropriate care and treatment:** this can be affected by factors such as the environmental conditions and facilities for handling, treatment and recovery, the skill and dedication of the carer and the availability of a suitable treatment. Compared with the human field, veterinary drugs and vaccines make up a small percentage of the profits made by pharmaceutical companies, and it is sometimes not economical for them to either develop new drugs or to maintain licensed products for some veterinary applications.
- **Unpleasant side effects associated with the treatment:** the distress and unpleasant taste associated with oral medications, the handling and restraint required for injections, the side effects from some drugs, pain, discomfort and confinement following surgery, and loss of function after surgical amputation or removal (McMillan, 2003) can influence treatment decisions.
- **Potential impact of treatment options on human health and environment:** some treatments for ectoparasites, such as sheep dips, are toxic for humans and wildlife.
- **Legal requirements:** animal welfare legislation to prevent suffering is present in many countries. For example, within the EU, any animal that appears to be ill or injured must be cared for appropriately without delay and, where an

animal does not respond to such care, veterinary advice must be obtained as soon as possible. Where necessary, sick or injured animals must be isolated in suitable accommodation with, where appropriate, dry and comfortable bedding. The owner or keeper of the animals must also maintain a record of any medicinal treatment given.

The main aim of treatment is to achieve rapid and permanent recovery, and this helps to reduce suffering. However, attention must also be given to the alleviation of suffering and feelings associated with sickness during disease states (Gregory, 1998). Unpleasant feelings and discomfort associated with disease may be relieved by medication, such as the use of analgesics, anti-emetics, laxatives, anxiolytics, antihistamines and corticosteroids (McMillan, 2003), or supportive therapy such as fluids and increased care/nursing.

8.8.3 Care of sick or injured animals

Sick animals require additional care and resources. Human tactile contact and talking to a sick animal can in some circumstances attenuate feelings of pain, fear and isolation (McMillan, 1999). Many animal units have a sick or hospital pen. These pens should be easily reached so that the animals can be checked regularly. Examples of extra resources that might be required include: extra environmental protection, additional bedding, increased floor space and easy access to water and food. Sick animals can be more susceptible to cold, and heating might be required. Although the use of a sick pen can isolate diseased animals from normal animals, and provide increased resources and protection from other animals, isolation of a social animal may not always be beneficial, and it should not be used as a convenient way of hiding sick animals and leaving them with inadequate care and treatment either to recover or die. When an animal is unable to rise (for example a 'downer' cow') the prospect for recovery can be greatly increased by providing high-quality care in the initial period of recumbency. Decisions should be prompt: if a sick animal with pain or discomfort does not respond to treatment or has an incurable condition, it should be euthanized as soon as possible, and if it cannot be moved or transported without causing more suffering, it should be euthanized where it is.

8.9 Role of the Veterinarian

Veterinary medicine comprises the prevention, diagnosis and treatment of animal disease or injury. The application of these veterinary skills is essential to the reduction of suffering associated with disease. In addition, the veterinary profession possesses a sufficiently broad scientific background to have animal welfare expertise. Professional ethics oblige veterinarians to work for the benefit of animal welfare, they are in regular contact with animals and are trusted by society as a source of practical and reliable advice (Edwards, 2004; Algers, 2008; Ladewig, 2008). Their role is key, whether in the clinic, on the farm treating sick animals and organizing prevention and disease control programmes, or as inspectors monitoring and enforcing regulations.

8.10 Conclusions

- Health and disease are important factors affecting welfare. Animals cared for appropriately and in accordance with welfare standards are more likely to be healthy, whereas animals kept in poor conditions are at greater risk of disease.
- Inferences about how a sick animal feels are based on behavioural and clinical observations, knowledge of its biology and our own experiences. This anthropomorphic approach is useful as many diseases are common to humans and animals, which often share clinical and pathological responses.
- Some diseases can cause pain and suffering, and make an animal feel ill. Pathological changes to sensitive tissues can cause damage, tension or pressure resulting in pain. Illness can be accompanied by feelings such as inappetence, thirst, fever, nausea, fatigue, fear and distress. In addition, physical impairments as a result of disease, such as reduced mobility, emaciation and reduced functioning of body systems can make an animal susceptible to other welfare issues.
- Infectious diseases, whether viral, bacterial or parasitic, are an important cause of suffering. Some only exert their malign effects when the pathogen interacts with predisposing factors such as genetics, age, nutrition, environment or management.
- Genetic selection for conformation or for maximum productivity, such as milk output, broiler

growth or egg number, can have deleterious effects on health in terms of normal functioning, metabolic diseases, lameness or bone fractures.

- Keeping large numbers of animals together requires meticulous attention to disease control, a written health plan, record keeping, biosecurity and appropriate preventive measures, such as vaccination, prophylactic medicines, parasiticides and an appropriate diet and environment. Good stockmanship is essential for welfare and requires professional competence, regular inspection and careful observation for signs of ill health. Any animal that appears to be ill or injured must be cared for appropriately and without delay. Where necessary, sick or injured animals must be provided with supportive care and suitable accommodation. Where an animal does not respond, veterinary advice should be obtained as soon as possible so that appropriate diagnosis, prognosis, treatment and/or other measures to reduce suffering and to prevent disease in other animals can be implemented.

References

Algers, B. (2008) Who is responsible for animal welfare? The veterinary answer. *Acta Veterinaria Scandinavica* 50 (Supplement 1), S11.

Andrews, A.H. (1992) Other clinical diagnostic methods. In: Moss, R. (ed.) *Livestock Health and Welfare.* Longman Scientific and Technical, Harlow, Essex, pp. 51–86.

Ashley, P.J. (2007) Fish welfare: current issues in aquaculture. *Applied Animal Behaviour Science* 104, 199–235.

Balsbaugh, R.K., Curtis, S.E., Meyer, R.C. and Norton, H.W. (1986) Cold resistance and environmental-temperature preference in diarrheic piglets. *Journal of Animal Science* 62, 315–326.

Bascom, S.S. and Young, A.J. (1998) A summary of the reasons why farmers cull cows. *Journal of Dairy Science* 81, 2299–2305.

Bayvel, A.C.D. (2004) Science-based animal welfare standards: the international role of the Office International des Epizooties. *Animal Welfare* 13 (Supplement 1), 163–169.

Blowey, R.W. (1998) Welfare aspects of foot lameness in cattle. *Irish Veterinary Journal* 51, 203–207.

Blunden, A., Murray, R., Dyson, S. and Schramme, M. (2005) Chronic foot pain in the horse – is it caused by bone or tendon pathology or what? *Research in Veterinary Science* 78, 41–42.

Bradshaw, R.H., Kirkden, R.D. and Broom, D.M. (2002) A review of the aetiology and pathology of leg weakness in broilers in relation to welfare. *Avian and Poultry Biology Reviews* 13, 45–103.

Brambell Committee (1965) *Report of the Technical Committee to Enquire into the Welfare of Animal Kept under Intensive Livestock Husbandry Systems.* Command Paper 2836, Her Majesty's Stationery Office, London.

Broom, D.M. (1988) The scientific assessment of poor welfare. *Applied Animal Behaviour Science* 20, 5–19.

Broom, D.M. (2007) Quality of life means welfare: How is it related to other concepts and assessed? *Animal Welfare* 16, 45–53.

Butterworth, A. (1999) Infectious components of broiler lameness: a review. *Worlds Poultry Science Journal* 55, 327–352.

Butterworth, A., Weeks, C.A., Crea, P.R. and Kestin, S.C. (2002) Dehydration and lameness in a broiler flock. *Animal Welfare* 11, 89–94.

Cheville, N.F. (2006) *Introduction to Veterinary Pathology.* Blackwell Publishing, Ames, Iowa.

Christiansen, S.B. and Forkman, B. (2007) Assessment of animal welfare in a veterinary context – a call for ethologists. *Applied Animal Behaviour Science* 106, 203–220.

Clarke, S.P. and Bennett, D. (2006) Feline osteoarthritis: a prospective study of 28 cases. *Journal of Small Animal Practice* 47, 439–445.

Cleveland-Nielsen, A., Christensen, G. and Ersboll, A.K. (2004) Prevalences of welfare-related lesions at post-mortem meat-inspection in Danish sows. *Preventive Veterinary Medicine* 64, 123–131.

Colditz, I.G. (2002) Effects of the immune system on metabolism: implications for production and disease resistance in livestock. *Livestock Production Science* 75, 257–268.

Crispin, S.M., Roger, P.A., O'Hare, H. and Binns, S.H. (2002) The 2001 foot and mouth disease epidemic in the United Kingdom: animal welfare perspectives. *Revue Scientifique et Technique de L'Office International Des Epizooties* 21, 877–883.

Danbury, T.C., Weeks, C.A., Chambers, J.P., Waterman-Pearson, A.E. and Kestin, S.C. (2000) Self-selection of the analgesic drug carprofen by lame broiler chickens. *Veterinary Record* 146, 307–311.

Dantzer, R., O'Connor, J.C., Freund, G.G., Johnson, R.W. and Kelley, K.W. (2008) From inflammation to sickness and depression: when the immune system subjugates the brain. *Nature Reviews Neuroscience* 9, 46–56.

Dawkins, M.S. (2008) The science of animal suffering. *Ethology* 114, 937–945.

Defra (Department for Environment, Food and Rural Affairs) (2003) *Code of Recommendations for the Welfare of Livestock: Cattle.* Defra Publications, London.

Defra (2008) Biosecurity guidance to prevent the spread of animal diseases. Available at: http://www.defra.gov.uk/foodfarm/farmanimal/diseases/documents/biosecurity_guidance.pdf (accessed 23 December 2010).

Duncan, I.J.H., Beatty, E.R., Hocking, P.M. and Duff, S.R.I. (1991) Assessment of pain associated with degenerative hip disorders in adult male turkeys. *Research in Veterinary Science* 50, 200–203.

Dyer, R.M., Neerchal, N.K., Tasch, U., Wu, Y., Dyer, P. and Rajkondawar, P.G. (2007) Objective determination of claw pain and its relationship to limb locomotion score in dairy cattle. *Journal of Dairy Science* 90, 4592–4602.

Dyson, S.J., Murray, R. and Schramme, M.C. (2005) Lameness associated with foot pain: results of magnetic resonance imaging in 199 horses (January 2001–December 2003) and response to treatment. *Equine Veterinary Journal* 37, 113–121.

Edwards, J.D. (2004) The role of the veterinarian in animal welfare — a global perspective. In: *Global Conference on Animal Welfare: An OIE Initiative, Proceedings*, Paris, 23–25 February 2004. OIE (World Organisation for Animal Health), Paris/ European Commission, Luxembourg, pp. 27–35.

Edwards, S.A. (2007) Experimental welfare assessment and on-farm application. *Animal Welfare* 16, 111–115.

Egenvall, A., Bonnett, B., Wattle, O. and Emanuelson, U. (2008) Veterinary-care events and costs over a 5-year follow-up period for warmblooded riding horses with or without previously recorded locomotor problems in Sweden. *Preventive Veterinary Medicine* 83, 130–143.

European Council (1998) Council Directive 98/58/EC of 20 July 1998 concerning the Protection of Animals kept for Farming Purposes. *Official Journal of the European Union* L221, 23–27.

FAWC (Farm Animal Welfare Council) (2009) *Five Freedoms*. Available at: http://www.fawc.org.uk/freedoms.htm (accessed July 2010).

Fèvre, E.M., Bronsvoort, B.M.d.C., Hamilton, K.A. and Cleaveland, S. (2006) Animal movements and the spread of infectious diseases. *Trends in Microbiology* 14, 125–131.

Fitzpatrick, J., Scott, M. and Nolan, A. (2006) Assessment of pain and welfare in sheep. *Small Ruminant Research* 62, 55–61.

Flower, F.C., Sedlbauer, M., Carter, E., von Keyserlingk, M.A.G., Sanderson, D.J. and Weary, D.M. (2008) Analgesics improve the gait of lame dairy cattle. *Journal of Dairy Science* 91, 3010–3014.

Fossum, O., Jansson, D.S., Etterlin, P.E. and Vagsholm, I. (2009) Causes of mortality in laying hens in different housing systems in 2001 to 2004. *Acta Veterinaria Scandinavica* 51, 1–28.

Green, L.E., Berriatua, E. and Morgan, K.L. (1997) The relationship between abnormalities detected in live lambs on farms and those detected at post mortem meat inspection. *Epidemiology and Infection* 118, 267–273.

Green, P. and Tong, J.M.J. (2004) The role of the veterinary surgeon in equine welfare cases. *Equine Veterinary Education* 16, 46–56.

Gregory, N.G. (1998) Physiological mechanisms causing sickness behaviour and suffering in diseased animals. *Animal Welfare* 7, 293–305.

Gregory, N.G. (2004) *Physiology and Behaviour of Animal Suffering*. Blackwell Publishing, Oxford, UK.

Gregory, N.G. and Wilkins, L.J. (1989) Broken bones in domestic fowl: handling and processing damage in end-of-lay battery hens. *British Poultry Science* 30, 555–582.

Griffin, J.F.T. (1989) Stress and immunity – a unifying concept. *Veterinary Immunology and Immunopathology* 20, 263–312.

Gross, W.B. and Siegel, P.B. (1997) Why some get sick. *Journal of Applied Poultry Research* 6, 453–460.

Gunnarsson, S. (2006) The conceptualisation of health and disease in veterinary medicine. *Acta Veterinaria Scandinavica* 48, 1–6.

Hansen, B. (2000) Acute pain management. *Veterinary Clinics of North America: Small Animal Practice* 30, 899–916.

Hardie, E.M. and Rawlings, C.A. (1983) Septic shock. 1. Patho-physiology. *Compendium on Continuing Education for the Practicing Veterinarian* 5, 369–376.

Hart, B.L. (1988) Biological basis of the behaviour of sick animals. *Neuroscience and Biobehavioral Reviews* 12, 123–137.

Holmes, P.H. (1987) Pathophysiology of parasitic infections. *Parasitology* 94, S29–S51.

Hu, H.H., MacAllister, C.G., Payton, M.E. and Erkert, R.S. (2005) Evaluation of the analgesic effects of phenylbutazone administered at a high or low dosage in horses with chronic lameness. *JAVMA – Journal of the American Veterinary Medical Association* 226, 414–417.

Huxley, J.N. and Whay, H.R. (2006) Current attitudes of cattle practitioners to pain and the use of analgesics in cattle. *Veterinary Record* 159, 662–668.

Julian, R.J. (2000) Physiological, management and environmental triggers of the ascites syndrome: a review. *Avian Pathology* 29, 519–527.

Julian, R.J. (2005) Production and growth related disorders and other metabolic diseases of poultry – a review. *Veterinary Journal* 169, 350–369.

Kaufmann-Bart, M. and Hoop, R.K. (2009) Diseases in chicks and laying hens during the first 12 years after battery cages were banned in Switzerland. *Veterinary Record* 164, 203–207.

Kelly, K.W. (1980) Stress and immune function: a bibliographic review. *Annales de Recherches Veterinaires/ Annals of Veterinary Research* 11, 445–478.

Kirkwood, J.K. (2007) Quality of life: the heart of the matter. *Animal Welfare* 16, 3–7.

Knowles, T.G., Kestin, S.C., Haslam, S.M., Brown, S.N., Green, L.E., Butterworth, A., Pope, S.J., Pfeiffer, D. and Nicol, C.J. (2008) Leg disorders in broiler chickens: prevalence, risk factors and prevention. *PLoS ONE* 3, e1545.

M.S. Cockram and B.O. Hughes

Kolkman, I., Aerts, S., Vervaecke, H., Vicca, J., Vandelook, J., de Kruif, A., Opsomer, G., Lips, D. (2010a) Assessment of differences in some indicators of pain in double muscled Belgian Blue cows following naturally calving vs Caesarean section. *Reproduction in Domestic Animals* 45, 160–167.

Kolkman, I., Opsomer, G., Aerts, S., Hoflack, G., Laevens, H., Lips, D. (2010b) Analysis of body measurements of newborn purebred Belgian Blue calves. *Animal* 4, 661–671.

Ladewig, J. (2008) The role of the veterinarian in animal welfare. *Acta Veterinaria Scandinavica* 50 (Supplement 1), S5.

Laven, R.A., Lawrence, K.E., Weston, J.F., Dowson K.R. and Stafford, K.J. (2008) Assessment of the duration of the pain response associated with lameness in dairy cows, and the influence of treatment. *New Zealand Veterinary Journal* 56, 210–217.

Lester, P. and Gaynor, J.S. (2000) Management of cancer pain. *Veterinary Clinics of North America: Small Animal Practice* 30, 951–966.

Ley, S.J., Livingston, A. and Waterman, A.E. (1989) The effect of chronic clinical pain on thermal and mechanical thresholds in sheep. *Pain* 39, 353–357.

Ley, S.J., Livingston, A. and Waterman, A.E. (1992) Effects of clinically occurring chronic lameness in sheep on the concentrations of plasma noradrenaline and adrenaline. *Research in Veterinary Science* 53, 122–125.

Ley, S.J., Waterman, A.E., Livingston, A. and Parkinson, T.J. (1994) Effect of chronic pain associated with lameness on plasma-cortisol concentrations in sheep – a field-study. *Research in Veterinary Science* 57, 332–335.

Livingston, A. (2000) Mechanism of action of nonsteroidal anti-inflammatory drugs. *Veterinary Clinics of North America: Small Animal Practice* 30, 773–781.

Lund, V. and Algers, B. (2003) Research on animal health and welfare in organic farming – a literature review. *Livestock Production Science* 80, 55–68.

Main, D.C.J. (2006) Offering the best to patients: ethical issues associated with the provision of veterinary services. *Veterinary Record* 158, 62–66.

Main, D.C.J., Webster, A.J.F. and Green, L.E. (2001) Animal welfare assessment in farm assurance schemes. *Acta Agriculturae Scandinavica, Section A – Animal Science* 51 (Supplement 30), 108–113.

Main, D.C.J., Kent, J.P., Wemelsfelder, F., Ofner, E. and Tuyttens, F.A.M. (2003) Applications for methods of on-farm welfare assessment. *Animal Welfare* 12, 523–528.

Mansa, S., Palmer, E., Grondahl, C., Lonaas, L. and Nyman, G. (2007) Long-term treatment with carprofen of 805 dogs with osteoarthritis. *Veterinary Record* 160, 427–430.

Mathews, K.A. (2000a) Nonsteroidal anti-inflammatory analgesics: Indications and contraindications for pain management in dogs and cats. *Veterinary Clinics of North America: Small Animal Practice* 30, 783–804.

Mathews, K.A. (2000b) Pain assessment and general approach to management. *Veterinary Clinics of North America: Small Animal Practice* 30, 729–755.

McEwen, S.A. (2006) Antibiotic use in animal agriculture: what have we learned and where are we going? *Animal Biotechnology* 17, 239–250.

McGeown, D., Danbury, T.C., Waterman-Pearson, A.E. and Kestin, S.C. (1999) Effect of carprofen on lameness in broiler chickens. *Veterinary Record* 144, 668–671.

McGuirk, B.J., Forsyth, R. and Dobson, H. (2007) Economic cost of difficult calvings in the United Kingdom dairy herd. *Veterinary Record* 161, 685–687.

McMillan, F.D. (1999) Effects of human contact on animal health and well-being. *Journal of the American Veterinary Medical Association* 215, 1592–1598.

McMillan, F.D. (2003) Maximizing quality of life in ill animals. *JAVMA – Journal of the American Animal Hospital Association* 39, 227–235.

Mellor, D.J. and Stafford K.J. (2004) Animal welfare implications of neonatal mortality and morbidity in farm animals. *Veterinary Journal* 168, 118–133.

Milne, C.E., Dalton, G.E. and Stott, A.W. (2008) Balancing the animal welfare, farm profitability, human health and environmental outcomes of sheep ectoparasite control in Scottish flocks. *Livestock Science* 118, 20–33.

Milne, M.H., Nolan, A.M., Cripps, P.J. and Fitzpatrick, J.L. (2003) Assessment and alleviation of pain in dairy cows with clinical mastitis. *Cattle Practice* 11, 289–293.

Munro, R. (2008) *Animal Abuse and Unlawful Killing: Forensic Veterinary Pathology*. Elsevier Saunders, Edinburgh, UK.

Newberry, R.C. (1995) Environmental enrichment – increasing the biological relevance of captive environments. *Applied Animal Behaviour Science* 44, 229–243.

Nolan, A.M. (2000) Pharmacology of analgesic drugs. In: Flecknell, P.A. and Waterman-Pearson, A.W.B. (eds) *Pain Management in Animals*. Saunders, London, pp. 21–52.

OIE (2009) Chapter 7.6: Killing of animals for disease control purposes. In: *Terrestrial Animal Health Code*. Available at: http://www.oie.int/eng/normes/mcode/en_chapitre_1.7.6.htm (accessed 23 December 2010).

Ortiz-Pelaez, A., Pritchard, D.G., Pfeiffer, D.U., Jones, E., Honeyman, P. and Mawdsley, J.J. (2008) Calf mortality as a welfare indicator on British cattle farms. *Veterinary Journal* 176, 177–181.

Owens, J.G., Kamerling, S.G., Stanton, S.R. and Keowen, M.L. (1995) Effects of ketoprofen and phenylbutazone on chronic hoof pain and lameness in the horse. *Equine Veterinary Journal* 27, 296–300.

Pastoret, P.P. (1999) Veterinary vaccinology. *Comptes Rendus de l'Academie des Sciences Serie Iii– Sciences de la Vie – Life Sciences* 322, 967–972.

Peterson, K.D. and Keefe, T.J. (2004) Effects of meloxicam on severity of lameness and other clinical signs of osteoarthritis in dogs. *JAVMA – Journal of the American Veterinary Medical Association* 225, 1056–1060.

Peterson, P.K., Chao, C.C., Molitor, T., Murtaugh, M., Strgar, F. and Sharp, B.M. (1991) Stress and pathogenesis of infectious-disease. *Reviews of Infectious Diseases* 13, 710–720.

Raj, M. (2008) Humane killing of nonhuman animals for disease control purposes. *Journal of Applied Animal Welfare Science* 11, 112–124.

Rauw, W.M., Kanis, E., Noordhuizen-Stassen, E.N. and Grommers, F.J. (1998) Undesirable side effects of selection for high production efficiency in farm animals: a review. *Livestock Production Science* 56, 15–33.

Refsdal, A.O. (2000) To treat or not to treat: a proper use of hormones and antibiotics. *Animal Reproduction Science* 60, 109–119.

Renberg, W.C. (2005) Pathophysiology and management of arthritis. *Veterinary Clinics of North America: Small Animal Practice* 35, 1073–1091.

Riddell, C., Helmbold, C.F., Singsen, E.P. and Matterson, L.D. (1968) Bone pathology of birds affected with cage layer fatigue. *Avian Diseases* 12, 285–297.

Rushen, J., de Passillé, A.M., von Keyserlingk, M.A.G. and Weary D.M. (2008) Health, disease, and productivity. In: Rushen, J., de Passillé, A.M., von Keyserlingk, M.A.G. and Weary, D.M. (eds) *The Welfare of Cattle.* Springer, Dordrecht, The Netherlands, pp. 15–42.

Rutherford, K.M.D. (2002) Assessing pain in animals. *Animal Welfare* 11, 31–53.

Sainsbury, D. (1998) *Animal Health: Health, Disease, and Welfare of Farm Livestock.* Blackwell Science, Oxford, UK.

Scott, H. (1999) Non-traumatic causes of lameness in the hindlimb of the growing dog. *In Practice* 21, 176–188.

Scott, P.R. (2005) The management and welfare of some common ovine obstetrical problems in the United Kingdom. *Veterinary Journal* 170, 33–40.

Scott, P.R., Sargison, N.D. and Wilson, D.J. (2007) The potential for improving welfare standards and productivity in United Kingdom sheep flocks using veterinary flock health plans. *Veterinary Journal* 173, 522–531.

Smith, H.A., Jones, T.C. and Hunt, R.D. (1972) *Veterinary Pathology.* Lea and Febiger, Philadelphia, Pennsylvania.

Smith, W.J. (1988) Lameness in pigs associated with foot and limb disorders. *In Practice* 10, 113–117.

Soulsby, E.J.L. (2007) Foreword. *Animal Welfare* 16 (Supplement 1), 1.

Stafford, K. (2006) *The Welfare of Dogs.* Springer, Dordrecht, The Netherlands.

Thiermann, A.B. (2004) The OIE process, procedures and international relations. *Global Conference on Animal Welfare: An OIE Initiative, Proceedings*, Paris, 23–25 February 2004. OIE, Paris/European Commission, Luxembourg, pp. 7–12.

Thoefner, M.B., Ersboll, B.K., Jansson, N. and Hesselholt, M. (2003) Diagnostic decision rule for support in clinical assessment of the need for surgical intervention in horses with acute abdominal pain. *Canadian Journal of Veterinary Research/Revue Canadienne de Recherche Veterinaire* 67, 20–29.

Thorp, B.H. (1994) Skeletal disorders in the fowl – a review. *Avian Pathology* 23, 203–236.

Thrusfield, M. (2005) *Veterinary Epidemiology.* Blackwell, Oxford, UK.

UFAW (Universities Federation for Animal Welfare) (2008) The Garden Bird Health Initiative. Available at: http://www.ufaw.org.uk/gbhi.php (accessed 23 December 2010).

Vedhara, K., Fox, J.D. and Wang, E.C.Y. (1999) The measurement of stress-related immune dysfunction in psychoneuroimmunology. *Neuroscience and Biobehavioral Reviews* 23, 699–715.

Vestergaard, K.S. and Sanotra, G.S. (1999) Relationships between leg disorders and changes in the behaviour of broiler chickens. *Veterinary Record* 144, 205–209.

Viñuela-Fernández, I., Jones, E., Welsh, E.M. and Fleetwood-Walker, S.M. (2007)Pain mechanisms and their implication for the management of pain in farm and companion animals. *The Veterinary Journal* 174, 227–239.

von Borell, E. and Sørensen, J.T. (2004) Organic livestock production in Europe: aims, rules and trends with special emphasis on animal health and welfare. *Livestock Production Science* 90, 3–9.

Waldenstedt, L. (2006) Nutritional factors of importance for optimal leg health in broilers: a review. *Animal Feed Science and Technology* 126, 291–307.

Webster, A.B. (2004) Welfare implications of avian osteoporosis. *Poultry Science* 83, 184–192.

Webster, A.J.F. (1992) Problems of feeding and housing: their diagnosis and control. In: Moss, R. (ed.) *Livestock Health and Welfare.* Longman Scientific and Technical, Harlow, Essex, UK, pp. 292–332.

Webster, A.J.F. (1995) Animal-welfare – who are our clients. *Irish Veterinary Journal* 48, 236–239.

Webster, A.J.F. (1998) What use is science to animal welfare? *Naturwissenschaften* 85, 262–269.

Webster, A.J.F. (2001) Farm animal welfare: the five freedoms and the free market. *Veterinary Journal* 161, 229–237.

Welsh, E.M. and Nolan, A.M. (1995) Effect of flunixin meglumine on the thresholds to mechanical stimulation in healthy and lame sheep. *Research in Veterinary Science* 58, 61–66.

Whay, H.R. (2007) The journey to animal welfare improvement. *Animal Welfare* 16, 117–122.

Whay, H.R., Waterman A.E., Webster A.J.F. and O'Brien J.K. (1998) The influence of lesion type on the duration of hyperalgesia associated with hindlimb lameness in dairy cattle. *Veterinary Journal* 156, 23–29.

Whay, H.R., Main, D.C.J., Green, L.E. and Webster, A.J.F. (2003) Assessment of the welfare of dairy cattle using animal-based measurements: direct observations and investigation of farm records. *Veterinary Record* 153, 197–202.

Whay, H.R., Webster, A.J.F. and Waterman-Pearson, A.E. (2005) Role of ketoprofen in the modulation of hyperalgesia associated with lameness in dairy cattle. *Veterinary Record* 157, 729–33.

Whitehead, C.C. (2004) Overview of bone biology in the egg-laying hen. *Poultry Science* 83, 193–199.

Whiting, T.L. (2003) Foreign animal disease outbreaks, the animal welfare implications for Canada: risks apparent from international experience. *Canadian Veterinary Journal/Revue Veterinaire Canadienne* 44, 805–815.

Willeberg, P. (1991) Animal welfare studies: epidemiological considerations. *Proceedings of the Society for Veterinary Epidemiology and Preventive Medicine*, 76–82.

Winter, A.C. (2008) Lameness in sheep. *Small Ruminant Research* 76, 149–153.

Zepeda, C., Salman M. and Ruppanner, R. (2001) International trade, animal health and veterinary epidemiology: challenges and opportunities. *Preventive Veterinary Medicine* 48, 261–271.

9 Behaviour

I. Anna S. Olsson, Hanno Würbel and Joy A. Mench

Abstract

Good stockpersons and animal managers have always used the behaviour of animals as a guide to their health and welfare. Behaviour also plays a key role in the scientific study of animal welfare, for two main reasons. First, it is one of the most easily observed indicators, and essential information can often be obtained from it using experience and a systematic approach, without the use of sophisticated equipment. Secondly, behaviour forms a bridge between the narrower concept of clinical health and the wider concept of animal welfare. Through behavioural methodologies, researchers can gain information about the motivation of animals and make inferences about their subjective experience. In this chapter, we focus on the relationships between behaviour and welfare. We discuss how the opportunity that animals have to perform different behaviours affects their welfare, and how changes in behaviour can be indicative of positive or negative emotional states. Abnormal behaviour is discussed in terms of its development and underlying causes, as well as its use as a welfare indicator. We present ways of using behavioural tests to gain information about animal welfare, focusing on tests that aim to assess the valence of emotional states. Some of the possible problems with using a behavioural approach to assess welfare are also raised in the chapter.

9.1 Introduction

The importance of behaviour as a means of assessing welfare in animals was made explicit by the Brambell Committee (1965, p. 10) in its landmark report on intensive farming practices:

> The scientific evidence bearing on the sensations and sufferings of animals is derived from anatomy and physiology on the one hand and from ethology, the science of animal behaviour, on the other ... we have been impressed by the evidence to be derived from the study of the behaviour of the animal. We consider that this is a field of scientific research in relation to animal husbandry which has not attracted the attention which it deserves and that opportunities should be sought to encourage its development.

Good stockpersons and animal managers have always used the behaviour of animals as a guide to their health and welfare. The Brambell Report, however, stimulated a more formal approach to the study of the behaviour of confined animals, which, in turn, stimulated controversy about what behaviour actually tells us about welfare. Despite this controversy, the advantage of using behaviour to assess welfare is that it is one of the most easily observed indicators. Behaviour, after all, is what animals do to change and control their environment, and thus provides information about their needs, preferences and internal states.

In this chapter, we focus on the relationships between behaviour and welfare. We discuss both normal and abnormal behaviour, provide examples of behaviour indicative of aversive states and present ways of using behavioural tests to gain information about animal welfare. Some of the possible problems with using a behavioural approach to assessing welfare are also raised in the chapter. Reflecting the focus of animal welfare research to date, the chapter primarily addresses animal welfare from the perspective of problems or their absence; research into positive experiences of animals is still scarce.

9.2 Normal Behaviour and Welfare

9.2.1 Natural behaviour and behavioural integrity

The study of natural behaviour can tell us what animals do when frightened, ill or in pain, as well as when they are healthy and are not restrained by lack of space or relevant resources. Comparisons of free-living and captive animals can show us which behaviours are absent in captivity, and provide clues as to which behaviours might be important for an animal to perform. Finally, ethological studies are vital to understanding the mechanisms that underlie puzzling behaviours, such as stereotypies, and their potential welfare significance.

To know what the behaviour of an animal 'means' in terms of its welfare, it is necessary to have a detailed knowledge of the behaviour characteristic of that species of animal. This is usually achieved by establishing an 'ethogram', which is more than merely a catalogue of behaviours (Banks, 1982), as it also contains information on the temporal, environmental and social context of behaviour. As most captive environments are highly artificial, and provide animals with only limited opportunities to perform their natural behaviours, observation of animals in more extensive environments is required in order to provide what Dawkins (1989, p. 77) describes as 'a baseline against which to compare behaviour in intensive systems'.

Ideally, such a baseline is established in near-natural conditions, even for domesticated animals. In addition, although the changes in behaviour caused by selection during domestication are mostly quantitative rather than qualitative in nature (Price, 2003), understanding those changes is also important for the assessment of animal welfare. Studying only the wild conspecifics is therefore helpful but not sufficient to understand captive species. It may even be relevant to study specific target strains or breeds of captive species for this purpose. 'A mouse is not just a mouse' argued Sluyter and van Oortmerssen (2000), based on studies of the natural behaviour and preferences of mice of three common laboratory strains that showed distinctly different adaptations to different environments.

Perhaps the most widely cited example illustrating how this approach can function in developing new housing systems is Stolba and Wood-Gush's (1989) work on domestic pigs allowed to range in two large forested enclosures (the 'Pig Park'), leading to the design of the Stolba family pen for pigs (Wechsler, 1996). Similar studies resulted in the development of aviary systems for laying hens (Fölsch et al., 1983) and group housing systems for laboratory rabbits (Stauffacher, 1992) (see Fig. 9.1). The intention behind these studies was to identify key stimuli that facilitate the expression of the natural behavioural repertoire. These (or adequate substitutes; Stauffacher, 1994), were then incorporated into the housing systems in order to allow animals to express their natural behavioural repertoire and hence maintain their 'behavioural integrity' (Würbel, 2009), assuming that this would also guarantee the animals' welfare. Implicitly, many guidelines for animals (e.g. the 2006 revision of *Appendix A to the European Convention for the Protection of*

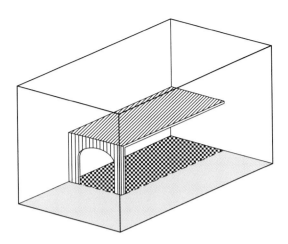

Fig. 9.1. Rabbit housing system taking natural behaviour into account. Compared with conventional cage housing for fattening rabbits, the raised solid-floor shelf provides shelter and an elevated place for observation, and encourages exercise. (After Stauffacher, 1992.)

Vertebrate Animals used for Experimental and other Scientific Purposes (Council of Europe, 2006) and the US *Guide for the Care and Use of Laboratory Animals* (2010), both of which provide recommendations on how to house animals used in research) are based on 'behavioural integrity' as a measure of animal welfare, in that they advocate allowing expression of a range of species-typical behaviour – a view going back to the Brambell Report (1965) (see also Chapter 2).

However, it must be realized that animals of most species can and do – to varying extents – adapt their behaviour to the environment and the situation in which they find themselves. Indeed, wild animals under natural conditions also show variation in behavioural activity under different circumstances. For example, in many deer species, both group size and activity pattern differ considerably between seasons, with rutting males reducing their food intake sometimes dramatically. Thus, differences in behaviour, whether found between animals in different natural environments or between animals in free-ranging and captive situations, do not necessarily imply anything about the welfare of those animals. Instead, they may simply show that animals are adaptable and exhibit behavioural plasticity. Under natural conditions, this helps them to adjust to changing environments and thereby to enhance their reproductive success and survival. For this reason, the 'naturalness' or 'integrity' of the behavioural repertoire has been heavily criticized as a measure of good welfare (e.g. Dawkins, 2008). Of greater importance for welfare is trying to understand what performing a particular natural behaviour actually means to the animal, a point illustrated by Fraser (1992, p. 100) with reference to three types of natural behaviour performed by pigs.

The point is … that a pig's natural behavior includes some things that they want to do (such as building a nest before farrowing), some things they want to do only if the conditions require it (such as panting and wallowing [under hot conditions]), and some things that they do not want to do at all (such as separation calls [emitted by piglets in the absence of the sow]). A simple comparison of animal behavior in confined and traditional environments may identify some interesting differences that merit further study, but this approach does not by itself allow us to say which of the differences are positive, negative or neutral in terms of animal welfare.

Distinguishing among these types of natural behaviour poses difficult (although not insurmountable) problems. Determining which behaviours animals 'want to do only when conditions require it' rests primarily on determining the functions of particular behaviours. In some cases, this is relatively simple. The wallowing referred to by Fraser above, for example, was seen in the pigs in the Pig Park only when ambient temperatures exceeded 18°C (Stolba and Wood-Gush, 1989), demonstrating that it is a thermoregulatory behaviour and therefore probably unnecessary for pigs in colder or climate-controlled conditions. In other cases, however, assessing the function (or functions) of a particular behaviour can be more complex (Duncan, 1981, p. 493):

It has been said that battery cages prevent wing-flapping … but perhaps the bird in a cage is not motivated to wing-flap. Wing-flapping is often described as the bird stretching its wings, but … it could also be a sexual signal or a social signal or an intention movement to fly. Until we know what wing-flapping is, what causes it, what function it serves, and how it develops and how it has evolved, we cannot say that caging prevents wing-flapping.

Because, at present, our understanding of the causes, functions and importance for welfare of many behaviours of species is still limited, 'behavioural integrity' based on ethograms established under extensive conditions can provide a useful starting point for ethical decision making if the goal is to 'give animals the benefit of the doubt' (Würbel, 2009). This is particularly true when behavioural integrity is defined in terms of: (i) natural behaviours that we know are strongly internally motivated (such as nest building in the pre-farrowing sow); and (ii) natural behavioural responses for which the eliciting stimuli are likely to be present in the animals' environment (such as wallowing at high temperatures in pigs housed without climate control). In these cases, 'behavioural integrity' will be fully consistent with one of Dawkins' (2008) criteria for good welfare: 'Do the animals have what they want?'

9.2.2 Behaviour indicative of disturbance, pain or distress

Wild, feral and captive animals show a number of behaviours characteristic of disturbance, fear or acute distress. Behaviours of this type (which fit Fraser's (1992) description of behaviours that animals 'do not want' to perform) include fleeing, avoidance, immobility, hiding and distress signals. Although the exact response will depend on the situation, behaviours of this sort are indicators of at least a short-term reduction in welfare.

The frequency and intensity of expression of some of these behaviours provides information about the distress experienced by the animal. Weary and Fraser (1995a, p. 1047) recorded the vocalizations given by piglets after separation from the sow:

'Non-thriving' and 'unfed' piglets called more and used more high-frequency calls, longer calls, and calls that rose more in frequency than their 'thriving' and 'fed' litter-mates … If a piglet's calls provide reliable information about its need for the sow's resources, then this calling can be used as a measure of its welfare.

Vocalizations may also be emitted more frequently and at a higher pitch in other animals when they are hungry, cold or in pain (Weary and Fraser, 1995b). In addition, specific types of vocalizations may be indicators of an animal's emotional state, although this relationship needs careful validation (Manteuffel et al., 2004). One of the best-studied vocalizations in this regard is the gakel call of chickens. Koene and Wiepkema (1991) found that hens gave this call when they were thwarted in their attempt to perform dust bathing behaviour. Subsequent studies showed that the same call is given in other situations where the hens are frustrated in their attempts to perform various behaviours, including feeding and drinking, and that the rate of calling is associated with the degree of frustration (Zimmerman et al., 2000).

When vocalizations are carefully validated, there is a potential for them to be used as diagnostic indicators of welfare in real-world settings. McCowan and Rommeck (2007) provided an example of how this approach could be used for assessing colonies of rhesus macaque monkeys. Managing aggression in mixed-sex macaque groups is a problem, because aggressive interactions can lead to severe injury. Analysis of vocalization patterns showed that certain calls were associated with particular types and levels of severity of aggression. The authors suggested that an automated acoustic monitoring system should be developed for continuous monitoring of aggression in these colonies, to enable managers to take corrective actions to decrease aggression or to intervene to prevent injury.

Because of their clinical and experimental importance, behaviours associated with pain, including vocalizations, have received much attention (see Chapter 5). Morton and Griffiths (1985) list signs of pain in laboratory animals, and also demonstrate an important consideration – that responses to pain, distress and disturbance are species specific. For example, guinea pigs squeal urgently and repetitively when in pain but rarely show aggression, while rats may squeal only when handled or when pressure is placed on the affected area but may also become aggressive. An understanding of species-characteristic responses to illness, pain and distress, and the environmental and social factors that modify those behaviours, is important for assessing the welfare of captive animals and taking appropriate action.

One of the best studied examples of behavioural pain assessment is methods of evaluating gait in lame dairy cattle (reviewed by Weary et al., 2006). Changes in gait can be assessed using a range of qualitative measures, including back arch, head bob and presence of a limp. Alone and in combination these behaviours can be scored reliably by the same and different observers, and can be used to identify animals with and without painful hoof injuries. These gait scores also improve when lame cows are given an analgesic, showing that changes in behaviour are at least partly due to the pain cows are experiencing (Flower et al., 2008). Most importantly, these measures can be used to evaluate improvements in barn design and management designed to help lame cows recover. For example, Bernardi et al. (2009) used gait scores to demonstrate that a simple modification to cubicle design (moving the neck rail partition out of the cow's way) allowed lame cows to recover (Fig. 9.2). While pain assessments in laboratory animals often

rely on clinical appearance, Roughan and Flecknell (2006) found that both experienced and inexperienced observers were more efficient in detecting post-surgical pain in rats (distinguishing between animals treated with analgesics or with saline post-surgery) when they used a behavioural approach especially developed to detect pain than when they used a clinical appearance scale.

9.2.3 Changes in behaviour

Morton and Griffiths (1985, p. 432) make another important point about using behaviour to evaluate welfare:

> Changes in the overall appearance of a group of animals, the way in which they interact and the deportment of an individual may indicate the first signs of abnormality.

One situation in which animals often show marked changes in behaviour is when they are sick. The reduced activity, according to Hart (1988, p. 133):

> can be viewed as representing an all-out effort to overcome the infectious disease, putting virtually all of the animal's resources into fending off the invading pathogens … Complete or partial anorexia, sleepiness, depression, lack of interest in drinking water and reduction of grooming activity, can be viewed as potentiating the fever or acute phase response by conserving energy … The same behavioral patterns might serve to protect a prey animal from predation while it is acutely ill.

The importance of observing behavioural changes in individuals is also demonstrated by studies of potentially painful procedures, for example declawing or tenectomy (cutting the tendons so that the claws remain retracted) of cats. These procedures are performed on cats to prevent them scratching and damaging furniture or other household items. Declawed or tenectomized cats show behavioural changes immediately post-surgery, including reductions in grooming and sternal resting behaviour, that are indicative of short-term pain (Cloutier et al., 2005). Conducting behavioural observations over a longer time frame can help to assess long-term pain and also evaluate alternative practices. Hemsworth et al. (2009) evaluated various alternatives to mulesing, a procedure performed on sheep to prevent blowfly infestation that involves surgical removal of the skin around the perineum. They compared mulesing with procedures that used clips or injections to stretch and enlarge, rather than

(a)

(b)

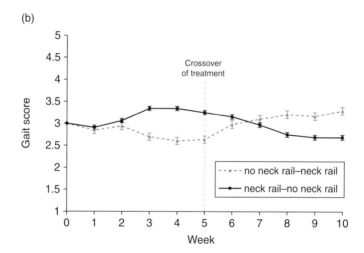

Fig. 9.2. Gait score as a measure of leg health in dairy cows improves when the neck rail is removed from the stalls in a loose housing system. The neck rail is designed to prevent cows from standing too far forward in the free stall, with the idea that this will help prevent cows from defecating and urinating in the stall. Unfortunately, preventing cows from standing fully in the stall increases the time they spend standing on wet concrete and manure slurry in the alley outside the stall. Bernardi *et al.* (2009) allocated two treatments to groups of cows using a crossover design. (a) Eight pens each containing 12 cows were either set up with the neck rail placed more centrally in the stall (the 'neck rail' treatment, with the rail positioned 130 cm from the end of the stall; seen in the nearest pen in the picture), or placed as far forward as possible so that cows would not contact the rail ('no neck rail' treatment; rail placed 190 cm from the end of the stall; seen in the second nearest pen). (b) Gait score was equal at the beginning of the experiment, but worsened (i.e. increased) over 5 weeks for cows in the 'neck rail treatment', and improved for cows in the 'no neck rail' treatment. When treatments were reversed after 5 weeks, this effect was also reversed. Behavioural observations confirmed that moving the neck rail forward allowed cows to spend more time standing fully in the stall, and reduced the time standing partially or fully in the alley. (Photo: Animal Welfare Program, UBC – University of British Columbia.)

I.A.S. Olsson *et al.*

remove, the skin. Lambs that had been mulesed still showed pronounced behavioural changes 3 weeks after the procedure, including reduced feeding and resting behaviours and abnormal walking. Lambs in the other two groups, however, were behaviourally similar to lambs that had not received any of the procedures, suggesting that the clips and injections are less painful alternatives than mulesing.

As Morton and Griffiths (1985) suggest, changes in social behaviours may also provide cues about welfare. Aggression is a normal part of the behavioural repertoire of social species. Aggression that causes injury clearly poses a welfare problem, and increases in this type of aggression can also indicate problems or inadequacies in the animal's environment – sometimes even inadequacies that are surprising because they are counter-intuitive. Howerton et al. (2008) gave group-housed male mice a highly preferred enrichment device, a running wheel. However, they found that this increased serious aggression and resulted in a destabilization of the social structure in the group. They hypothesized that this might be owing to the running wheel compartmentalizing the cage and thus creating territories within it that the mice attempted to defend. In a subsequent study (Howerton and Mench, in preparation) these researchers showed that providing rigid enrichments that could compartmentalize the cage, such as wheels and plastic tunnels, did indeed tend to increase aggression (although this depended upon the particular strain of mice studied), while providing destructible enrichments such as nesting material did not (Fig. 9.3).

Increases even in non-injurious aggression, however, may also be associated with reduced welfare. Duncan and Wood-Gush (1971) deprived chickens of food for 24 h and then provided them with food under a perspex cover. They found an increase in aggression, and suggested that the frustration associated with being unable to access the food was associated with this increase. Laboratory mice subject to a chronic stress protocol, in which unpleasant and uncontrollable environmental events occurred at unpredictable intervals, were more aggressive both when confronted with an unfamiliar mouse and when interacting with their cage mates (Mineur et al., 2003).

9.2.4 Suppression of normal behaviour

When animals are disturbed, ongoing behaviour is often temporarily suppressed and replaced by more appropriate behaviours. Under conditions of higher

(a)

(b)

Fig. 9.3. Commercial enrichments used to study the effects of enrichment on aggression in male mice. (a) Destructible enrichments included a Shepherd Shack® (left), which is a shelter-type structure made of recycled paper, and FiberCore Eco-Bedding© paper nesting material (right). (b) Rigid enrichments included a polycarbonate tunnel (left) and a Bio-Serv ®Fast-Trac (right) – a polycarbonate shelter structure (igloo) with an affixed running wheel. Providing rigid enrichments increased inter-male aggression in some strains of mice, while providing destructible enrichments did not have this effect. (Photo: Christopher Howerton.)

intensity or longer lasting stress, however, certain behaviours may be suppressed even after a disturbance has ended. Behaviours associated with reproduction are one example in some species (Moberg, 1985, p. 264):

> The expression of both male and female reproductive behavior appears to be modified by stress, especially social stress. Although the physiological mechanisms involved have not been conclusively demonstrated, at least in the female the response of the adrenal axis to stress seems to be to prevent the gonadal steroids from eliciting sexual behavior.

This was demonstrated in a study by Ehnert and Moberg (1991), in which social isolation or transportation stress delayed or blocked oestrus in ewes that had been injected with oestradiol to stimulate oestrus. Reduced self-grooming activity (Yalcin *et al.*, 2008) accompanied by deteriorating fur condition (Mineur *et al.*, 2003) has been observed in chronically stressed mice. Suppression of certain behaviour may therefore provide evidence that the animal has experienced significant stress. Behaviours associated with exploration and play (Chapter 3) may also be affected by stress. In one study of exploration (Arnsten *et al.*, 1985, p. 803):

> Naive rats were exposed to one of three stressors (restraint, tailpinch pressure, high intensity white noise) or to control procedures, and were observed in a novel environment ... The average time an animal spent per contact with stimuli in the environment was decreased significantly with stress.

So the primary effect of stress in this study was to change the character and range of exploratory activity. A similar reaction was found in healthy and diseased calves feeding from an automated milk feeder. While both groups of calves paid the same number of visits to the feeder to drink, there was a significant reduction in non-nutritive 'exploratory' visits by the diseased calves (Svensson and Jensen, 2007).

In contrast, play behaviour decreases or disappears entirely under conditions where there is food shortage, drought (Lee, 1984) or other stressors. For example, dairy calves given a reduced milk allowance showed less play running than fully fed calves (Krachun *et al.*, 2010). Weaning, which is known to be stressful to calves, also decreased play. However, the decrease was greater when the calves were weaned at 7 rather than 13 weeks of age, suggesting that the decreased food energy that accompanied early weaning also affected play behaviour. Play and diverse exploration may thus be 'luxuries' that are dispensed with during periods of stress and resource shortage.

9.2.5 Normal behaviour out of context

In a conflict or frustrating situation, behavioural activities may appear which are out of context and similarly irrelevant. An example of such an activity, referred to as displacement behaviour, is preening shown by hungry domestic fowl in response to being presented with food covered by perspex. Duncan and Wood-Gush (1972, p. 68) reported that these preening movements differed from the normal situation in that they were of shorter duration and primarily directed at areas easy to reach. But some displacement activities are seemingly 'indistinguishable in form or orientation from the same behaviour patterns in normal contexts ... and their identification [has to be] based almost exclusively on a contextual analysis' (Maestripieri *et al.*, 1992, p. 968).

Displacement behaviour can therefore be difficult to characterize, and this has caused some controversy regarding its usefulness as a category. In a review of displacement activities in primates, however, Maestripieri *et al.* (1992, p. 967) concluded that:

> Displacement activities tend to occur in situations of psycho-social stress and their frequency of occurrence is affected by anxiogenic and anxiolytic drugs. In the light of this evidence, it is suggested that displacement activities can be used as indicators of emotional states.

In interpreting displacement behaviours (as well as frustration-induced behaviours such as aggression) in terms of welfare, it is important to realize (Dawkins, 1980, pp. 75–76) that:

> Conflict is a widespread occurrence in animals, and, more importantly, that conflict behaviour patterns may be adaptive, enabling an animal to cope with the conflict and often eventually to resolve it ... while conflict and frustration do not always indicate suffering, they sometimes certainly seem to ... [particularly during] prolonged or intense occurrences.

So, as for other potentially challenging situations, the welfare consequences depend on the intensity and the duration of the situation in which displacement activities occur.

9.3 Abnormal Behaviour and Welfare

9.3.1 Understanding abnormal behaviour

Animals in captivity sometimes perform behaviour patterns or sequences of behaviour that differ fundamentally from the behaviour of free-living animals. The differences may be in the form of the behaviour (e.g. crib biting in horses (McGreevy *et al.*, 1995), somersaulting in rodents (Würbel, 2006)), its intensity (e.g. over-grooming (barbering) in mice, Garner *et al.*, 2004; pecking-induced cannibalism in chickens, Dixon *et al.*, 2008), variability (e.g. pacing in captive carnivores, Clubb and Mason, 2007) or orientation (e.g. wood chewing in horses, Nicol, 1999; feather pecking in chickens, Dixon *et al.*, 2008). Such behaviour is commonly termed 'abnormal behaviour', but its adaptive significance and implications for welfare are often controversial. One controversy relates to the meaning of abnormal behaviour. Abnormal literally means 'away from the norm', that is statistically rare or different from a reference population (normally conspecifics living free or under naturalistic conditions). However, as Dawkins (1980, p. 77) pointed out:

> The moment a behaviour is labelled 'abnormal' … it can become almost impossible not to assume the animal doing it must be suffering, because 'abnormal' is such an emotionally loaded word.

This is because in a second, colloquial sense 'abnormal' means 'pathological', assuming that the behaviour reflects some form of dysfunction caused by damage or illness (Mason, 1991). Indeed, in some cases, abnormal behaviour is the immediate behavioural expression of some specific dysfunction, as in the case of waltzing (circling) in mice caused by genetic inner ear damage (Lee *et al.*, 2002) or limping in cattle caused by severe foot lesions (Weary *et al.*, 2006). Clearly, however, not all behaviour that is abnormal in the first sense is also pathological. Instead, it may represent an 'adaptive modification' to an environment that differs from the species' natural environment, as in the case of prematurely weaned calves sucking milk from nipple drinkers or rats kept in operant boxes pressing levers for food rewards. In many cases, though, adaptation may fail because either the normal performance of the behaviour is compromised or reaching the intended goal is thwarted. In such cases, the animals may develop an aberrant form of the behaviour (e.g. horse-like standing up in tethered cows, Chaplin and Munksgaard, 2001) or redirect the behaviour towards inappropriate objects (e.g. bar biting in hungry pigs, Terlouw and Lawrence, 1993) or pen mates (e.g. cross-sucking in prematurely weaned calves, de Passillé *et al.*, 2010) or even towards themselves (e.g. self-biting in primates, Novak *et al.*, 2006; self-plucking of feathers in parrots, Garner *et al.*, 2006).

In most cases where behaviour is compromised or thwarted, abnormal behaviour originates from normal behavioural responses aimed at overcoming these restrictions, but over time this develops, often with characteristic changes in performance. Thus, the behaviour may become increasingly repetitive with time and less variable in form and orientation, and may sometimes become elicited by a wider range of conditions (a process called emancipation, Mason, 1991) or even persist under conditions under which it would not normally develop (e.g. jumping in bank voles, Cooper *et al.*, 1996). Owing to its repetitive nature, such behaviour is commonly termed 'stereotypic behaviour' or 'abnormal repetitive behaviour' and covers a wide range of different forms, including true stereotypies, compulsive behaviour, tics, dyskinesias, etc. (Garner, 2006; Mills and Lüscher, 2006). Because most abnormal behaviour in captive animals is stereotypic in nature, and because most research on the causation and welfare implications of abnormal behaviour has been devoted to stereotypic behaviour, we will now consider the evidence that links abnormal behaviour with poor welfare based primarily on research into stereotypic behaviour.

9.3.2 Stereotypic behaviour is linked to poor housing conditions

In ethology, repetitive, invariant behaviour patterns without obvious goal or function are traditionally termed 'stereotypies' (Ödberg, 1978; Mason, 1991). However, in the human medical literature, the term stereotypy implies specific clinical features and, therefore, the terms '(abnormal) stereotypic behaviour' or 'abnormal repetitive behaviour' should be used as umbrella terms instead (Garner, 2006; Mills and Lüscher, 2006).

Stereotypic behaviour is generally most prevalent in conditions believed to be aversive to animals. These include physical confinement (e.g. weaving in chained elephants, Mason and Veasey, 2010), lack of critical resources (e.g. digging in gerbils without a burrow, Wiedenmayer, 1997), lack of stimulation (e.g. pacing by leopards kept 'off exhibit' in small, barren quarters, Mallapur and Cheelam, 2002), social isolation (e.g. self-injurious behaviour in primates, Novak *et al.*, 2006), fear (e.g. pacing by leopard cats, *Felis bengalensis*, housed near natural predators, Carlstead *et al.*, 1993), and frustration (e.g. oral stereotypic behaviour in food-restricted pigs, Bergeron *et al.*, 2006). These accounts paint a picture confirmed by meta-analysis. Mason and Latham (2004) reviewed the literature to assess the relationship between stereotypic behaviour and welfare based on other reported indicators, such as behavioural avoidance, alarm calls, infant mortality, etc. The data revealed a consistent pattern: stereotypic behaviour was significantly associated with signs of poor welfare (Fig. 9.4a). So treatments that induce the most stereotypic behaviour are typically also poorer for welfare. For example, self-injurious behaviours in laboratory rhesus macaques

(*Macacca mulatta*) are predicted by the number of aversive medical procedures (e.g. blood sampling) that they have previously experienced (Novak *et al.*, 2006). The few exceptions seem to occur when poor conditions induce inactivity instead. For example, extreme cold or unpredictable electric shock does not induce stereotypic behaviour, but rather huddling or crouching (Mason, 1991).

9.3.3 Individual differences have more paradoxical links with welfare

Despite living in the same poor conditions, typically some individuals display high levels of stereotypic behaviour, but others little or none. Paradoxically, highly stereotypic individuals sometimes seem to fare better than non-stereotypic conspecifics. For example, farmed mink (*Mustela vison*) with high levels of pacing had larger litters and lower infant mortality (Jeppesen *et al.*, 2004), as did caged African striped mice (*Rhabdomys*) with high levels of jumping and/or somersaulting (Jones *et al.*, 2010). Again, this picture was confirmed by meta-analyses. Thus, among animals from the same treatment groups, those with high levels of stereotypic behaviour appeared to cope better with the poor conditions (Fig. 9.4b).

This suggests that the poorest welfare occurs in animals who fail to develop stereotypic behaviour when this is the population norm. Possible explanations include that performing stereotypic behaviour has beneficial consequences (discussed in Section 9.3.5), or that the failure to develop stereotypic behaviour despite poor conditions reflects even greater impairments in welfare. These may be in the form of depression-like states leading to inactivity (Cabib, 2006; Novak *et al.*, 2006), extreme fear causing hiding or 'freezing', or clinical conditions such as arthritis that make activity painful and aversive. As Fig. 9.4b shows, however, exceptions to this pattern are common, and individual variation in the expression of stereotypic behaviour within populations does not generally correlate with individual variation in welfare.

9.3.4 Other welfare issues associated with abnormal behaviour

Sometimes the link between abnormal behaviour and welfare problems is far more straightforward. This is particularly the case when physical damage

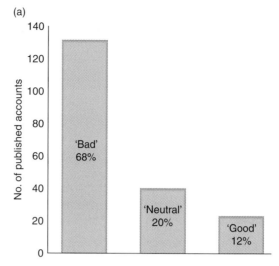

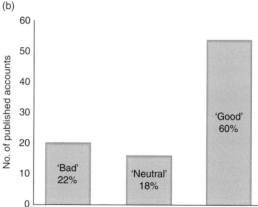

Fig. 9.4. How stereotypic behaviour co-varies with other welfare measures. Accounts of stereotypic behaviour that reported additional welfare measures were scored as to whether they associated relatively prevalent or frequent stereotypic behaviour with decreased ('bad'), unchanged ('neutral') or improved ('good') welfare relative to low-/non-stereotyping controls. (a) Results from 196 reports where stereotypers and low-/non-stereotyping controls came from different treatment groups (e.g. different housing or feeding regimens). (b) Results from 90 reports where stereotypers and low-/non-stereotyping controls came from within the same population/treatment group. The resulting patterns are significantly different from chance, and also significantly different from each other. (Modified from Mason and Latham, 2004.)

is a result. Often the animals suffering physical damage are not the ones behaving abnormally, but others that share the same living space, as in the case of feather pecking (Fig. 9.5) and cannibalism

I.A.S. Olsson *et al.*

Fig. 9.5. Some abnormal behaviour gives rise to welfare problems in the individuals subject to the behaviour. This is the case with feather pecking, which is painful when the feathers are pulled out and leaves the victim with a damaged plumage and sometimes also skin injuries. (Photo: Anne Larsen, Swedish University of Agricultural Sciences.)

in poultry, tail biting and belly nosing in pigs, or over-grooming (barbering) in mice. In these cases, the fact that the receiver may be injured renders these behaviours not only indicators but also even causes of poor welfare, regardless of whether the behaviour itself reflects pathology or suffering in the actor. The impact of such behaviours can be significant; cannibalism is one of the main causes of death among laying hens housed in non-cage systems (Fossum *et al.*, 2009).

9.3.5 What causes abnormal stereotypic behaviour?

Understanding the causation of abnormal stereotypic behaviour can help to elucidate the underlying or associated welfare problems. Three main reasons are thought to be responsible for their development (Mason and Turner, 1993; Würbel, 2006). The first involves sustained eliciting stimuli, possibly in combination with the mechanisms of habit formation (Mason and Turner, 1993). The second involves progressive pathological changes in the neural substrate underlying behavioural control, which leads to 'perseverative behaviour' (Garner, 2006). The third possibility involves reinforcement through reward in the form of some coping effect (Würbel, 2006).

Sustained eliciting stimuli

Ethologists have long explained abnormal stereotypic behaviour in terms of sustained attempts to perform highly motivated behaviour thwarted by captivity (Rushen *et al.*, 1993; Chapter 7). Likely evidence includes the observations that oral stereotypic behaviour by food-restricted sows originates from foraging and feeding behaviour (Terlouw *et al.*, 1993); that stereotypic bar mouthing by laboratory mice (*Mus musculus*) develops from repeated escape attempts at locations where external stimuli enter the cage, or where exit from the cage is possible (Würbel *et al.*, 1996; Nevison *et al.*, 1999); and that stereotypic digging in caged gerbils (*Meriones unguiculatus*) is triggered by cage corners that resemble unfinished burrow entrances (Wiedenmeyer, 1997). Sustained repetition may thus occur because motivationally salient factors elicit the recurrent performance of specific normal actions. As motivational frustration is aversive (see Chapter 7), this helps to explain why stereotypic behaviour is associated with poor welfare. However, the same explanation may also apply to non-stereotypic abnormal behaviour. So, if the elicited normal behaviour patterns naturally comprise 'fixed action patterns' and are performed in environments which themselves do not vary, or have been repeated often enough to become predictable habits, the result is likely to be a string of similar repetitions, while if the elicited normal behaviour patterns are not inherently unvarying, the environment requires them to be performed more flexibly (e.g. if they are directed at a mobile cage mate).

Impaired behavioural control

Researchers from neuroscience and related disciplines (Turner *et al.*, 2001; Wang *et al.* 2009) explain abnormal repetitive behaviour in terms of central nervous system (CNS) pathologies similar to those known to underlie the stereotypic behaviours of humans with schizophrenia or autism (Turner, 1997), subjects treated with psychostimulant drugs such as amphetamine (Robbins *et al.*, 1990) and severely maternally deprived primates (Latham and Mason, 2008). In these subjects, stereotypic behaviour is associated with specific forms of forebrain dysfunction that compromise the ability to inhibit inappropriate responses, resulting in perseveration – 'the continuation or recurrence of an activity without the appropriate stimulus' (Sandson and Albert, 1987) – and behavioural inflexibility. In particular, the functioning of neural loops running between the basal ganglia and regions of the cortex is believed to be compromised. These loops 'translate intention into action' (Graybiel, 1998) and their malfunction can have a variety of effects, including persistently recurrent plans or goals, exaggerated responses to motivationally important cues, poor abilities to cease repeating motor patterns or switch between different behaviours (Garner, 2006). In captive animals, such impairments may be caused by the chronic stress imposed by socially or physically impoverished environments (Novak et al., 2006; Tanimura et al., 2008), possibly in combination with impaired brain development in such environments (Würbel, 2001, 2006). Evidence in caged animals includes correlations between levels of stereotypic behaviour and perseverative behaviour, e.g. in extinction tasks (Garner and Mason, 2002; Vickery and Mason, 2005; Garner, 2006).

Dysfunction of behavioural control can also account for properties of abnormal stereotypic behaviour that are hard to explain in purely motivational terms, especially that behaviours are self-damaging, or that adverse early experiences (e.g. being prematurely removed from the mother, or raised in captivity rather than the wild) cause lasting propensities to develop stereotypic behaviour, and that these behaviours can sometimes be elicited by multiple stressors or persist under conditions that would not normally induce their development (Würbel, 2006).

However, it is worth noting that the differences in perseveration associated with stereotypic behaviour need not indicate pathology: the tendency to perseverate varies naturally across normal individuals and increases with age as part of normal senescence. Instead of pathology, perseveration and the underlying neurobiological correlates might, therefore, reflect normal differences in personality or 'style' that do, none the less, explain individual differences in the predisposition of animals to develop stereotypic behaviour under adverse conditions.

Coping with adversity

Despite being defined as functionless, the hypothesis that stereotypic behaviour may have rewarding properties was formulated early on (reviewed by Mason, 1991; Würbel, 2006). Empirical evidence for this hypothesis includes studies: with calves, in which repetitively sucking dry rubber teats and other objects was found to have physiological consequences, including insulin release, that may contribute to satiety (de Passillé et al., 1993); with rhesus monkeys, where bouts of self-injurious biting were found to be associated with reductions in various indices of physiological stress (Novak et al., 2006); and with wheel running, which is such a powerful reinforcer for rodents that they will lever-press to perform it (Latham and Würbel, 2006). These findings suggest that stereotypic captive animals should not be physically prevented from performing such behaviour, as this may further reduce welfare by preventing possible coping effects. However, whether coping effects play a widespread role in maintaining these behaviours is unknown; and even if they do, these benefits do not generally compensate for the welfare-compromising effects of poor housing conditions, as Fig. 9.4a makes clear.

How do these explanations fit together?

In principle, explanations in terms of frustrated motivations can combine with explanations in terms of either CNS pathology or coping, whereas CNS pathology and coping appear to be alternative hypotheses, although each may apply to a different subset of abnormal stereotypic behaviour. It seems plausible that frustrated motivations are involved whenever poor housing conditions give rise to the development of stereotypic behaviour. Frustrated motivation may explain the form and initial performance of the behaviour, while CNS changes or coping effects may account for its further development, including emancipation from the originally eliciting stimuli and persistence under conditions where the behaviour would not normally develop (Würbel, 2006).

Researchers are still some way from complete understanding of the causation and development of stereotypic behaviour. However, appreciating these different types of processes/explanations generates testable hypotheses that can aid future progress. Such progress is also needed with respect to the implications of stereotypic behaviour in terms of animal well-being. Measures of cognitive bias (see Section 9.4) may be a particularly promising approach in this respect, as they allow us to investigate the relationship between the performance of stereotypic behaviour and the valence of associated emotional states. Preliminary evidence in starlings indicates that stereotypic behaviour may be associated with pessimistic emotional states (Brilot et al., 2010). However, given the paradoxical relationship between the levels of stereotypic behaviour of individuals and welfare discussed above, more research on different species and different forms of stereotypic behaviour are needed before general conclusions can be drawn. Nevertheless, as Mason (2006, p. 345) put it:

> If we want to reduce stereotypic behaviour and improve welfare simultaneously, improving captive conditions will usually be the key.

9.4 Measuring and Testing Behaviour in Animal Welfare Research

The focus of this chapter is on the behaviour that animals show in the environment in which they live, quantified using ethological approaches. However, animal welfare researchers also draw important information from more specifically developed approaches, which include different types of behavioural tests – where the reaction of an animal is studied in a specific testing situation – as well as evaluation of qualitative aspects of behaviour.

The first specific behaviour test used in animal welfare research was the study of preferences and then, later, of operant responses, to evaluate animals' use of or need for different housing environments or resources. This is the subject of Chapter 12 and will not be discussed further here. A second set of behaviour tests measures the tendency of animals to react fearfully or aggressively in set situations. These include exposing animals to novel environments and/ or novel individuals (humans as well as unfamiliar conspecifics). When such tests are used to study differences between animals that have been subject to different treatments, such as different housing environments or husbandry practices, they can provide information about how these treatments affect the welfare of the animals. Chapter 6 provides many examples of this, as well as a critical discussion of these tests. Behavioural tests can also be used to make predictions about what will happen in a real-life situation. In this way, a test can provide information on which one can base management or breeding decisions. D'Eath (2002, p. 281) tested 2-month-old pigs in a resident–intruder test before mixing groups of unfamiliar pigs and concluded that:

> Weight and aggressiveness of individual pigs interact to influence the process and outcomes of aggressive behaviour at mixing. Both affected the severity of aggression as evidenced by increased lesions. Weight affected involvement and success in the initial and intense fighting that occurs at mixing, whereas pigs with high aggressiveness scores (measured in a resident-intruder test) showed more persistence in their aggressive behaviour. They were more involved in

bullying after initial fighting, and initiated further aggression the following day.

The third and most recent group of approaches aims to provide measures of animals' emotional states. Here, we will describe studies of cognitive bias (Mendl, Paul and co-workers) and qualitative behaviour assessment (Wemelsfelder and co-workers) in some detail, whereas the cognitive appraisal approach (Boissy and co-workers) is covered in Chapter 6.

Human psychology research suggests that changes in cognitive information processing, generally referred to as 'cognitive bias', can be reliable indicators of whether individuals perceive stimuli or events as positive or negative, so-called 'emotional valence' (Mineka and Sutton, 1992; Mathews et al., 1995; Warda and Bryant, 1998). An individual's emotional valence affects various cognitive processes, such as attention, memory and judgement. Thus, people in negative emotional states show enhanced attention to threatening stimuli, are more likely to recall negative memories, and make more negative judgements about future events or ambiguous stimuli (Paul et al., 2005; Mendl et al., 2009).

Research has recently been conducted to investigate the possibility that such 'pessimistic' judgement biases also occur in animals. If they do, they could provide more accurate measures of the emotional states of animals than more traditional behavioural and physiological measures. Researchers trained rats on a discrimination task in which one tone predicted a positive event (food) and another tone predicted a negative event (white noise). Once the rats had reliably learned this discrimination, the experimenter presented them with unreinforced ambiguous tones with characteristics interspersed between the positive and negative tones. Rats with a recent history of chronic mild stress were more likely to respond to these ambiguous cues as if they predicted the negative event than rats not subjected to such stress (Harding et al., 2004). Work using other species (including rats, dogs, rhesus monkeys and starlings) has shown similar results (Mendl et al., 2009).

There are still some uncertainties in these research findings. For example, human studies have shown that different states that have similar valence, such as depression, anxiety, fear and anger, do not always have the same effects on cognitive function (Lerner and Keltner, 2000), and that some effects may reflect the personality characteristics of individuals rather than their actual emotional states (Mineka and Zinbarg, 1998; Mogg and Bradley,

2005). Moreover, the term 'cognitive bias' does not imply that the underlying cognitive processes involve either the subjective experience of emotion, or conscious thought processes. Nevertheless, Mendl *et al.* (2009, p. 163) conclude that:

> Measures of attention, memory and judgement biases … offer a number of advantages to behavioural and physiological indicators of emotion. These include specifically measuring emotional valence; providing general *a priori* predictions, based on the findings of human studies, for how cognitive performance and emotion co-vary and hence allowing tests to be readily applied across different species; and allowing the measurement of positive affective states.

Qualitative behavioural assessment differs from the kinds of tests we have discussed so far in this section, in that it involves assessing observers' interpretations of an animal's behaviour rather than directly quantifying behaviours themselves. In a qualitative behavioural assessment, observers rate animals with respect to various traits or attributes. These ratings can be numerical – for example, a score of 1 to 5 for a trait like fearfulness, with 1 representing the lowest level of fearfulness and 5 the highest. Alternatively, observers can be allowed to develop their own descriptors (such as 'bored', 'nervous', 'attentive') to characterize the animal, a method known as free-choice profiling (Wemelsfelder *et al.*, 2001). The use of observer ratings to evaluate animal behaviour is not new – for example, qualitative rating systems have been used by animal breeders for many years to select animals that have a particular temperament or other desired behavioural characteristics. It is only recently, however, that qualitative methodologies have been used for animal welfare assessment. Wemelsfelder *et al.* (2001, p. 209) suggest that the strength of this technique is that it is an integrative and dynamic approach that provides much more information about the animal than separate pieces of behavioural data can, particularly information about the animal's 'style' and emotional expression:

> The qualitative assessment of behaviour is based upon the integration by the observer of many pieces of information that in conventional quantitative approaches are recorded separately, or are not recorded at all. This may include incidental behavioural events, subtle details of movement and posture, and aspects of the context in which behaviour occurs. In summarizing such details … qualitative behavioural assessment specifies not so much what an animal does, but how it does it.

This method has now been used to study several aspects of welfare, including the responses of foals to humans (Minero *et al.*, 2009), the social behaviour of dairy cows in loose housing systems (Rousing and Wemelsfelder, 2006) and the responses of horses to novel environments (Napolitano *et al.*, 2008).

The introduction of new methods such as those described here raises the question of validation – a question which also applies to more established methods. The many different aspects which need to be considered before we can with some confidence say that a particular method measures what it is supposed to measure in a reliable manner are discussed in detail in Chapter 6.

9.5 Conclusions

- Behaviour is one of the most easily observed and non-invasive indicators of welfare: it provides information about the needs, preferences and internal states of animals. The study of normal behaviour can tell us what animals do when they are frightened, frustrated, distressed, ill or in pain, as well as when they have abundant resources and are free from predation.
- Normal behaviour can be divided into three categories, behaviour that animals want to do, behaviour that they want to do but only when external circumstances require it, and behaviour that they don't want to do.
- No one set of behavioural responses can indicate reduced welfare, though changes in the frequencies of individual and social behaviours, or in the suppression of behaviours, can provide clues about welfare problems. Behaviour that occurs out of context may also indicate disturbance.
- Responses to aversive experiences vary with situation, with species and with strain of animal. Their interpretation requires knowledge of the species-characteristic behavioural repertoire, the functions of behaviour patterns, the contexts in which they occur and the behaviour that is characteristic of groups and individual animals.
- Abnormal behaviour appears to be closely related to adverse situations where animals are frustrated or restricted in the performance of highly motivated behaviour. However, individual variation in abnormal behaviour does not necessarily reflect individual variation in welfare state. Thus, in some cases, a high level of abnormal behaviour is associated with better coping abilities or even appears to act to improve

welfare and, in other cases, abnormal behaviour persists even when the environment is improved, indicating underlying developmental changes.

- Several behavioural approaches aiming to provide measures of animal emotional state have been developed specifically by animal welfare researchers. The cognitive bias approach relies on the idea that emotional state influences how facts are interpreted, so that individuals in a negative mood tend to make a pessimistic assessment of ambiguous facts. Qualitative behaviour assessment is based on the idea that emotional state is visible in *how* animals express different behaviours, such as through their posture. Such methods may give access to information not available through more established methods. However, it is important to validate both new and previously used methods.

Acknowledgements

Thanks to Georgia Mason, University of Guelph, for general input and revision of the section on stereotypic behaviour and to Dan Weary, University of British Columbia, for providing material on lameness in dairy cows.

References

Arnsten, A.F.T., Berridge, C. and Segal, D.S. (1985) Stress produces opioid-like effects on investigatory behavior. *Pharmacology Biochemistry and Behavior* 22, 803–809.

Banks, E.M. (1982) Behavioral research to answer questions about animal welfare. *Journal of Animal Science* 54, 434–446.

Bergeron, R., Badnell-Waters, A., Lambton, S. and Mason, G. (2006) Stereotypic oral behaviour in captive ungulates: foraging, diet and gastro-intestinal function, In: Mason, G. and Rushen, J. (eds) *Stereotypic Behaviour in Captive Animals: Fundamentals and Applications to Welfare*, 2nd edn. CAB International, Wallingford, UK, pp. 19–57.

Bernardi, F., Fregonesi, J., Winckler, C., Veira, D.M., von Keyserlingk, M.A.G. and Weary, D.M. (2009) The stall-design paradox: neck rails increase lameness but improve udder and stall hygiene. *Journal of Dairy Science* 92, 3074–3080.

Brambell Committee (1965) *Report of the Technical Committee to Enquire into the Welfare of Animal Kept under Intensive Livestock Husbandry Systems*. Command Paper 2836, Her Majesty's Stationery Office, London.

Brilot, B.O., Asher, L. and Bateson, M. (2010) Stereotyping starlings are more 'pessimistic'. *Animal Cognition* 13, 721–731.

Cabib, S. (2006) The neurophysiology of stereotypy II – the role of stress. In: Mason, G. and Rushen, J. (eds) *Stereotypic Behaviour in Captive Animals: Fundamentals and Applications to Welfare*, 2nd edn. CAB International, Wallingford, UK, pp. 227–255.

Carlstead, K., Brown, J.L. and Seidensticker, J. (1993) Behavioural and adrenocortical responses to environmental changes in leopard cats (*Felis bengalensis*). *Zoo Biology* 12, 321–331.

Chaplin, S. and Munksgaard, L. (2001) Evaluation of a simple method for assessment of rising behaviour in tethered dairy cows. *Animal Science* 72, 191–197.

Cloutier, S., Newberry, R.C., Cambridge, A.J. and Tobias, K.M. (2005) Behavioural signs of postoperative pain in cats following onychectomy or tenectomy surgery. *Applied Animal Behaviour Science* 92, 325–335.

Clubb, R. and Mason, G.J. (2007) Natural behavioural biology as a risk factor in carnivore welfare: how analysing species differences could help zoos improve enclosures. *Applied Animal Behaviour Science* 102, 303–328.

Cooper, J.J., Odberg, F., Nicol, C.J. (1996) Limitations on the effectiveness of environmental improvement in reducing stereotypic behaviour in bank voles (*Clethrionomys glareolus*). *Applied Animal Behaviour Science* 48, 237–248.

Council of Europe (2006) *Appendix A to the European Convention for the Protection of Vertebrate Animals used for Experimental and other Scientific Purposes: Guidelines for Accomodation and Care of Animals*. Strasbourg, France.

Dawkins, M.S. (1980) *Animal Suffering: The Science of Animal Welfare*. Chapman and Hall, London.

Dawkins, M.S. (1989) Time budgets in red junglefowl as a baseline for the assessment of welfare in domestic fowl. *Applied Animal Behaviour Science* 24, 77–80.

Dawkins, M.S. (2008) The science of animal suffering. *Ethology* 114, 937–945.

de Passillé, A.M.B., Christopherson, R. and Rushen, J. (1993) Nonnutritive sucking by the calf and postprandial secretion of insulin, CCK, and gastrin. *Physiology and Behavior* 54, 1069–1073.

de Passillé A.M., Sweeney B. and Rushen, J. (2010) Cross-sucking and gradual weaning of dairy calves *Applied Animal Behaviour Science* 124, 11–15.

D'Eath, R.B. (2002) Individual aggressiveness measured in a resident-intruder test predicts the persistence of aggressive behaviour and weight gain of young pigs after mixing. *Applied Animal Behaviour Science* 77, 267–283.

Dixon, L.M., Duncan, I.J.H. and Mason, G. (2008) What's in a peck? Using fixed action pattern morphology to identify the motivational basis of abnormal feather-pecking behaviour. *Animal Behaviour* 76, 1035–1042.

Duncan, I.J.H. (1981) Animal rights – animal welfare: a scientist's assessment. *Poultry Science* 60, 489–499.

Duncan, I.J.H. and Wood-Gush, D.G.M. (1971) Frustration and aggression in the domestic fowl. *Animal Behaviour* 19, 500–504.

Duncan, I.J.H. and Wood-Gush, D.G.M. (1972) An analysis of displacement preening in the domestic fowl. *Animal Behaviour* 20, 68–71.

Ehnert, K. and Moberg, G.P. (1991) Disruption of estrous behavior in ewes by dexamethasone or management-related stress. *Journal of Animal Science* 69, 2988–2994.

Flower, F.C., Sedlbauer, M., Carter, E., von Keyserlingk, M.A.G., Sanderson, D.J. and Weary, D.M. (2008) Analgesics improve the gait of lame dairy cattle. *Journal of Dairy Science* 91, 3010–3014.

Fölsch, D.W., Dolf, C., Ehrbar, H., Bleuler, T. and Teijgeler, H. (1983) Ethologic and economic examination of aviary housing for commercial laying flocks. *International Journal for the Study of Animal Problems* 4, 330–335.

Fossum, O., Jansson, D.S., Etterlin, P.E. and Vågsholm, I. (2009) Causes of mortality in laying hens in different housing systems in 2001 to 2004. *Acta Veterinaria Scandinavica* 51(3) doi:10.1186/1751-0147-51-3.

Fraser, D. (1992) Role of ethology in determining farm animal well-being In: Guttman, H.N., Mench, J.A. and Simmonds, R.C. (eds) *Science and Animals: Addressing Contemporary Issues*. SCAW (Scientists Center for Animal Welfare), Bethesda, Maryland, pp. 95–102.

Garner, J.P. (2006) Perseveration and stereotypy – systems-level insights from clinical psychology. In: Mason, G. and Rushen, J. (eds) *Stereotypic Animal Behaviour: Fundamentals and Applications to Welfare*, 2nd edn. CAB International, Wallingford, UK, pp. 121–152.

Garner, J.P. and Mason, G.J. (2002) Evidence for a relationship between cage stereotypies and behavioural disinhibition in laboratory rodents. *Behavioural Brain Research* 136, 83–92.

Garner, J.P., Dufour, B., Gregg, L.E., Weisker, S.M. and Mench, J.A. (2004) Social and husbandry factors affecting the prevalence and severity of barbering ('whisker trimming') by laboratory mice. *Applied Animal Behaviour Science* 89, 263–282.

Garner, J.P., Meehan, C.L., Famula, T.R. and Mench, J.A. (2006) Genetic, environmental, and neighbor effects on the severity of stereotypies and feather picking in orange-winged Amazon parrots (*Amazona amazonica*): an epidemiological study. *Applied Animal Behaviour Science* 96, 153–168.

Graybiel, A.M. (1998) The basal ganglia and chunking of action repertoires. *Neurobiology of Learning and Memory* 70, 119–136.

Harding, E.P., Paul, E.S. and Mendl, M. (2004) Animal behavior – cognitive bias and affective state. *Nature* 427, 312.

Hart, B.L. (1988) Biological basis of the behavior of sick animals. *Neuroscience and Biobehavioral Reviews* 12, 123–137.

Hemsworth, P.H., Barnett, J.L., Karlen, G.M., Fisher, A.D., Butler, K.L. and Arnold, N.A. (2009) Effects of mulesing and alternative procedures to mulesing on the behaviour and physiology of lambs. *Applied Animal Behaviour Science* 117, 20–27.

Howerton, C.L., Garner, J.P. and Mench, J.A. (2008) Effects of a running wheel-igloo enrichment on aggression, hierarchy linearity, and stereotypy in group-housed male CD-1 (ICR) mice. *Applied Animal Behaviour Science* 115, 90–103.

Jeppesen, L.L., Heller, K.E. and Bildsoe, A. (2004) Stereotypies in female farm mink (*Mustela vison*) may be genetically transmitted and associated with higher fertility due to effects on body weight. *Applied Animal Behaviour Science* 86, 137–143.

Jones, M.A., van Lierop, M., Mason, G. and Pillay, N. (2010) Increased reproductive output in stereotypic captive *Rhabdomys* females: potential implications for captive breeding. *Applied Animal Behaviour Science* 123, 63–69.

Koene, P. and Wiepkema, P.R. (1991) Pre-dustbathing vocalizations as an indicator of a 'need' in domestic hens. In: Boehnke, E. and Molkenthin, V. (eds) *Proceedings of the International Conference on Alternatives in Animal Husbandry*, July 1991, University of Kassel, Witzenhausen. Agrarkultur Verlag, Witzenhausen, Germany, pp. 95–103.

Krachun, C., Rushen, J. and de Passillé, A.M. (2010) Play behaviour in dairy calves is reduced by weaning and by a low energy intake. *Applied Animal Behaviour Science* 122, 71–76.

Latham, N. and Mason, G. (2008) Maternal deprivation and the development of stereotypic behaviour. *Applied Animal Behaviour Science* 110, 84–108.

Latham, N. and Würbel, H. (2006) Wheel running: a common rodent stereotypy In: Mason, G. and Rushen, J. (eds) *Stereotypic Behaviour in Captive Animals: Fundamentals and Applications to Welfare*, 2nd edn. CAB International, Wallingford, UK, pp. 91–92.

Lee, J.W., Ryoo, Z.Y., Lee, E.J., Hong, S.H., Chung, W.H., Lee, H.T., Chung, K.S., Kim, T.Y., Oh, Y.S. and Suh, J.G. (2002) Circling mouse, a spontaneous mutant in the inner ear. *Experimental Animals* 51, 167–171.

Lee, P.C. (1984) Ecological constraints on the social development of vervet monkeys. *Behaviour* 91, 245–262.

Lerner, J.S. and Keltner, D. (2000) Beyond valence: toward a model of emotion-specific influences on judgement and choice. *Cognition and Emotion* 14, 473–493.

Maestripieri, D., Schino, G., Aureli, F. and Troisi, A. (1992) A modest proposal – displacement activities as an indicator of emotions in primates. *Animal Behaviour* 44, 967–979.

Mallapur, A. and Cheelam, R. (2002) Environmental influences on stereotypy and the activity budget of Indian leopards (*Panthera pardus*) in four zoos in southern India. *Zoo Biology* 21, 585–595.

Manteuffel, G., Puppe, B. and Schon, P.C. (2004) Vocalization of farm animals as a measure of welfare. *Applied Animal Behaviour Science* 88, 163–182.

Mason, G.J. (1991) Stereotypies: a critical review. *Animal Behaviour* 41, 1015–1037.

Mason, G.J. (2006) Stereotypic behaviour in captive animals: fundamentals and implications for animal welfare and beyond. In: Rushen, J. and Mason, G. (eds) *Stereotypic Animal Behaviour: Fundamentals and Applications to Welfare*, 2nd edn. CAB International, Wallingford, UK, pp. 325–356.

Mason, G.J. and Latham, N.R. (2004) Can't stop, won't stop: is stereotypy a reliable animal welfare indicator? *Animal Welfare* 13 (Supplement 1), 57–69.

Mason, G.J. and Turner, M.A. (1993) Mechanisms involved in the development and control of stereotypies. In: Bateson, P.P.G., Klopfer, P.H. and Thompson, N.K. (eds) *Perspectives in Ethology, Volume 10: Behavior and Evolution.* Plenum Press, New York, pp. 53–85.

Mason, G.J. and Veasey, J. (2010) What do population-level welfare indices suggest about the well-being of zoo elephants? *Zoo Biology* 29, 256–273.

Mathews, A., Mogg, K., Kentish, J. and Eysenck, M. (1995) Effect of psychological treatment on cognitive bias in generalized anxiety disorder. *Behaviour Research and Therapy* 33, 293–303.

McCowan, B. and Rommeck, I. (2007) Bioacoustic monitoring of aggression in group-housed rhesus macaques (*Macaca mulatta*). *American Journal of Primatology* 69 (Supplement 1), 47 (abstract).

McGreevy, P.D., Cripps, P.J., French, N.P., Green, L.E. and Nicol, C.J. (1995) Management factors associated with stereotypic and redirected behavior in the thoroughbred horse. *Equine Veterinary Journal* 27, 86–91.

Mendl, M., Burman, O.H.P., Parker, R.M.A. and Paul, E.S. (2009) Cognitive bias as an indicator of animal emotion and welfare: emerging evidence and underlying mechanisms. *Applied Animal Behaviour Science* 118, 161–181.

Mills, D. and Lüscher, A. (2006) Veterinary and pharmacological approaches to abnormal repetitive behaviour. In: Mason, G. and Rushen, J. (eds) *Stereotypic Animal Behaviour: Fundamentals and Applications to Welfare*, 2nd edn. CAB International, Wallingford, UK, pp. 286–324.

Mineka, S. and Sutton, S.K. (1992) Cognitive biases and the emotional disorders. *Psychological Science* 3, 65–69.

Mineka, S. and Zinbarg, R. (1998) Experimental approaches to the anxiety and mood disorders, In: Adair, J.G., Belanger, D. and Dion, K.L. (eds) *Advances in Psychological Science: Vol. 1. Social, Personal and Cultural Aspects.* Psychology Press, Hove, UK, pp. 429–454.

Minero, M., Tosi, M.V., Canali, E. and Wemelsfelder, F. (2009) Quantitative and qualitative assessment of the response of foals to the presence of an unfamiliar human. *Applied Animal Behaviour Science* 116, 74–81.

Mineur, Y.S., Prasol, D.J., Belzung, C. and Crusio, W.E. (2003) Agonistic behavior and unpredictable chronic mild stress in mice. *Behavior Genetics* 33, 513–519.

Moberg, G.P. (1985) Influence of stress on reproduction: measure of well-being, In: Moberg, G.P. (ed.) *Animal Stress.* American Physiological Society, Bethesda, Maryland, pp. 245–267.

Mogg, K. and Bradley, B.P. (2005) Attentional bias in generalized anxiety disorder versus depressive disorder. *Cognitive Therapy and Research* 29, 29–45.

Morton, D.B. and Griffiths, P.H.M. (1985) Guidelines on the recognition of pain, distress and discomfort in experimental animals and an hypothesis for assessment. *Veterinary Record* 116, 431–436.

Napolitano, F., De Rosa, G., Braghieri, A., Grasso, F., Bordi, A. and Wemelsfelder, F. (2008) The qualitative assessment of responsiveness to environmental challenge in horses and ponies. *Applied Animal Behaviour Science* 109, 342–354.

National Research Council (2010) *Guide for the Care and Use of Laboratory Animals.* National Academy Press, Washington, DC.

Nevison, C., Hurst, J. and Barnard, C. (1999) Why do male ICR(CD-1) mice perform bar-related (stereotypic) behaviour? *Behavioural Processes* 47, 95–111.

Nicol, C.J. (1999) Understanding equine stereotypies. *Equine Veterinary Journal* 31 (S28), 20–25.

Novak, M.A., Meyer, J.S., Lutz, C. and Tiefenbacher, S. (2006) Social deprivation and social separation: developmental insights from primatology. In: Mason, G. and Rushen, J. (eds) *Stereotypic Behaviour in Captive Animals: Fundamentals and Applications to Welfare*, 2nd edn. CAB International, Wallingford, UK, pp. 153–189.

Ödberg, F.O. (1978) Abnormal behaviours (stereotypies). Introduction to the Round Table. In: Garsi, J. (ed.) *Proceedings of the First World Congress of Ethology Applied to Zootechnics.* Industrias Graficas Espana, Madrid, pp. 475–480.

Paul, E.S., Harding, E.J. and Mendl, M. (2005) Measuring emotional processes in animals: the utility of a cognitive approach. *Neuroscience and Biobehavioral Reviews* 29, 469–491.

Price, T.D., Qvarnstrom, A. and Irwin, D.E. (2003) The role of phenotypic plasticity in driving genetic evolution. *Proceedings of the Royal Society of London Series B – Biological Sciences* 270, 1433–1440.

Robbins, T., Mittleman, G., O'Brien, J. and Winn, P. (1990) The neurobiological significance of stereotypy induced by stimulant drugs. In: Cooper, S.J. and Dourish, C.T. (eds) *The Neurobiology of Stereotyped Behaviour.* Clarendon Press, Oxford, UK, pp. 25–63.

Roughan, J.V. and Flecknell, P.A. (2006) Training in behaviour-based post-operative pain scoring in rats – an evaluation based on improved recognition of analgesic requirements. *Applied Animal Behaviour Science* 96, 327–342.

Rousing, T. and Wemelsfelder, F. (2006) Qualitative assessment of social behaviour of dairy cows housed in loose housing systems. *Applied Animal Behaviour Science* 101, 40–53.

Rushen, J., Lawrence, A. and Terlouw, C. (1993) The motivational basis of stereotypies. In: Lawrence, A. and Rushen, J. (eds) *Stereotypic Behaviour: Fundamentals and Applications to Welfare*. CAB International, Wallingford, UK, pp. 41–64.

Sandson, J. and Albert, M. (1987) Perseveration in behavioural neurology. *Neurology* 37, 1736–1741.

Sluyter, F. and Van Oortmerssen, G.A. (2000) A mouse is not just a mouse. *Animal Welfare* 9, 193–205.

Stauffacher, M. (1992) Rabbit breeding and animal welfare – new housing concepts for laboratory and fattening rabbits. *Deutsche Tierärztliche Wochenschrift* 99, 9–15.

Stauffacher, M. (1994) Ethologische Konzepte zur Entwicklung tiergerechter Haltungssysteme und Haltungsnormen für Versuchstiere. *Tierärztliche Umschau* 49, 560–569.

Stolba, A. and Wood-Gush, D.G.M. (1989) The behavior of pigs in a semi-natural environment. *Animal Production* 48, 419–425.

Svensson, C. and Jensen, M.B. (2007) Short communication: identification of diseased calves by use of data from automatic milk feeders. *Journal of Dairy Science* 90, 994–997.

Tanimura, Y., Yang, M.C. and Lewis, M.H. (2008) Procedural learning and cognitive flexibility in a mouse model of restricted, repetitive behavior. *Behavioural Brain Research* 189, 250–256.

Terlouw, A.B. and Lawrence, E.M.C. (1993) Long-term effects of food allowance and housing on the development of stereotypies in pigs. *Applied Animal Behaviour Science* 38, 103–126.

Terlouw, E.M.C., Wiersma, A., Lawrence, A.B. and Macleod, H.A. (1993) Ingestion of food facilitates the performance of stereotypies in sows. *Animal Behaviour* 46, 939–950.

Turner, C., Presti, M., Newman, H., Bugenhagen, P., Crnic, L. and Lewis, M.H. (2001) Spontaneous stereotypy in an animal model of Down syndrome: Ts65Dn mice. *Behavior Genetics* 31, 393–400.

Turner, M. (1997) Towards an executive dysfunction account of repetitive behaviour in autism. In: Russell, J. (ed.) *Autism as an Executive Disorder*. Oxford University Press, New York, pp. 57–100.

Vickery, S. and Mason, G. (2005) Stereotypy and perseverative responding in caged bears: further data and analysis. *Applied Animal Behaviour Science* 91, 247–260.

Wang, L., Simpson, H. and Dulawa, S. (2009) Assessing the validity of current mouse genetic models of obsessive–compulsive disorder. *Behavioural Pharmacology* 20, 119–133.

Warda, G. and Bryant, R.A. (1998) Cognitive bias in acute stress disorder. *Behaviour Research and Therapy* 36, 1177–1183.

Weary, D.M. and Fraser, D. (1995a) Calling by domestic piglets – reliable signals of need. *Animal Behaviour* 50, 1047–1055.

Weary, D.M. and Fraser, D. (1995b) Signaling need – costly signals and animal-welfare assessment. *Applied Animal Behaviour Science* 44, 159–169.

Weary, D.M., Niel, L., Flower, F.C. and Fraser, D. (2006) Identifying and preventing pain in animals. *Applied Animal Behaviour Science* 100, 64–76.

Weary, D.M., Huzzey, J.M. and von Keyserlingk, M.A.G. (2009) Board-Invited Review: Using behavior to predict and identify ill health in animals. *Journal of Animal Science* 87, 770–777.

Wechsler, B. (1996) Rearing pigs in species-specific family groups. *Animal Welfare* 5, 25–35.

Wemelsfelder, F., Hunter, E.A., Mendl, M.T. and Lawrence, A.B. (2001) Assessing the 'whole animal': a free choice profiling approach. *Animal Behaviour* 62, 209–220.

Wiedenmayer, C. (1997) Causation of the ontogenetic development of stereotypic digging in gerbils. *Animal Behaviour* 53, 461–470.

Wood-Gush, D.G.M., Duncan, I.J.H. and Fraser, D. (1975) Social stress and welfare problems in agricultural animals. In: Hafez, E.S.E. (ed.) *The Behaviour of Domestic Animals*. Baillière Tindall, London, pp. 183–200.

Würbel, H. (2001) Ideal homes? Housing effects on rodent brain and behaviour. *Trends in Neurosciences* 24, 207–211.

Würbel, H. (2006) The motivational basis of caged rodents' stereotypies. In: Mason, G. and Rushen, J. (eds) *Stereotypic Behaviour in Captive Animals: Fundamentals and Applications to Welfare*, 2nd edn. CAB International, Wallingford, UK, pp. 86–120.

Würbel, H. (2009) Ethology applied to animal ethics. *Applied Animal Behaviour Science* 118, 118–127.

Würbel, H., Stauffacher, M. and von Holst, D. (1996) Stereotypies in laboratory mice: quantitative and qualitative description of the ontogeny of 'wiregnawing' and 'jumping' in Zur:ICR and Zur:ICR nu. *Ethology* 102, 371–385.

Yalcin, I., Belzung, C. and Surget, A. (2008) Mouse strain differences in the unpredictable chronic mild stress: a four-antidepressant survey. *Behavioural Brain Research* 193, 140–143.

Zimmerman, P.H., Koene, P. and van Hooff, J. (2000) The vocal expression of feeding motivation and frustration in the domestic laying hen, *Gallus gallus domesticus*. *Applied Animal Behaviour Science* 69, 265–273.

10 Physiology

DOMINIQUE BLACHE, CLAUDIA TERLOUW AND SHANE K. MALONEY

Abstract

The assessment of animal welfare using physiological parameters depends on the measurement of change(s) in biological function(s) in response to stressors. This complex biological response is commonly called the 'stress response' and has been repeatedly conceptualized since homeostasis was first formulated. This chapter first examines the changing concept of stress, describing how it has evolved from a reference to the physiological responses involved in adaptation to the environment, to the animal's state when it is challenged beyond its behavioural and physiological capacity to adapt to its environment. Next, the interactions between physical stress and emotional stress, as well as the interaction between emotional status and physiological responses to stress, are considered. Examples of different sources of stress and their biological consequences are then provided to illustrate the interplay between biological systems which constitutes the response to stress. We discuss the limitations of the physiological parameters that are currently used, and some that could potentially be used, to assess the response to stress. We conclude that the assessment of animal welfare using physiological parameters is useful, but is still evolving owing to both technological advances and progressive change in the concept of stress.

10.1 Introduction

Physiological measurements are important tools for the assessment of animal stress or welfare. Physiological changes underlie the adaptive responses of the animal to any perturbation, providing the behavioural and physiological adjustments that work to maintain homeostasis. The interpretation of physiological data can be difficult because there are several facets to the concept of stress, and also a lack of universal agreement as to what stress is in biological terms. Nevertheless, physiological stress data can be interpreted in concert with other physiological and/or behavioural parameters, while at the same time taking into account the genetic, environmental and temporal contexts. But much work is still needed to better understand the relationship between the external and internal environment, the physiological responses, and the emotional state of an animal as components of stress and welfare.

10.2 The Concept of Stress and its Relationships with Physiology

Originally introduced by Selye (1936), the term *stress* has been the centre of much controversy, partly because the word can be used as a noun, a verb or an adjective to qualify the input, the response or the outcome of the response to challenge (Engel, 1985; Levine and Ursin, 1991; Le Moal, 2007).

10.2.1 Stress in physics and in biology

Rather than concluding that the term stress is meaningless (Engel, 1985), it is informative to look at its original meaning and trace the development of its use in biology. The word stress was originally adopted in biology from its use in the field of mechanical engineering. In mechanical engineering, stress quantifies the load placed on a system. Stress causes *strain* in the system, where strain is a measure of the magnitude of the deformation of the system. For example, in engineering terms, stress is the load placed on the end of a cable, while strain is the stretch induced in the cable in response to the load. In biology, stress is the intensity and duration of the challenge faced by the organism, while strain is the alteration in physiological system(s) that is induced. In biological systems, the situation is more complex than in mechanical engineering because homeostatic responses, using mechanisms such as negative feedback, tend to minimize the deviations that are induced by stress in the regulated variables that define homeostasis. Thus, thermal stress ensues when the ambient temperature increases, but the induced increase in body core temperature is limited by the activation of homeothermic defences. In panting animals, thermal strain can be assessed as the increase in body core temperature or the increase in respiration rate. From the above discussion, it is evident that the use in biology of both the term strain and the term stress – mirroring their original definitions in mechanical engineering – would help to prevent a great deal of confusion. However, it is probably not realistic to expect to change terminology that has been used for over 60 years. It would possibly be akin to expecting the entire world to adopt the metric system, which has not happened since its creation at the end of the 18th century and the establishment of the international system of units in 1960.

Several characteristics of materials determine the shape of the stress–strain relationship. The *strength* of a material determines the initial slope of the stress–strain relationship. It is worth remarking on another inheritance from mechanical engineering in the stress–strain relationship: that is, the convention that stress (the cause) be plotted on the *y*-axis and strain (the effect) be plotted on the *x*-axis. The stronger the material is, the steeper the slope of the line to point A (Fig. 10.1a). In biology, a well-insulated homeotherm will display less strain in response to a cold challenge than a less well-insulated homeotherm; a fleeced sheep will be stronger (resist cold stress better) and will display a steeper stress–strain relationship than a shorn sheep.

In material physics, point A is known as the elastic limit, because beyond that point the material begins to deform. In biology, point A reflects the point beyond which normal regulatory feedback mechanisms can no longer maintain a variable within its biological boundaries (Fig. 10.1a). For example, during heat exposure, panting, sweating and peripheral vasodilation may be sufficient to maintain normal body temperature up to point A. When the heat stress increases further, the animal becomes hyperthermic. *Toughness*, the area under the curve, is the ability of a material to cope with deformation challenge without collapsing; the material displays some plasticity and stretches in response to additional load. In biological systems, toughness refers to the ability of the system to cope with regulated variables outside the normal range, such as hyperthermia. Hyperthermia is not immediately lethal, and animals can cope with such challenges. Repeated exposure to temperature changes, within a certain range, may help animals

to acclimate/acclimatize, i.e. to increase toughness (Ringer, 1971). For example, routine heat exposure increases the sweating capacity in humans (Henane and Valatx, 1973). When stress on the system increases beyond point B, the material will start to deform. The system can recover from a challenge up to a certain level of strain (point C), the breakpoint of the material in physics. Finally, *ductility* (length of the curve, Fig. 10.1b) refers to the ability to recover from challenges beyond point A.

The shape of the relationship between stress and strain varies between different physiological systems because the values of toughness, elasticity (equivalent to mechanical ductility) and strength vary between systems. The variation of these three parameters is reflected in the safety factors built into physiological systems (Diamond, 1993). For example, in humans, the safety factor of the length of the small intestine is two, meaning that the length of intestine can be reduced by 50% without any measurable adverse consequences. Some enzymatic systems, such as the arginine transporter in the cat's intestine, have a safety factor as high as ten. The safety factors of various physiological systems allow the individual to cope with different challenges, and presumably reflect the importance of particular systems to fitness. Owing to differences in safety factors, stress–strain curves have different shapes for different physiological systems (Fig. 10.1b). For example, the reproductive system is less ductile (less elastic), with less strength, than metabolic systems. So, in response to food restriction, the system controlling metabolism follows a Curve 1 shape, while the reproductive system will follow more of a Curve 3 shape. This means that the reproductive system will be compromised by energy restriction before metabolism. Evolution and natural selection have produced animals with stress–strain relationships that are sufficient to cope with the challenges presented to those animals in their normal lives. As will be illustrated below, the resistance of an individual to stress, that is, the shape of the stress–strain relationships of its physiological systems, not only depends on its genetic background, but also adapts according to the nature and intensity of the challenges faced by that individual throughout its life.

From the above point of view, homeostasis is not compromised if the response to challenge is within the elastic range of the physiological system (normal homeostatic function; up to point A in Fig. 10.1a). When the stress increases further, either

D. Blache *et al.*

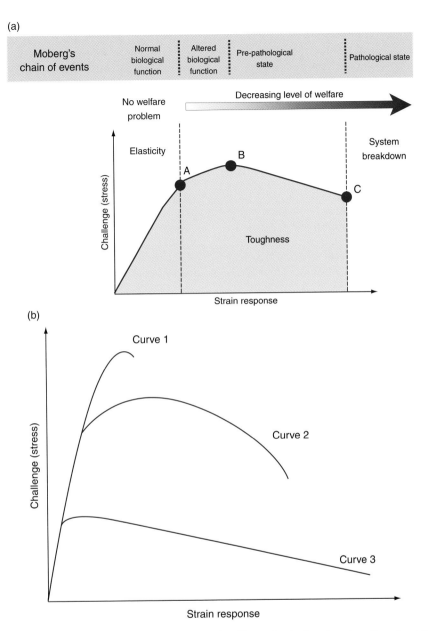

Fig. 10.1. (a) Relationship between stress and strain as defined in mechanical engineering. From the origin to point A (yield strength or elastic limit), the system is plastic and there is no deformation. When the strain is greater than point B (tensile strength), the system can recover from the stress, but it is submitted to some degree of deformation because the strain exceeds the elastic limit. Point C denotes the fracture or break-point and is when the system fails and recovery is not possible (as when glass breaks). The top bar aligns the chain of events described by Moberg (2000) using the concept of stress and strain in mechanical engineering. Each of the different levels of strain in the response to stress coincides with a part of the stress–strain curve. (b) Variation in the shape of the stress–strain curve according to different values of the three parameters of strength, ductility (elasticity) and toughness. NB: In biology, the *x* and *y* axes would be reversed, because, by convention, the dependent variable (the response that varies as a function of the stimulus or independent variable) is always presented on the *y*-axis, while the independent variable is presented on the *x*-axis.

in intensity or duration, the adaptive response will need additional physiological investment (i.e. strain increases more per unit of additional stress) and the effort required to maintain homeostasis progressively increases (up to point B). When the stress exceeds the physiological adaptive capacities of the animal (up to point C) the ability to maintain homeostasis decreases even faster. Beyond point C, the system will no longer function and will also not recover, and ultimately the organism dies. So applying the concepts used in physics to biological systems would suggest that from a physiological viewpoint, animal welfare is compromised from point B onwards, that is, when an animal's physiological functions have insufficient capacity to maintain homeostasis in an appropriate time frame. This is very close to the concept of stress proposed by Fraser *et al.* (1975), who stated that stress is the animal's state when it is challenged beyond its behavioural and physiological capacity to adapt to its environment. The main difficulty with such a definition is to determine exactly when the animal starts to have difficulties adapting (between point A and point B in Fig. 10.1a) and thus when the animal's welfare is compromised (Moberg, 2000; Terlouw, 2005).

10.2.2 The evolution of the concept of stress

The main theories of stress have recently been excellently reviewed by Pacák and Palkovits (2001) and will be only briefly described here. Figure 10.2 provides a schematic of how the different theories are related to one another. The concept of stress was first dealt with by Charles Darwin when he elaborated the theory of natural selection (Darwin, 1866). The concept of selection pressure embodies the notion that the capacity to adapt to environmental stressors determines which animals and plants survive and reproduce (Darwin, 1866). Claude Bernard (1878) is considered the first scientist to have recognized that animals need to maintain their *milieu interieur* (internal environment) stable and constant regardless of changes in the external milieu, as illustrated by this quotation from his lessons on the phenomena of life (Bernard, 1878, p. 113): 'Constancy and stability of the internal environment is the condition that life should be free and independent'. Using the terms presented above, stress would represent the external pressures that impact on the internal environment, while strain would represent the deviations from constancy of the milieu interieur. Cannon (1914) showed that physical and emotional disturbances trigger many of the same physiological responses during the *fight or flight* response, including activation of the activity of the autonomic nervous system (ANS) and the secretion of catecholamines. The important role of glucocorticosteroids (depending on species, mainly cortisol and corticosterone) in the response to stress was first recognized by Selye (1936), who found that many different noxious stimuli provoke a rise in plasma glucocorticosteroids secreted by the hypothalamo–pituitary–adrenal axis (HPA axis; see Section 10.3.2). Because of the apparent generality of the physiological response, Selye formulated his classic theory of the 'General Adaptation Syndrome'. He called negative stress 'distress' and positive stress 'eustress'. In the terms used by Selye, Point A in Fig. 10.1a demarks the progression from *eustress* to *stress* (Selye, 1950).

From the end of the 1970s to the beginning of the 1990s, the concept of stress was often discussed in terms of adaptation. First, Henessy *et al.* (1979) formulated the *psychoneuroendocrine hypothesis* in which neurohormones have an important role in feedback on brain functions, including memory formation. Later, Munck and Guyre (1986) proposed that glucocorticoids modulate the strain of physiological systems, because some systems are stimulated by glucocorticoids while others are inhibited. Moreover, depending on their level, circulating glucocorticoids may either inhibit or stimulate a specific system. The complex actions of glucocorticoids in the response to stress can be distinguished as permissive, stimulatory, suppressive or preparative (Sapolsky *et al.*, 2000). In addition, the role of environmental context as a modulator of the response to stress was outlined by Krantz and Lazar (1987) and later by Levine and Ursin (1991), who showed that *adaptive biological responses* may occur during long periods of repeated exposure to stress. The interaction of the animal with its environment and the importance of the behavioural component of the response to stress was then formulated (Weiner, 1991).

In the 1990s, Chrousos and Gold (1992) and Goldstein (1995) defined stress as *a state of disharmony* or of *threatened homeostasis*. Moberg (2000) suggested that coping with stress may cause a transfer of 'biological resources' away from non-stress functions or activities, resulting in the development of a pre-pathological or even a pathological

D. Blache *et al.*

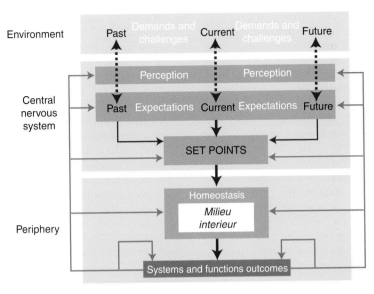

Fig. 10.2. Illustration of the physiological processing involved in an animal's response to challenge, to repeated challenges, or to different simultaneous or sequential challenges. The solid dark blue arrows represent the pathways by which the environmental challenges are processed by the organism. The dashed arrows represent the process used to evaluate whether there is a mismatch between present or anticipated future challenges and the organism's past, present and future expectations. The degree of mismatch depends on the perception of the challenge(s) and those past, present and future expectations. The light blue arrows represent the pathways by which the response(s) to challenge(s) of the different systems and functions affect (by feedback, feedforward and integrative mechanisms) the processes of: (i) the cascade of information processing illustrated by the dark blue arrows (by altering perceptions or expectations); (ii) set points for various physiological systems; and (iii) homeostasis. The behavioural component of the response(s) is included in the light blue arrows.

state (Moberg, 2000; point B in Fig. 10.1a). According to Moberg (2000, p. 2), '*distress*' refers to 'the biological state where the stress response [response to stress] has a deleterious effect on the individual's welfare' and 'our challenge is to determine when stress becomes distress'. At the same time, the notion that ability to meet the energetic costs of coping with environmental challenge might underlie the success or failure of the response led to the development of the concept of *allostasis*, literally 'stability through change'. First coined by Sterling (1988) this concept was further developed by McEwen (1998), essentially to explain modern human pathogenesis. Allostasis involves mechanisms that allow anticipating demands and is, therefore, different from homeostasis, but does not exclude the basic notion of homeostasis as originally described by Cannon (1929). Similarly, rheostasis incorporates the ability of the organism to trade off homeostasis between competing systems (Mrosovsky, 1990). While Bernard's (1878) and Cannon's (1929) concepts of homeostasis focus on

feedback loops between afferent measurements of physiological variables and efferent control of effectors that serve to regulate those variables, allostasis enshrines the pre-eminent role of the brain in adjusting the set point for feedback loops according to a plethora of inputs, including learning and emotion, which allow the system to anticipate (McEwen and Wingfield, 2007). Lately, allostasis has been proposed as a useful concept to help assess animal welfare (Korte *et al.*, 2007):

> The concept of animal welfare based on allostasis is a better alternative that incorporates recent scientific developments in behavioural physiology and neurobiology ... Good animal welfare is characterised by broad predictive and behavioural capacity to anticipate environmental challenges ... Good animal welfare is guaranteed when the regulatory range of allostatic mechanisms matches the environmental demands.

It is necessary to point out that an important aspect of animal welfare is the emotional component of the exposure to stress. Several of the above cited concepts, from the psychoneuroendocrine hypothesis to the concept of allostasis, consider that physiological responses to stress can be modulated, or even caused, by psychological factors or emotion. As a heuristic, the stress concept is constantly under development. This development has itself been affected by the importance of physiological measurements in contributing to the assessment of the response to stress and animal welfare.

10.3 Physiological Parameters to Assess Animal Welfare

10.3.1 Technical considerations

As indicated above, animals may be exposed to different types of stressors. These may be of physical origin, due, for example, to temperature variations or food restriction, or of emotional origin, due, for example, to human presence, isolation, changes in environment or the presence of unfamiliar or dominant animals. In addition, physical stress and the associated physiological responses may influence the emotional status of the animal. It is often difficult to determine which part of the response to stress is due to physiological adjustments and which part reflects the individual's emotional status.

In this section, we will discuss the physiological systems for which measurable and validated physiological parameters exist, or are likely to exist in the near future. Table 10.1 provides a summary of the physiological systems involved in the adaptive response to emotional and physical stress, the possible indicators of their activity, the techniques used to measure them, and their relevance and limitations in the assessment of welfare. Our list is not exhaustive, but it covers the principal systems. As indicated earlier, the measures do not reflect the stress response but the physiological strain induced by stress, i.e. the quantitative measure of the animal's reaction to the perceived challenge. To interpret such data in terms of animal welfare, it is important to consider them in concert with other measurements while taking into account the genetic, environmental, and temporal context.

The utility of physiological variables as welfare indicators can be affected by the level of invasiveness of the method used to obtain them, and the specificity and sensitivity of the method used to quantify them. Particularly, the baseline against which a change is to be compared needs to be of good quality. The relevant physiological systems should not be disturbed by stimuli other than those tested in the experiment. For example, human presence may be a source of emotional stress and the act of handling animals to obtain samples can induce changes in many of the physiological systems of interest. New technologies allow the use of remote telemetric recording of physiological parameters such as body or skin temperature, respiration rate or heart rate (Marchant-Forde *et al.*, 2004). Remote sensing is becoming increasingly popular in the field of animal welfare, including for non-domesticated animals. However, the usefulness of the data will depend on temporal resolution and the detection resolution of the parameter – for example, measuring core temperature to 1°C resolution will not be as useful as measuring to 0.1°C resolution because many stressors will only alter core temperature by a few tenths of a degree (Ropert-Coudert and Wilson, 2005).

Many physiological parameters can be measured in body fluids. Metabolites and hormones can be measured in blood samples using various methods. Venipuncture by needle is the simplest method, but it requires training of the operator and handling and restraint of the animals, and may be painful. Chronically indwelling venous catheters allow remote blood sampling while limiting disturbance, but often require surgical intervention (Säkkinen *et al.*, 2004; also see an example in sheep, Fig. 10.3a). Alternatively, automatic blood sampling devices can be used, but the animals need to be able to bear the weight of the equipment and need to be habituated to the device (Cook *et al.*, 2000; Goddard *et al.*, 1998; Fig. 10.3b).

In addition to blood plasma, other biological fluids can be sampled, such as cerebrospinal fluid (CSF), saliva, urine or faeces (Owens *et al.*, 1984; Cook *et al.*, 1996; Cavigelli, 1999; Strawn *et al.*, 2004; Foury *et al.*, 2007). A major benefit of using the last three fluids is the low degree of invasiveness required for sample collection. The use of saliva, urine and faeces is very popular in non-domestic species, especially wild animals, because the technique is non-invasive and sometimes also does not require the capture or handling of an animal.

10.3.2 Physiological indicators of stress

The hypothalamo–pituitary–adrenal axis

The secretion of glucocorticoids (GCs) is the most commonly used physiological measure of an animal's response to stress (Broom and Johnson,

Table 10.1. Systems and the known parameters that can or could be used to assess the level of animal strain (response to stress of physical or emotional origin). Biological media in which the parameters can be sampled and the techniques to measure them are summarized (not exhaustively). Information is given on the relevance and limitations of each system in relation to its value in the assessment of animal welfare. For example, hormone analysis can be obtained only within a few hours to a few days after collection of the sample, and hence even longer after the event of interest (lagged results).

Systems/functions	Parameters[a]	Biological media[b]	Techniques[c]	Assessment of animal welfare (AW)	
				Relevance[d]	Limitation/s
Hypothalamo–pituitary–adrenal axis (HPA axis)	Glucocorticoids (cortisol, corticosterone or their metabolites), ACTH, CRH, prolactin, vasopressin	Blood, urine, saliva, CSF, faeces, milk	Automated sampling devices, EIA, ELISA, RIA HPLC	Very well studied and accepted as a standard measure of the animal response to almost any physiological, physical, or psychological challenge; lagged results	Changes not always linked to a decrease in welfare; HPA axis can be activated by sample collection; periodical secretion in some species (pulse and diurnal variation); expensive for some hormones in some media; need for validation
Sympatho–adrenal axis	Adrenaline, noradrenaline	Blood, urine, saliva	ELISA, RIA GC, HPLC	Good measurement of the fight or flight response; lagged results	Very rapid response, not always easy to detect because of the very short half-life of catecholamines
	Electrical activity	Neural tissue	Electrophysiology, fMRI	Precise measure of the activation of specific parts of the ANS	fMRI is non-invasive, electrophysiology is invasive and both techniques are only applicable to caged (and restrained) animals
Circulation	Heart rate	Whole body	External monitor, internal logger ± transmitter	Indication of the activation of the ANS and of the strain on the cardiovascular system	Difficult to differentiate between response to increased activity and response to stress; non-invasive (counting heart beat manually) to invasive depending on where the loggers are implanted; expensive if using live (real time) transmitters
	Blood pressure	Whole body	External monitor, internal logger ± transmitter	Indication of the activation of the ANS; non-invasive (external blood pressure monitor) to invasive depending on whether and where the loggers are implanted	Same limitations as above; not easy in most animals because of difficulties keeping the animal steady when using an external monitor; expensive if using live transmitters

Continued

Table 10.1. Continued.

Systems/functions	Parameters[a]	Biological media[b]	Techniques[c]	Relevance[d]	Limitation/s
					Assessment of animal welfare (AW)
Ventilation	Breathing rate	Whole body	External monitor, internal logger ± transmitter, visual count	Indicator of the level of general activity of the organism in response to either psychological or physical stress; non-invasive (counting respiration rate) to invasive depending on where the loggers are implanted	Difficult to differentiate between response to increased activity and the response to stress; expensive if using live transmitters
	Blood chemistry, blood cells	Blood	Blood gas monitor	Indicative of the capacity of the organism to mount a response; slightly invasive; relatively cheap	Same limitations as above; lagged results; need for species-specific references
Thermoregulation	Temperature	Skin, rectum, whole body, core (abdominal)	External monitor, internal logger/transmitter	Cheap, practical method for short- and long-term monitoring of thermal strain; expensive if using live transmitters	Thermoregulation is affected by a large array of stimuli; variability between individuals and very species specific
	Hormones (thyroid hormone, catecholamines, cortisol)	Blood	ELISA, RIA	Indicative of the thermal load and the capacity of the organism to respond to further insult.	Slightly invasive; lagged results; same as above
Osmoregulation	Urine excretion Biochemical parameters: pH, urea, acid uric, ammonia, taurine	Urine	Urine collection in metabolic cage, urine bag or cup collection	Indicative of adequate drinking and feeding; cheap	Difficult to obtain individual and clear samples
	Water intake		Water meter	Indicative of adequate drinking and feeding; cheap but can be costly if large number of animals, or under extensive conditions	Difficult to obtain individual data; need to be interpreted within environmental context

D. Blache *et al.*

	Blood, saliva, urine	ELISA, RIA Automated assays for blood parameters and electrolytes	Indicative of the strain of the regulatory systems; slightly invasive	Blood parameters can be measured using automated assays; need for specialized assays for hormones; multiple hormones should be analysed together and some hormones are not only specific to water/salt balance
Blood parameters: packed cell volume, plasma osmolality, plasma creatine kinase, electrolyte (Na^+, K^+, Cl^-) concentrations. Saliva flow and composition (Na^+, HPO_4^-) Hormones: vasopressin, atrial natriuretic peptide, aldosterone, catecholamines, relaxin, ghrelin				
Pain				
Peripheral component				
Hormones and transmitters such as substance P, serotonin, prostaglandins	Blood, saliva	EIA, ELISA, RIA HPLC	Large array of indicators, some very specific to the different components of pain-signalling systems	Need for validation before true interpretation because activation of some parts of the signalling system might not transcribe into sensation; lagged results
Central component				
Opioids, neurotransmitters; electrical activity	Neural tissue, blood, CSF	ELISA, RIA Electrophysiology (EEG)	Possibility to locate and quantify pain sensation in distinct areas of the brain including cortical structures	Lagged results; invasive and expensive; only applicable to caged or restrained animals
Immunity				
Load				
Parasite number	Blood, faeces, infected tissue such as intestine, skin	Microscopy, culture	Non-invasive when collection in cage to slightly invasive when individual collection or blood or tissue collection	Difficult for free-range animals; requires good identification to follow temporal change; costly and lagged results
Immune response				
Hormones, blood cells, cytokines	Blood, lymphoid tissue	ELISA, RIA Cell count ± IHC	Indicative of the pre-pathological and pathological status of an animal; good monitoring of recovery	Cheap to very expensive

Continued

Table 10.1. Continued.

Systems/functions	Parameters[a]	Biological media[b]	Techniques[c]	Assessment of animal welfare (AW)	
				Relevance[d]	Limitation/s
Reproduction	Reproductive success		Production of a fertile progeny	Cheap if animals can be easily identified and followed up; indicator of strongly compromised AW	Difficult in free range animals; need to know the biology of reproduction for each species to differentiate between natural regulation of reproduction and AW problems; quite insensitive, long-term measurement on larger groups
	Hormones: prolactin, LH, FSH	Blood, saliva, faeces, CSF	ELISA, RIA GC, HPLC	Decrease of the reproductive activity indicates a decrease of AW	Slightly invasive for blood sampling; difficult in free-range animals; requires multiple blood samples in time; need to know the endocrinology of reproduction for each species to differentiate between natural regulation of reproduction and AW problems
Excitable tissues					
Muscle	Markers of oxidative stress, lactate and enzymatic activity	Muscle tissue, blood	ELISA, RIA GC, HPLC	Good indicator of fatigue and exhaustion	Slightly invasive in case of blood sampling. Invasive for muscle sampling
	Hormone (see bioenergetics)	Blood, saliva, faeces	ELISA, RIA GC, HPLC	Indicative of energy metabolism at muscle levels	Slightly invasive for blood sampling
	Electrical activity	Muscle tissue	Electrophysiology	Good indicator of muscle fatigue and muscle damage	Invasive or non-invasive electrophysiology
Central nervous system	Neurohormones (dopamine, adrenaline, noradrenaline, serotonin, opioids, etc.)	Neural tissue, CSF, blood	ELISA, RIA GC, HPLC	Indicative of the mental state and feelings and emotion such as pleasure, pain, and fear, and activation of the integrated pathways linking perception to action	Invasive; need validation for specificity of the changes in relation to emotional state

D. Blache *et al.*

Electrical activity	Neural tissue, whole brain	CT-scan, fMNR Electrophysiology	True measurements of the brain activity and potentially indicative of pleasure and pain; these technologies have the potential to become very good indicators of AW if interpretation of local activity is studied	Need for mapping and testing of the brain activity to interpret the readings; very expensive, requires immobilization of the animals; not practical yet
Bioenergetics Metabolites	Glucose, NEFA, urea, β-OH butyrate, lipids, triglycerides	Blood, urine, faeces	ELISA, RIA GC, HPLC Portable monitor	Good indicators of energy metabolism (anabolism, catabolism) and its regulation	Slightly invasive for blood sampling; difficult to interpret because of the complex interaction between systems and controlling factors; need to assess multiple parameters
Metabolic hormones	Insulin, glucagon, IGF-1, GH, leptin, T3 and T4, neuropeptide Y, cholecystokinin, etc.	Blood, specific tissues	ELISA, RIA GC, HPLC	Good indicators of energy metabolism and its regulation; indicative of the energy balance and energy partitioning; also indicative of the drive to feed and level of intake; can be interpreted in the assessment of appetite	Slightly invasive for blood sampling; expensive for some hormones; lagged results; difficult to interpret because of the complex interaction between systems and controlling factors; metabolic hormones are not only affected by intake or digestion but also by the energy balance between intake, reserve and expenditure; need to assess multiple hormonal parameters
Physical parameters	Live weight (mass)	Whole animal	Scale	Cheap, practical method, non-invasive; gross assessment of AW	Not reliable without other measurements such as fat mass, body conformation, age; extreme natural variations in some species
	Growth rate (mass over time)	Whole animal	Scale	Cheap, practical method, non-invasive; gross assessment of AW	Not very informative without reference data for each species
	Body condition	Back	Body condition scoring	Cheap, practical method, non-invasive; gross assessment of AW	Consistent and reliable only with trained operators; extreme natural variations in some species

Continued

Table 10.1. Continued.

Systems/functions	Parameters[a]	Biological media[b]	Techniques[c]	Assessment of animal welfare (AW)	
				Relevance[d]	Limitation/s
		Back	Ultrasonic scanning	Standardized method, non-invasive; estimated assessment of body reserves	Not very precise, consistency and reliability only with trained operators; extreme natural variations in some species
		Whole animal	X-ray CT	Very precise, reliable technique; measure of whole body reserves, so very good indicator of energy reserves	Expensive, requires immobilization of the animal; extreme natural variations in some species
		Whole animal	Electrical impedance	Easy to use, non-invasive; relatively good indicator of whole body reserves	Need for validation in different species and breeds

Abbreviations:

[a]ACTH = adrenocorticotrophic hormone; CRH = corticotrophin-releasing hormone; FSH = follicle-stimulating hormone; GH = growth hormone; IGF-1 = insulin-like growth factor 1; LH = luteinizing hormone; NEFA = non-esterified fatty acids; T3 and T4 = triiodothyronine and thyroxine.

[b]CSF = cerebrospinal fluid.

[c]CT = computerized tomography; EEG = electroencephalogram; EIA = enzyme immunoassay; ELISA = enzyme-linked immunosorbent assay; fMRI = functional magnetic resonance imaging; GC = gas chromatography; HPLC = high performance liquid chromatography; IHC = immunohistochemistry; RIA = radioimmunoassay.

[d]ANS = autonomic nervous system.

D. Blache *et al.*

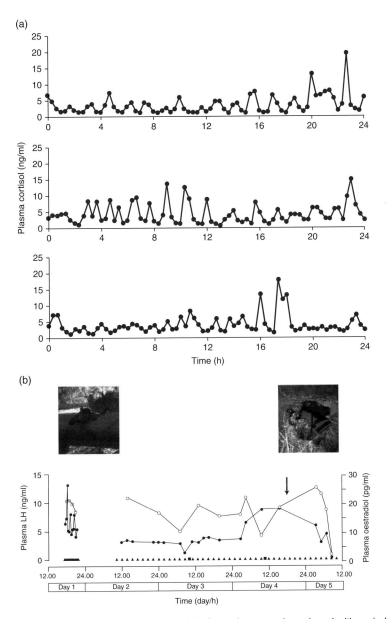

Fig. 10.3. (a) Profiles of plasma concentrations of cortisol in three sheep, each equipped with an indwelling catheter during a 24 h sampling period at a frequency of one sample every 20 min. The differences in the pattern of secretion of cortisol illustrate the inherent variability of any episodic (or pulsatile) endocrine system. Technically, this variability requires the assessment of the episodic system to take place over long periods of time with the appropriate frequency of sampling. Here, human contact was minimal and did not induce disturbances of the hypothalamo–pituitary–adrenal (HPA) axis. (Unpublished data from Blache and Maloney). (b) Study of an emu wearing automatic blood sampling equipment (ABSE), attached using a specially designed harness. Below, profiles of plasma concentrations of LH (luteinizing hormone; closed circles) and oestradiol (open circles) in a female emu bled without human contact via the ABSE. The female emu was trained to human contact and to carry the ABSE. The profiles showed pulsatility of LH and secretion of LH and oestradiol while the female was laying (arrow). The triangles indicate times when the ABSE was programmed to take samples, and the squares indicate when the ABSE was reloaded and reprogrammed. The success rate of obtaining a sample was 60%. (Data from Van Cleeff, 2002.)

1993; Moberg and Mench, 2000; Mormède et al., 2007). Increases in plasma GCs indicate that the hypothalamo–pituitary–adrenal (HPA) axis has been activated. The HPA axis is a pituitary dependent neuroendocrine system (Fig. 10.4). Neurons located in the paraventricular nucleus (PVN) of the hypothalamus produce and release corticotrophin-releasing hormone (CRH) into the portal blood vessels of the median eminence. CRH reaches its receptors on the cells in the anterior pituitary and stimulates the release of adrenocorticotrophic hormone (ACTH) into the systemic blood. ACTH then reaches the adrenal glands, where the release of GCs is stimulated. GCs have a negative feedback effect on the secretion of both ACTH and CRH.

The HPA axis may be activated as part of the physiological responses to stress, related to the level of feed and water intake, temperature or activity of the immune system. Emotional stress may also increase HPA activity (Moberg et al., 1980). However, factors unrelated to stress may also influence HPA activity, such as age (Munck and Guyre, 1986; Sapolsky et al., 2000). Some

challenges activate the HPA axis for a short time, often called the *acute response to stress*, whereas other challenges can result in activation of the HPA axis over longer time periods, which is called a *chronic response to stress* (Moberg and Mench, 2000). Owing to feedback mechanisms, the levels of GCs rarely remain high for more than several days. The HPA response to a stressor depends on many aspects, for example, the type of stressor, the duration of exposure, the genetic background of the animal, and the difference between expected and actual outcomes of a physiological response (Moberg and Mench, 2000), as discussed in the previous section.

Both the universality and the variation in duration of HPA activation make it difficult to establish a threshold above which animal welfare is affected. After extensive work on pigs and poultry, Barnett and Hemsworth (1990) proposed that an increase of 40% in the concentration of plasma GCs from the baseline is indicative of a 'non-beneficial stress of activation'. Later, the threshold value of 40% was questioned because while the HPA axis does

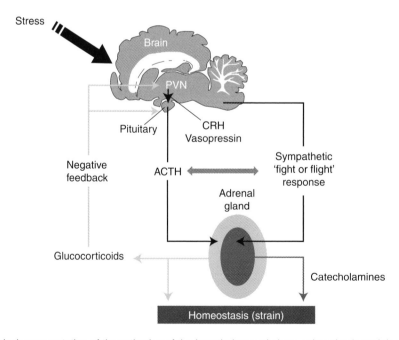

Fig. 10.4. Classical representation of the activation of the hypothalamo–pituitary–adrenal axis and the sympatho–adrenal axis by stress. Both systems are activated in response to many stressors, and these are the most studied responses to stress. Note the interaction between the two systems and the feedback loop of the glucocorticoids on the brain structures (memory structures are not indicated for clarity). ACTH = adrenocorticotrophin hormone; CRH = cortiocotrophin-releasing hormone; PVN = paraventricular nucleus of brain.

D. Blache *et al.*

not always respond to apparent negative stimuli, it does respond to apparently pleasurable stimuli such as mating, and its activation does not necessarily have a negative effect on the physiology of the individual (Sapolsky et al., 2000). Further, the response to a given stimulus varies between individuals and species. The interpretation of the HPA response therefore needs to be considered in a broader context of time and consequences. Moberg proposed that activation of the HPA axis that leads to a pre-pathological state is indicative of a decrease in animal welfare (Moberg and Mench, 2000). To integrate both the magnitude and the temporal component of the HPA activation, CRH, ACTH or cortisol may be administered to animals to measure changes in cortisol and/or CRH and ACTH concentrations (Gupta et al., 2004). For example, the cortisol response to ACTH administration was depressed in individually housed, compared with group-housed, bulls (Ladewig and Smidt, 1989) and sows (Von Borell and Ladewig, 1989). Similarly, the cortisol and ACTH responses to the administration of CRH were lower in individually housed horses than they were in horses housed in pairs (Visser et al., 2008). These changes illustrate the adaptive capacity of the HPA axis to supposed chronic stimulation resulting from socially or physically inadequate housing conditions.

The interpretation of GC levels should also take into account other considerations (Morméde et al., 2007). In most species, the secretion of GC exhibits both circadian and circannual patterns, independent of external stimuli (apart, perhaps, from light, which is probably driving the seasonal changes). In addition, the secretion of GC is pulsatile, with a rapid increase in concentrations and a slow return to baseline levels, with each pulse lasting about 90 min. Alteration in the diurnal rhythm of GC secretion has been observed in humans with depression (Turek, 2007), indicating that in some situations, changes in diurnal patterns might allow the assessment of changes in animal welfare.

The sympatho–adrenal system

Activation of the sympatho–adrenal system in response to emotional or physical stress may be assessed by measurements of plasma catecholamine (adrenaline and noradrenaline, also called epinephrine and norepinephrine) levels, or of heart rate (HR) and its variability. Plasma noradrenaline is principally released by the nerve endings of the sympathetic nervous system and, to a lesser extent, by the adrenal medulla, which is a specialized end organ of the sympathetic nervous system. In contrast, adrenaline is almost exclusively released by the adrenal medulla. Variations in adrenaline and/or noradrenaline have been used to assess welfare in animals in response to either psychological challenges (such as social isolation; Lefcourt and Elsasser, 1995) or physical challenges such as husbandry procedures (e.g. branding, Lay et al., 1992; handling in fish, Ashley, 2007) or routine laboratory procedures (such as catheterization; Lucot et al., 2005). In contrast, a role for plasma dopamine in the assessment of animal welfare seems limited, and it is considered to be essentially a precursor of noradrenaline.

HR reflects the activity of the cardiac branches of both the sympathetic and the parasympathetic nervous systems (Von Borell et al., 2007b). Like other physiological measurements, HR is affected by a large number of factors, including some not necessarily related to welfare, such as exercise and digestion. Therefore, the interpretation of changes in HR may benefit from the use of other physiological and/or behavioural measurements, or appropriate controls. For example, HR in young calves increased when they were hot-iron or freeze branded, but the changes were the same in a control group that was not branded. The changes in HR were thus at least in part due to the capture and restraint in a crush, and not to branding per se (Lay et al., 1992).

Heart rate variability (HRV), which is the variability in the time interval between consecutive heart beats (IBI, inter-beat interval, i.e. the interval between R waves of the QRS complex on a standard electrocardiogram), is an alternative indicator of cardiac stimulation. HRV reflects the balance between the activity of the sympathetic and parasympathetic branches (vagus nerve) of the ANS (Crawford et al., 1999) and may be an additional measurement in the context of stress research; various reports have shown that certain situations can result in differences in HRV while not changing average HR (Porges, 1995; Mohr et al., 2002; Von Borell et al., 2007a). HRV has been used to assess welfare in horses, cattle, sheep, goats, pigs and dogs, among others. However, depending on the nature of the stressor and on the animal, HRV can either increase or decrease, possibly as a function of the animal's interpretation of a stressor (see Chapter 8; von Borell, 2007a; Greiveldinger et al., 2007). After exposure to a stressor, changes in cardiac function are apparent before any alteration of behaviour, possibly reflecting the rapidity of autonomic responses compared with the time needed to prepare a behavioural response (Von Borell et al., 2007a). Because of the large number of factors affecting HR and HRV, it is useful to obtain resting values (when the animal is stationary and not exposed to any challenge), and to take into account circadian variation, season, and age and metabolic state of the animal (Von Borell et al., 2007a).

Neurotransmitters

Neurotransmitters, such as serotonin, dopamine, opioids and orexin, are central to the integrated response to stress and could represent a more direct measurement of strain status than peripheral measurements. Serotonin is a central neurotransmitter involved in maintaining the adaptive physiology of cognitive and emotional processes. Mood disorders in humans are often associated with changes in serotonin systems in the brain (Dayan and Huys, 2008). Brain serotonin systems also seem to be central in the response to social challenge in fish and rats (Caramaschi et al., 2007; Cubitt et al., 2008). Changes in other neurotransmitters, such as dopamine and noradrenaline, have been linked to aggression (Van Erp and Miczek, 2000; Chichinadze, 2004). However, the classic techniques used to collect data are very invasive.

More recently developed non-invasive brain scanning techniques, such as functional magnetic resonance imaging (fMRI) and positron emission tomography (PET) scanning, may provide valuable information on the involvement of brain areas in the response to stress. These techniques are used most often in humans and require controlled conditions (e.g. Damasio et al., 2000), but can be used for non-human animals. For example, visualization of central pathways activated in response to noxious stimuli by fMRI has increased our knowledge of the physiological differences and similarities between humans and other animals in sensations such as pain and pleasure (Derbyshire, 2008; Leknes and Tracey, 2008).

Stress-induced hyperthermia

In humans, emotional stimuli, such as those resulting from watching movies or boxing matches or participating in exams, can result in increased core temperature (Kleitman and Jackson, 1950; Renbourn, 1960; Marazziti et al., 1992; Briese, 1995). Similarly, placing laboratory animals into novel situations leads to an increase in core temperature, comparable to but shorter than the fever response observed after immune challenge (Bouwknecht et al., 2007). Capture hyperthermia is a major problem for wildlife translocation in Africa, and was assumed to be a result of intense muscular activity during flight. But recent research has shown that intense hyperthermia occurs even in animals that do not run, as reported in a paper with the informative title 'Hyperthermia in captured impala (Aepyceros melampus): a fright not flight response' (Meyer et al., 2008).

Like the fever response, stress induced hyperthermia (SIH) appears to involve prostaglandin, vasopressin and interleukin-based mechanisms (Singer et al., 1986; Kluger et al., 1987; Terlouw et al., 1996). In rodents, there is individual variation in the core temperature increase induced by stressors, such as handling or injection, but the response is consistent within individuals (Vinkers et al., 2008). The SIH response correlates well with other traditional measures of the response to stress, such as HPA activity (Groenink et al., 1994; Spooren et al., 2002; Veening et al., 2004). Repeated exposure of experimental animals to the same emotional stressor can in some cases lead to an attenuation of the response, and the habituation evidenced by a reduced SIH correlates with habituation of the corticosterone response (Barnum et al., 2007).

Reproduction

Indicators of reproductive capacity are often used to assess animal welfare, probably because reproduction is such an essential part of production performance. The production of viable young is a rather inefficient measure of animal welfare because it is an a posteriori measurement in that the production or not of offspring depends on conditions weeks or months previously. But reproductive capacity can also be determined on short time scales by assessing some of the hormones that control reproduction. Such assessments of reproductive hormones are common in large animals, and several reproductive hormones have been found to be altered by physical and/or emotional stress. The secretion of luteinizing hormone (LH) and follicle-stimulating hormone (FSH) are decreased by challenges such as transport, changes in social group structure, low feed intake or exposure to high temperature (Willmer *et al.*, 2000). Prolactin has a role in the control of parental behaviour such as nesting, seasonal reproduction and lactation, and also responds to stress (Fava and Guaraldi, 2006). In rats, the prolactin response diminished when exposure to a stressor was repeated over several days, suggesting habituation of this response (Kant *et al.*, 1985). The authors tested stressors with both a strong physical component (running) and a strong emotional component (restraint), as well as foot shock. Interestingly, the habituation of the prolactin response was stressor specific, that is, repeated exposure to each of the three types of stressor caused habituation of the prolactin response to that stressor, but not to any of the other stressors (Kant *et al.*, 1985). Recently, prolactin was injected into chronically stressed mice, and shown to prevent the detrimental effects of stress on neurogenesis in the hippocampus, an important area in memory formation (Torner *et al.*, 2009). Thus, prolactin might be of interest in assessing the longer term effects of stress.

Finally, reproduction is influenced by other factors, such as photoperiod and nutrition (Jordan, 2003; Blache *et al.*, 2007), and these should be taken into account in the interpretation of the results.

Production performance

Physical parameters such as growth rate, body mass or body weight and body condition can indicate the overall metabolic status of an animal. Both physical stress and emotional stress have an effect on resource allocation and on body condition (Elsasser *et al.*, 2000; Broom, 2008). Body condition is a measure of the level of adiposity and muscle deposition of an animal, and is an indicator of fitness and body reserve status that can be measured either by palpation or by using specific techniques, such as double X-ray (Jones *et al.*, 2002). Estimations of body condition score were found to correlate well with leptin levels (Blache *et al.*, 2000), and so may be a good indicator of energy reserves.

Non-stress-related factors may also influence feeding behaviour and body condition, and should be considered in any assessment of animal welfare. For example, season may also play a role (Fig. 10.5a). Animals may decrease food intake, leading to reduced live weight, during the reproductive period. Male emus can lose up to 25% of their bodyweight during breeding, because they do not feed during the 8 weeks of incubation (see Fig. 10.5b). Some species exhibit increased eating before the breeding period, or before migration or hibernation.

10.3.3 Examples of causes of stress and associated measurements

Emotional stress

Many studies have been purposely designed to try to understand the role of emotions in the response to stress. Many of them use, in addition to behavioural measurements, indicators of activation of the HPA axis and/or the sympatho–adrenal axis, but other measurements, such as pathological status, or metabolic activity, have also been used.

Behavioural tests, mainly developed for laboratory rodents and adapted for larger animals, are used to investigate how, depending on their genetic background or prior experience, animals respond to their environment (for review, see Forkman, 2007). For example, compared with Large White pigs, Duroc pigs touched an unfamiliar person more often and had a higher heart rate in a human exposure test (Terlouw and Rybarczyk, 2008). There are also attempts to understand to which aspect of the context the animal reacts. The degree of controllability and predictability of the stressor seems to influence the response to stress. Early studies by Weiss (1971a,b) showed that receiving electrical shocks was more stressful, in terms of the development of stomach ulcers, if the shocks were unpredictable or uncontrollable. More recent work on sheep has found that unpredictable events cause a stronger behavioural and cardiac response than predictable events (Greiveldinger *et al.*, 2007). The familiarity and suddenness of a stressor can also affect the reaction of sheep to those stressors (Désiré *et al.*, 2004). The relationship between emotional stress and animal welfare is discussed in greater detail in Chapter 6.

Further efforts have been made to understand the cause(s) of individual differences in the response to emotional stress. Studies using rats and mice genetically selected for a particular behavioural response (active avoidance, aggression) have revealed a genetic basis for *coping style*, which is classified as proactive or reactive (Koolhaas *et al.*, 1999). Individuals expressing different behavioural responses display different physiological and neuroendocrine responses. The proactive coping style is

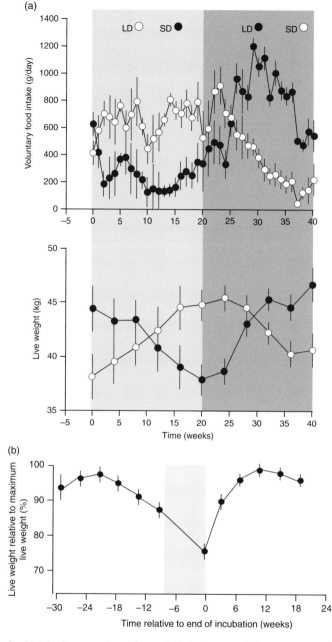

Fig. 10.5. (a) Voluntary food intake (top panel) and live weight (bottom panel) in two groups of four male emus maintained on artificial photoperiod for two periods of 20 weeks (shaded areas). One group (open circles) was exposed to a long-day photoperiod (LD, 14 h light:10 h dark) then to a short-day photoperiod (SD, 10 h light:14 h dark), while the other group (closed circles) was exposed to SD then to LD. The males were not in contact with the females, so the changes in voluntary intake were not the result of a shift in time budget due to expression of courtship or mating behaviour. The food intake response to photoperiod is an illustration of the interaction between genetic and feedforward processes. (Data adapted from Blache and Martin, 1999; Blache *et al.*, 1999.) (b) Change in live weight relative to maximum live weight observed before the beginning of incubation in male emus (*n* = 8). Data are aligned to end of incubation (time = 0 weeks). Shaded area indicates incubation period. (From Van Cleeff, 2002.)

D. Blache *et al.*

characterized by low activity and reactivity of the HPA axis, low parasympathetic reactivity and high sympathetic activity, while the reactive animals have the opposite pattern of physiological responses (Koolhaas *et al.*, 1999). Similarly, individuals show consistent differences in their behavioural and physiological responses to stress compared with conspecifics (Fig. 10.6). Those differences persist across different situations and are due to differences in genetic background as well as to earlier experience (Boissy and Bouissou, 1988; Van Reenen *et al.*, 2005; Terlouw and Rybarczyk, 2008). Comparisons between behavioural and physiological responses may help to understand the motivation underlying these responses. For example, calves that had more contact with an unfamiliar object had lower cortisol levels after exposure to the object, but also after an open-field test. These calves were identified as less fearful (Van Reenen *et al.*, 2005). Similarly, in pigs the reactivity to humans predicted the reactivity to the slaughter procedure as indicated by metabolic activity in the post-mortem muscle, suggesting that the presence of humans is a stressor during the slaughter procedure (Terlouw, 2005; Terlouw and Rybarczyk, 2008).

Pain

Assessment of pain is essential in the assessment of animal welfare and a large number of physiological parameters are available to assess pain. These are briefly summarized in Table 10.1, and the topic of pain and animal welfare is covered in detail in Chapter 5.

Thermal stress

In mammals and birds, thermoregulation is central to the maintenance of homeostasis and the correct functioning of all physiological systems, mainly because biochemical systems function optimally over a limited temperature range. The thermoneutral zone is the range of ambient temperatures within which an endotherm can regulate its temperature without elevating its metabolic rate or evaporative water loss (IUPS Thermal Commission, 2001). Outside the thermoneutral zone, endotherms are exposed to heat or cold stress. Cold stress results in the appropriate stimulation of heat production through increased metabolic rate, including via shivering. Just as the capacity of individuals to engage in sustained exercise varies, so too does an animal's ability to sustain the elevated metabolic rate required to achieve heat balance in the cold. As such, the adaptive capacity of the animal during cold stress depends on other factors, such as its health status and energy stores. For example, in sheep, exposure to cold weather in the first few days after shearing can result in significant mortality (Lynch *et al.*, 1980).

Heat stress may be of greater concern than cold stress because the 'safety margin' between normal core temperature and lethal hyperthermia is smaller than that for lethal hypothermia. The physiological responses to heat stress include a reduction in food intake, an increase in respiration rate, water consumption, sweating (in animals able to), heart rate, and skin and body temperature and scrotal skin temperature. The best indicator of heat stress is core temperature, because the strain of thermoregulation

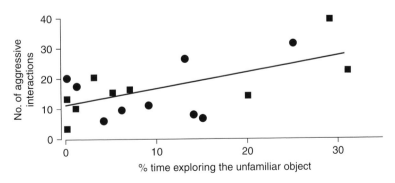

Fig. 10.6. Illustration of consistency in reactivity across different situations. There was a positive correlation ($r = 0.59$; $p = 0.01$) between the time spent exploring an unfamiliar object in a test conducted during the rearing period and the number of aggressive interactions during mixing before slaughter, for Large White (squares) and Duroc (circles) pigs. (Data from Terlouw and Rybarczyk, 2008.)

becomes a problem for animal welfare when the animal cannot regulate its body temperature using the mechanisms described above. Therefore, because evaporative heat loss is the main defence against hyperthermia, an animal exposed to heat stress but with access to drinking water has better welfare than an animal that becomes dehydrated.

Under heat stress, animals grow or reproduce less well than they do in their thermoneutral zone. For example, heat stress has a dramatic effect on the reproductive system of ruminants. In cattle, heat stress alters the duration of oestrus, colostrum quality, conception rate, uterine function, endocrine status, follicular growth and development, luteolytic mechanisms, early embryonic development and fetal growth (De Rensis and Scaramuzzi, 2003). Hormonal systems are also affected by exposure to high ambient temperature. In sheep, an excessive heat load can lead to a *heat syndrome* which is characterized by an activation of the HPA axis and the secretion of catecholamines, and a decrease in insulin and thyroid hormone, leading to lipolysis and protein anabolism, and ultimately to tissue damage (Marai *et al.*, 2007). Heat stress also increases the plasma concentration of prolactin (Parrott *et al.*, 1996). Although the underlying mechanism is not understood, it could be related to vasopressin systems, which are involved in the control of prolactin release as well as in body temperature (Matthews and Parrott, 1994; Terlouw *et al.*, 1996). Thus, in addition to body core temperature, measurement of the activation of these different systems can provide an indication of the strain experienced by an animal exposed to temperatures higher than the upper limit of its thermoneutral zone.

Osmotic stress

The regulation of an adequate volume of body water and of the osmolarity of the intracellular and extracellular compartments is essential for survival (McKinley *et al.*, 2004). Osmoregulation is very important for aquatic animals. In fish, both chronic and acute stress, for example during transport, crowding and handling, induce dramatic changes in the osmoregulatory system (Iwama *et al.*, 1997). The response to stress in fish, as in other animals, is primarily mediated by the changes in plasma adrenaline and cortisol, which are concomitant with changes in electrolyte disturbance in the blood leading to osmotic stress (McDonald and Milligan,

1997). In fish, the physiological markers of osmoregulation are a very good indicator of the level of welfare (Ashley, 2007).

Osmoregulation may be influenced not only by increased water loss during heat exposure, but also by other challenges, such as changes in salt or water intake. The physiological responses to water restriction stimulate a reduction in water loss and, in humans and presumably in other animals, the perception of thirst. If water is available, the perception of thirst stimulates an increase in water intake. Hence, simple behaviours such as drinking and urination are potential indicators of animal welfare status. The degree of dehydration, hypovolaemia (decrease in blood volume) and hypervolaemia (fluid overload) indicate the degree of osmoregulatory strain. The physiological indicators of these conditions include blood parameters such as packed cell volume, concentrations of electrolytes (Na^+, K^+, Cl^-) and osmolality, which may be obtained with automated assays (but need to be validated for each species). However, these are all only indirect measures of the osmotic state of the intracellular or extracellular fluid compartments. Saliva plays an important role in osmoregulation because animals can excrete and recycle electrolytes (Na^+, HPO_4^-). For example, dehydration decreases the salivary flow and increases saliva osmolality. Urine composition also reflects the strain on kidney function. The renal excretion of nitrogenous compounds (urea, uric acid, nitrogen) reflects not only the elimination of nitrogenous end products but also the regulation of acid–base status and the osmolality of body fluids (King and Goldstein, 1985). The activity of the osmoregulatory system can be assessed using a number of hormones that control both water intake and kidney function.

Variability in body water and its regulation also has implications for the interpretation of changes in the blood levels of hormones or electrolytes, because when the extracellular volume changes, the concentrations of hormones or electrolytes are directly affected. Further, the adaptation to extreme conditions can result in variability between species and breeds in the volume change that a species can withstand. The use of these parameters as indicators of animal welfare therefore requires good knowledge of the animal's biology. For example, ruminants can use their rumens as a water reserve, allowing certain goat breeds (black Bedouin, Barmer) to survive in the desert without drinking for up to 4 days (Silanikove, 1994). These goats can tolerate a water loss of up to 40% of their live weight, while in other species a loss of 20% (farmed sheep), 18% (cattle), or 15% (monogastric mammals) is often lethal (Willmer et al., 2000). Some non-mammalian vertebrates, such as desert frogs, seabirds, and crocodilians also have specific osmoregulatory mechanisms (Bentley, 1998). When fish are in hypo-osmotic water, prolactin is important for maintaining a low water permeability of the transport epithelia (Nishimura, 1985). In the assessment of welfare relative to strain on the osmoregulatory system, it is essential to take into account that osmoregulation is a very resilient system in most species because of its central role in survival.

Energy depletion

The capacity of any biological system to respond to challenge depends to a large degree on its energy reserves. An example is the modern dairy cow. During the first few weeks of lactation, a dairy cow can face an energy deficit because the sum of the energy required to support her maintenance requirements plus milk production exceeds her energy intake, resulting in her having to use her energy reserves. Energy balance can be measured as the difference between the energy used in metabolic processes and the energy available (the difference between intake in feed and output in faeces and urine); this difference accounts for the use of energy reserves, which can be estimated using changes in live weight and body reserves. However, none of these measures are easy to obtain because they require the use of metabolic crates or metabolic chambers and several days of data collection (Blaxter, 1989; Lighton, 2008). Information on the regulatory systems controlling energy balance can be obtained by measuring the main regulatory hormones, such as insulin and glucagon, as well as growth hormone, cortisol and catecholamines. The level of body reserves in adipose tissue correlates well with the circulating level of the adipose-derived hormone leptin. The level of activation of these hormonal systems may serve as an indicator of the strain on the bioenergetics of the animal during long-term challenges such as underfeeding, exercise or milk production. For example, plasma concentrations of insulin, IGF-1 (insulin-like growth factor 1) and leptin fall sharply during the first few weeks after parturition in dairy cows,

when they are in a state of negative energy balance (Ingvartsen and Boisclair, 2001). At that time, dairy cows are very susceptible to pathogens and often develop mastitis or lameness (Ingvartsen et al., 2003). Cold stress may also influence hormonal systems related to energy status. In merino sheep, plasma leptin levels decreased after shearing and remained low for up to 14 days (Fig. 10.7). However, the relevance of these systems as indicators of animal welfare during short-term challenge remains to be established.

Fatigue

Muscle function depends on the capacity of the organism to provide the muscle with enough energy substrate (glucose and fatty acids) and oxygen. Indicators of energy use in blood (oxygen, glucose, lactate) or in the muscle itself (including glycogen and lactate), may help to assess muscle fatigue. Such measurements have been used in the assessment of animal welfare during long-term haulage of livestock, during which the animals have difficulty in resting (Warriss et al., 1993; Knowles et al., 1997). In addition, enzymes in the pathways of energy utilization in muscle, such as creatine kinase and lactate dehydrogenase, leak from the muscle tissue into the general circulation and their plasma levels have also been used as markers of muscle fatigue in transport studies (Knowles et al., 1996; Grandin, 2007; Barton Gade, 2008).

The immune system and the stress of infection

The immune system has evolved to limit and/or prevent disease by identifying and destroying pathogens and tumour cells. Pathogenic organisms stimulate the immune system, and the recognition of infection and mobilization of defences are a challenge to the physiology of animal. The capacity of an animal to mount an appropriate immune response is influenced by its genetics (including gender) and capacity to mobilize reserves (Colditz, 2008), the season (Nelson et al., 2002), the social environment (Bartolomucci, 2007) and the animal's emotional status (Ader, 2000). In contrast, stressors that induce an HPA-axis response induce some degree of immunosuppression. For example, in laboratory rodents, a large number of studies have shown that social stress (presence of intruder,

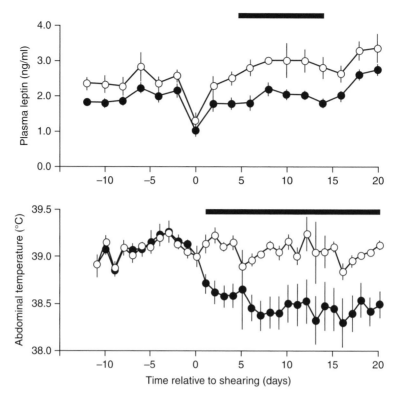

Fig. 10.7. Plasma concentration of leptin (top panel) and abdominal temperature (bottom panel) in merino sheep (*n* = 5) either mock shorn (no blade on the clippers; open circles) or shorn (closed circles) before and after mock shearing or shearing (day 0) during winter in Western Australia (ambient temperature, max. = 15 °C, min. = 4 °C). For 14 days after shearing the concentration of leptin was lower in the shorn sheep than in the control (mock shorn) sheep, while the abdominal temperature remained lower in the shorn sheep until the experiment ended 20 days after shearing (blue bars). Note the decrease in leptin in both groups resulting from the shearing procedure. The sample on day 0 was taken 10 min after (mock) shearing. (Unpublished data from Chen, Maloney and Blache.)

isolation) induces a change in cellular immunity that reduces lymphocyte proliferation (Bartolomucci, 2007). In fact, a large array of indicators, which include specific signals (cytokines), blood cells and internal temperature (fever), can be used to assess the level of strain on the immune system. In addition, parasite load can be used to assess the level of stress applied to the immune system.

10.4 Limitations of the Physiological Assessment of Welfare

Physiological parameters are often considered to be objective measurements of animal welfare or the response to stress because they are simple, scientifically well-defined measurements and, in general, their 'normal' range is known or can be determined.

However, in practice, their measurement can be difficult because the act of measurement may negatively influence the welfare of the animals, for example in the case of invasive procedures such as biopsy or intensive blood sampling. The choice of parameters may also be difficult. Different physiological parameters give complementary information about animal welfare while, at the same time, they are often interdependent. The behavioural context should also be taken into account. For example, an animal showing no overt behaviour may simply be resting, or it may be behaviourally inhibited owing to high arousal. Similarly, changes in physiological variables, such as an increase in heart rate, may be a response to stress, or may be provoked by supposedly pleasant events, such as food delivery (Bloom *et al.*, 1975) or sexual

D. Blache *et al.*

activity. Finally, the interpretation should take into account that physiological parameters are controlled by several, sometimes many, inputs and that the physiological response to a given stimulus varies not only between species but also between breeds and individuals. An integrated approach is needed to study potential sources of stress and their impact on animal welfare. When individual differences in behavioural and physiological responses are taken into account, the different strategies by means of which individual animals respond to stress and the adaptive capacity of the individual may be evaluated (Boissy and Bouissou, 1995; Greiveldinger *et al.*, 2007).

10.5 Conclusions

- The concept of stress has evolved in parallel with our increasing understanding of animal and human biology.
- The assessment of animal welfare is currently based on the measurement of a few isolated parameters, such as the activation of the hypothalamo–pituitary–adrenal axis, or changes in heart rate or respiration rate, and so measure few facets of the complex response to stress.
- The threshold of activation, duration of activation and amplitude of responses of any particular physiological system can be modified by its interaction with other physiological systems, and therefore should not be considered individually.
- Many stressors have an emotional as well as a physical component, and both components activate physiological and emotional responses.
- In the light of new theoretical concepts of stress and adaptation, and increased knowledge of the physiological systems involved in the response to stress, the assessment of animal welfare needs to evolve further towards measures that will integrate the emotional and physiological dimensions of the various responses to stressors of physical and emotional origin.
- The measurement of multiple physiological parameters is necessary to better understand the integrated response to stress.
- The brain has a central role in the integration of signals from both the environment and the body, and controls physiological systems and behaviour either directly or indirectly.
- Part of physiological control involves the modification of the set points of systems,

whether these are directly or indirectly influenced by the stressor.
- When assessing animal welfare based on physiological measures of the response to stress, it is important to consider the fact that the brain may modify set points to optimize the adaptive response to future challenges.
- Allostasis is a potential framework in which to develop integrative physiological measures of animal welfare.

References

Ader, R. (2000) On the development of psychoneuroimmunology. *European Journal of Pharmacology* 405, 167–176.

Ashley, P.J. (2007) Fish welfare: current issues in aquaculture. *Applied Animal Behaviour Science* 104, 199–235.

Barnett, J.L. and Hemsworth, P.H. (1990) The validity of physiological and behavioural measures of animal welfare. *Applied Animal Behaviour Science* 25, 177–187.

Barnum, C.J., Blandino, P. Jr, and Deak, T. (2007) Adaptation in the corticosterone and hyperthermic responses to stress following repeated stressor exposure. *Journal of Neuroendocrinology* 19, 632–642.

Bartolomucci, A. (2007) Social stress, immune functions and disease in rodents. *Frontieres in Neuroendocrinology* 28, 28–49.

Barton Gade, P. (2008) Effect of rearing system and mixing at loading on transport and lairage behaviour and meat quality: comparison of outdoor and conventionally raised pigs. *Animal* 2, 902–911.

Bentley, P.J. (1998) *Comparative Vertebrate Endocrinology*, 3rd edn. Cambridge University Press, Cambridge, UK.

Bernard, C. (1878) *Leçons sur les Phénomènes de la Vie Communs aux Animaux et aux Végétaux* (Dastre, A., ed.). Baillière, Paris.

Blache, D. and Martin, G.B. (1999) Day length affects feeding behaviour and food intake in adult male emus (*Dromaius novaehollandiae*). *British Poultry Science* 40, 573–578.

Blache, D., Malecki, I.A., Williams K.M. and Martin, G.B. (1997) Responses of juvenile and adult emus (*Dromaius novaehollandiae*) to artificial photoperiod. In: Kawashima, S. and Kikuyama, S. (eds) *Proceedings of the XIII International Congress of Comparative Endocrinology*. International Society for Avian Endocrinology, Yokohama, Japan, pp. 445–450.

Blache, D., Tellam, R., Chagas, L.M., Blackberry, M.A., Vercoe, P.V. and Martin, G.B. (2000) Level of nutrition affects leptin concentrations in plasma and cerebrospinal fluid in sheep. *Journal of Endocrinology* 165, 625–637.

Blache, D., Chagas, L.M. and Martin, G.B. (2007) Nutritional inputs into the reproductive neuroendocrine control system – a multidimensional perspective. In: Juengel, J.I., Murray, J.F. and Smith, M.F. (eds) *Reproduction in Domestic Ruminants VI*. Nottingham University Press, Nottingham, UK, pp. 123–139.

Blaxter, K.L. (1989) *Energy metabolism in animals and man*. Cambridge University Press, Cambridge, UK.

Bloom, S.R., Edwards, A.V., Hardy, R.N., Malinowska, K. and Silver, M. (1975) Cardiovascular and endocrine responses to feeding in the young calf. *Journal of Physiology* 253, 135–155.

Boissy, A. and Bouissou, M.-F. (1988) Effects of early handling on heifers' subsequent reactivity to humans and to unfamiliar situations. *Applied Animal Behaviour Science* 20, 259–273.

Boissy, A. and Bouissou, M.-F. (1995) Assessment of individual differences in behavioural reactions of heifers exposed to various fear-eliciting situations. *Applied Animal Behaviour Science* 46, 17–31.

Bouwknecht, A.J., Olivier, B. and Paylor, R.E. (2007) The stress-induced hyperthermia paradigm as a physiological animal model for anxiety: a review of pharmacological and genetic studies in the mouse. *Neuroscience and Biobehavioral Reviews* 31, 41–59.

Briese, E. (1995) Emotional hyperthermia and performance in humans. *Physiology and Behavior* 58, 615–618.

Broom, D.M. (2008) Consequences of biological engineering for resource allocation and welfare. In: Rauw, W. (ed.) *Resource Allocation Theory Applied to Farm Animal Production*. Oxford University Press, Oxford, UK, pp. 261–274.

Broom, D.M. and Johnson, K.G. (1993) *Stress and Animal welfare*. Chapman and Hall, London.

Cannon, W.B. (1914) The emergency function of the adrenal medulla in pain and the major emotions. *American Journal of Physiology* 33, 356–372.

Cannon, W.B. (1929) Organization for physiological homeostasis. *Physiological Reviews* 9, 399–431.

Caramaschi, D., De Boer, S.F. and Koolhaas, J.M. (2007) Differential role of the 5-HT1A receptor in aggressive and non-aggressive mice: an across-strain comparison. *Physiology and Behavior* 90, 590–601.

Cavigelli, S.A. (1999) Behavioural patterns associated with faecal cortisol levels in free-ranging female ring-tailed lemurs, *Lemur catta*. *Animal Behaviour* 57, 935–944.

Chichinadze, K. (2004) Motor and neurochemical correlates of aggressive behavior in male mice *Neurophysiology* 36, 262–269.

Chrousos, G.P. and Gold, P.W. (1992) The concepts of stress and stress system disorders. Overview of physical and behavioral homeostasis. *The Journal of the American Medical Association* 267, 1244–1254.

Colditz, I.G. (2008) Allocation of resources to immune responses. In: Rauw, W. (ed.) *Resource Allocation Theory Applied to Farm Animal Production*. Oxford University Press, Oxford, UK, pp. 192–209.

Cook, C.J., Mellor, D.J., Harris, P.J., Ingram, J.R. and Matthews, L.R. (2000) Hands-on and hands-off measurements of stress. In: Moberg, G.P. and Mench, J.A. (eds) *The Biology of Animal Stress: Basic Principles and Implications for Animal Welfare*. CAB International, Wallingford, UK, pp. 123–146.

Cook, N.J., Schaefer, A.L., Lepage, P. and Jones, S.M. (1996) Salivary vs serum cortisol for the assessment of adrenal activity in swine. *Canadian Journal of Animal Science* 76, 329–335.

Crawford, M.H., Bernstein, S.J., Deedwania, P.C., Dimarco, J.P., Ferrick, K.J., Garson, A., Green, L.A., Greene, H.L., Silka, M.J., Stone, P.H. and Tracy, C.M. (1999) ACC/AHA guidelines for ambulatory electrocardiography: executive summary and recommendations. *Circulation* 100, 886–893.

Cubitt, K.F., Winberg, S., Huntingford, F.A., Kadri, S., Crampton, V.O. and Øverli, Ø. (2008) Social hierarchies, growth and brain serotonin metabolism in Atlantic salmon (*Salmo salar*) kept under commercial rearing conditions. *Physiology and Behavior* 94, 529–535.

Damasio, A.R., Grabowski, T.J., Bechara, A., Damasio, H., Ponto, L.L.B., Parvizi, J. and Hichwa, R.D. (2000) Subcortical and cortical brain activity during the feeling of self-generated emotions. *Nature Neuroscience* 3, 1049–1056.

Darwin, C. (1866) *On the Origin of Species by Means of Natural Selection, or, the Preservation of Favoured Races in the Struggle of Life*, 4th ed. John Murray, London.

Dayan, P. and Huys, Q.J.M. (2008) Serotonin, inhibition, and negative mood. *PLoS Computational Biology* 4, e4.

Derbyshire, S.W.G. (2008) Assessing pain in animals. In: Bushnell, C. and Basbaum, I.A. (eds) *The Senses: A Comprehensive Reference*. Elsevier, Oxford, UK, pp. 969–974.

De Rensis, F. and Scaramuzzi, R.J. (2003) Heat stress and seasonal effects on reproduction in the dairy cow: a review. *Theriogenology* 60, 1139–1151.

Désiré, L., Veissier, I., Despres, G. and Boissy, A. (2004) On the way to assess emotions in animals: do lambs (*Ovis aries*) evaluate an event through its suddenness, novelty, or unpredictability? *Journal of Comparative Psychology* 118, 363–374.

Diamond, J. (1993) Evolutionary physiology. In: Nobel, D. and Boyd, C.A.R. (eds) *The Logic of Life*. Oxford University Press, Oxford, UK, pp. 89–111.

Elsasser, T.H., Klasing, K.C., Filipov, N. and Thompson, F. (2000) The metabolic consequences of stress: targets for stress and priorities of nutrient use. In: Moberg, G.P. and Mench, J.A. (eds) *The Biology of*

Animal Stress: Basic Principles and Implications for Animal Welfare. CAB International, Wallingford, UK, pp. 77–110.

Engel, B.T. (1985) Stress is a noun! No, a verb! No, an adjective! In: Field, T.M., McCabe, P.M. and Schneiderman, N. (eds) *Stress and Coping.* Lawrence Erlbaum Associates, Hillsdale, New Jersey, pp. 3–12.

Fava, M. and Guaraldi, G.P. (2006) Prolactin and stress. *Stress and Health* 3, 211–216.

Forkman, B., Boissy, A., Meunier-Salaun, M.-C., Canali, E. and Jones, R.B. (2007) A critical review of fear tests used on cattle, pigs, sheep, poultry and horses. *Physiology and Behavior* 91, 531–565.

Foury, A., Geverink, N.A., Gil, M., Gispert, M., Hortós, M., Furnols, M.F.I., Carrion, D., Blott, S.C., Plastow, G.S. and Mormède, P. (2007) Stress neuroendocrine profiles in five pig breeding lines and the relationship with carcass composition. *Animal* 1, 973–982.

Fraser, D. Ritchie, J.S.D. and Faser, A.F. (1975) The term "stress" in a veterinary context. *British Veterinary Journal* 131, 653–662.

Goddard, P.J., Gaskin, G.J. and Macdonald, A.J. (1998) Automatic blood sampling equipment for use in studies of animal physiology. *Animal Science* 66, 796–775.

Goldstein, D.S. (1995) *Stress, Catecholamines, and Cardiovascular Disease.* Oxford University Press, New York.

Grandin, T. (2007) *Livestock Handling and Transport,* 3rd edn. CAB International, Wallingford, UK.

Greiveldinger, L., Veissier, I. and Boissy, A. (2007) Emotional experience in sheep: predictability of a sudden event lowers subsequent emotional responses. *Physiology and Behavior* 92, 675–683.

Groenink, L., Van Der Gugten, J., Zethof, T., Van Der Heyden, J. and Olivier, B. (1994) Stress-induced hyperthermia in mice: hormonal correlates. *Physiology and Behavior* 56, 747–749.

Gupta, S., Earley, B., Ting, S.T.L., Leonard, N. and Crowe, M.A. (2004) Technical note: effect of corticotropin-releasing hormone on adrenocorticotropic hormone and cortisol in steers. *Journal of Animal Science* 82, 1952–1956.

Henane, R. and Valatx, J.L. (1973) Thermoregulatory changes induced during heat acclimatization by controlled hyperthermia in man. *The Journal of Physiology* 230, 255–271.

Henessy, J.W. and Levine, S. (1979) Stress, arousal, and the pituitary adrenal system: a psychoendocrine hypothesis. *Progress in Psychobiology and Physiological Psychology* 8, 133–178.

Ingvartsen, K.L. and Boisclair, Y.R. (2001) Leptin and the regulation of food intake, energy homeostasis and immunity with special focus on periparturient ruminants. *Domestic Animal Endocrinology* 21, 215–250.

Ingvartsen, K.L., Dewhurst, R.J. and Friggens, N.C. (2003) On the relationship between lactational performance and health: is it yield or metabolic imbalance that cause production diseases in dairy cattle? A position paper. *Livestock Production Science* 83, 277–308.

IUPS (International Union of Physiological Sciences) Thermal Commission (2001) Glossary of terms for thermal physiology: third edition. *Japan Journal of Physiology* 51, 245–280.

Iwama, G., Pickering, A., Sumpter, J. and Schreck, C. (1997) *Fish Stress and Health in Aquaculture.* Cambridge University Press, Cambridge, UK.

Jones, H.E., Lewis, R.M., Young, M.J. and Wolf, B.T. (2002) The use of X-ray computer tomography for measuring the muscularity of live sheep. *Animal Science* 75, 387–399.

Jordan, E.R. (2003) Effects of heat stress on reproduction. *Journal of Dairy Science* 86 (Supplement), E104–E114.

Kant, J.G., Eggleston, T., Landman-Roberts, L., Kenion, C.C., Driver, G.C. and Meyerhoff, J.L. (1985) Habituation to repeated stress is stressor specific. *Pharmacology Biochemistry and Behavior* 22, 631–634.

King, P. and Goldstein, L. (1985) Renal excretion of nitrogenous compounds in vertebrates. *Renal Physiology* 8, 261–278.

Kleitman, N. and Jackson, D.P. (1950) Body temperature and performance under different routines. *Journal of Applied Physiology* 3, 309–328.

Kluger, M.J. O'Reilly, B., Shope, T.R. and Vander, A.J. (1987) Further evidence that stress hyperthermia is a fever. *Physiology and Behavior* 39, 763–766.

Knowles, T.G., Warriss, P.D., Brown, S.N., Kestin, S.C., Edwards, J.E., Perry, A.M., Watkins, P.E. and Phillips, A.J. (1996) Effects of feeding, watering and resting intervals on lambs transported by road and ferry to France. *Veterinary Record* 139, 335–339.

Knowles, T.G., Warriss, P.D., Brown, S.N., Edwards, J.E., Watkins, P.E. and Philips, A.J. (1997) Effects on calves less than one month old of feeding or not feeding them during road transport of up to 24 hours. *Veterinary Record* 140, 116–124.

Koolhaas, J.M., Korte, S.M., De Boer, S.F., Van Der Veght, B.J., Van Reenen, C.G., Hopster, H., De Jong, I.C. and Blokhuis, H.J. (1999) Coping styles in animals: current status in behavior and stress-physiology. *Neuroscience and Biobehavioral Reviews* 23, 925–935.

Korte, S.M. Olivier, B. and Koolhaas, J.M. (2007) A new animal welfare concept based on allostasis. *Physiology & Behavior* 92, 422–428.

Krantz, D.S. and Lazar, J.D. (1987) Behavioral factors in hypertension. In: Julius, S. and Bassett, D.R. (eds) *The Stress Concept: Issues and Measurements.* Elsevier, New York, pp. 43–58.

Ladewig, J. and Smidt, D. (1989) Behavior, episodic secretion of cortisol and adrenocortical reactivity in bulls subjected to tethering. *Hormones and Behavior* 23, 344–360.

Lay, D., Friend, T., Bowers, C., Grissom, K. and Jenkins, O. (1992) A comparative physiological and behavioral study of freeze and hot-iron branding using dairy cows. *Journal of Animal Science* 70, 1121–1125.

Lefcourt, A.M. and Elsasser, T.H. (1995) Adrenal responses of Angus × Hereford cattle to the stress of weaning. *Journal of Animal Science* 73, 2669–2676.

Leknes, S. and Tracey, I. (2008) A common neurobiology for pain and pleasure. *Nature Reviews Neuroscience* 9, 314–320.

Le Moal, M. (2007) Historical approach and evolution of the stress concept: a personal account. *Psychoneuroendocrinology* 32, S3–S9.

Levine, S. and Ursin, H. (1991) What is stress? In: Brown, M.R., Koob, G.F. and Rivier, C. (eds) *Stress: Neurobiology and Neuroendocrinology*. Dekker, New York, pp. 3–21.

Lighton, J.R.B. (2008) *Measuring Metabolic Rates: A Manual for Scientists*. Oxford University Press, Oxford, UK.

Lucot, J.B., Jackson, N., Bernatova, I. and Morris, M. (2005) Measurement of plasma catecholamines in small samples from mice. *Journal of Pharmacological and Toxicological Methods* 52, 274–277.

Lynch, J.J., Mottershead, B.E. and Alexander, G. (1980) Sheltering behaviour and lamb mortality amongst shorn Merino ewes lambing in paddocks with a restricted area of shelter or no shelter. *Applied Animal Ethology* 6, 163–174.

Marai, I.F.M., El-Darawany, A.A,. Fadiel, A. and Abdel-Hafez, M.A.M. (2007) Physiological traits as affected by heat stress in sheep – a review. *Small Ruminant Research* 71, 1–12.

Marazziti, D., Dimuro, A. and Castrogiovanni, P. (1992) Psychological stress and body temperature changes in humans. *Physiology and Behavior* 52, 393–395.

Marchant-Forde, R.M., Marlin, D.J. and Marchant-Forde, J.N. (2004) Validation of a cardiac monitor for measuring heart rate variability in adult female pigs: accuracy, artefacts and editing. *Physiology and Behavior* 80, 449–458.

Matthews, S.G. and Parrott, R.F. (1994) Centrally administered vasopressin modifies stress hormone (cortisol, prolactin) secretion in sheep under basal conditions, during restraint and following intravenous corticotrophin-releasing hormone. *European Journal of Endocrinology* 130, 297–301.

McDonald, G. and Milligan, L. (1997) Ionic, osmotic and acid–base regulation in stress. In: Iwama, G., Pickering, A., Sumpter, J. and Schreck, C. (eds) *Fish Stress and Health in Aquaculture*. Cambridge University Press, Cambridge, UK, pp. 119–144.

McEwen, B.S. (1998) Stress, adaptation, and disease. Allostasis and allostatic load. *Annals of the New York Academy of Sciences* 840, 33–44.

McEwen, B.S. and Wingfield, J.C. (2007) Allostasis and allostatic load. In: Fink, G. (ed.) *Encyclopedia of Stress*. Academic Press, New York, pp. 135–141.

McKinley, M.J., Cairns, M.J., Denton, D.A., Egan, G., Mathai, M.L., Uschakov, A., Wade, J.D., Weisinger, R.S. and Oldfield, B.J. (2004) Physiological and pathophysiological influences on thirst. *Physiology and Behavior* 81, 795–803.

Meyer, L., Fick, L., Matthee, A., Mitchell, D. and Fuller, A. (2008) Hyperthermia in captured impala (*Aepyceros melampus*): a fright not flight response. *Journal of Wildlife Diseases* 44, 404–416.

Moberg, G.P. (2000) Biological response to stress: implications for animal welfare. In: Moberg, G.P. and Mench, J.A. (eds.) *The Biology of Animal Stress: Basic Principles and Implications for Animal Welfare*. CAB International, Wallingford, UK, pp. 1–21.

Moberg, G.P. and Mench, J.A. (2000) *The Biology of Animal Stress: Basic Principles and Implications for Animal Welfare*. CAB International, Wallingford, UK.

Moberg, G.P., Anderson, C.O. and Underwood, T.R. (1980) Ontogeny of the adrenal and behavioral responses of lambs to emotional stress. *Journal of Animal Science* 51, 138–142.

Mohr, E., Langbein, J. and Nürnberg, G. (2002) Heart rate variability – a noninvasive approach to measure stress in calves and cows. *Physiology and Behavior* 75, 251–259.

Mormède, P., Andanson, S., Auperin, B., Beerda, B., Guemene, D., Malmkvist, J., Manteca, X., Manteuffel, G., Prunet, P., Van Reenen, C.G., Richard, S. and Veissier, I. (2007) Exploration of the hypothalamic–pituitary–adrenal function as a tool to evaluate animal welfare. *Physiology and Behavior* 92, 317–339.

Mrosovsky, N. (1990) *Rheostasis: The Physiology of Change*. Oxford University Press, Oxford, UK.

Munck, A. and Guyre, P.M. (1986) Glucocorticoid physiology, pharmacology, and stress. In: Chrousos, G.P., Loriaux, D.L. and Lipsett, M.B. (eds) *Steroid Hormone Resistance*. Plenum Press, New York, pp. 81–96.

Nelson, R., Demas, G.E., Klein, S.L. and Kriegsfeld, L.J. (2002) *Seasonal Patterns of Stress, Immune Function, and Disease*. Cambridge University Press, Cambridge, UK.

Nishimura, H. (1985) Endocrine control of renal handling of solutes and water in vertebrates. *Renal Physiology* 8, 279–300.

Owens, P., Smith, R., Green, D. and Falconer, L. (1984) Effect of hypoglycemic stress on plasma and cerebrospinal fluid immunoreactive beta-endorphin in conscious sheep. *Neuroscience Letters* 49, 1–6.

Pacák, K. and Palkovits, M. (2001) Stressor specificity of central neuroendocrine responses: implications for stress-related disorders. *Endocrine Reviews* 22, 502–548.

Parrott, R.F., Lloyd, D.M. and Goode, J.A. (1996) Stress hormone responses of sheep to food and water deprivation at high and low ambient temperatures. *Animal Welfare* 1, 45–56.

Porges, S.W. (1995) Cardiac vagal tone: a physiological index of stress. *Neuroscience and Biobehavioral Reviews* 19, 225–233.

Renbourn, E.T. (1960) Body temperature and pulse rate in boys and young men prior to sporting contests. A study of emotional hyperthermia: with a review of the literature. *Journal of Psychosomatic Research* 4, 149–175.

Ringer, R.K. (1971) Adaptation of poultry to confinement rearing systems. *Journal of Animal Science* 32, 590–598.

Ropert-Coudert, Y. and Wilson, R.P. (2005) Trends and Perspectives in animal-attached remote sensing. *Frontiers in Ecology and the Environment* 3, 437–444.

Säkkinen, H., Tornbeg, J., Goddard, P.J., Eloranta, E., Ropstad, E. and Saarela, S. (2004) The effect of blood sampling method on indicators of physiological stress in reindeer (*Rangifer tarandus tarandus*). *Domestic Animal Endocrinology* 26, 87–98.

Sapolsky, R.M., Romero, L.M. and Munck, A.U. (2000) How do glucocorticoids influence stress responses? Integrating permissive, suppressive, stimulatory, and preparative actions. *Endocrine Reviews* 21, 55–89.

Selye, H. (1936) A syndrome produced by diverse nocuous agents. *Nature* 138, 32.

Selye, H. (1950) *The Physiology and Pathology of Exposure to Stress: A Treatise Based on the Concepts of the General-Adaptation-Syndrome and the Diseases of Adaptation.* Acta, Inc. Medical Publishers, Montreal.

Silanikove, N. (1994) The struggle to maintain hydration and osmoregulation in animals experiencing severe dehydration and rapid rehydration: the story of ruminants. *Experimental Physiology* 79, 281–300.

Singer, R., Harker, C.T., Vander, A.J. and Kluger, M.J. (1986) Hyperthermia induced by open-field stress is blocked by salicylate. *Physiology and Behavior* 36, 1179–1182.

Spooren, W.P.J.M., Schoeffter, P., Gasparini, F., Kuhn, R. and Gentsch, C. (2002) Pharmacological and endocrinological characterisation of stress-induced hyperthermia in singly housed mice using classical and candidate anxiolytics (LY314582, MPEP and NKP608). *European Journal of Pharmacology* 435, 161–170.

Sterling, P. and Eyer, J. (1988) Allostasis: a new paradigm to explain arousal pathology. In: Fisher, S. and Reason, J. (eds) *Handbook of Life Stress, Cognition, and Health.* Wiley, New York, pp. 629–649.

Strawn, J.R., Ekhator, N.N., Horn, P.S., Baker, D.G. and Geracioti, T.D.J. (2004) Blood pressure and cerebrospinal fluid norepinephrine in combat-related post-traumatic stress disorder. *Psychosomatic Medicine* 66, 757–759.

Terlouw, C. (2005) Stress reactions at slaughter and meat quality in pigs: genetic background and prior experience. A brief review of recent findings: product quality and livestock systems. *Livestock Production Science* 94, 125–135.

Terlouw, E.M.C. and Rybarczyk, P. (2008) Explaining and predicting differences in meat quality through stress reactions at slaughter: the case of Large White and Duroc pigs. *Meat Science* 79, 795–805.

Terlouw, E.M., Kent, S., Cremona, S. and Dantzer, R. (1996) Effect of intracerebroventricular administration of vasopressin on stress-induced hyperthermia in rats. *Physiology and Behavior* 60, 417–424.

Torner, L., Karg, S., Blume, A., Kandasamy, M., Kuhn, H.-G., Winkler, J., Aigner, L. and Neumann, I.D. (2009) Prolactin prevents chronic stress-induced decrease of adult hippocampal neurogenesis and promotes neuronal fate. *Journal of Neuroscience* 29, 1826–1833.

Turek, F.W. (2007) From circadian rhythms to clock genes in depression. *International Clinical Psychopharmacology* 22, S1–S8.

Van Cleeff, J. (2002). Reproductive activity alters fat metabolism patterns in the male emu (*Dromaius novaehollandiae*). PhD thesis, The University of Western Australia, Crawley, Western Australia.

Van Erp, A.M.M. and Miczek, K.A. (2000) Aggressive behavior, increased accumbal dopamine, and decreased cortical serotonin in rats. *Journal of Neuroscience* 20, 9320–9325.

Van Reenen, C.G., O'Connell, N.E., Van Der Werf, J.T.N., Korte, S.M., Hopster, H., Jones, R.B. and Blokhuis, H.J. (2005) Responses of calves to acute stress: individual consistency and relations between behavioral and physiological measures. *Physiology and Behavior* 85, 557–570.

Veening, J.G., Bouwknecht, J.A., Joosten, H.J.J., Dederen, P.J., Zethof, T.J.J., Groenink, L., Van Der Gugten, J. and Olivier, B. (2004) Stress-induced hyperthermia in the mouse: c-fos expression, corticosterone and temperature changes. *Progress in Neuro-Psychopharmacology and Biological Psychiatry* 28, 699–707.

Vinkers, C.H., Van Bogaert, M.J.V., Klanker, M., Korte, S.M., Oosting, R., Hanania, T., Hopkins, S.C., Olivier, B. and Groenink, L. (2008) Translational aspects of pharmacological research into anxiety disorders: the stress-induced hyperthermia (SIH) paradigm. *European Journal of Pharmacology* 585, 407–425.

Visser, E.K., Ellis, A.D. and Van Reenen, C.G. (2008) The effect of two different housing conditions on the welfare of young horses stabled for the first time. *Applied Animal Behaviour Science* 114, 521–533.

Von Borell, E. and Ladewig, J. (1989) Altered adrenocortical response to acute stressors or ACTH (1-24) in intensively housed pigs. *Domestic Animal Endocrinology* 6, 299–309.

Von Borell, E., Dobson, H. and Prunier, A. (2007a) Stress, behaviour and reproductive performance in female cattle and pigs. *Hormones and Behavior* 52, 130–138.

Von Borell, E., Langbein, J., Despres, G., Hansen, S., Leterrier, C., Marchant-Forde, J., Marchant-Forde, R., Minero, M., Mohr, E., Prunier, A., Valance, D. and Veissier, I. (2007b) Heart rate variability as a measure of autonomic regulation of cardiac activity for assessing stress and welfare in farm animals – a review. *Physiology and Behavior* 92, 293–316.

Warriss, P.D., Kestin, S.C., Brown, S.N., Knowles, T.G., Wilkins, L.J., Edwards, J.E., Austin, S.D. and Nicol, C.J. (1993) The depletion of glycogen stores and indices of dehydration in transported broilers. *British Veterinary Journal* 149, 391–398.

Weiner, H. (1991) Behavioral biology of stress and psychosomatic medicine. In: Brown, M.R., Koob, G.F. and Rivier, C. (eds) *Stress: Neurobiology and Neuroendocrinology*. Dekker, New York, pp. 23–51.

Weiss, J.M. (1971a) Effects of punishing the coping response (conflict) on stress pathology in rats. *Journal of Comparative and Physiological Psychology* 77, 14–21.

Weiss, J.M. (1971b) Effects of coping behavior with and without a feedback signal on stress pathology in rats. *Journal of Comparative and Physiological Psychology* 77, 22–30.

Willmer, P., Stone, G. and Johnston, I. (2000) *Environmental Physiology of Animals*. Blackwell Publishing, Oxford, UK.

D. Blache *et al.*

11 Preference and Motivation Research

David Fraser and Christine J. Nicol

Abstract

A major approach within animal welfare science involves research on the preferences of animals for options such as ambient temperature, illumination and types of flooring, and on the strength of animals' motivation to perform certain types of behaviour, obtain certain resources, and avoid unpleasant features of their housing and handling. Various cautions need to be taken into account in such research. The preferences of an animal are likely to vary with the animal's condition, the time of day, environmental conditions and the animal's on-going behaviour; therefore, preference experiments must be comprehensive enough to identify the relevant sources of variation. Experiments must also avoid confounding preference with familiarity. Preference research is most useful if it identifies the factors underlying the preferences that animals show. To draw inferences about animal welfare from preference research generally requires that we establish how strongly an animal prefers a chosen option or is motivated to perform a type of behaviour (such as nest-building or exploration) that a given environment allows. Various measures of motivation strength have been used, mostly based on determining how hard an animal will work, or what other benefits it will forego, to obtain a preferred option. The conditions preferred by an animal will often promote its welfare. However, preferences may not correspond to welfare if the choices fall outside the animals' sensory, cognitive and affective capacities, or if animals are asked to make choices very unlike those that would have occurred in the environment where the species evolved. There is great scope for incorporating knowledge of animals' preferences and motivations into the design of animal housing and management, and for integrating preference research with other indicators of animal welfare.

11.1 Introduction

Research on the preferences of animals is commonly used as a tool in the study of animal welfare. Fundamental to this research is the assumption that animals make choices that are in their own best interests, and that allowing animals to live as they prefer will ensure a high level of animal welfare. But under what conditions do the choices made by an animal serve as a reliable guide to its welfare? And how can we best incorporate the preferences of animals into housing and management systems?

11.2 The Early Use of Preference Testing

The naturalistic study of animal behaviour is an important precursor to the study of environmental preferences. The fact that birds perch on branches or wires, and that mice burrow in fields or walls, provides information about the environments that these animals prefer. If collected in a systematic and quantitative way – for example by identifying the sizes of branches that birds do and do not use for perching – such observations can be a starting point for designing animal environments and a source of hypotheses for more controlled research on animals' preferences (Dawkins et al., 2003).

Similarly, traditional laboratory studies of behaviour have provided significant insights into the preferences of animals, although such work was often done as basic research on animal motivation. For example, Barnett et al. (1971) monitored the movements of rats and mice in residential mazes where food, water, nesting material and other resources were available in different compartments (Fig. 11.1). Such methods have now been applied more specifically to questions about animal welfare.

The formal proposal to study the preferences of animals as a component of animal welfare research arose in the classic essay, 'The assessment of pain and distress in animals', by British ethologist W.H. Thorpe (1965). Thorpe recounted the story of a group of African buffalo that had to be relocated to Nairobi National Park. In preparation for release, they were kept in paddocks like those used for domestic cattle. After their release they

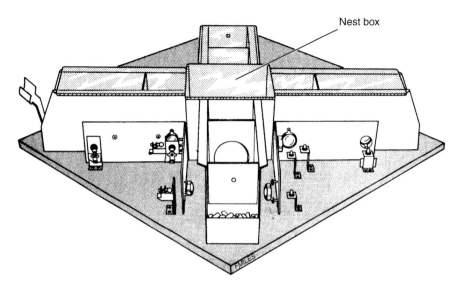

Nest box

Fig. 11.1. A residential maze used to monitor the movements of rats or mice among chambers that gave the animals access to resources such as food, nesting material and wood for gnawing. (From Barnett *et al.*, 1971. This material is reproduced with permission of John Wiley & Sons, Inc.)

continued to return and tried to re-enter the paddocks toward nightfall. Thorpe wrote:

> The natural assumption is that the unfamiliar National Park, reeking of lion, leopard and other dangerous and uncomfortable neighbours, must have seemed a very unfriendly place; far inferior to the luxurious though restricted quarters they had become used to inhabiting!

Thorpe concluded that because these animals had experienced a range of living conditions, we could legitimately 'ask' them which they preferred.

The first actual experiment using preference research to resolve an animal welfare issue arose from a debate over the flooring used in cages for laying hens. The Brambell Committee (1965) had recommended that fine-gauge 'chicken wire', which was used in some of the early cage designs, should be replaced with a heavier wire mesh. To obtain the hens' own view, Hughes and Black (1973) observed where hens chose to be when kept in cages with different sections, each floored in a different material. The hens actually showed no strong preferences, but their overall preference was for, rather than against, the fine-gauge 'chicken wire' that the Committee had deemed unsuitable. In reporting this finding, Hughes and Black expressed enthusiasm for the potential of research on the preferences of animals. They wrote:

> We feel that this type of experiment offers a new approach to animal welfare; objective assessment of animals' preferences should ultimately make subjective value judgements superfluous.

Other early preference research tried to answer broader questions. For example, Dawkins (1977) used the approach of asking whether hens prefer battery cages to large pens or outdoor runs. In one experiment, she gave hens free access to cages and to larger pens for 12 h, and she observed the amount of time that the birds spent in each environment. Perhaps surprisingly, the hens spent considerable time in the cages. Dawkins then did a series of trials in a T-maze, where turning in one direction caused hens to spend the next 5 min in a battery cage, while turning in the other direction led to 5 min in an outdoor run. With this procedure, where discrete choices had clear consequences, the hens tended to select the outdoor run. While fully recognizing the preliminary nature of her results, Dawkins, too, expressed optimism about the approach:

> I hope that one day such work will enable us to take an objective and humane judgment on life in a battery cage.

Since these early examples, environmental preference research and related methods have been

D. Fraser and C.J. Nicol

Table 11.1. Selected examples of the use of preference and motivation research to study animals' preferences, motivations and aversions.

Variable	Species	Reference
Preferences for:		
Ambient temperature	Piglets	Morrison *et al.*, 1987
Illumination level	Pigs	Baldwin and Start, 1985
	Gerbils	van den Broek *et al.*, 1995
	Cattle	Phillips and Morris, 2001
Social contact	Pigs	Matthews and Ladewig, 1994
	Sows	Kirkden and Pajor, 2006
	Rats	Patterson-Kane *et al.*, 2004
Bedding	Pigs	Fraser, 1985
	Horses	Hunter and Houpt, 1989
	Rodents	Blom *et al.*, 1993
	Cattle	Tucker *et al.*, 2003
Flooring	Pigs	Farmer and Christison, 1982
	Sows	Phillips *et al.*, 1996
	Cattle	Telezhenko *et al.*, 2007
Nesting materials	Mice	van de Weerd *et al.*, 1998
Dust-bathing materials	Hens	van Liere *et al.*, 1990
Preferred design features of:		
Loading ramps	Pigs	Phillips *et al.*, 1989
Roosts	Hens	Muiruri *et al.*, 1990
Motivation to perform behaviour:		
Swimming	Mink	Mason *et al.*, 2001
Roosting	Hens	Olsson and Keeling, 2002
Use of nest boxes	Hens	Cooper and Appleby, 1996
Dust-bathing	Hens	Widowski and Duncan, 2000
Exercise	Mice	Sherwin, 1998
Rest	Cattle	Jensen *et al.*, 2005
Aversion to:		
Noise and vibration	Pigs	Stephens *et al.*, 1985
Heat and vibration	Chickens	Abeyesinghe *et al.*, 2001
Polluted air	Pigs	Jones *et al.*, 1999
Gases used for euthanasia	Rodents	Leach *et al.*, 2004
Electro-immobilization	Sheep	Rushen, 1986
Rough handling	Cattle	Pajor *et al.*, 2000b

used for an impressive variety of purposes in animal welfare science (Table 11.1). Topics include preferences for environmental variables such as temperature and illumination, preferred design features of items such as loading ramps and nest boxes, the strength of animals' motivation to perform behaviour such as swimming or roosting, and the strength of their aversions to features such as noise and rough handling.

Despite its widespread use, however, preference research has remained a controversial tool. In fact, the ink had barely dried on the earliest reports of preference experiments when debate broke out over what we can actually conclude from the technique. Initially Duncan (1978), and subsequently many others, provided criticisms of preference research, and these stimulated major changes in how preference research is conducted and interpreted (reviewed by Duncan, 1992; Fraser *et al.*, 1993; Fraser, 2008). In this chapter, we briefly review some of the methodological and conceptual issues that have been raised and that now give us a more nuanced understanding of the role of preference and motivation research in understanding animal welfare.

11.3 Ensuring that Experiments Accurately Identify Animals' Preferences

The most basic concern over preference research is that the experiments performed should accurately identify the animals' true preferences. This requires attention to several points about how we design and conduct the research.

11.3.1 Recognizing the complexity of animals' preferences

One criticism of the early preference research is that some of the questions asked were too simple. On the surface, it might seem reasonable to ask whether pigs prefer pens with straw bedding or pens with bare concrete floors. However, research designed to answer this question showed that pigs sometimes prefer straw, sometimes avoid straw and sometimes are indifferent to straw, depending on a variety of factors. In one study, for example, pigs had free access to two identical pens, one with straw and one with bare concrete flooring (Fig. 11.2). The pigs strongly preferred the straw pen when they were actively rooting on the floor; this indicated that the straw was an effective stimulus for the pigs' natural exploration and foraging behaviour. However, the pigs were relatively indifferent to the presence of straw when they were using a feed or water dispenser. Finally, the pigs clearly chose to lie on the straw when the room was cool, but preferred to rest on the bare concrete

rather than the straw when the room was hot, presumably because of the thermal insulation provided by the straw (Fraser, 1985). Moreover, other research has shown that sows' preference for straw increases sharply just before parturition when they normally make nests (Arey, 1992). In view of this complexity, we need to ask not whether pigs prefer straw, but how their preference is influenced by features of the environment and by the animals' condition and behaviour. Even when the ultimate objective is to decide what kind of housing is best on average for a certain type of animal, research methods that ignore relevant variables are likely to give confusing and contradictory results.

More complex questions about the preferences of animals generally require more complex research methods. First, experiments must be of sufficient duration to monitor preferences under a range of fluctuations in both the environment and the animal's condition. Brief tests in a T-maze, as used in the 1970s, have largely given way to methods such as continuous monitoring over periods of days or weeks in 'closed-economy' studies where animals cannot obtain the resource under investigation outside the experiment itself. For example, Sherwin and Nicol (1996) monitored the movements of mice for 9 days in an apparatus where the animals

Fig. 11.2. Preference research to test the preferences of pigs for different substrates. Two adjoining pens, each equipped with food and water, were identical except that one had a bare concrete floor while the other was bedded with straw. (Reproduced from Fraser, 1996.)

D. Fraser and C.J. Nicol

had to walk through tunnels to reach four different resources: food, shelter, additional space and visual contact with another mouse (Fig. 11.3).

Secondly, experiments should be designed so that variability in the animals' responses can be interpreted. In preference research, animals are rarely completely consistent in the choices they make. For example, Lindberg and Nicol (1996) allowed hens to choose between larger and smaller pens. The birds showed a significant preference for larger enclosures, but almost all birds selected the smaller enclosure at least once.

Why do such exceptions occur? Perhaps animals simply make mistakes, for instance by forgetting which passage in a T-maze leads to the preferred reward. However, minority choices could be important to the animals, for example because animals are motivated to monitor all available environments occasionally. Laboratory mice will work to gain entry to empty spaces outside their home pens, regardless of the quantity of additional space offered, and regardless of the quality, size and presence of enrichments and companions in the home pen (Sherwin and Nicol, 1997; Sherwin, 2004, 2007). Sherwin (2007) argued that this behaviour is most consistent with a strong need to gather information.

The occasional mistake, or time spent patrolling or monitoring, need not affect the overall conclusions drawn from preference experiments, but these forms of behaviour need to be distinguished from genuine minority preferences. Perhaps hens really do want to spend a certain proportion of their time in a small enclosed area. Nicol (1986) suggested that the distinction between genuine minority preference and monitoring behaviour could be investigated by offering two versions of the minority-choice environment side by side: one free to enter and the other involving a cost. If the animal had an occasional need for the environment it could take the freely available option, but if it was strongly motivated to patrol or monitor all nearby environments it should continue to choose both available versions.

11.3.2 Separating preference from familiarity

In the study by Hughes and Black (1973) described earlier, the hens in the experiment had previous experience of chicken-wire flooring. When the birds showed a mild preference for this type of flooring, were they merely choosing what was familiar to them, rather than demonstrating a preference typical of hens in general?

The previous experience of animals can affect the results of preference experiments in several ways. In the simplest cases, animals may show a temporary avoidance of, or attraction to, unfamiliar options. For example, Dawkins (1980) noted that hens housed in cages tended, in initial preference trials, to select a cage over an outdoor run, but a few minutes of exposure to the run was enough to overcome this initial reaction. In other cases, preferences may undergo longer-term change as the animals gain experience of the different options. For example, Phillips et al. (1996) housed sows for 3 weeks in a preference apparatus where the animals could choose to be on different types of flooring. During the first few days in the apparatus, the sows strongly preferred concrete flooring (which was familiar) to alternative products made from metal or plastic. However, this preference waned over several weeks as the animals gained experience of the different materials. In this case, the animals may have needed prolonged exposure to become confident in walking and going through their normal postural changes on unfamiliar surfaces.

In some cases, familiarity has only minor effects. It has been proposed that the substrate preferences

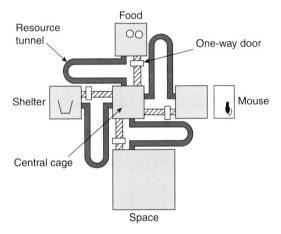

Fig. 11.3. A 'closed economy' apparatus for testing the motivation of mice to access four different resources: food, shelter, extra space and visual contact with another mouse. From the central cage, mice could walk to the different resources via the 'resource tunnels' and then return via the short tunnels with one-way doors. To vary the difficulty of accessing resources, shallow water was added to sections of the resource tunnels. The mice remained in the apparatus for 9 days of testing. (From Sherwin and Nicol, 1996.)

of chickens are 'fixed' during a sensitive period of early development through a process analogous to young birds 'imprinting' on their mothers (Vestergaard and Baranyiova, 1996). To test this idea, Nicol *et al.* (2001) raised hens in wire-floored cages but exposed them to wood shavings for 10-day periods at different times during their development. As adults, the birds' preferences for dust-bathing substrates were influenced to some extent by the age at which they had previously experienced shavings, but their foraging behaviour seemed to be entirely influenced by the current availability of different substrates, regardless of previous exposure.

In summary, familiarity can have various effects – including little or no effect – on animals' preferences, and needs to be included as a feature investigated in preference research. Especially in the case where animals learn about the options only when the preference experiment is being conducted, the experiments need to continue until stable results have been obtained.

11.3.3 Understanding the basis for preferences

Experiments that compare different products or materials (types of flooring or bedding) are of more general value if they also identify the features that underlie the animals' preferences. For example, Farmer and Christison (1982) established the preferences of young pigs for a variety of flooring products, and also measured many properties of the products, including the amount of traction they offered, the degree of heat loss through the material and the abrasiveness of the surface. Statistical analysis then showed that weaned pigs tended to choose high-traction floors, whereas very young piglets chose floors that did not conduct heat away from the body (Christison and Farmer, 1983). By identifying the factors underlying the preferences, the research could provide guidance on products that were not part of the initial research.

An alternative approach to identifying underlying factors was used by Phillips *et al.* (1988) to identify the features that make ramps acceptable to pigs. Ramps are widely used in loading pigs, but pigs baulk at walking up certain ramps for reasons that were not well understood. Phillips *et al.* placed young pigs in a small pen that gave free access to four ramps that the pigs could climb

at will. The research consisted of a series of experiments, each one comparing a different design feature: one experiment used ramps with four different slopes; another compared four levels of illumination; and so on. The pigs showed clear preferences for ramps with shallow slopes (up to about 25°) versus steeper slopes, and for ramps with closely spaced footholds (horizontal bars to provide secure footing) versus ramps with widely spaced footholds. In contrast, they showed no preferences based on ramp width, level of illumination or the openness of the side walls. By studying the various factors in turn, the research thus identified the design features that are important to the animals.

11.4 Other Clarifications

Even when research has established valid information about animals' preferences, several other issues need to be clarified to give useful insights into animal welfare.

11.4.1 Motivation to perform behaviour or obtain an outcome?

When we see gerbils selecting a floor where they can dig a burrow, or sows working to obtain straw before farrowing, are they motivated to perform a certain type of behaviour (burrowing, nest-building) or to have the outcome (a burrow, a nest) that the behaviour commonly produces? The question is important for animal welfare: if the animal merely wants the outcome of the behaviour, then we could provide this feature in the environment rather than attempting to accommodate the behaviour.

The question of nest-building by sows was put to an experimental test by Arey *et al.* (1991), who observed six sows making nests before farrowing. The researchers quantified the time the animals spent in the major nest-building activities of rooting or pawing the substrate and carrying straw to the site. They also recorded the physical features of the nests the sows built. They then provided six other sows with a preformed nest constructed to match the nests that the previous sows had built. Instead of just accepting the preformed nest and settling down to farrow, the sows still went through elaborate nest-building. The researchers concluded that the motivation to build a nest is not eliminated by providing a suitable preformed nest,

D. Fraser and C.J. Nicol

and they proposed that 'farrowing accommodation should therefore enable sows to perform nest building'. Laying hens show similar results: they are motivated to build a nest, but they shun preformed nests provided for them (Hughes *et al.*, 1989).

The answer may be different in the case of digging by Mongolian gerbils. In studying this behaviour, Wiedenmayer (1997) observed that caged gerbils often dig in the bedding, sometimes in short bouts throughout the cage and sometimes in sustained bouts of digging directed at the edges and corners of the cage. However, when gerbils were kept in a cage fitted with an artificial burrow system below the cage floor, they did fewer short bouts of digging and did not do sustained digging at all. In this case, the motivation (at least for the sustained digging) appeared to be to have access to a burrow system rather than to dig. Hence, the solution to the welfare problem would be to provide a suitable environment rather than to accommodate the behaviour.

11.4.2 Taking immediate cues into account

When a hen in a motivation experiment works to obtain litter for dust-bathing, has the sight of the litter triggered the motivation, or would the bird be motivated to dust-bathe even if no litter was visible?

This kind of question was explored by Warburton and Mason (2003), who trained mink to push against weighted doors for access to tunnels leading to four rewards: food, a bath where they could swim, a toy (a red plastic ball), and 'unpredictable social contact' in a cage where another mink was sometimes in close proximity. The mink were tested under two conditions: when the rewards were several metres away and could not be seen from the home cages, and when the rewards were close by and visible but still accessed via tunnels of similar length. The mink generally showed a high level of motivation to enter the tunnel leading to food, and a medium level of motivation to enter the tunnels leading to the bath and social contact, whether they could see these rewards or not. Motivation for the ball, however, seemed to depend strongly on the sight of the reward. When the mink could see the ball from the home cage, then they worked fairly hard to enter the tunnel that (they had learned) would lead to it; but if the ball was not visible, they showed little motivation to enter the tunnel.

11.4.3 Ensuring that questions about preferences are within the cognitive capacity of animals

One of the earliest objections to preference research was that animals may not be able to weigh up the longer-term consequences of their choices, and would therefore make decisions based only on their current state. Thus, Duncan (1978) observed that hens would repeatedly enter nest boxes for laying, even if they then remained trapped in the box for many hours without food or water. This allows two possible explanations: that the strong motivation to enter a nest box outweighed the known risk of future confinement, or that the hens lacked the ability to anticipate the future consequence of their actions (Špinka *et al.*, 1998).

To study whether pigs can anticipate such future consequences, Špinka *et al.* (1998) allowed pigs to choose to enter one of two feeding stalls whose characteristics were signalled by visual cues and position. Each stall provided food, but the animals were confined in one stall for 30 min and in the other for 240 min. Under these conditions, the pigs learned to avoid the longer duration of confinement. In a more complex experiment, however, mice showed little evidence of anticipating future consequences. Warburton and Nicol (1998) trained mice to press a lever up to 80 times to enter a cage with food, or allowed them to enter the food cage freely but required up to 80 lever presses to return to the home cage. If the mice had to work to enter the food cage, then they made relatively few visits, but if they had to do the same amount of work to leave the food cage (so that the overall cost of obtaining food was the same in both cases) then they made many more visits. Thus the mice appeared to respond to the immediate cost of entry rather than the overall (immediate plus future) cost of a round trip.

Evidence of 'self-control' – or choosing to wait for a larger reward rather than accepting a smaller, more immediate reward – provides further evidence of anticipating the future. In several experiments, chickens were allowed to choose between two such options. With a sufficiently extreme difference, they showed self-control by selecting a larger and slightly delayed reward: specifically, they learned to select 22 s of access to food after 6 s of delay, rather than 3 s access to food after 2 s delay. In less extreme cases, however, the birds acted 'impulsively' by choosing a smaller but more immediate reward (Abeyesinghe *et al.*, 2005). Just how far into the future chickens can plan remains to be discovered. Duncan's (1978) observation suggests that there are limits on the extent to which chickens can act rationally when their immediate needs are pressing, but the results of Špinka *et al.* (1998) suggest that results may be quite different with other species or testing methods.

11.5 Assessing the Strength of Animals' Preferences

A preference shown by an animal in a choice experiment may be a weak preference, such as a preference for grapes over cherries, or a strong preference, such as a preference to live in a house rather than a dungeon. Denying animals access to their preferred options presumably affects their welfare more if the preference is strong. Thus, in addition to establishing what an animal prefers, we also need some indication of the strength of the preference or the animal's level of motivation for the preferred option (Dawkins, 1983).

The issue is particularly important because of a peculiar feature of preference research methods. In most other types of experiments – such as comparisons of growth rate on different diets – the difference between the treatments is compared against extraneous variation (such as the natural variation between animals and between pens) to test whether the effect of the treatment is statistically significant. In preference experiments, however, the different options are usually presented to the same animals at the same time and in almost the same place. Consequently, preference research generally minimizes extraneous variation such that even mild preferences may be statistically significant (Fraser *et al.*, 1993).

11.5.1 Assessing motivation strength

As the simplest approach to establishing preference strength, experimenters have tried to determine whether a preferred option is sufficiently rewarding that an animal will learn to perform an 'instrumental' response (such as pressing a lever or opening a door) to gain access to it. For example, having established that hens prefer floors with litter instead of bare wire, Dawkins and Beardsley (1986) tested whether hens would learn to peck a key or break a photo beam to gain access to a cage with litter. If the animal can be trained in such a way, then the motivation for that option is assumed to be more than trivial.

'Trade-offs' among alternative environmental features may also be informative. For example, to test how strongly mice prefer bedding material, van de Weerd *et al.* (1998) kept mice in a complex cage where they could choose between a compartment with nesting material or a compartment with other environmental features, such as a nest box (which mice generally prefer) and a metal grid floor (which mice generally avoid). Most mice chose to spend their time in the compartment with nesting material, even if it was presented in a compartment with no nest box and a metal grid floor. The researchers concluded that mice were highly motivated to have nesting material, and that this contributes significantly to their welfare.

Another way of examining trade-offs is to allow animals to work simultaneously for two different resources at different workloads. Pedersen *et al.* (2005) allowed pigs to obtain straw by pressing one panel, and peat by pressing a second panel. The number of presses required on each panel was systematically varied to find the point at which the pigs would take equal amounts of both substrates. This equality was observed at a point when straw was relatively cheap (9 presses) and peat relatively expensive (39 presses); the result demonstrated a stronger motivation for peat.

The strength of motivation can also be studied by asking animals to choose between different activities. For example, Dawkins (1983) proposed that we might assess the strength of a hen's motivation for dust-bathing by 'titrating' it against its motivation to eat. To do this, Dawkins trained hens to enter two cages from a common choice-point. One cage contained litter (to permit dust-bathing) but no food, while the other contained food but no litter. Dawkins then required the hens to choose between the two cages after 0, 3, or 12 h of food deprivation. The results suggested that the hens' motivation for dust-bathing, under the given conditions, was about as strong as their motivation to eat when food had been withheld for 3 h.

11.5.2 Elasticity of demand

As a more elaborate option for comparing motivation strength, Dawkins (1983) proposed an approach that she described in the language of microeconomics (Fig. 11.4). She noted that

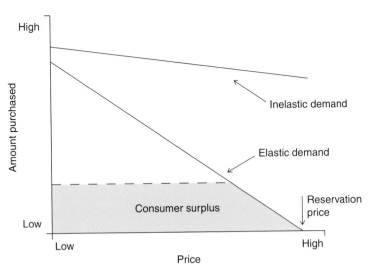

Fig. 11.4. An idealized illustration of 'demand curves' for commodities with elastic and inelastic demand. For commodities with inelastic demand (such as bread), the amount purchased remains fairly constant over a wide range of prices. For commodities with elastic demand (like fine wine), the amount purchased falls with increasing price. The slope of the line (i.e. how steeply the line falls with increasing price) is commonly used as a quantitative measure of elasticity. Two other measures of willingness to pay are also shown. Reservation price is the highest price paid for the commodity. In animal studies, consumer surplus can be calculated as the area to the left of the demand curve, and applies to a given quantity of the commodity. (From Fraser, 2008.)

consumers purchase some commodities (such as bread) in roughly the same amounts even if the price increases or if people's incomes fall. These commodities are said to have 'inelastic demand' and are sometimes called 'necessities' because consumers are willing to pay more and more of their income to obtain them. For other commodities (such as fine wine) consumption falls if the price increases or if incomes decline. Such commodities are said to have 'elastic demand' and may be regarded as 'luxuries' because consumers will forego them more readily. Applying this thinking to animals, Dawkins (1983, 1990) proposed that elasticity of demand could be used to compare how important different options are to animals.

To apply this concept, a commodity such as food can be provided as a reward for some work ('price') that the animal has to perform, and the price can then be varied experimentally to determine the 'price elasticity' of the demand. Alternatively, the animal can be given a limited amount of time ('income') to access various resources; then the amount of time can be varied to determine the 'income elasticity' of the demand. The assumption is that the animal should devote more and more

effort or available time (thus demonstrating inelastic demand) to maintain a given level of reward if that reward is very important.

11.5.3 The debate about measures of demand

Although elasticity of demand provides an intuitively appealing approach, many technical and conceptual problems arise over how to generate valid results. One basic issue is how to parcel out different rewards into discrete amounts (Mason *et al.*, 1998; Kirkden *et al.*, 2003). Food can easily be divided into small quantities so as to allow the many measurements needed to generate a demand curve. But how should we parcel out rewards such as bedding or opportunities for dust-bathing, mating or rest? Such decisions can affect the elasticity of demand. For example, Jensen *et al.* (2005) allowed dairy heifers to lie down in a bedded stall for 9h a day, and then required them to press a panel 10–50 times in order to obtain additional rest periods. As long as each reward consisted of 30–80min of uninterrupted rest, the heifers showed relatively inelastic demand, but the demand was much more elastic if the rewards were limited to only 20 min of uninterrupted rest. Jensen *et al.* concluded that the heifers had an inelastic demand for rest of up to 12–13 h per day, but that short rewards gave an incorrect picture of the animals' motivation.

A second concern arises over the selection of appropriate instrumental responses. Animals may find it easier to associate certain tasks with certain rewards (Young *et al.*, 1994; Sumpter *et al.*, 1999). For example, it is natural for hens to find food by pecking and to enter a new area by walking; but if a hen is required to peck a key in order to enter a new area, will the results reflect the bird's true motivation, or is the task simply too unnatural for the type of reward? More natural tasks such as walking or pushing against weighted doors are especially useful if the energetic costs imposed by the tasks can be quantified (Olsson and Keeling, 2002; Champion and Matthews, 2007).

As an even more fundamental concern, there is probably no single demand curve that characterizes a given commodity for a given animal. Hens, for example, may have an inelastic demand for a nest box at the time of laying an egg, but not at other times.

These and other issues have generated a large literature on the use of demand elasticity and related measures (Dawkins, 1990; Mason *et al.*, 1998; Kirkden *et al.*, 2003). Some of the problems involved can be addressed by sufficiently detailed research that determines how elasticity of demand is influenced by the size of the reward, level of deprivation, time of day and other variables (Jensen *et al.*, 2004), but such comprehensive research is rarely undertaken.

Fortunately, there are simpler measures of motivation that raise fewer technical problems (see Fig. 11.4). One, called 'reservation price', is the maximum price that an animal will pay for a given reward. For example, by varying the price one can simply determine the maximum price that a hen will pay for a pellet of food, or for a nest box or for a perch where she can roost. The value of these resources will fluctuate depending on the level of deprivation and other factors, but if the hen will pay a very high price at any given time, then the resource must be important to the hen. Moreover, this measure can be applied even to resources (such as mating opportunities) that do not lend themselves to being divided into parcels as would be needed to generate a demand curve. A second measure, called 'consumer surplus', is

effectively the area to the left of the demand curve for a given amount of reward (see Fig. 11.4). Consumer surplus is used by economists as a measure of the importance of a good or service when assessing human welfare. Resources are viewed as more important if they involve a larger consumer surplus.

Seaman *et al.* (2008) provided an illustration of the different measures of motivation. They kept laboratory rabbits in a central area that provided access to four chambers, one containing food, one with a resting platform, one allowing social contact and one empty cage (as a control). The rabbits could access the different chambers by opening doors that were fitted with different weights to increase the cost. The results (Fig. 11.5) allowed different measures of demand – reservation price, consumer surplus and expenditure rate – to be calculated. All three measures ranked the resources similarly as food > social contact > resting platform > empty cage.

11.6 Measures of Avoidance

Measures of preference and motivation can also be applied to situations that animals avoid such as unpleasant temperatures, noise, or management practices that are expected to cause fear or pain.

Such approaches have played a key role in studies designed to find more humane ways of killing animals. Of the millions of rats used each year by scientists, most are killed at the end of the experiments. One of the most common methods is to put the animals in a sealed chamber containing a lethal level of carbon dioxide gas, although inhalant anaesthetic gases can also be used. The various gases or gas combinations normally cause unconsciousness followed by death, but are they so unpleasant that the animal will experience distress before losing consciousness? Leach *et al.* (2004) tested rats in an apparatus consisting of two similar chambers separated by a short tunnel with a plastic flap door. A gas (carbon dioxide or anaesthetic gases) was introduced into one of the chambers, and the animal's behaviour was observed for 3 min, starting when it first entered that chamber. When the chamber contained 25% carbon dioxide (a concentration that would cause some loss of motor control after 30 s) the rats strongly avoided it, spending only 2 s of the 3 min in the chamber. When halothane was used, also at a concentration that would cause similar loss of

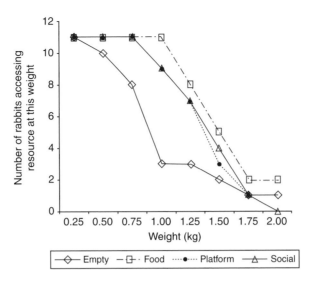

Fig. 11.5. An example of the use of demand theory to assess the motivation of animals for different resources. In this study, rabbits opened doors fitted with different weights to access four resources: food, social contact, a resting platform or an empty cage (as a control). As the weight increased, fewer and fewer rabbits entered the empty cage, but they expended more and more effort to access the other options. Only at weights above 1 kg did any rabbits fail to access food within the time allowed. In this example, the number of animals accessing the resource (rather than the number of rewards obtained) is used as the measure of demand. (From Seaman *et al.*, 2008.)

motor control in 30 s, the rats showed much less avoidance, spending an average of 43 s in the test chamber. Through this and similar research Leach *et al.* (2004) concluded that killing rats by carbon dioxide is likely to cause 'considerable pain and distress', and that less aversive options should be used instead.

In this experiment, rats could demonstrate that they found carbon dioxide unpleasant simply by escaping from it. Where such a direct approach is not feasible, it may be possible instead to see whether animals will learn to avoid a situation after repeated exposure. When working with dairy cattle, some handlers move the animals by negative methods such as shouting, hitting and using electric prods. Rough handling is a known cause of impaired welfare and reduced productivity, but which procedures do the cattle find most unpleasant? To understand this better, Pajor *et al.* (2000b) moved cattle into a 10-m raceway that ended in a small pen where a handler applied one of the common treatments. These included shouting at the cow, slapping its rump with an open hand, twisting its tail, applying an electric prod, or doing nothing as a control treatment. After several repetitions of the procedure, control animals moved fairly readily

along the raceway, but experimental animals took longer and needed increasing amounts of force to make them walk to the end. After four exposures, the animals that received shouting were the slowest and hardest to move, but by the eighth and ninth repetitions, the animals that received electric prods were more difficult to move than all the others. The results suggested that cattle perceive the electric prod as the most unpleasant treatment, and that shouting can be more aversive than mild physical methods such as slapping the cow on the rump.

Here, as elsewhere, instrumental tasks need to be chosen carefully. For example, if a treatment produces fear, the animal may simply freeze or attempt to escape and thus be incapable of performing certain instrumental tasks. Rutter and Duncan (1991) attempted to test fear in chickens using a two-chamber apparatus. The fear-producing stimulus was a balloon that could be inflated rapidly; to avoid it, the birds simply needed to move into the opposite chamber when they heard a signal. Instead of doing this, however, the chickens 'froze' when the signal was given, and they did not learn to avoid the stimulus. The authors concluded that a different procedure – involving a more passive response – was needed to study fear in these birds.

11.7 Applying Knowledge of Animals' Preferences and Motivations

The ultimate aim of most of the above research is to improve the welfare of animals, especially through improvements in housing and handling. Some research – such as knowledge of animals' preferences for bedding materials, temperatures and flooring products – is relatively easy to incorporate into practice. Other research presents a greater challenge. For example, it seems clear that hens are motivated to obtain sufficient space to perform comfort behaviours, to lay in nest-boxes, and to perch at night for resting. The evidence that hens are strongly motivated to perform dust-bathing for feather care is not so clear, although the behaviour is almost certainly something that hens enjoy doing if they have the chance (Widowski and Duncan, 2000; Weeks and Nicol, 2006). However, accommodating these types of behaviour in efficient, high-health environments has required a good deal of research and ingenuity (Appleby and Hughes, 1995; Tauson, 2002; Blokhuis et al., 2007).

Another approach in environmental design is to create environments where the animals themselves can make choices in their daily lives (Wathes, 1997). Especially in intensive production systems, many environments are simple and uniform, and provide little scope for animals to exercise their preferences. However, simple devices such as radiant heat lamps or warmed areas allow broiler chickens or piglets to select their own thermal environment. Free-access, two-level pens for pigs allow animals to choose their microclimate (warmer above, cooler below) and their immediate social companions (Phillips and Fraser, 1987). Free-access stalls for pregnant sows or dairy cattle allow animals either to mingle or to select individual resting areas (Phillips, 1997). Similar principles hold for animals in rescue shelters and laboratory animal housing. For example, cats in rescue shelters often appear fearful and agitated in conventional, featureless cages, but simple modifications can give cats the choice of perching on a raised shelf or hiding from view (Gourkow and Fraser, 2006). Beyond these direct benefits, simply having some control over a situation can in itself be beneficial for animal welfare (Bassett and Buchanan-Smith, 2007).

Environments that permit choice can also allow for important individual differences. Sometimes, genetic differences between strains or breeds can result in animals having different preferences. For example, mice of the strain ICR-CD(1) actively investigate and explore novel items placed regularly in their cages, whereas the seemingly more anxious C57/Bl/6 mice avoid them (Nicol et al., 2008). Even within the same breed, individuals may differ. For example, sows naturally wean their young by gradually increasing the time they spend away from the piglets, but some sows reduce their contact time quite quickly beginning in the second week of lactation, while others remain in close contact with the young for several weeks (Pajor et al., 2000a). On many commercial farms, sows are penned permanently with the young, and are then separated abruptly when the piglets reach a standard age. 'Get-away' pens allow sows to control the weaning process by choosing how much time to stay with the young, and they accommodate the large individual differences between sows in the age at which they prefer to wean their piglets (Bøe, 1991).

11.8 Clarifying the Link between Preferences and Welfare

Even when we have accurately identified an animal's preferences and assessed their strength, can we assume that allowing animals to live as they prefer will necessarily improve their welfare? To understand this question, we need to be clear on what we mean by animal welfare (Fraser, 2008). If we take the view that animal welfare involves three broad elements – the basic health of animals, their affective states (especially freedom from negative states such as pain and fear), and the ability of animals to live in a manner for which they are adapted – then we can ask the question about each of these elements.

There are many examples of animals choosing environments that allow them to use their adaptations. Thus, we see hens selecting litter where they can perform their natural behaviour of dust-bathing for feather care, sows working to obtain straw for nest-building, and rats selecting social environments where they can exercise the social behaviour that is natural for the species.

Animals also seem likely to choose environments that promote positive affective states, at least in the short term (Bechara and Damasio, 2005). Thus animals can be expected to choose comfortable lying surfaces, social environments where they will not be afraid of aggression, and foods that they find palatable.

To some degree, we can expect animals' preferences to favour good health. Applying evolutionary logic, we would expect the preferences and motivations of animals to function as adaptations that would allow animals to maximize their reproductive success (which for many species is likely to involve survival and good health) in the environment in which the species evolved. Hence, we would expect to find a good correlation between preferences and health as long as the environment presents the types of choices that the species would have made in the environment in which it evolved; but in artificial environments, this relationship may well break down.

The simplest examples of such a breakdown arise when animals are exposed to potential dangers or benefits that are beyond their sensory and affective capacity. Many fish species successfully avoid being harmed by certain aquatic pollutants such as copper simply by swimming away from contaminated water (Giattina and Garton, 1983). However, fish generally fail to avoid certain other contaminants (phenol, selenium) even at levels that cause serious damage or death (Giattina and Garton, 1983; Hartwell *et al.*, 1989). Presumably the fish never evolved or developed the capacity to detect and avoid these contaminants, and in these cases their preferences fail to protect their health.

A similar limitation may occur if a choice requires a cognitive skill that the animal does not possess. Rats rapidly learn to avoid a poisoned food on the basis of its flavour, but not if colour or pellet size is its distinguishing feature (McFarland, 1985). In this case, the rats presumably can detect the distinguishing stimuli, but do not readily associate feelings of illness with the visual properties of the food.

Preferences may also fail to promote good health if the options available have features that would be adaptive under more natural conditions but are not adaptive under the animal's current circumstances. For example, a preference for energy-rich food may be adaptive under natural conditions where energy is needed, but may lead to obesity in captivity where food is abundant.

Finally, genetic selection for specific production traits may alter the preferences of animals in ways that are not consistent with basic health. Genetic selection of meat-producing chickens for very rapid growth has led to birds with voracious appetites. If these birds are kept for breeding they cannot be fed the amounts of food that they would prefer or the birds would become obese.

In summary, the preferences of an animal will probably correspond to its short-term affective states such as pain, fear and discomfort, and animals are likely to prefer environments that allow those types of natural behaviour that they are highly motivated to perform. Whether the preferences of animals promote their health and positive affect in the longer term will depend on how well the options correspond to the kind of choices the animals are adapted to make.

If preferred environments improve animal welfare, then we should expect a close correspondence between preferences and other presumed indicators of animal welfare. For example Mason *et al.* (2001), having found that mink strongly prefer to have access to a pool where they can swim, also found increased levels of the stress-related hormone cortisol in mink that were prevented from swimming; and Fisher *et al.* (2008) found that the motivation of sheep to avoid hot conditions corresponded with the measured physiological cost of remaining in those conditions. However, Hurst *et al.* (1997) found evidence that singly housed rats showed signs of seeking social interaction, even though the stress-related hormone corticosterone, and some types of organ pathology, were higher in group-housed animals.

Further investigation is clearly needed of the links between preferences and other indicators of animal welfare. In a recent experiment, hens were housed for a series of 5-week periods in three environments: wire-floored pens, pens with shavings and enriched pens with shavings, peat, perches and nest-boxes (Nicol *et al.*, 2009). After every two 5-week exposures, the birds were tested to see which of the two environments they preferred. When housed in the more preferred environment, the birds showed lower body temperature, blood glucose and heterophil:lymphocyte ratio, plus improved feed digestibility. They also showed more self-grooming and low reaction to novelty. Conversely, many measures commonly used as 'indicators' of animal welfare, including plumage condition and corticosterone concentration, were not correlated with the environmental preferences of the birds. Such studies provide a start at combining methods of animal welfare research that have been conducted separately for too long.

11.9 Conclusions

- Measuring animals' preferences and motivations provides an animal-centred way of finding out what is important for their welfare.

- Measuring the preferences and motivations of animals is not simple; the influences of familiarity, information gathering, learning and other factors must be considered in designing appropriate tests.
- Animals are likely to choose things that are good for their long-term welfare in environments that are very like those where the species evolved. However, the preferences of domesticated animals in artificial environments may not always be a guide to their best long-term interests.
- Motivation strength can be measured in tests that manipulate the amount of energy, effort or time that animals must expend to access different resources. Different measures of demand, drawn from economic theory, include consumer surplus, reservation price and elasticity.
- Benefits to animal welfare can be achieved by housing animals in situations that they prefer and that accommodate behaviour they are motivated to perform. Welfare can also be improved by situations where they can express their individual preferences for physical or social environments.
- New research is examining the extent to which animals prefer environments in which other indicators of their health and physical condition suggest good welfare.

Acknowledgements

Portions of this chapter are based on *Understanding Animal Welfare: The Science in its Cultural Context* (Fraser, 2008); we are grateful to Wiley-Blackwell and the Universities Federation for Animal Welfare for allowing us to rework some of that material here and to reproduce Figure 11.4. We are also grateful to Wiley-Blackwell for permission to reproduce Figure 11.1; to Lab Animal magazine for permission to reproduce Figure 11.2; and to Elsevier Science for permission to reproduce Figures 11.3 and 11.5.

References

Abeyesinghe, S., Wathes, C.M., Nicol, C.J. and Randall, J.M. (2001) The aversion of broiler chickens to concurrent vibrational and thermal stressors. *Applied Animal Behaviour Science* 73, 199–216.

Abeyesinghe, S.M., Nicol, C.J., Hartnell, S.J. and Wathes, C.M. (2005) Can domestic fowl show self-control? *Animal Behaviour* 70, 1–11.

Appleby, M.C. and Hughes, B.O. (1995) The Edinburgh Modified Cage for laying hens. *British Poultry Science* 36, 707–718.

Arey, D.S. (1992) Straw and food as reinforcers for prepartal sows. *Applied Animal Behaviour Science* 33, 217–226.

Arey, D.S., Petchey, A.M. and Fowler, V.R. (1991) The preparturient behaviour of sows in enriched pens and the effect of pre-formed nests. *Applied Animal Behaviour Science* 31, 61–68.

Baldwin, B.A. and Start, I.B. (1985) Illumination preferences of pigs. *Applied Animal Behaviour Science* 14, 233–243.

Barnett, S.A., Smart, J.L. and Widdowson, E.M. (1971) Early nutrition and the activity and feeding of rats in an artificial environment. *Developmental Psychobiology* 4, 1–15.

Bassett, L. and Buchanan-Smith, H.M. (2007) Effects of predictability on the welfare of captive animals. *Applied Animal Behaviour Science* 102, 223–245.

Bechara, A. and Damasio, A.R. (2005) The somatic marker hypothesis: a neural theory of economic decision. *Games and Economic Behavior* 52, 336–372.

Blokhuis, H.J., van Niekerk, T.F., Bessei, W., Elson, A., Guemene, D., Kjaer, J.B., Levrino, G.A.M., Nicol, C.J., Tauson, R., Weeks, C.A. and de Weerd, H.A.V. (2007) The LayWel project: welfare implications of changes in production systems for laying hens. *World's Poultry Science Journal* 63, 101–114.

Blom, H.J.M., Baumans, V., van Vorstenbosch, C.J.A.H.V., van Zutphen, L.F.M. and Beynen, A.C. (1993) Preference tests with rodents to assess housing conditions. *Animal Welfare* 2, 81–87.

Bøe, K. (1991) The process of weaning in pigs: when the sow decides. *Applied Animal Behaviour Science* 30, 47–59.

Brambell Committee (1965) *Report of the Technical Committee to Enquire into the Welfare of Animal Kept under Intensive Livestock Husbandry Systems.* Command Paper 2836, Her Majesty's Stationery Office, London.

Champion, R.A. and Matthews, L.R. (2007) An operant-conditioning technique for the automatic measurement of feeding motivation in cows. *Computers and Electronics in Agriculture* 57, 115–122.

Christison, G.I. and Farmer, C. (1983) Physical characteristics of perforated floors for young pigs. *Canadian Agricultural Engineering* 25, 75–80.

Cooper, J.J. and Appleby, M.C. (1996) Demand for nest boxes in laying hens. *Behavioural Processes* 36, 171–182.

Dawkins, M.[S.] (1977) Do hens suffer in battery cages? Environmental preferences and welfare. *Animal Behaviour* 25, 1034–1046.

Dawkins, M.S. (1980) *Animal Suffering.* Chapman and Hall, London.

D. Fraser and C.J. Nicol

Dawkins, M.S. (1983) Battery hens name their price: consumer demand theory and the measurement of ethological 'needs'. *Animal Behaviour* 31, 1195–1205.

Dawkins, M.S. (1990) From an animal's point of view: motivation, fitness, and animal welfare. *Behavioral and Brain Sciences* 13, 1–9, 54–61.

Dawkins, M.S. and Beardsley, T. (1986) Reinforcing properties of access to litter in hens. *Applied Animal Behaviour Science* 15, 351–364.

Dawkins, M.S., Cook, P.A., Whittingham, M.J., Mansell, K.A. and Harper, A.E. (2003) What makes free-range broiler chickens range? *In situ* measurement of habitat preference. *Animal Behaviour* 66, 151–160.

Duncan I.J.H. (1978) The interpretation of preference tests in animal behaviour. *Applied Animal Ethology* 4, 197–200.

Duncan, I.J.H. (1992) Measuring preferences and the strength of preferences. *Poultry Science* 71, 658–663.

Farmer, C. and Christison, G.I. (1982) Selection of perforated floors by newborn and weanling pigs. *Canadian Journal of Animal Science* 62, 1229–1236.

Fisher, A.D., Roberts, N., Matthews, L.R. and Hinch, G.R. (2008) Does a sheep's motivation to avoid hot conditions correspond to the physiological cost of remaining in those conditions? In: Boyle, L., O'Connell, N. and Hanlon, A. (eds) *Proceedings of the 42nd Congress of the ISAE [International Society for Applied Ethology] 'Applied Ethology: Addressing Future Challenges in Animal Agriculture'*, University College Dublin, Ireland, 5–9 August 2008. Wageningen Academic Publishers, Wageningen, The Netherlands, p. 45 (abstract).

Fraser, D. (1985) Selection of bedded and unbedded areas by pigs in relation to environmental temperature and behaviour. *Applied Animal Behaviour Science* 14, 117–126.

Fraser, D. (1996) Preference and motivational testing to improve animal well-being. *Laboratory Animals* 25, 27–31.

Fraser, D. (2008) *Understanding Animal Welfare: The Science in its Cultural Context*. Wiley-Blackwell, Oxford, UK.

Fraser, D., Phillips, P.A. and Thompson, B.K. (1993) Environmental preference testing to assess the well-being of animals – an evolving paradigm. *Journal of Agricultural and Environmental Ethics* 6 (Supplement 2), 104–114.

Giattina, J.D. and Garton, R.R. (1983) A review of the preference–avoidance responses of fishes to aquatic contaminants. *Residue Reviews* 87, 43–90.

Gourkow, N. and Fraser, D. (2006) The effect of housing and handling practices on the welfare, behaviour and selection of domestic cats (*Felis sylvestris catus*) by adopters in an animal shelter. *Animal Welfare* 15, 371–377.

Hartwell, S.I., Jin, J.H., Cherry, D.S. and Cairns, J. Jr (1989) Toxicity versus avoidance response of golden shiner, *Notemigonus crysoleucas*, to five metals. *Journal of Fish Biology* 35, 447–456.

Hughes, B.O. and Black, A.J. (1973) The preference of domestic hens for different types of battery cage floor. *British Poultry Science* 14, 615–619.

Hughes, B.O., Duncan, I.J.H. and Brown, M.F. (1989) The performance of nest building by domestic hens – is it more important than the construction of a nest. *Animal Behaviour* 37, 210–214.

Hunter, L. and Houpt, K.A. (1989) Bedding material preferences of ponies. *Journal of Animal Science* 67, 1986–1991.

Hurst, J.L., Barnard, C.J., Nevison, C.M. and West, C.D. (1997) Housing and welfare in laboratory rats: welfare implications of isolation and social contact among caged males. *Animal Welfare* 6, 329–347.

Jensen, M.B., Pedersen, L.J. and Ladewig, J. (2004) The use of demand functions to assess behavioural priorities in farm animals. *Animal Welfare* 13 (Supplement 1), 27–32.

Jensen, M.B., Pedersen, L.J. and Munksgaard, L. (2005) The effect of reward duration on demand functions for rest in dairy heifers and lying requirements as measured by demand functions. *Applied Animal Behaviour Science* 90, 207–217.

Jones, J.B., Webster, A.J.F. and Wathes, C.M. (1999) Trade-off between ammonia exposure and thermal comfort in pigs and the influence of social contact. *Animal Science* 68, 387–398.

Kirkden, R.D. and Pajor, E.A. (2006) Motivation for group housing in gestating sows. *Animal Welfare* 15, 119–130.

Kirkden, R.D, Edwards, J.S.S. and Broom, D.M. (2003) A theoretical comparison of the consumer surplus and the elasticities of demand as measures of motivational strength. *Animal Behaviour* 65, 157–178.

Leach, M.C, Bowell, V.A., Allan, T.F. and Morton, D.B. (2004) Measurement of aversion to determine humane methods of anaesthesia and euthanasia. *Animal Welfare* 13 (Supplement 1), 77–S86.

Lindberg, A.C. and Nicol, C.J. (1996) Space and density effects on group size preferences in laying hens. *British Poultry Science* 37, 709–721.

Mason, G. [J.], McFarland, D. and Garner, J. (1998) A demanding task: using economic techniques to assess animal priorities. *Animal Behaviour* 55, 1071–1075.

Mason, G.J., Cooper, J. and Clarebrough, C. (2001) Frustrations of fur-farmed mink. *Nature* 410, 35–36.

Matthews, L.R. and Ladewig, J. (1994) Environmental requirements of pigs measured by behavioural demand functions. *Animal Behaviour* 47, 713–719.

McFarland, D. (1985) *Animal Behaviour: Psychology, Ethology and Evolution*. Longman Scientific and Technical, Harlow, UK.

Morrison, W.D., Bate, L.A., McMillan, I. and Amyot, E. (1987) Operant heat demand of piglets housed on four different floors. *Canadian Journal of Animal Science* 67, 337–341.

Muiruri, H.K., Harrison, P.C. and Gonyou, H.W. (1990) Preferences of hens for shape and size of roosts. *Applied Animal Behaviour Science* 27, 141–147.

Nicol, C.J. (1986) Non-exclusive spatial preference in the laying hen. *Applied Animal Behaviour Science* 15, 337–350.

Nicol, C.J., Lindberg, A.C., Phillips, A.J., Pope, S.J., Wilkins, L.J. and Green, L.E. (2001) Influence of substrate exposure during rearing on feather pecking, foraging and dustbathing in adult laying hens. *Applied Animal Behaviour Science* 73, 141–156.

Nicol, C.J., Brocklebank, S., Mendl, M. and Sherwin, C.M. (2008) A targeted approach to developing environmental enrichment for two strains of laboratory mouse. *Applied Animal Behaviour Science* 110, 341–353.

Nicol, C.J., Caplen, G., Edgar, J. and Browne, W.J. (2009) Associations between welfare indicators and environmental choice in laying hens. *Animal Behaviour* 78, 413–424.

Olsson, I.A.S. and Keeling, L.J. (2002) The push-door for measuring motivation in hens: laying hens are motivated to perch at night. *Animal Welfare* 11, 11–19.

Pajor, E.A., Kramer, D.L. and Fraser, D. (2000a) Regulation of contact with offspring by domestic sows: temporal patterns and individual variation. *Ethology* 106, 37–51.

Pajor, E.A., Rushen, J. and de Passillé, A.M.B. (2000b) Aversion learning techniques to evaluate dairy cattle handling practices. *Applied Animal Behaviour Science* 69, 289–102.

Patterson-Kane, E.P., Hunt, M. and Harper, D. (2004) Short communication: rat's demand for group size. *Journal of Applied Animal Welfare Science* 7, 267–272.

Pedersen, L.J., Holm, L., Jensen, M.B. and Jorgensen, E. (2005) The strength of pigs' preferences for different rooting materials using concurrent schedules of reinforcement. *Applied Animal Behaviour Science* 94, 31–48.

Phillips, C.J.C. and Morris, I.D. (2001) A novel operant conditioning test to determine whether dairy cows dislike passageways that are dark or covered with excreta. *Animal Welfare* 10, 65–72.

Phillips, P.A. (1997) A two-level system for housing dry sows. In: Bottcher, R.W. and Hoff, S.J. (eds) *Livestock Environment V: Proceedings of the Fifth International Symposium, American Society of Agricultural Engineers* (ASAE) [now American Society of Agricultural and Biological Engineers, ASABE]. ASABE, St. Joseph, Michigan, pp. 266–272.

Phillips, P.A. and Fraser, D. (1987) Design, cost and performance of a free-access two-level pen for growing–finishing pigs. *Canadian Journal of Agricultural Engineering* 29, 193–195.

Phillips, P.A., Thompson, B.K. and Fraser, D. (1988) Preference tests of ramp designs for young pigs. *Canadian Journal of Animal Science* 68, 41–48.

Phillips, P.A., Thompson, B.K. and Fraser, D. (1989) The importance of cleat spacing in ramp design for young pigs. *Canadian Journal of Animal Science* 69, 483–486.

Phillips, P.A., Fraser, D. and Thompson, B.K. (1996) Sow preference for types of flooring in farrowing crates. *Canadian Journal of Animal Science* 76, 485–489.

Rushen, J. (1986) Aversion of sheep to electro-immobilization and physical restraint. *Applied Animal Behaviour Science* 15, 315–324.

Rutter, S.M. and Duncan, I.J.H. (1991) Shuttle and one-way avoidance as measures of aversion in the domestic fowl. *Applied Animal Behaviour Science* 30, 117–124.

Seaman, S.C., Waran, N.K., Mason, G. and D'Eath, R.B. (2008) Animal economics: assessing the motivation of female laboratory rabbits to reach a platform, social contact and food. *Animal Behaviour* 75, 31–42.

Sherwin, C (1998) The use and perceived importance of three resources which provide caged laboratory mice the opportunity for extended locomotion. *Applied Animal Behaviour Science* 55, 353–367.

Sherwin, C.M. (2004) The motivation of group-housed laboratory mice, *Mus musculus*, for additional space. *Animal Behaviour* 67, 711–717.

Sherwin, C.M. (2007) The motivation of group-housed laboratory mice to leave an enriched laboratory cage. *Animal Behaviour* 73, 29–35.

Sherwin, C.M. and Nicol, C.J. (1996) Reorganization of behaviour in laboratory mice, *Mus musculus*, with varying cost of access to resources. *Animal Behaviour* 51, 1087–1093.

Sherwin, C.M. and Nicol, C.J. (1997) Behavioural demand functions of caged laboratory mice for additional space. *Animal Behaviour* 53, 67–74.

Špinka, M., Duncan, I.J.H. and Widowski, T.M. (1998) Do domestic pigs prefer short-term to medium-term confinement? *Applied Animal Behaviour Science* 58, 221–232.

Stephens, D.B., Bailey, K.J., Sharman, D.F. and Ingram, D.L. (1985) An analysis of some behavioural effects of the vibration and noise components of transport in pigs. *Quarterly Journal of Experimental Physiology* 70, 211–217.

Sumpter, C.E., Temple, W. and Foster, T.M. (1999) The effects of differing response types and price manipulations on demand measures. *Journal of the Experimental Analysis of Behavior* 71, 329–354.

Tauson, R. (2002) Furnished cages and aviaries: production and health. *World's Poultry Science Journal* 58, 49–63.

Telezhenko, E., Lidfors, L. and Bergsten, C. (2007) Dairy cow preferences for soft or hard flooring when standing or walking. *Journal of Dairy Science* 90, 3716–3724.

Thorpe, W.H. (1965) The assessment of pain and distress in animals. In: Brambell Committee (1965) *Report of the Technical Committee to Enquire into the Welfare of Animal Kept under Intensive Livestock Husbandry Systems.* Command Paper 2836, Her Majesty's Stationery Office, London, pp. 71–79.

Tucker, C.B., Weary, D.M. and Fraser, D. (2003) Effects of three types of freestall surfaces on preferences and stall usage by dairy cows. *Journal of Dairy Science* 86, 521–529.

van de Weerd, H.A., van Loo, P.L.P, van Zutphen, L.F.M., Koolhaas, J.M. and Baumans, V. (1998) Strength of preference for nesting material as environmental enrichment for laboratory mice. *Applied Animal Behaviour Science* 55, 369–382.

van den Broek, F.A.R., Klompmaker, H., Bakker, R. and Beynen, A.C. (1995) Gerbils prefer partially darkened cages. *Animal Welfare* 4, 119–123.

van Liere, D.W., Kooijman, J. and Wiepkema, P.R. (1990) Dustbathing behaviour of laying hens as related to quality of dustbathing material. *Applied Animal Behaviour Science* 26, 127–141.

Vestergaard, K.S. and Baranyiova, E. (1996) Pecking and scratching in the development of dust perception in young chicks. *Acta Veterinaria Brno* 65, 133–142.

Warburton, H. and Mason, G. (2003) Is out of sight out of mind? The effects of resource cues on motivation in mink, *Mustela vison. Animal Behaviour* 65, 755–762.

Warburton, H.J and Nicol, C.J. (1998) Position of operant costs affects visits to resources by laboratory mice. *Animal Behaviour* 55, 1325–1333.

Wathes, C.M. (1997) Engineering choices into animal environments. In: Forbes, J.M., Lawrence, T.L.J., Rodway, R.G. and Varley, M.A. (eds) *Animal Choices.* Occasional Publication, British Society of Animal Science (BSAS) No. 20., BSAS, Edinburgh, UK, pp. 67–73.

Weeks, C.A. and Nicol, C.J. (2006) Behavioural needs, priorities and preferences of laying hens. *World's Poultry Science Journal* 62, 296–307.

Widowski, T.M. and Duncan, I.J.H. (2000) Working for a dustbath: are hens increasing pleasure rather than reducing suffering? *Applied Animal Behaviour Science* 68, 39–53.

Wiedenmayer, C. (1997) Causation of the ontogenetic development of stereotypic digging in gerbils. *Animal Behaviour* 53, 461–470.

Young, R.J., Macleod, H.A. and Lawrence, A.B. (1994) Effect of manipulandum design on operant responding in pigs. *Animal Behaviour* 47, 1488–1490.

12 Practical Strategies to Assess (and Improve) Welfare

ANDREW BUTTERWORTH, JOY A. MENCH AND NADJA WIELEBNOWSKI

Abstract

There has been a growing interest in developing and implementing animal welfare assessment schemes on farm, in zoos, in experimental situations and even in the wild. Scientists, inspection bodies and politicians are also starting to consider seriously the use of animal welfare outcome-based measures (OBMs) such as behaviour or physical condition as a progression from resource-based measures (RBMs) for such schemes. Measures based directly on the animals can provide good indicators under a variety of conditions, as welfare is a characteristic of the individual animal, not just of the system in which animals are kept. Many modern farms, zoos, aquaria and other facilities are keen to identify welfare-monitoring tools that can be readily applied and allow for rapid assessment to provide timely feedback for management decisions. Improvements in welfare can be achieved through the combination of: (i) measurement; (ii) analysis of risk and environmental factors; (iii) provision of information resulting from the assessment; and (iv) promotion of positive change by supporting management decisions. In principle, individual measures can also be combined to give aggregate scores that can be presented to the producer, the animal keeper or the consumer. This requires the attribution of weighted values to the measures used to assess the impact of each measure with respect to animal welfare. Four questions arise about any approach that assesses the animals themselves: is it practical, is it valid (providing 'real' information about welfare), is it repeatable and is it robust (not influenced by weather, etc.)? Yet animal experience cannot be reduced to a mechanistic assessment: animals are variable, living and sentient beings and this must be realistically addressed in practical assessment systems. A multi-pronged strategy involving various RBMs and OBMs is most likely to provide the capacity for comprehensive welfare monitoring. Such a strategy may include regularly updated guidelines for species care, accreditation standards, longitudinal, multi-disciplinary and multi-institutional studies, and assessment tools for continuous welfare monitoring. Two case studies are discussed in detail: welfare assessments of broiler chickens on farms and of clouded leopards in zoos. A laboratory animal application is then described related to the implementation of humane end points for research.

12.1 Introduction

Many studies have focused on developing measures to assess animal welfare in an experimental setting. However, using such approaches *in situ*, whether on a farm, in a laboratory or at a zoo, may pose challenges. These challenges vary from one setting to another, but they may include the following: the animals may be kept in large groups, making observations of individual animals difficult; only minimal (or no) direct contact with the animals may be possible, making assessment methods that involve testing or closely examining the animals impractical; and only limited time may be available for observation. Developing methods for such *in situ* assessment, however, has become a high priority area to address public concerns about animal welfare.

Many countries now have laws regulating how animals can be housed or treated, and these generally require inspections for enforcement (Chapter 18).

In addition, numerous trade groups, including producers, distributors, retailers and chain restaurants, have recently been involved with creating or endorsing certification schemes for on-farm assessment which are either focused on animal welfare or include elements of animal welfare (Mench, 2004; Veissier *et al.*, 2008). Lastly, there is increasing emphasis on the use of welfare check sheets to evaluate the welfare of laboratory animals during and after experimental procedures in order to establish humane end points and refine experimental procedures (Morton, 2000), and to perform welfare evaluations of zoo animals to improve their housing and care (Goulart *et al.*, 2009).

In this chapter we will discuss the kinds of methods that are being developed and validated to assess animal welfare *in situ* and how those can lead to practical improvements in welfare, focusing on farm and zoo animals.

12.2 What Should We 'Measure' to Assess Welfare?

Much previous work on welfare monitoring systems has focused on 'what' or 'how much' of different resources are given to animals (so-called resource-based measures, or RBMs), and this approach is the basis of much existing legislation. Examples of RBMs are space provided per animal or the type and quantity of food given. The sorts of questions which are generally asked are:

- Are the animals properly fed and supplied with water?
- Are the animals properly housed?
- Are the animals provided with appropriate veterinary care?
- Are the animals given sufficient space to express a range of behaviours?

However, there is a quiet evolution (rather than a revolution) happening now, in different parts of the world, that involves scientists, inspection bodies and politicians starting to consider seriously the use of animal welfare outcome-based measures (OBMs), such as behaviour or physical condition, as a progression from (predominately) resource-based measures. Animals differ in their experience, temperament and the way in which their genetic make-up interacts with their environment. The influence of management and the stockperson can also dramatically influence not only measures of growth and reproduction, but also an animal's experience of a particular situation. Thus resource- or management-based measures (such as breeding strategies, health plans, etc.) may be a poor guarantee of good animal welfare in a particular situation.

Research scientists have for some time suggested that OBMs could provide better indicators of animal welfare, as welfare is a characteristic of the individual animal, not just of the system in which animals are kept. They have further suggested that welfare assessment could be based on OBMs supported by RBMs and management-based measures to identify risk factors, which are those factors used to help diagnose the actual causes of welfare problems. Once the cause or causes are identified, strategies can then be put into place to reduce or eliminate the problems.

Let us consider an example. A dairy farmer has a problem with lameness in his or her dairy cows. A structured assessment will tell him/her how many lame cows he actually has when referenced to some standard scoring system. Once the farmer has a tool to gauge whether the amount of lameness is increasing (or decreasing) then this can be a barometer against which to judge practical steps he may make to reduce the problem. If this OBM is combined with, for example, information on the type of flooring and the farmer's hoof-care strategy, this could be used to help decide on remedial solutions. In the case of cattle lameness, the problem can be both an economic cost (lameness in dairy cows costs the farmer in terms of lost productivity) and also a cost to the animals in terms of disability or discomfort, so targeted improvement may help both the farmer and the animal. To be viable, remedial strategies must satisfy both welfare and economic requirements, and they must be practicable, i.e. affordable and easy to implement by the farmer and/or breeding company. When information like this is linked to economic information that the farmer is likely to share with his veterinarian or advisor, then the combined use of OBMs (as a barometer of success or failure) linked with resource and environment information, may be a powerful tool to help and support farmers and to promote best management of animal health and welfare.

In developing welfare-assessment schemes, as many different potential welfare 'factors' as possible should be taken into account, and these should, where possible, be based on scientific knowledge of animal welfare rather than on conjecture based purely on anthropomorphized positions. Some single OBMs have been suggested as being capable of providing an integrated assessment of animal welfare; these include corticosteroids, acute-phase proteins and longevity (Barnett and Hemsworth, 1990; Hurnik, 1990; Geers *et al.*, 2003). However, none of these single measures can cover all of the dimensions of welfare, and it seems probable that several measures are necessary to obtain a comprehensive view of any particular animal's welfare (Dawkins, 1990; Webster, 1998; Rutter, 1998). Recognizing the difficulties of 'single-measure approaches', grouped measures have been used to advise farmers (Sørensen, 2001) in 'branded' welfare certification schemes (e.g. the Freedom Food Scheme; Main *et al.*, 2001), in comparing systems to provide information in the creation of legislation (Bracke *et al.*, 2002) or in checking compliance with legislative requirements (Keeling and Svedberg, 1999); see Table 12.1.

Table 12.1. An example of farm measures for broiler chickens, arranged in categories, including both resource- and outcome-based measures (RBMs and OBMs). Italicized entries indicate that farm measures are not available. (From the Welfare Quality® project; www.welfarequality.net.)

Welfare category	Aim	Measure/criterion
Good feeding	1. Absence of prolonged hunger	*This criterion is measured at the slaughterhouse – thin and emaciated birds can be detected when they can be more easily seen without feathers (after slaughter) and where large numbers can be scored as they pass by on the slaughter line*
	2. Absence of prolonged thirst	Drinker space
Good housing	3. Comfort around resting	Plumage cleanliness, litter quality
		Dust sheet test – a simple measure of aerial (and potentially inspired) dust
	4. Thermal comfort	Panting, huddling
	5. Ease of movement	Stocking density
Good health	6. Absence of injuries	Lameness, hock burn, foot pad dermatitis
	7. Absence of disease	On-farm mortality, culls on farm
	8. Absence of pain induced by management procedures	*This criterion is not applied on farm but at the slaughterhouse – the effectiveness of stunning during slaughter operations can be assessed at slaughter*
Appropriate behaviour	9. Expression of social behaviours	*As yet, no measure is developed*
	10. Expression of other behaviours	Cover on the range, free range (applicable only to birds with access to range)
	11. Good human–animal relationship	Fear of humans as measured by the avoidance distance test
	12. Positive emotional state	Qualitative behaviour assessment (QBA) (see Chapter 9)

Alongside, and in combination with OBMs, resource- or management-based measures should probably be included in a welfare assessment as a basis for the identification of 'causes of poor welfare', or because they can help to identify risk factors for welfare problems.

12.3 Developing and Testing Measures

When a new method for assessment is proposed, four fundamental questions quickly emerge – is it practical (how long will it take, how much will it cost?), does it tell you something 'real' about the animal's welfare (is it valid?), can two or more assessors give the same answer or score when assessing the same animal (is it repeatable?) and is the measure influenced by weather, season, time of day, etc. (is it robust?).

To be not only 'useful' but also to engender confidence – from the public, the farmer or institution housing the animals and the assessor – these four conditions should be 'reasonably met'. In order to meet these criteria 'reasonably', it is not

realistic to consider that animal experience can be reduced to a mechanistic assessment alone – animals are variable, living and sentient beings. We cannot simply plug in formulae and expect to understand how animals see and respond to their world, and so assessment of animal welfare needs to accommodate and include these 'animal' characteristics. Practically speaking, it may be possible to combine RBMs and OBMs. For example, if some animals are lame (assessed using an OBM) it may be possible to predict lameness in others if the floor condition is poor (an RBM). However, the philosophy of OBM-based assessment is that, if there is an OBM that fulfils the conditions described above (practical, valid, repeatable, robust) then this should be used in preference to the RBM alone because individual animals may vary in their responses. For example, a given floor condition may be very good for one animal but very difficult for another – measuring the animal response accommodates this individual variation.

Additionally, it is clear that some simple questions, for example, 'is the animal thirsty?', are not

A. Butterworth *et al.*

so easy to answer in a farm, zoo or laboratory situation. There is no currently feasible OBM for dehydration which can be carried out on farm. A blood measure could be carried out at the slaughterhouse, or it could be theoretically 'possible' to offer water to animals in a choice test – but these are both considered impractical as day-to-day tests, so a combination of resource-based (drinker availability, access to water) and management-based measures (how water supply is actually managed) must be used.

12.4 Applying New Measures

There are a number of stages required in applying an OBM. First, a standardized description is needed. Then a description of the practical application of the method and of any linked RBMs is required. Finally, the ways in which the information that is produced from the use of the OBM and RBMs can be used to promote and support management decisions and practices to create improvements in welfare. The four steps involved are:

Step 1. Measure.
Step 2. Analyse risk factors.
Step 3. Inform.
Step 4. Support management decisions to create improvements in welfare.

We will discuss each of these steps using examples drawn from welfare assessments of farm animals (broiler chickens) and zoo animals (clouded leopards).

12.4.1 Lameness in broiler chickens

The occurrence of leg disorders which lead to lameness can be high in rapidly growing broilers (Sanotra, 2000). This has been highlighted as a major welfare concern (FAWC 1992, 1998; European Commission 2000). Studies have shown that lame birds with high gait scores negotiate an obstacle course more rapidly after the administration of a non-steroidal anti-inflammatory drug, and preference testing reveals that they have a greater tendency than non-lame birds to select feed treated with analgesics (Danbury *et al.*, 2000). These studies demonstrate that birds with lameness may be experiencing pain as well as impaired mobility. Leg weakness may be due to skeletal deformity or to joint or bone infections (Mench *et al.*, 2001; Butterworth *et al.*, 2002).

Let us say that you are a chicken farmer, and you have 100,000 birds on your farm, and 0.8% of these birds are lost through lameness-related culling or do not achieve full body weight because of lameness. So 0.8% of your final productivity is lost or only partially achieved; in an industry where margins per bird and per flock are very small, this can be a significant loss. If you could reduce this percentage of compromised birds, this would not only result in a tangible increase in animal welfare, but potentially a measurable improvement in profit.

Step 1. Measure

Gait scoring: this is a recognized method for assessing the walking ability of birds and was developed by Kestin *et al.* (1992). Birds are given a gait score from 0 to 5, with 5 being the poorest. This gait-scoring system has been validated by others and been found to be practical and repeatable (Berg and Sanotra, 2001; Garner *et al.*, 2002). Written descriptions of the gait scores can be used for the assessment:

Gait Score 0. Normal, dextrous and agile.
Gait Score 1. Slight abnormality, but difficult to define.
Gait Score 2. Definite and identifiable abnormality.
Gait Score 3. Obvious abnormality, affects ability to move.
Gait Score 4. Severe abnormality, only takes a few steps.
Gait Score 5. Incapable of walking.

However, training using video footage and on-farm evaluation of the performance of the assessor is considered to be more robust, giving assurance as to the uniformity, repeatability and validity of the assessor's performance.

For gait scoring, 250 birds are selected at random within the house by reference to a randomized location identifier. Birds are selected from four locations, in groups of up to 80 birds, by corralling at each location using a hinged catching pen (Fig. 12.1). Each bird is individually encouraged to walk out of the pen (Fig. 12.2) and is scored as it does so (Fig. 12.3).

Step 2. Analyse risk factors

While on the farm measuring gait, a variety of information can be collected about the farm and its management practices that can help in analysing

Fig. 12.1. Gait scoring of broilers: penning. Approximately 60 birds are penned in an area of the house. Birds should not be herded into the pen. The pen should be placed quietly around a group of birds with minimal disturbance.

the risk factors for lameness. Information that might be of importance includes:

1. Breeder flock information, including genotype/strain, breeder history and age.
2. Hatchery information, including hatchery, distance/time transported and hatchery vaccination programme for chicks.
3. General information, including number and weight of chicks placed, sex, time of year, age at assessment and slaughter.
4. Specific husbandry practices, including stocking density, brooding conditions, nutritional profile, litter substrate, feeder and drinker design/type, lighting programme, age of house, construction details, ventilation, diseases and medication history, vaccination programme and water source.
5. Performance information, including growth, mortality and culls for leg problems and other reasons.
6. Processing plant information, including scratches, breast blisters, incidence of hock burn and foot burn, condemnations and weight at slaughter.
7. Background information about the management, including: ratio of stockworker to animal

numbers, and age and training/qualifications of stockworkers.
8. Background information about the site/company, including size of houses, number of birds on site and biosecurity measures.

In addition to collecting farm information, ten birds with poor gait score can be culled from the flock, and post-mortems conducted. This would help to determine the main pathological causes of lameness for the company and the farm as a specific risk assessment, to enable them to understand the causes of lameness within their own business.

The broiler company can (in this specific example) analyse the OBM and RBM information, and:

1. Find out the prevalence and severity of leg disorders in the flocks.
2. Make comparisons between 'good' and 'poor' farms (with respect to lameness) within the company to help identify management, house environment, feeding, medication, stockmanship and genotype factors which differ between these farms.
3. Investigate the influence of specific factors on different farms, for example, the use of water.

A. Butterworth *et al.*

Farms with increased water use per bird (in equal weather conditions) may have systematic problems with leaking drinkers. Small amounts of water leaked chronically into the litter can severely reduce litter quality and affect leg health.

4. Carry out an investigation of the bacteriological pathologies linked with lameness and identify whether these bacteria may have come from the hatchery, the transportation or other lapses in farm biosecurity.

In general, significant improvements in profitability, as well as an improvement in overall bird welfare, can be made if lameness is tackled within a company. Inspection bodies in some countries are now beginning to focus on leg health issues as a marker for company welfare performance.

Step 3. Inform, and Step 4. Support management decisions to create improvements in welfare

The farmer can be informed about the extent (and perhaps the 'type') of lameness on the farm, and, with time, and after analysis, a pattern of risk factors may emerge which allow him/her to make decisions which can reduce lameness. Factors which, in real farm experience, have been shown to be risk factors for lameness include: growth rate, the age of the birds at slaughter, the use of whole cereals in the diet, the type of feed, the quality and application of biosecurity, litter condition and

Fig. 12.2. Gait scoring of broilers: scoring. One person can quietly move individual birds out of the pen using a cane (about 1 m in length). The birds should not be hurried, pushed or lifted. Each bird is gait scored as it leaves the pen.

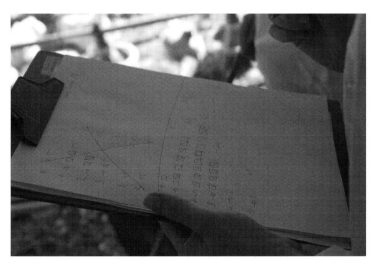

Fig. 12.3. Gait scoring of broilers: recording. A simple tabulation of the numbers of birds scored in each gait score category is created. It is recommended that approximately 250 birds are scored to give a significant sample size.

genotype of the birds grown. The following have also been altered in successful manipulations to reduce lameness on farms in the UK: the sex of the birds (overall, levels of leg weakness are higher in males, and for farms experiencing problems with lameness, selection of the sex of the birds reared can have an effect on overall lameness), levels of feed restriction, the lighting pattern and light intensity, bird activity levels and stocking density (Butterworth, personal observation).

12.4.2 Animal welfare assessment in zoos

Over the past decade, zoos have increasingly been putting efforts into identifying appropriate measures and processes to assess the welfare of zoo animals. For example, in 2000, the Association of Zoos and Aquaria in the United States (AZA) established an Animal Welfare Committee (AWC) to start addressing the increasing need for systematic and scientific assessment of zoo animal welfare. One of the first tasks of this committee was to initiate the compilation of AZA care guidelines, or best practices, officially termed care manuals, for all zoo-held species. While only a handful of these Animal Care Manuals (ACMs) have so far been completed, at least 150 drafts are currently in various stages of the development and review process, and will be published over the next few years (see AZA, 2010).

The goal of ACMs is to facilitate the development of consistent animal care practices across taxa for the wide variety of species regularly held in zoos and aquaria. ACMs are also intended to help identify knowledge gaps and describe future research approaches for filling existing gaps (Barber, 2009). Once available on the AZA web site, these documents can clearly provide a great resource for zoo professionals. But, similarly to the situation for farm animal assessment, the ACMs can only provide very general resource-based and management-based welfare guidelines that allow zoos and aquaria to set the stage for good welfare (Barber, 2009), but naturally cannot provide an assessment of the actual welfare experienced by individual animals. In addition to ACMs, the AZA Accreditation Standards (AZA, 2009) themselves are intended to set the stage for good potential welfare at least, at the species level, by regularly (every 5 years) assessing various components of necessary care, management and housing at accredited institutions.

However, all of the above measures to ensure good animal welfare represent RBMs, and will ideally need to be linked with indicators or measures of welfare obtained on individual animals and specific outcomes (OBMs) to ensure that care recommendations are truly effective at the individual animal level. A more holistic approach to animal welfare assessment that includes both RBM and OBM assessments is therefore recommended for zoo animals, in a manner similar to current trends for farm animals.

A number of studies attempting to assess various aspects of individual zoo animal welfare and welfare outcomes based on variations in management or housing have been conducted during recent years (Wielebnowski *et al.*, 2002; Shepherdson *et al.*, 2004; Carlstead and Brown, 2005; Moreira *et al.*, 2007). These studies, while providing highly valuable information, have also further highlighted the complexity of animal-focused assessments. To validate such measures properly and to ensure their universal applicability, longitudinal data collection is usually necessary, and large samples sizes are generally needed. This requires studies to be carried out across multiple institutions (i.e. across zoos and aquaria), as the most common sample sizes for many species at any given zoo or aquarium are one to three animals. In addition, most of the data collection on zoo animals has to be conducted non-invasively (e.g. usually without direct access, or with only very limited access to the animals) and with respect to the fact that individual animals are being viewed by zoo guests, commonly on a daily basis (so keeper routines, animal visibility and guest events need to be taken into account when designing studies and experiments). Furthermore, the large-scale longitudinal data collection required for such studies is naturally very time intensive, and some of the data analyses, such as physiological assessments, may also be quite costly. All of these challenges combined make it difficult to conduct such studies on a regular basis to arrive at welfare assessments for the wide variety of species and individuals housed in zoos and aquaria. Thus, this much more intensive research approach can, like the above-mentioned RBMs, only represent one part of an overall welfare assessment strategy that ultimately needs to include a variety of both RBMs and OBMs for effective assessments across individuals, species and institutions.

A. Butterworth *et al.*

12.4.3 The welfare of clouded leopards

An example of using the four-step approach described above to improve the welfare of zoo animals is the process undertaken to evaluate the husbandry of captive clouded leopards (*Neofelis nebulosa*). Clouded leopards (Fig. 12.4) are difficult to manage and breed in captivity. Behavioural problems, such as fur plucking, excessive pacing or hiding, and severe male–female aggression abound in this species. These husbandry problems are presently one of the major obstacles to establishing a self-sustaining captive population. Animal managers and keepers have often felt that clouded leopards may be particularly prone to stress, possibly linked to various elements of their captive environment.

Step 1. Measure

A multi-institutional study was carried out using faecal hormone monitoring in combination with behavioural and keeper surveys to assess whether certain management variables may correlate with increased adrenal activity in this species. Scientists were able to validate an assay for faecal glucocorticoid metabolites for clouded leopards. They then monitored faecal glucocorticoid concentrations in 74 clouded leopards (37 males and 37 females) at

Fig. 12.4. Clouded leopard (*Neofelis nebulosa*). (Photo: Jim Schultz, Chicago Zoological Society.)

12 zoological institutions, and obtained keeper ratings and other pertinent survey information for each animal (Wielebnowski *et al.*, 2002). The following are some examples of questions addressed to keepers for ratings on a scale from 1 to 5: how frequently does the animal pace?; how frequently does the animal hide?; and how frequently does the animal show self-injuring behaviours (such as tail biting or fur plucking)? Other survey information collected included enclosure size, enclosure height, climbing structures, hiding places, diet, keeper interaction type (e.g. 'hands off', 'hands on'), rearing type (hand versus mother reared), number of keepers taking care of an animal and time spent per day on animal care.

Step 2. Analyse risk factors

Analyses showed higher average faecal corticoid concentrations to be positively correlated with the occurrence of self-injurious behaviours (fur plucking), as well as with the frequency of pacing and hiding. Furthermore, analysis of various husbandry factors showed that higher enclosures (accessible height available to the cat by using climbing structures) and increased time spent by the keeper with the animal were associated with lower corticoid concentrations, while visibility of potential predators (tigers and other large cats) and public display were associated with higher concentrations (Wielebnowski *et al.*, 2002).

Based on these results, it was decided to design an experiment in which two of the above-mentioned variables, available enclosure height and public display, could be systematically varied while leaving all other measures unaltered. These follow-up experiments were conducted on 12 clouded leopards (six males and six females) at four zoological institutions. Regular quantitative behavioural observations and non-invasive faecal hormone monitoring were used to track changes in behaviour and adrenal activity in individuals before and after the implementation of enclosure changes. Initial baseline behavioural and hormone data, before enclosure changes, was collected for 2–3 months. Changes in enclosure height were made by the addition of climbing structures to allow the animals increased access to higher areas of the enclosure ($n = 8$, 4 males and 4 females). Changes for 'public display' were made by adding additional hiding spaces ($n = 4$, 2 males and 2 females). Following the changes, data were collected for another 2–3 months.

Results indicated that both the addition of climbing structures and the addition of hiding spaces resulted in significant reductions of faecal glucocorticoid concentrations for the majority of study individuals (ten out of 12) (Shepherdson *et al.*, 2004; Fig. 12.5). Recommendations for future housing and exhibit design were made accordingly, based on the results obtained from the initial large scale study and the follow-up experiments. For example, institutions now regularly try to maximize available enclosure height for clouded leopards, add additional climbing and hiding spaces, and try to avoid locating clouded leopards in close proximity to other large cats.

Step 3. Inform, and Step 4. Support management decisions to create improvements in welfare

The results of these studies were reported at conferences, through publications (e.g. Wielebnowski *et al.*, 2002; Shepherdson *et al.*, 2004) and to the clouded leopard Species Survival Plan (SSP)

Committee that helps to facilitate the breeding and management of clouded leopards in North American zoos. The clouded leopard SSP and the clouded leopard husbandry manual produced through the SSP serve as key information resources for all managers and keepers working with this species in zoos worldwide. Therefore, the integration of new management information into current husbandry manuals and the distribution of this information through the SSP is most effective. The expansion of available enclosure height, in particular, has been increasingly implemented in new exhibits for this species, and other factors (such as increased hiding spaces) are now also regularly considered as part of species management; see Thailand Clouded Leopard Consortium (Smithsonian Conservation Biology Institute, 2010) or The Clouded Leopard Project (2010). Successful outcomes have already been identified at some institutions and these can be measured in improved animal health and reproductive success, and a significant reduction or elimination of potentially harmful behaviours such as fur plucking and pacing. Using combined measures and

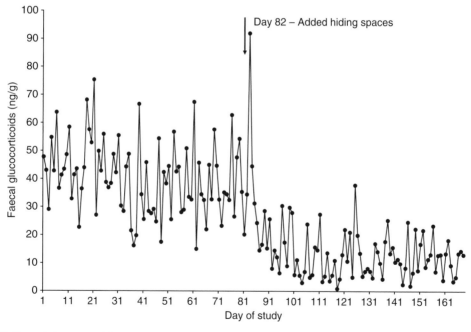

Fig. 12.5. Faecal glucocorticoid metabolite concentrations monitored in a clouded leopard (*Neofelis nebulosa*) before and after the addition of more hiding spaces to its enclosure. Hiding spaces were added on day 82; the highest peak hormone value was observed on the following day. This peak value is likely to reflect a stress response to the initial disturbance caused by the changes; but, subsequently, faecal corticoid concentrations declined significantly in response to the (previous) addition of more hiding spaces. (Shepherdson *et al.*, 2004.)

A. Butterworth *et al.*

data from multiple institutions was of course key to the success of this study, as well as the buy-in of all stakeholders (clouded leopard SSP members and animal managers) from the start so that information on outcomes could be distributed widely and effectively and applied to future management decisions.

12.5 Turning Assessments into Scores

12.5.1 Farm animal inspection programmes

Assurance schemes in a number of countries have the potential to influence and improve farm animal welfare through their programmes of inspection to standards. Additionally, the 'branding' or linking of assurance-scheme labelling and information can facilitate favourable market responses. Higher welfare animal products, sometimes carrying the label of an assurance scheme, are generally sold to consumers as 'quality items' or as 'best' or 'finest', thereby generating a higher price (Buller *et al.* 2009). When welfare assessments are used in this way, there is often interest in using the OBMs, RBMs and management-based measures to create an overall welfare 'score' that can then be used in characterizing the product from the animal, particularly given that most assurance schemes in fact make demands above (sometimes far above) the legal minimums in particular countries, where these exist.

There are multiple levels to such scoring, and multiple methods for determining them. The first level would be the results for each measure, provided initially as 'benchmarking' or 'initial position' scores – which tell an individual farmer how he or she is positioned at the very start of the assessments, and also allows her to compare herself with some known baseline values, and also with her farming peers. The individual scores for each measure made can be fed back to the farmer for her own information, so that she can see how she did, and also to support any areas that she might wish to develop or focus on (if she has a problem). For example, she might find that the assessment shows that she has an unusually high number of tail-bitten pigs when compared with her previous assessments. She might use the periodic information provided by an assessment as a barometer for any changes she chooses to make. If they cause reductions in the 'area of concern', then she can use this information to assess the degree of 'success' that

she has achieved – and, conversely, if there is no improvement or even an increase, she may be able to use the risk factor assessment to target other areas of change. Von Borell *et al.* (2001) created a recording system for pig housing based on hazard analysis and critical control points (HACCP), with the potential to feed back results of the effects of altered management at these critical control points. The concept of feeding back individual 'non-compliant' assessment points is the basis of all existing farm assurance and certification schemes, for example those related to food safety, and so this approach – detect areas of concern and then request rectification of specific points – is familiar to many farmers, and is seen as the basis of a reasonable approach to ensuring compliance with a recognized 'standard'.

Some schemes, such as the Red Tractor Farm Assured Dairy Scheme in the UK, have two categories of standards, those that 'must' be met (within a given time frame) and those that 'should' be met (i.e. strongly recommended but not imposed) (NDFAS, 2004). However, no scheme yet uses such a structured approach to promote animal welfare improvement using predominantly OBMs. Most schemes, including the welfare-oriented schemes such as the RSPCA Freedom Food Scheme, still rely heavily on RBMs, with inclusion of some key animal performance and animal-based criteria, such as absence of injuries caused by the environment or evidence of failure to treat disease without delay (examples which originate in legal requirements).

The second level of 'scoring' would be to combine the individual assessment results, perhaps into linked groupings, for example, in the area of housing. A combined score for related areas is possible, but only if:

1. The individual measures can be combined to give aggregate scores which are considered meaningful, and credible, by the producer and the consumer.
2. The process of combining scores does not devalue the overall meaning of the assessment information – for example, by compensating for a very poor score in one area, with a number of reasonable scores in other areas. If this occurs, then the power of discrimination can be lost, and the credibility of the combined score brought into doubt. Spoolder *et al.* (2003) indicated that when welfare scores are simply summated to give an overall score, a large

welfare disadvantage can be compensated for by a number of minor advantages. This effect can be limited if minimal requirements – below which specific scores cannot be 'permitted'– are set.

Whay *et al.* (2003b) assessed dairy calf welfare on 45 farms by using 19 measures, including health parameters, nutrition and general appearance. The farms were ranked on the basis of each measure from best (rank 1) to worst (rank 45). Each farm could thus be ranked, measure by measure. By summing these ranks it was possible to compare the farms within the 45 farms visited. It is also possible to 'sum the scores' produced, to give an absolute value which is independent of the number of farms. This is the basis of a number of animal welfare studies (Keeling and Svedberg, 1999; Bracke *et al.*, 2002).

Once raw 'values' for measures have been made, the 'natural' next step is to ascribe different weighted values to these measures, to give impact factors for each measure with respect to animal welfare (Bracke *et al.* 2002). This type of weighting system is seen in the Austrian Tiergerechtheitsindex (TGI) (Bartussek, 1999). Weighted sums of scores appear intuitive and the principle is usually readily understood by users.

In the Welfare Quality® project (Welfare Quality®, 2004) a large integrated European Union (EU)-wide research project which has brought together over 30 institutes from across the EU, animal and social scientists have worked together to outline the areas of animal welfare concern, and then to create an assessment system – a set of measures which can be used on the farm and at the slaughterhouse – to assess a range of health, behaviour, injury, disease, comfort, feeding, management, human–animal interactions, emotional state and housing factors that can affect animal welfare (Blokhuis *et al.*, 2003; Botreau *et al.*, 2007; Butterworth *et al.*, 2008). After a farm assessment, the results for individual measures can be fed back to the farmer, and aggregated scores can then be created using combined weighted sums. By combining resource-based information with animal-based measures, the potential exists to provide a powerful tool for informing the farmer of the welfare status of his animals, to enable him to see how his farm compares with other farms, and also to support improvements and management decisions. The farmer can receive a result for each single assessment measure, for example, how many thin cows

he has. He could also be given the information collected together into grouped data, for example, for animal body health measures, measures linked with environmental factors such as skin lesions, hock and foot damage, or behavioural measures. Finally, it is possible to combine all measurement results to give an aggregated overall score – and it may be possible in the future that retailers and consumers would make purchasing decisions for some animal products based on combined welfare-assessment scores.

But what do farmers actually think about the use of OBMs and what are some of the potential problems associated with these kinds of assessment schemes? Some recent focus group interview work by Buller *et al.* (2009) found the following concerns voiced:

1. Who carries the cost of assessing OBMs?
2. How will they work in terms of periodicity and seasonality of assessment?
3. Difficulties (time, money and the accurate identification of causes of failure) in achieving compliance.
4. Potential for mistrust, among producers and others, of the pertinence of certain animal behaviour assessments.
5. Can reduction of specific animal-based assessments to a single farm-based algorithm actually give a meaningful 'single score' which is useful to producers, retailers and advisors?
6. The difficulty of employing OBMs in conveying 'positive' information to consumers, as most OBMs are based on potentially negative characteristics (injuries, hunger, avoidance, etc.).

Resource-based measures lend themselves more readily to 'objective' and numerical assessment procedures (and therefore to algorithms) than do OBMs, which can be seen as more inherently 'subjective', open to challenge and displaying lower confidence levels, and therefore less acceptable to farmers as the basis for determining compliance failure.

12.5.2 Zoo animal welfare scores

Some of the newest efforts in zoo animal welfare assessments also involve the development of rapid, low-cost scoring tools that can serve as first indicators of potential problems and as a guide for continuous improvements. Whitham and Wielebnowski (2009) have recently outlined a

A. Butterworth *et al.*

process for the development and use of individual welfare score sheets, which, if validated successfully, has the potential to be applicable to most zoo species and institutions. The score sheet development is based on the Delphi technique employed by Whay *et al.* (2003a), and the resulting score sheets are intended to quantify the qualitative assessments made by keepers on a daily basis for the rapid use and assessment of animal welfare. The key is that score sheets produced for each species will be short, easy to complete, user friendly and available electronically, so that input and output can be rapid and easily accessible. Score sheet results will be graphed regularly, and a software program, still in its developmental stages at the moment, will be used to flag potential problem cases or cases in which welfare has been improving, so that timely feedback and management changes can occur. Such a system would allow more rapid responses to potentially significant changes in an individual animal's welfare state. It would also allow for continuous improvements in welfare status, owing to an adjustable scale focused on each individual, which will allow welfare scoring from poor to good on a variety of key measures, rather than solely on the avoidance of poor welfare. Used in combination with the currently available resource-based approaches, this would result in a more holistic assessment of welfare at zoos and aquaria. In addition, in cases where distinct welfare problems are identified, more in-depth longitudinal, multi-institutional and multi-disciplinary studies can then be designed to address these questions specifically and to help inform species-specific care manuals and accreditation processes for further enhancement of individual animal welfare.

12.6 A Laboratory Animal Application: Humane Endpoints

This chapter has presented two detailed examples from farm animal production and zoo animal keeping, respectively. Practical strategies to assess and improve welfare are equally important for the use of animals in scientific research – however, the nature of their use presents some particular challenges and limitations. In most other applications, the main limiting factors in promoting good animal welfare are probably economic constraints. The market determines how high the price of animal products or the zoo entrance ticket can be before the consumer/visitor goes somewhere else, and this sets a limit on how many costly measures can be implemented. Economic constraints also affect laboratory animal care, but in this area it is often the scientific use of animals that presents the most challenging hurdles. Not all research with animals results in animal suffering of course, but when the research requires that animals develop pathologies, this will be accompanied by welfare problems. The more severe the disease, the more animal welfare will, in general, be compromised.

The particular ethical concern over experiments causing animals to die from the pathologies they develop has led to the development of the concept of humane end points. Humane end points mean that earlier, less severe clinical signs are used to determine the point at which animals are euthanized (or sometimes treated) rather than awaiting spontaneous death (for an extensive overview of the application of humane end points, see a special issue of *ILAR Journal* entitled *Humane Endpoints for Animals Used in Biomedical Research and Testing* (ILAR Journal, 2000)). A crucial component in the correct application of humane end points is the establishment of a welfare assessment protocol specific for the experiment in question. This should lay down how frequently the animals need to be inspected (related to how rapidly animal health is expected to deteriorate), which parameters should be measured (including both general measures of health and well-being such as body weight and temperature, as well as disease-specific parameters such as paralysis) and criteria for when to take action. In practice, this is usually translated into score sheets to be filled out by the person responsible for animal monitoring (Morton, 2000). Humane end points are applicable in a wide range of research and testing situations, including tumour research, infectious disease research, vaccine testing and tests of systemic toxicity. In many countries, a research project in which animals are expected to become severely ill will not be given ethical approval unless a protocol for humane end points is established. In regulatory testing of substances, the implementation of humane end points has led to the replacement of the infamous LD_{50} test (Lethal Dose 50, meaning the dose at which half of the animals die within 14 days of a single exposure) by tests in which most animals are euthanized at an earlier end point (OECD, 2002).

12.7 Conclusions

- Existing animal welfare assessment schemes tend to assess welfare by examination of the provision of housing or resources (resource-based measures, RBMs), rather than looking at outcomes for the animals themselves (outcome-based measures, OBMs).

- Research scientists have for some time suggested that OBMs could provide valid indicators of animal welfare, as welfare is a characteristic of the individual animal, not just of the system in which animals are kept.

- The sorts of questions being asked are: Are the animals properly fed and supplied with water? Are the animals properly housed? Are the animals healthy? Can the animals express a range of behaviours and emotional states?

- Many modern farms, zoos, aquaria and other facilities are keen to identify welfare monitoring tools that can be readily applied and allow for rapid assessment to provide timely feedback for management decisions.

- To implement effective use of animal-based assessment methods on farm, it is necessary to adopt the following steps: Step 1, Measure (RBMs and OBMs); Step 2, Analyse risk factors; Step 3, Inform (producer, purchaser); Step 4, Support management decisions to create improvements in welfare.

- It may be possible to create a range of 'scores'. The individual measures can be combined to give aggregate scores which can be presented to the producer and the consumer. This requires the attribution of weighted values to the measures in order to assess the impact of each measure with respect to animal welfare.

- A multi-pronged strategy involving various RBMs and OBMs is most likely to provide the capacity for comprehensive welfare monitoring of farm animal, zoo animal and laboratory animal welfare in the future. Such a strategy may include regularly updated species care guidelines, accreditation standards, longitudinal, multi-disciplinary and multi-institutional studies, and rapid assessment tools for continuous welfare monitoring.

Acknowledgment

Anna Olsson's generous contribution of Section 12.6 is gratefully acknowledged.

References

AZA (Association of Zoos and Aquariums) (2009) *The Accreditation Standards and Related Policies*, 2008 edn. AZA, Silver Springs, Maryland.

AZA (2010) Animal Care Manuals. Available at: www.aza.org/animal-care-manuals/ (accessed 30 December 2010).

Barber, J.C.E. (2009) Programmatic approaches to assessing and improving animal welfare in zoos and aquariums. *Zoo Biology* 28, 519–530.

Barnett, J.L. and Hemsworth, P.H. (1990) The validity of physiological and behavioural measures of animal welfare. *Applied Animal Behaviour Science* 25, 177–187.

Bartussek, H. (1999) A review of the animal needs index (ANI) for the assessment of animals' well-being in the housing systems for Austrian proprietary products and legislation. *Livestock Production Science* 61, 179–192.

Berg, C. and Sanotra, G.S. (2001) Kartlaggning av forekomsten av benfel hos svenska slaktkycklingar – en pilotstudie. (Survey of the prevalence of leg weakness in Swedish broiler chickens – a pilot study, with translation). *Svensk Veterinartidning* 53, 5–13.

Blokhuis, H.J., Jones, R.B., Geers, R., Miele, M. and Veissier, I. (2003) Measuring and monitoring animal welfare: transparency in the food product quality chain. *Animal Welfare*, 12, 445–455.

Botreau, R., Veissier, I., Butterworth, A., Bracke, M.B.M. and Keeling, L. (2007) Definition of criteria for overall assessment of animal welfare. *Animal Welfare* 16, 225–228.

Bracke, M.B.M., Spruijt, B.M., Metz, J.H.M. and Schouten, W.G.P. (2002) Decision support system for overall welfare assessment in pregnant sows A: Model structure and weighting procedure. *Journal of Animal Science* 80, 1819–1834.

Buller, H., Roe, E., Bull, J., Dockes, A.C., Kling-Eveillard, F. and Godefroy, C. (2009) *Constructing Quality: Negotiating Farm Animal Welfare in Food Assurance Schemes in the UK and France.* Welfare Quality® Report 4.1.1.2 D 4.17, March 2009. Project Office Welfare Quality®, Lelystad, The Netherlands.

Butterworth, A., Weeks, C.A., Crea, P.R. and Kestin, S. C. (2002) Dehydration and lameness in a broiler flock. *Animal Welfare* 11, 89–94.

Butterworth, A., Veissier, I., Manteca, J.X. and Blokhuis, H.J. (2008) Welfare trade. European Union 15. *Public Service Review* 18:54:39, pp. 456–459.

Carlstead, K. and Brown, J.L. (2005) Relationships between patterns of fecal corticoid excretion and behavior, reproduction, and environmental factors in captive black (*Diceros bicornis*) and white (*Ceratotherium simum*) rhinoceros. *Zoo Biology* 24, 215–232.

A. Butterworth *et al.*

Danbury, T.C., Weeks, C.A., Chambers, J.P., Waterman-Pearson, A.E. and Kestin, S.C. (2000) Self-selection of the analgesic drug carprophen by lame broiler chickens. *Veterinary Record* 146, 307–311.

Dawkins, M.S. (1990) From an animal's point of view: motivation, fitness, and animal welfare. *Behavioural and Brain Sciences* 13, 1–61.

European Commission (2000) *The Welfare of Chickens Kept for Meat Production (Broilers). Report of the Scientific Committee on Animal Health and Animal Welfare. Adopted 21 March 2000.* Available at: http://ec.europa.eu/food/fs/sc/scah/out39_en.pdf (accessed 29 December 2010).

FAWC (Farm Animal Welfare Council) (1992) *Report on the Welfare of Broiler Chickens*. FAWC, Surbiton, UK.

FAWC (1998) *Report on the Welfare of Broiler Breeders*. FAWC, Surbiton, UK.

Garner, J.P., Falcone, C., Wakenell, P., Martin, M. and Mench, J.A. (2002) Reliability and validity of a modified gait scoring system and its use in assessing tibial dyschondroplasia in broilers. *British Poultry Science*, 43, 355–363.

Geers, R., Petersen, B., Huysmans, K., Knura-Deszczka, S., De Becker, M., Gymnich, S., Henot, D., Hiss, S. and Sauerwein, H. (2003) On-farm monitoring of pig welfare by assessment of housing, management, health records and plasma haptoglobin. *Animal Welfare* 12, 643–647.

Goulart, V.D., Azevedo, P.G., van de Schepop, J.A., Teixeira, C.P., Barcante, L., Azevedo, C.S. and Young, R.J. (2009) GAPs [Gaps] in the study of zoo and wild animal welfare. *Zoo Biology* 28, 561–573.

Hurnik, J.F. (1990) World's Poultry Science Association Invited Lecture. Animal welfare: ethical aspects and practical considerations. *Poultry Science* 69, 1827–1834.

ILAR Journal (2000) *Humane Endpoints for Animals Used in Biomedical Research and Testing. ILAR Journal* 41(2). Available [online only] at: http://dels-old.nas.edu/ilar_n/ilarjournal/41_2/index.shtml (accessed 30 December 2010).

Keeling, L. and Svedberg, J. (1999) Legislation banning conventional battery cages in Sweden and a subsequent phase-out programme. In: Kunisch, M. and Eckel, H. (eds) *Proceedings of the Congress 'Regulation of Animal Production in Europe'*, 9–12 May, Wiesbaden, Germany. KTBL – Schriften-Vertrieb im Landwirtschaftsverlag, Darmstadt, Germany, pp. 73–78.

Kestin, S.C., Knowles, T.G., Tinch, A.E. and Gregory, N.G. (1992) Prevalence of leg weakness in broiler chickens and its relationship with genotype. *Veterinary Record* 131, 190–194.

Main, D.C.J., Webster, F. and Green, L.E. (2001) Animal welfare assessment in farm assurance schemes. *Acta Agriculturae Scandinavica, Section A – Animal Science* Supplementum 30, 108–113.

Mench, J.A. (2004) Assessing animal welfare at the farm and group level: a United States perspective. *Animal Welfare* 12, 493–503.

Mench, J.A., Garner, J.P. and Falcone, C. (2001) Behavioural activity and its effects on leg problems in broiler chickens. In: Oester, H. and Wyss, C. (eds) *Proceedings of the 6th European Symposium on Poultry Welfare*, Zollikofen. World's Poultry Science Association, Swiss Branch, Zollikofen, Switzerland, pp. 152–156.

Moreira, N., Brown, J.L., Moraes, W., Swanson, W.F. and Monteiro-Filho, E.L.A. (2007) Effect of housing and environmental enrichment on adrenocortical activity, behavior, and reproductive cyclicity in the female tigrina (*Leopradus tigrinus*) and margay (*Leopardus wiedii*). *Zoo Biology* 26, 441–460.

Morton, D. (2000) A systematic approach for establishing humane endpoints. *ILAR Journal* 41(2). Available at: http://dels-old.nas.edu/ilar_n/ilarjournal/41_2/Systematic.shtml (accessed 29 December 2010).

NDFAS (National Dairy Farm Assured Scheme) (2004) *National Dairy Farm Assured Scheme: Standards and Guidelines for Assessment*, 3rd edn. Available at: http://www.ndfas.org.uk/uploadeddocuments/NDFAS_3rd_Edition.doc (accessed 29 December 2010).

OECD (Organisation for Economic Co-operation and Development) (2002) OECD Test Guideline 401 was deleted in 2002: A Major Step in Animal Welfare: OECD Reached Agreement on the Abolishment of the LD_{50} Acute Toxicity Test. Available at: http://www.oecd.org/document/52/0,2340,en_2649_34377_2752116_1_1_1_1,00.html (accessed 29 December 2010).

Rutter, S.M. (1998) Assessing the welfare of intensive and extensive livestock. In: *Proceedings of the EC Workshop 'Pasture Ecology and Animal Intake'*, 24–25 September 1996, Dublin, Ireland, pp. 1–9.

Sanotra, G.S. (2000) *Leg Problems in Broilers: A Survey of Conventional Production Systems in Denmark*. Dyrenes Beskyttelse, Frederiksberg, Denmark.

Shepherdson, D.J., Carlstead, K.C. and Wielebnowski, N. (2004) Cross-institutional assessment of stress responses in zoo animals using longitudinal monitoring of fecal corticoids and behavior. *Animal Welfare* 13 (Supplement 1), 105–113.

Smithsonian Conservation Biology Institute (2010) Thailand Clouded Leopard Consortium. Available at: http://nationalzoo.si.edu/SCBI/ReproductiveScience/ConsEndangeredCats/CloudedLeopards/consortium.cfm (accessed 30 December 2010).

Sørensen, P. (2001) Breeding strategies in poultry for genetic adaptation to the organic environment. In: *Proceedings of the 4th NAHWOA Workshop*, 24–27 March 2001, Wageningen. Network for Animal Health and Welfare in Organic Agriculture (NAHWOA), Wageningen, The Netherlands, pp. 51–61.

Spoolder, H., De Rosa, G., Horning, B., Waiblinger, S. and Wemelsfelder, F. (2003) Integrating parameters to assess on-farm welfare. *Animal Welfare* 12, 529–534.

The Clouded Leopard Project (2010) Available at: http://www.cloudedleopard.org/default.aspx (accessed 30 December 2010).

Veissier, I., Butterworth, A., Bock, B. and Roe, E. (2008) European approaches to ensure good animal welfare *Applied Animal Behaviour Science* 113, 279–297.

Von Borell, E., Bockisch, F.J., Buscher, W., Hoy, S., Krieter, J., Muller, C., Parvizi, N., Richter, T., Rudovsky, A., Sundrum, A. and Van Den Weghe, H. (2001) Critical control points for on-farm assessment of pig housing. *Livestock Production Science* 72, 177–184.

Webster, A.J.F. (1998) What use is science to animal welfare? *Naturwissenschaften* 85, pp. 262–269.

Welfare Quality® (2004) Welfare Quality®: Science and society improving animal welfare in the food quality chain, EU funded project FOOD-CT-2004-506508.

Details available at: www.welfarequality.net (accessed 30 December 2010).

Whay, H.R., Main, D.C.J., Green, L.E. and Webster, A.J.F. (2003a) Animal-based measures for the assessment of welfare state of dairy cattle, pigs and laying hens: consensus of expert opinion. *Animal Welfare* 12, 205–217.

Whay, H.R., Main, D.C.J., Green, L.E. and Webster, A.J.F. (2003b) An animal-based welfare assessment of group-housed calves on UK dairy farms. *Animal Welfare* 12, 611–617.

Whitham, J.C. and Wielebnowski, N. (2009) Animal-based welfare monitoring: using keeper ratings as an assessment tool. *Zoo Biology* 28, 545–560.

Wielebnowski, N.C., Fletchall, N., Carlstead, K., Busso, J.M. and Brown, J.L. (2002) Noninvasive assessment of adrenal activity associated with husbandry and behavioral factors in the North American clouded leopard population. *Zoo Biology* 21, 77–98.

13 Physical Conditions

BIRTE L. NIELSEN, MICHAEL C. APPLEBY
AND NATALIE K. WARAN

Abstract

The physical environment of an animal is sometimes altered if it is found to cause problems for animal welfare. These changes are commonly quite specific (making changes to space, food, water, aspects of housing design such as flooring, or to other environmental factors such as air quality) and may be effective in preventing injuries or disease. However, such measures may not be implemented in practice (usually for economic reasons), and where implemented may cause other problems, as when concern for hygiene leads to animals being kept in barren conditions. Numerous ways have also been tried to diversify feeding methods in order to improve animal welfare, but specific changes to the environment such as these often have widespread effects, some of which may be detrimental. For example, inclusion of novel pen structures meant to enrich the environment may lead to increased aggression. A more general approach is therefore appropriate. One area where this is particularly relevant is handling and transport, when animals encounter environments that are wholly new to them. For environments where animals spend more time, several studies have attempted a 'biological approach' in which a biological functioning is considered while avoiding simplistic assumptions of 'natural is best'. We consider as examples systematic tests of environmental enrichment for pigs, novel designs for loose housing of lactating sows and their litters, and furnished cages for laying hens. Stringent tests of every design feature and their interactions are necessary to produce commercial designs from such studies.

13.1 Introduction

The physical environment in which animals live is critical to their welfare. It is relevant to observe, then, that the environment of most animals, throughout the world, is influenced to a greater or lesser degree by humans. The principles of this chapter therefore apply to all animals, including those for which human influence is indirect or slight. However, it is mostly concerned with animals for which we have a major influence on the physical environment, such as farm, zoo, companion and laboratory animals. Indeed, the welfare of animals managed by humans has commonly been evaluated based on provisions of space, food, water, housing design (for example flooring) and other environmental factors (for example air quality), although increasing emphasis is now being put on indicators of welfare based on the direct consequences for the animals. Veissier *et al.* (2008b, p. 295) write:

> There are moves in some certification schemes to try and start to include animal-based measures, for example, the incidence of lameness in cattle and sheep, of foot pad lesions in poultry, or the comfort behaviours of animals in cubicles and stalls.

These developments are discussed in Chapter 2, Understanding Animal Welfare, and Chapter 12, Practical Strategies to Assess (and Improve) Welfare. However, it is also important to know how physical factors affect those outcomes.

Two points are pertinent to evaluation of the effects of physical conditions on animal welfare. First, an understanding of the problems is essential if solutions are to be found, particularly if these are to be general, and not specific to one situation. This chapter should therefore be read in conjunction with previous chapters that considered underlying problems. Secondly, many of the problems and many of the solutions relate to behaviour – more so than to other factors relevant to welfare. The extent to which the performance of behaviour is itself important for welfare is discussed in Chapters 7 and 9. Jensen and Toates (1993, p. 177) comment as follows:

> In considering ethological needs, it is essential to understand the consequences to the animals of not being able to perform a certain behaviour ... Ultimately, what should guide our welfare considerations is the degree of suffering an animal will experience if a certain behaviour is impaired.

So the emphasis on behaviour does not imply that other factors are less important. Rather, it occurs because behaviour is the interface between the animal and most aspects of its environment. Thus, injury is not itself a behavioural welfare problem, but it occurs or is prevented partly through behaviour. An example is the finding that cows are likely to need more than one hour of exercise a day both to fulfil their behavioural needs and to stay in good health (Veissier *et al.*, 2008a) – to be fit enough to avoid injury. For those welfare problems that are not mediated by behaviour, such as some diseases, behaviour is important as a symptom in detecting the problem and in assessing the effectiveness of any cure.

Assessment of effectiveness is vital, partly because there may be different solutions available for the same problem. We cannot simply assume that our supposed solution has worked. Furthermore, solving some problems creates others. For example, Marashi *et al.* (2003) compared standard cages used for male laboratory mice with more varied environments by adding wooden scaffolding and transparent plastic covers in either a standard cage or a much larger cage where extra sources of food and water had also been added. They found that environmental enrichment led to an increase in aggressive behaviour in the strain of mice they studied (p. 288):

> The increased aggression of males in both kinds of enriched cages may have resulted because the animals were beginning to defend territories, facilitated by the provision of structural elements ... Interestingly, the degree of enrichment as well as the size of the enriched cages seems to be of minor importance in causing agonistic behavior, since no significant differences existed in frequency and duration of aggressive encounters observed in either enclosure.

Sometimes, then, one or a few changes to the environment turn out to be inadequate or inappropriate. It may be necessary to make other associated changes and perhaps compromise on certain characteristics that cause problems. We shall consider specific changes first: those that have attempted to tackle a specific problem or to promote a specific advantage. One important example that illustrates many of the principles will then be considered in more detail, namely, methods of feeding. After this, we shall consider whole environments; initially the special case of environments for handling and transport – because during these potentially stressful procedures the environment is new to the animals – and then general approaches to environmental design. We should also point out

that changes to the environment made for reasons unconnected with welfare will have effects, and we shall consider some such changes in passing.

13.2 Specific Changes

A considerable amount is known about environmental design and management in relation to reduction or avoidance of specific physical welfare problems of injury and disease. This is particularly true in agriculture, where such problems reduce profit. For example, in the prevention of injury, softness of the floor is important for both pigs (Lewis *et al.*, 2005) and cattle (Frankena *et al.*, 2009). However, the relationship between floor type and lameness may not be straightforward (Telezhenko *et al.*, 2007, p. 3720):

> The majority of cows preferred to walk and stand on soft rubber flooring rather than on concrete floors ... Yet, lame cows within a group apparently did not have a greater preference than nonlame cows for soft flooring.

Small changes to the housing design may also reduce stereotypic behaviour. Cooper *et al.* (2000) found that providing horses with additional fields of view simply by opening more windows in the stable reduced the incidence of stereotypic weaving in horses that had been known to perform this behaviour for at least 2 years.

Stocking density is another specific physical condition with great influence on animal welfare. For example, the number of broilers per square metre can affect the litter moisture which, in turn, affects the prevalence and severity of foot pad lesions of the birds (Dozier *et al.*, 2006), although this and other effects of stocking density interact with other aspects of the environment such as ventilation, which may be even more important (Dawkins *et al.*, 2004). Estévez (2007) summarizes the negative impact of high stocking density in terms of increased hock and foot pad damage, more scratches and bruising, poorer gait, reduced movement and an increased frequency of disturbance. Despite this, she concludes (p. 1265):

> However, results overwhelmingly suggest that while stocking density has major consequences for the health and welfare of broilers, the quality of the environment, which has been largely underestimated, is far more relevant. Advances in broiler welfare will be difficult to achieve unless some criteria for environmental quality are also established.

B.L. Nielsen *et al.*

This suggests that current scientific endeavours to develop more animal-based welfare parameters (such as hock and foot pad damage) should not ignore physical features such as stocking density and the provision of loose substrate, but encompass them in biologically meaningful ways.

There are two major reservations to make about measures to reduce injury and disease. Sometimes measures known to be effective are not used, for various reasons, including economics, as emphasized in Chapter 8 on Health and Disease. In contrast, some measures may be adopted irrespective of their deleterious effects on other aspects of welfare. Most obviously, concern for hygiene has led to some animals being kept in barren, highly sanitized conditions. This is one reason that otherwise suitable roughage materials are not used for growing pigs, sows, hens or broiler breeders, and that some zoo animals are kept on washed concrete floors. The concern for hygiene is often exaggerated and may have unwanted side effects. Cleaning the cages of laboratory mice is known to cause increased aggression, and van Loo et al. (2000, p. 291) examined whether some olfactory cues could affect aggression in male mice:

Animals whose cages were cleaned with transfer of nesting material showed lower levels of agonistic behaviours and higher latencies to first agonistic encounters than those whose cages were cleaned either completely or with transfer of saw dust.

Burn and Mason (2008) also found reduced pup cannibalism in rats when cages were cleaned less often. Results such as these led Olsson et al. (2003, p. 258) to conclude that:

More basic research into the behavioural priorities of rodents is crucial for developing improved housing systems, and ethological studies are important to evaluate new systems and ensure that these result in the intended improvements of welfare and biological functioning of the animals.

Where animals are kept in barren conditions, the need has arisen for making changes to ameliorate the associated problems. For example, when mink are kept in simple cages, abnormal behaviour such as stereotypies and tail chewing may develop. Hansen et al. (2007) provided mink with occupational material consisting of pulling ropes, balls and tubes that the mink could enter, as well as access to either one or two cages, thereby increasing available space (p. 73):

Females in enriched cages performed less stereotypies and had lower stress hormone levels than females in standard cages (Figure [13.1]) … The positive results are achieved independently of the number of cages and confirm previous findings that concluded that environmental enrichments are more important for the welfare of mink than the size of the cage.

In the case of other welfare problems associated with behaviour, methods of prevention are more debatable, as mentioned in the introduction. For example, Baxter (1983) suggested that it might be possible to provide farrowing sows, usually kept in barren pens, with an environment which makes it unnecessary for them to perform nesting behaviour. However, Arey et al. (1991, p. 61) found that this suggestion was not borne out:

When pre-formed nests were presented to six sows on the day before farrowing, straw removal was reduced, straw carrying and pawing remained the same. Rooting, lying and the duration of nest building increased. Nest building is highly motivated behaviour in sows and the performance of the activities themselves appears to have a significant role in reducing the motivation. Nesting material appears to be a key factor both in the regulation and performance of the behaviour.

Nest building has also been found to be important for the birth process in mink (Malmkvist and Palme, 2008, p. 279):

Access to straw for nest building is beneficial for the progress of parturition, whereas the feedback from an artificial nest alone had no such effect.

Space restrictions, with associated effects on behaviour, are sometimes applied not because they are necessary in themselves but to achieve some other result. Metabolism crates are small barren enclosures used for collection of urine and faeces, and in experiments requiring surgical procedures, animals are housed so as to restrict movement and social interaction, which may cause damage, for example to fitted catheters (Fig. 13.2). However, although group housing may not be possible, a more thoughtful approach may allow the result to be achieved without complete isolation. Herskin and Jensen (2000) compared the behaviour of pigs housed at weaning in groups or fully isolated in individual metabolism crates to that of pigs housed with restricted physical contact with other pigs. They found (p. 246 and Fig. 13.3) that:

Long-term effects of isolation may be ameliorated by provision of limited contact with littermates through a wire mesh … [After 13 days] partly isolated piglets only differed from group-housed by decreased

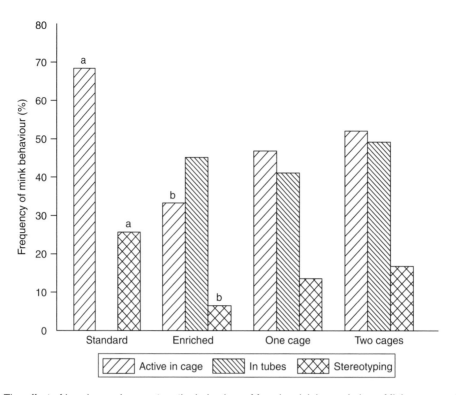

Fig. 13.1. The effect of housing environment on the behaviour of female mink housed alone. Mink were provided with either standard cages or cages enriched with occupational materials (such as resting tubes), and with access to either one or two cages, and the frequency of three types of behaviours was measured under each of these conditions. Enrichment affected behaviour independently of the number of cages, as indicated by the different letters (a, b). (Data from Hansen *et al.*, 2007.)

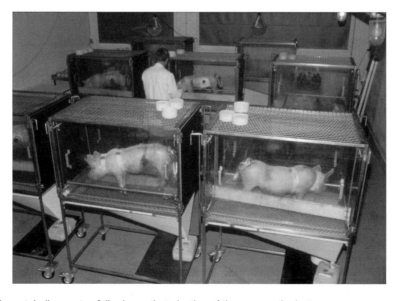

Fig. 13.2. Pigs in metabolism crates following catheterization of the pancreatic duct.

B.L. Nielsen *et al.*

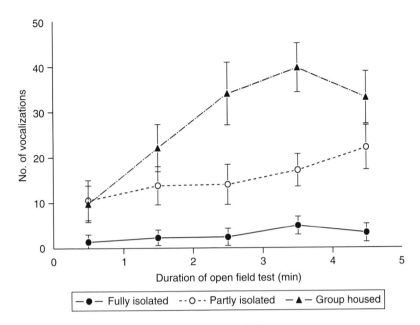

Fig. 13.3. Vocalizations (as means ± SEM) by pigs during an open field test were compared between pigs from three different housing environments. Pigs were housed at weaning either in groups, fully isolated in individual metabolism crates, or partly isolated with restricted physical contact with other pigs. (Data from Herskin and Jensen, 2000.)

frequency of play, whereas fully isolated piglets showed a more pronounced decrease in play behaviour as well as an increased frequency of pawing.

Subtle changes to the physical environment, such as pushing metabolism crates against each other, resulted in a behavioural response resembling that of social housing. Thus, the effects of a given physical environment cannot be estimated without due consideration to the social environment (which is dealt with in Chapter 14). It has already been emphasized that a specific change to the environment may have widespread effects. Similarly, provision for one aspect of animal behaviour or welfare may affect many other aspects. We shall now consider in more detail one such area that influences, among other factors, nutrition, use of time, use of space and social interactions, namely, feeding methods.

13.3 Feeding Methods

Feeding behaviour is highly controlled by humans when animals are maintained in captive environments. Despite the benefits of such control over the intake of food, there are also clear disadvantages

when animals have difficulties in adapting to environments different from those in which their behaviour evolved. So although the effectiveness of a production environment is often judged in terms of the efficiency of food intake, the design of the feeding system may lead to problems that compromise the individuals within the system in some way. For example Martin and Edwards (1994, p. 64) state that:

> In modern husbandry systems, pregnant sows typically receive all of their food for 24 h in one or two small concentrate meals. This food allowance primarily supplies requirements for maintenance, with only a small additional allowance for growth of maternal tissue and conceptus. The level of feed provided is well below the voluntary intake of the animal and most of the observed aggression in stable groups is related to feeding.

When the environment is modified to solve such problems, the aim of the change is generally to encourage animals to feed in a more natural manner. The first step to any change is therefore an investigation of the probable causes of the problems. Broiler breeders, which are feed restricted during rearing to achieve a body weight suitable for reproduction, are sometimes fed commercially using scatter feeding with relatively small pellets to encourage active foraging. The idea is that prolonged feeding behaviour improves satiety and ensures a more even distribution of feed among the birds. However, de Jong *et al.* (2005, p. 74) found that:

> Both scattered feeding and [feeding twice a day] as well as a combination of these two feeding strategies do not significantly improve broiler breeder welfare during rearing, as behavioural indicators of hunger and frustration appear not to be reduced.

Sometimes, however, modification of the environment may not only lead to reduction in undesirable food-related behaviour, but may also positively affect production-related measures. This is the case when increased feeding space leads to a decrease in aggression and to more time spent feeding, especially for the low-ranking individuals, as for example in goats (Jørgensen *et al.*, 2007) and cows (Huzzey *et al.*, 2006). Similarly, protection while drinking milk from a teat bucket reduces competition between calves (Jensen *et al.*, 2008, p. 1611):

> In contrast [to a short barrier], a long barrier (that extended beyond the calves' heads) doubled the latency of the calves to switch teats, greatly reduced the frequency of switching, and eliminated the number of calves displacing other calves during a feeding event.

Modifications can range from a total change in environment, as when changing from indoor to outdoor housing and allowing foraging, to small changes in the way that food is offered. For example, the diet itself can be modified to make it more palatable using flavourings, or be made more bulky to increase gut fill. Access to straw can reduce the risk of tail biting in pigs (Moinard *et al.*, 2003), feather pecking in hens (van Krimpen *et al.*, 2005) and oral stereotypies in cattle (Tuyttens, 2005). An understanding of the way that food preferences develop can also aid effective feeding. Some ruminants are naturally neophobic, which presents a problem when they are offered novel foodstuffs during fattening. This may be overcome by repeated exposure and/or addition of familiar flavourings (Launchbaugh *et al.*, 1997) to the foodstuffs. Conversely, some animals have a natural tendency to sample feeds separately, and this has been used in the development of methods of choice feeding. For example, selenium-deficient laying hens choose a mixed diet with high selenium concentration (Zuberbuehler *et al.*, 2002). Furthermore, Görgülü *et al.* (2008, p. 45) found that:

> Lactating goats were able to [select] a diet to meet their nutritional requirements when they were offered feed ingredients on [a] self selection basis, even ... when they were subjected to time restriction for feeding.

This approach has been extended in some studies in which animals are required to work for access to their choice of food. Various feeding devices have been designed with the aim of increasing species-specific feeding behaviour and decreasing behavioural management problems such as stereotypies. Fernandez *et al.* (2008) were partly successful in reducing oral sterotypies in giraffes by changing the design of the feeders (p. 208):

> For Betunia, the female that spent the most time performing stereotypic licking during the baseline, each increasingly complex feeder decreased licking significantly more than the previous manipulation.

However, feeding enrichment devices may not always solve feeding problems, and it is important to evaluate quantitatively the effects of such devices on a variety of behaviour patterns.

It is known that animals will work (e.g. press a lever) to obtain food, even though identical food can be obtained freely from a nearby dish (Inglis *et al.*, 1997). This response has been utilized to stimulate the natural foraging behaviour of captive wild animals, for example by feeding salmon and apples frozen into ice blocks to grizzly bears (Trudelle-Schwarz *et al.*, 2004). There is some evidence to suggest that the simultaneous presence of free food is important to avoid frustration (Lindberg and Nicol, 1994, p. 225):

> Hens are clearly willing to use operant feeders, whether because of behavioural 'need' or [for] information-gathering ... but in the absence of conventional ad libitum feeders it seems that arousal levels and frustration became too high, resulting in feather pecking and loss of condition. It is concluded that operant feeders are unsuitable for use by themselves.

There is also the problem of habituation, when an animal becomes accustomed to the device with the result that usage declines over time. This may be avoided or delayed by using feeders that deliver food in a pattern unpredictable in time and space, as in this example of feeding enrichment for captive red foxes (Kistler *et al.*, 2009, p. 262):

> The dispenser consisted of a plastic tub, with a distributor placed inside to partition the food into small portions and an analogue time device. The mechanism activating the distributor was started by the timer. On activation, the distributor released a

B.L. Nielsen *et al.*

small amount of food, which fell onto a fast rotating disk and from there was dispersed with a radius of about six meters from the dispenser.

It is often difficult to see whether there are feeding problems among group-housed animals. Individual feeding behaviour may be overlooked in farming environments where feed intake or growth is measured at group level. However, the advent of technologies such as transponders, which automatically register individual visits to the food trough, allows caretakers to monitor the feeding behaviour of each animal. The solution to many feeding problems lies not only in knowledge of species-specific feeding behaviour and requirements, but also in a sound knowledge of the range of behaviour patterns that the animal needs to perform.

13.4 Handling and Transport

Handling or transporting animals involves changes to their whole environment, or at least to many of its most important aspects, to the extent that welfare is often compromised in all the areas indicated by the Five Freedoms of the Farm Animal Welfare Council (FAWC, 1992). This occurs even though welfare problems such as fear during handling or injuries during transport are often also problems for the owners. Of course, methods of handling and transport are influenced by experience and knowledge of animals – consider, for example, the knowledge involved in using a dog to herd sheep. Grandin's book (2007) is a particularly important collection of papers on the topic of transport and handling and, more recently, long-distance transport has been covered by Appleby and co-editors (2008). Hutson (1993, p. 133) writes about the behavioural principles of sheep handling and lists what he regards as important characteristics of the animals with respect to handling:

These four characteristics of the sheep – vision, flocking behaviour, following behaviour and intelligence – form the basis of all behavioural principles of sheep handling ... [Concerning design, I] recommended that the most crucial design criterion was to give sheep a clear, unobstructed view towards the exit, or towards where they are meant to move. This often becomes more evident by taking a sheep's eye view of the facility.

Prior knowledge of the biology and behaviour of a species will not always be enough to plan a system or to predict its effects on the animals. To give just one example, which could probably not have been predicted, Matthews (1993, p. 269) reports that 'deer remain settled in padded crushes for up to 60 min provided the shoulders of the animal are well restrained'. So empirical work is necessary. As with other work on welfare, this has involved the two complementary approaches of describing the conditions and their effects on the animals, and of investigating the animals' perceptions of those conditions. One particularly clear example of the descriptive approach is work done on the conditions experienced by broilers transported by lorry (Mitchell and Kettlewell, 2004, p. 150):

Prior to and during transit, birds may be exposed to a variety of potential stressors, including the thermal demands of the transport microenvironment, acceleration, vibration, motion, impacts, fasting, withdrawal of water, social disruption and noise ... Each of these factors and their various combinations may impose stress upon the birds, but it is well recognized that thermal challenges and in particular heat-stress constitute the major threat to animal well-being and productivity.

Mitchell and Kettlewell (2004) have developed an index to express the interaction of temperature and relative humidity in causing heat stress (Fig. 13.4).

Work on animal perceptions of situations involving handling or transport has included preference or aversion testing and operant techniques. Thus, MacCaluim et al. (2003) found that, when given a choice, broilers did not avoid vibration but significantly avoided thermal stressors. Grigor et al. (1998) examined the relative aversiveness to deer of different transport stressors (p. 260):

The results ... suggest that, compared with the control deer, restraint and transport were the most aversive treatments, followed by human proximity and visual isolation.

Gibbs and Friend (1999) found no strong preference of horses for a specific orientation, i.e. facing backwards or forwards, during transport. Stephens and Perry (1990, p. 50) used an operant approach to test the response of pigs to vibration and noise in a transport simulator:

All the pigs learned to press the switch panel which turned the transport simulator off. The animals usually began to make responses during the first training session of 30 min and by the fourth session they kept the apparatus switched off for about 75% of the time ... These experiments clearly demonstrate that young growing pigs find the vibration to be aversive

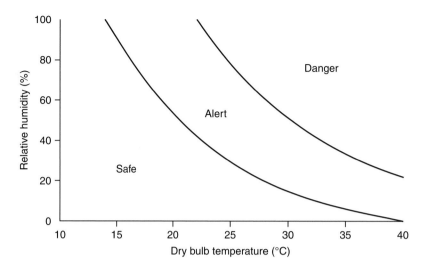

Fig. 13.4. Thermal comfort zones for broilers during transport, indicating the combined effect of relative humidity and temperature in defining safe, alert and danger zones. (Adapted from Mitchell and Kettlewell, 2004.)

and that the pigs responded behaviourally to terminate the vibration of their pen.

This sort of work is being used to identify new and improved methods of handling and transport which have a less negative impact on welfare than those currently in use. To summarize (Gonyou, 1993, p. 17):

> The ease of animal handling is determined by a combination of the animal's previous experience, the facilities and personnel involved, and the normal behavioural characteristics of the species. All these factors should be considered in the management of animals which will be handled and transported.

It is important to bear in mind that implementation of welfare-oriented changes to transport and handling procedures is not just limited by unanswered questions or lack of suitable methods. It also depends on the practicality, expense and relative priority given to welfare. Specific guidelines on handling during transport have in recent years been developed by a number of official bodies, such as the Council of Europe, the European Union (EU) and the OIE (World Organisation for Animal Health), and have been adopted in many countries. This topic is discussed in Chapter 19 (International Issues). Indeed, this whole area of management requires more than just better methods, as it raises such issues as to what extent pre-slaughter transport is necessary. Science suggests that, not only for animal welfare but also in terms of restriction of disease spread, sustainability and food safety, there should

be reduced transport of animals, as well as of animal feed and food from animals within and between countries (Appleby, 2003; Appleby *et al.*, 2008).

13.5 General Approaches

We have emphasized that specific changes to the environment are rarely specific in their effects. For example, although Marashi *et al.* (2003) found an increase in male aggression when mouse cages were enriched, they also found evidence of improved welfare in terms of positive social behaviour and more play behaviour.

The theme of this book is that animal welfare is increasingly an explicit aspect of any housing compromise. Environmental enrichment has probably been the most widely used concept in achieving this, at least since Hebb (1947) observed that rats he had taken home seemed more intelligent than those he kept in the laboratory; and since enrichment was used in 1958 to improve the learning ability of 'bright and dull rats' (Cooper and Zubek, 1958). Environmental enrichment has been attempted in many contexts, including farms, laboratories and zoos, as we have previously shown. Sometimes, the term environmental enrichment is used by researchers interested in the plasticity of behaviour with no welfare improvement in mind. Würbel and Garner (2007, p. 3) argue that:

> Debate about whether *enrichment in general* improves animal welfare is obsolete, as it springs from the inconsistent use of the term.

B.L. Nielsen *et al.*

However, while the concept of environmental enrichment is a general one, too often the actual changes made have been specific rather than part of a general approach, and success has been variable. In extreme cases, the enrichment may cause injury, as happened to a dog that swallowed vinyl covering placed as part of the enrichment (Veeder and Taylor, 2009). In other words, the concept has often been applied uncritically, as Newberry (1995, p. 230) points out:

> The term 'enrichment' implies an improvement. However, the term is frequently applied to types of environmental change ... rather than the outcome ... I define environmental enrichment as an improvement in the biological functioning of captive animals resulting from modifications to their environment.

Duncan and Olsson (2001, p. 73) are even more sceptical of the use of the term:

> We contend that the term 'enrichment' is used misleadingly most of the time. First, it is used to describe improvements to the environment that supply the animals with basic needs. However, 'enrichment' actually means the process of making richer and not the process of alleviating poverty. Therefore, describing these manipulations as 'enrichment' is fraudulent, creating the impression that the environments already cater for basic needs and that the described manipulation enhances the animals' lives even further. Second, it is often used to describe any increase in environmental complexity regardless of the effect on the animal's quality of life.

Newberry (1995, p. 235) also stresses the importance of a biological approach:

> Enrichment attempts will fail if the environmental modifications have little functional significance to the animals, are not sufficiently focused to meet a specific goal, or are based on an incorrect hypothesis regarding the causation and mechanisms underlying a problem.

So what is involved in a biological approach? Description of an environment as promoting natural behaviour does not necessarily imply that welfare has been increased. Van de Weerd and Day (2009, p. 3) review environmental enrichment in pigs and state in their introduction that:

> It is often not possible to use the behaviour of the animal in the wild as a benchmark because a lot of this natural behaviour is not documented and may be highly variable and dependent on local environmental conditions ... Therefore, a more workable approach is to specify target behaviour that is functional and adaptive in specific environments rather than

'natural'. Enhancing this target behaviour then becomes the objective of the enrichment programme.

Widowski and Duncan (2000) found that although hens are willing to work for access to a substrate in which to perform dust bathing, they are not necessarily willing to work harder when deprived of dust-bathing substrate. The authors concluded that the motivation to perform dust bathing in hens is opportunistic, so that when the opportunity to perform this behaviour arises, it leads to a state of pleasure rather than a reduction in suffering. Access to dust-bathing substrate would thus conform to Duncan and Olsson's (2001) request, that environmental enrichment should supply more than just basic requirements.

In considering further what may be involved in a biological approach to environmental design, and the success of such an approach, we shall consider examples in which an integrated biological approach has resulted in systematic investigation of enrichments for pigs and for laboratory mice, and the development of two housing systems for lactating sows with piglets, and for laying hens.

Van de Weerd et al. (2003) examined no less than 74 different objects to assess their potential enrichment value for pigs of different ages over a period of 5 days. The aim was to identify the common characteristics of the 28 objects used most intensively by the animals (p. 113):

> The emergence of the main characteristics (especially 'ingestible', 'odorous', 'chewable', 'deformable' and 'destructible') may be explained by motivations such as exploration and foraging, which are important behavioural systems for pigs.

They continue (p. 116):

> The analysis yielded two different sets of characteristics for days 1 and 5, indicating a separation between the characteristics which determined the initial attractiveness of an object (when a pig was first exposed to it) and the characteristics which kept the animal's attention over a number of days.

Among the initial characteristics of importance were odorous, deformable and chewable, whereas after 5 days, the important characteristics included ingestible and destructible. Two of the most intensively used enrichment objects for the duration of the test were lavender straw with whole peanuts and coconut halves suspended on a string. A similar approach has been used to identify appropriate enrichment for laboratory mice (Nicol et al., 2008).

The housing of sows has developed, such that increasing emphasis in later years has been placed on freedom of movement. This has led to changes in national and EU legislation in which, from 2013, gestating (pregnant) sows can no longer be confined as is currently common practice. In most countries, however, sows are still housed in crates when farrowing and suckling their young. Farrowing crates have been deemed necessary to prevent the sows from lying on their piglets and crushing them, although there is no consistent evidence that crating sows does prevent them from crushing the piglets. This problem does not exist under natural conditions, where wild and feral sows build elaborate nests of branches, twigs and softer materials in which the piglets are born and remain for the first period of life. There are now efforts to develop suitable farrowing pens, which allow more freedom for the sow without compromising the welfare of the piglets. In a large Danish project, a number of pen features were tested to develop a welfare-improved farrowing environment that could be incorporated in pens of the same dimensions as those on existing farms. Among the tested features were access to straw for nest-building material, a covered area to give the sow an impression of isolation, floor heating and sloping walls (Fig. 13.5). The latter allowed the sows to lie down while sliding against a wall,

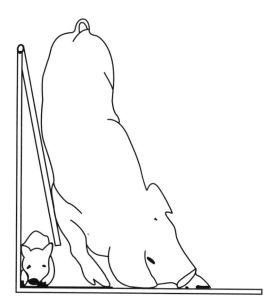

Fig. 13.5. Drawing showing the principle of a sloping wall to assist lying behaviour in sows while reducing the risk of crushing of the piglets. (From Moustsen, 2006.)

which they prefer (Damm *et al.*, 2006), and which also reduced the risk of piglet crushing. Floor heating was found to have positive effects on the welfare of piglets (Malmkvist *et al.*, 2006, p. 100):

> Our results indicate that floor heating (around 33.5°C) during the first 48 h after parturition is favourable for the neonate pig, since it resulted in earlier recovery of piglet body temperature, reduced latency to first suckle and lowered mortality.

The welfare of the sows was also investigated, as heat stress during parturition could be detrimental to the sows (Damgaard *et al.*, 2009, p. 142):

> The findings of this experiment suggest that floor heating from 12 h after onset of nest building and until 48 h after birth did not compromise physiological and immunological parameters, water intake and body temperature in loose-housed sows.

Furthermore, Pedersen *et al.* (2007, p. 8) stated that:

> Sows neither avoided nor preferred farrowing on a heated floor. However, sows that farrowed in the activity area of the pen did gradually change their lying site to the resting area after farrowing and did so particularly when the resting area was heated.

These studies present an example where the welfare of two types of animals – the sow and her piglets – with quite different environmental demands needs to be catered for in the same relatively small area. It is important to test various features simultaneously to allow for interactive effects between them, because without such specifications, producers may 'cut corners' and save costs where they could and the performance of the system would become less reliable. Also, it is worth emphasizing that this project was conducted in collaboration between scientists working on animal behaviour and welfare, and the pig industry, which greatly increases the likelihood of subsequent commercial application.

Whatever the approach to designing a system, if it is going to be developed commercially, every feature of it has to be justified by exhaustive 'testing to destruction'. Appleby and colleagues attempted to do this in the development of the Edinburgh Modified Cage for laying hens by adopting what they called (Appleby, 1993, p. 67) 'a stage-by-stage, systematic approach to cage design'. Thus (pp. 72–74):

> The Edinburgh project started with a pilot trial which compared prototype cage designs with conventional controls. These designs all included perches and nest boxes; some also had dust baths … A series of trials then examined facilities separately, to consider their

use by the birds and their effects … Current trials demonstrate that these possibilities can be combined in practical designs for groups of four or five birds, with few production problems and with benefits for welfare compared to conventional cages.

These trials formed the basis for furnished or enriched cages, which in 2012 will replace the conventional battery cage for caged laying hens within the EU. This is an example where systematic testing of biologically relevant feasible structures led to international adoption of a housing design that provides the hen with more behavioural opportunities.

13.6 Compromise

While welfare is increasingly considered in environmental design for animals, it is never the sole criterion and rarely the primary one. This is often reflected in legislation on the housing of livestock, where concerns for the welfare of the animals are weighed against feasibility and cost, as the keeping of the animals is, in the first place, intended for economic gain. However, improved health and welfare through appropriate housing is more often than not associated with increased output, either in quality or quantity. For example, Leone and Estévez (2008, p. 18) showed this to be the case for broiler breeders given environmental complexity:

In this experiment we were able to improve reproductive performance, specifically fertility, hatchability, and egg production, by providing environmental enrichment in the form of cover panels.

In other contexts, such as zoos, economics may not be the primary aim but there are still other criteria in addition to welfare. Providing captive chimpanzees with an artificial termite mound from which they could extract food items with twig-type tools led Nash (1982, p. 211) to conclude that:

The artificial mound provides the chimpanzees with a stimulating and rewarding activity, interest and enjoyment for the public, and an opportunity for researchers to study tool use under more controlled conditions than are possible in the field.

Where compromise is necessary, it should not be regarded as wholly negative, because economics and other similar factors cannot be divorced from environmental design, and can be used to establish priorities between varied demands for both humans and animals. This subject is discussed further in Chapter 17 (Economics).

13.7 Conclusions

- The physical environment has significant impacts on the welfare of an animal.
- Design of the physical environment should be based on knowledge of species- and age-specific requirements, both in terms of physiological and behavioural needs.
- There is no general principle that 'natural is best' but certain aspects of natural environments may be important, such as controllability, predictability (or conversely stimulation such as variability) and complexity.
- Different solutions may exist to a given problem, just as certain changes to the physical environment may solve more than one problem. Conversely, some changes to the physical environment may not solve the underlying problem, or may create other welfare problems.

References

Appleby, M.C. (1993) Should cages for laying hens be banned or modified? *Animal Welfare* 2, 67–80.

Appleby, M.C. (2003) Farm disease crises in the UK: lessons to be learned. In: Salem, D.J. and Rowan, A.N. (eds) *The State of the Animals II: 2003*. Humane Society Press, Washington, DC, pp. 149–158.

Appleby, M.C., Cussen, V., Garcés, L., Lambert, L.A. and Turner, J. (eds) (2008) *Long Distance Transport and Welfare of Farm Animals*. CAB International, Wallingford, UK.

Arey, D.S., Petchey, A.M. and Fowler, V.R. (1991) The preparturient behaviour of sows in enriched pens and the effect of pre-formed nests. *Applied Animal Behaviour Science* 31, 61–68.

Baxter, M.R. (1983) Ethology in environmental design. *Applied Animal Ethology* 9, 207–220.

Burn, C.C. and Mason, G.J. (2008) Effects of cage-cleaning frequency on laboratory rat reproduction, cannibalism, and welfare. *Applied Animal Behaviour Science* 114, 235–247.

Cooper, J.J., McDonald, L. and Mills, D.S. (2000) The effect of increasing visual horizons on stereotypic weaving: implications for the social housing of stabled horses. *Applied Animal Behaviour Science* 69, 67–83.

Cooper, R.M. and Zubek, J.P. (1958) Effects of enriched and restricted early environments on the learning-ability of bright and dull rats. *Canadian Journal of Psychology* 12, 159–164.

Damgaard, B.M., Malmkvist, J., Pedersen, L.J., Jensen, K.H., Thodberg, K., Jørgensen, E. and Juul-Madsen, H.R. (2009) The effects of floor heating on body temperature, water consumption, stress response and immune competence around parturition in loose-housed sows. *Research in Veterinary Science* 86, 136–145.

Damm, B.I., Moustsen, V., Jørgensen, E., Pedersen, L.J., Heiskanen, T. and Forkman, B. (2006) Sow preferences for walls to lean against when lying down. *Applied Animal Behaviour Science* 99, 53–63.

Dawkins, M.S., Donnelly, C.A. and Jones, T.A. (2004) Chicken welfare is influenced more by housing conditions than by stocking density. *Nature* 427, 342–344.

de Jong, I.C., Fillerup, M. and Blokhuis, H.J. (2005) Effect of scattered feeding and feeding twice a day during rearing on indicators of hunger and frustration in broiler breeders. *Applied Animal Behaviour Science* 92, 61–76.

Dozier, W.A. III, Thaxton, J.P., Purswell, J.L., Olanrewaju, H.A., Branton, S.L. and Roush, W.B. (2006) Stocking density effects on male broilers grown to 1.8 kilograms of body weight. *Poultry Science* 85, 344–351.

Duncan, I.J.H. and Olsson, I.A.S. (2001) Environmental enrichment: from flawed concept to pseudo-science. In: Garner, J.P., Mench, J.A. and Heekin, S.P. (eds) *Proceedings of the 35th International Congress of the ISAE [International Society for Applied Ethology]*, Davis. The Center for Animal Welfare at UC [University of California] Davis, California, p. 73 (abstract).

Estévez, I. (2007) Density allowances for broilers: where to set the limits? *Poultry Science* 86, 1265–1272.

FAWC (Farm Animal Welfare Council) (1992) FAWC updates the five freedoms. *Veterinary Record* 131, 357.

Fernandez, L.T., Bashaw, M.J., Sartor, R.L., Bouwens, N.R. and Maki, T.S. (2008) Tongue twisters: feeding enrichment to reduce oral stereotypy in giraffe. *Zoo Biology* 27, 200–212.

Frankena, K., Somers, J.G.C.J., Schouten, W.G.P., van Stek, J.V., Metz, J.H.M., Stassen, E.N. and Graat, E.A.M. (2009) The effect of digital lesions and floor type on locomotion score in Dutch dairy cows. *Preventive Veterinary Medicine* 88, 150–157.

Gibbs, A.E. and Friend, T.H. (1999) Horse preference for orientation during transport and the effect of orientation on balancing ability. *Applied Animal Behaviour Science* 63, 1–9.

Gonyou, H.W. (1993) Behavioural principles of animal handling and transport. In: Grandin, T. (ed.) *Livestock Handling and Transport*. CAB International, Wallingford, UK, pp. 11–20.

Görgülü, M., Mustafa, B., Şahin, A., Serbester, U., Kutlu, H.R. and Şahinler, S. (2008) Diet selection and eating behaviour of lactating goats subjected to time restricted feeding in choice and single feeding system. *Small Ruminant Research* 78, 41–47.

Grandin, T. (ed.) (2007) *Livestock Handling and Transport*, 3rd edn. CAB International, Wallingford, UK.

Grigor, P.N., Goddard, P.J. and Littlewood, C.A. (1998) The relative aversiveness to farmed red deer of transport, physical restraint, human proximity and social isolation. *Applied Animal Behaviour Science* 56, 255–262.

Hansen, S.W., Malmkvist, J., Palme, R. and Damgaard, B.M. (2007) Do double cages and access to occupational materials improve the welfare of farmed mink? *Animal Welfare* 16, 63–76.

Hebb, D.O. (1947) The effects of early experience on problem-solving at maturity. *American Psychologist* 2, 306–307.

Herskin, M.S. and Jensen, K.H. (2000) Effects of different degrees of social isolation on the behaviour of weaned piglets kept for experimental purposes. *Animal Welfare* 9, 237–249.

Hutson, G.D. (1993) Behavioural principles of sheep handling. In: Grandin, T. (ed.) *Livestock Handling and Transport*. CAB International, Wallingford, UK, pp. 127–146.

Huzzey, J.M., DeVries, T.J., Valois, P. and von Keyserlingk, M.A.G. (2006) Stocking density and feed barrier design affect the feeding and social behavior of dairy cattle. *Journal of Dairy Science* 89, 126–133.

Inglis, I.R., Forkman, B. and Lazarus, J. (1997) Free food or earned food? A review and fuzzy model of contra-freeloading. *Animal Behaviour* 53, 1171–1191.

Jensen, M.B., de Passillé, A.M., von Keyserlingk, M.A.G. and Rushen, J. (2008) A barrier can reduce competition over teats in pair-housed milk-fed calves. *Journal of Dairy Science* 91, 1607–1613.

Jensen, P. and Toates, F.M. (1993) Who needs 'behavioural needs'? Motivational aspects of the needs of animals. *Applied Animal Behaviour Science* 37, 161–181.

Jørgensen, G.H.M., Andersen, I.L. and Bøe, K.E. (2007) Feed intake and social interactions in dairy goats – the effects of feeding space and type of roughage. *Applied Animal Behaviour Science* 107, 239–251.

Kistler, C., Hegglin, D., Würbel, H. and König, B. (2009) Feeding enrichment in an opportunistic carnivore: the red fox. *Applied Animal Behaviour Science* 116, 260–265.

Launchbaugh, K.L., Provenza, F.D. and Werkmeister, M.J. (1997) Overcoming food neophobia in domestic ruminants through addition of a familiar flavor and repeated exposure to novel foods. *Applied Animal Behaviour Science* 54, 327–334.

Leone, E.H. and Estévez, I. (2008) Economic and welfare benefits of environmental enrichment for broiler breeders. *Poultry Science* 87, 14–21.

Lewis, E., Boyle, L.A., O'Doherty, J.V., Brophy, P. and Lynch, P.B. (2005) The effect of floor type in farrowing crates on piglet welfare. *Irish Journal of Agricultural and Food Research* 44, 69–81.

Lindberg, A.C. and Nicol, C.J. (1994) An evaluation of the effect of operant feeders on welfare of hens maintained on litter. *Applied Animal Behaviour Science* 41, 211–227.

MacCaluim, J.M., Abeyesinghe, S.M., White, R.P. and Wathes, C.M. (2003) A continuous-choice assessment of the domestic fowl's aversion to concurrent transport stressors. *Animal Welfare* 12, 95–107.

B.L. Nielsen *et al.*

Malmkvist, J. and Palme, R. (2008) Periparturient nest building: implications for parturition, kit survival, maternal stress and behaviour in farmed mink (*Mustela vison*). *Applied Animal Behaviour Science* 114, 270–283.

Malmkvist, J., Pedersen, L.J., Damgaard, B.M., Thodberg, K., Jørgensen, E. and Labouriau, R. (2006) Does floor heating around parturition affect the vitality of piglets born to loose housed sows? *Applied Animal Behaviour Science* 99, 88–105.

Marashi, V., Barnekow, A., Ossendorf, E. and Sachser, N. (2003) Effects of different forms of environmental enrichment on behavioral, endocrinological, and immunological parameters in male mice. *Hormones and Behavior* 43, 281–292.

Martin, J.E. and Edwards, S.A. (1994) Feeding behaviour of outdoor sows: the effects of diet quantity and type. *Applied Animal Behaviour Science* 41, 63–74.

Matthews, L.R. (1993) Deer handling and transport. In: Grandin, T. (ed.) *Livestock Handling and Transport.* CAB International, Wallingford, UK, pp. 253–272.

Mitchell, M.A. and Kettlewell, P.J. (2004) Transport and handling. In: Weeks, C.A. and Butterworth, A. (eds) *Measuring and Auditing Broiler Welfare.* CAB International, Wallingford, UK, pp. 145–160.

Moinard, C., Mendl, M., Nicol, C.J. and Green, L.E. (2003) A case control study of on-farm risk factors for tail biting in pigs. *Applied Animal Behaviour Science* 81, 333–355.

Moustsen, V.A. (2006) *Skrå Liggevægge i Stier til Løsgående Diegivende Søer* [Sloped Walls in Pens for Loose Housed, Lactating Sows]. Meddelelse No. 755, Dansk Svineproduktion, Den Rullende Afprøvning, Copenhagen, Denmark, 13 pp.

Nash, V.J. (1982) Tool use by captive chimpanzees at an artificial termite mound. *Zoo Biology* 1, 211–221.

Newberry, R.C. (1995) Environmental enrichment: increasing the biological relevance of captive environments. *Applied Animal Behaviour Science* 44, 229–243.

Nicol, C.J., Brocklebank, S., Mendl, M. and Sherwin, C.M. (2008) A targeted approach to developing environmental enrichment for two strains of laboratory mice. *Applied Animal Behaviour Science* 110, 341–353.

Olsson, I.A.S., Nevison, C.M., Patterson-Kane, E.G., Sherwin, C.M., van de Weerd, H.A. and Würbel, H. (2003) Understanding behaviour: the relevance of ethological approaches in laboratory animal science. *Applied Animal Behaviour Science* 81, 245–264.

Pedersen, L.J., Malmkvist, J. and Jørgensen, E. (2007) The use of a heated floor area by sows and piglets in farrowing pens. *Applied Animal Behaviour Science* 103, 1–11.

Stephens, D.B. and Perry, G.C. (1990) The effects of restraint, handling, simulated and real transport in the pig (with reference to man and other species). *Applied Animal Behaviour Science* 28, 41–55.

Telezhenko, E., Lidfors, L. and Bergsten, C. (2007) Dairy cow preferences for soft or hard flooring when standing or walking. *Journal of Dairy Science* 90, 3716–3724.

Trudelle-Schwarz, R.M., Newberry, R.C., Robbins, C.T. and Alldredge, J.R. (2004) Contrafreeloading in grizzly bears. In: Hänninen, L. and Valros, A. (eds) *Proceedings of the 38th International Congress of the ISAE [International Society for Applied Ethology]*, Helsinki, Finland. ISAE, p. 53 (abstract).

Tuyttens, F.A.M. (2005) The importance of straw for pig and cattle welfare: a review. *Applied Animal Behaviour Science* 92, 261–282.

van de Weerd, H.A. and Day, J.E.L. (2009) A review of environmental enrichment for pigs housed in intensive housing systems. *Applied Animal Behaviour Science* 116, 1–20.

van de Weerd, H.A., Docking, C.M., Day, J.E.L., Avery, P.J. and Edwards, S.A. (2003) A systematic approach towards developing environmental enrichment for pigs. *Applied Animal Behaviour Science* 84, 101–118.

van Krimpen, M.M., Kwakkel, R.P., Reuvekamp, B.F.J., van der Peet-Schwering, C.M.C., den Hartog, L.A. and Verstegen, M.W.A. (2005) Impact of feeding management on feather pecking in laying hens. *World's Poultry Science Journal* 61, 663–685.

van Loo, P.L.P., Kruitwagen, C.L.J.J., Van Zutphen, L.F.M., Koolhaas, J.M. and Baumans, V. (2000) Modulation of aggression in male mice: influence of cage cleaning regime and scent marks. *Animal Welfare* 9, 281–295.

Veeder, C.L. and Taylor, D.K. (2009) Injury related to environmental enrichment in a dog (*Canis familiaris*): gastric foreign body. *Journal of the American Association for Laboratory Animal Science* 48, 76–78.

Veissier, I., Andanson, S., Dubroeucq, H. and Pomies, D. (2008a) The motivation of cows to walk as thwarted by tethering. *Journal of Animal Science* 86, 2723–2729.

Veissier, I., Butterworth, A., Bock, B. and Roe, E. (2008b) European approaches to ensure good animal welfare. *Applied Animal Behaviour Science* 113, 279–297.

Widowski, T.M. and Duncan, I.J.H. (2000) Working for a dustbath: are hens increasing pleasure rather than reducing suffering? *Applied Animal Behaviour Science* 68, 39–53.

Würbel, H. and Garner, J.P. (2007) *Refinement of Rodent Research Through Environmental Enrichment and Systematic Randomization.* National Centre for Replacement, Refinement and Reduction of Animals in Research (NC3Rs), London. Available at: http://www.nc3rs.org.uk/downloaddoc.asp?id=506&page=395&skin=0 (accessed 30 December 2010).

Zuberbuehler, C.A., Messikommer, R.E. and Wenk, C. (2002) Choice feeding of selenium-deficient laying hens affects diet selection, selenium intake and body weight. *Journal of Nutrition* 132, 3411–3417.

14 Social Conditions

FRANCISCO GALINDO, RUTH C. NEWBERRY AND MIKE MENDL

Abstract

In this chapter we describe how the constrained structure of managed social groups may lead to the occurrence of animal welfare problems and discuss potential solutions. Solving socially induced welfare problems is particularly challenging because it is usually more difficult to predict how animals will respond to each other than it is to predict how they will respond to changes in the physical environment. One approach is to design social environments which allow the expression of the types of behaviour and social organization that seem to be important for stable social structures and harmony in natural groups. For example, housing animals in relatively stable groups, and allowing them the freedom to separate themselves from the main group, to select which individuals they associate with and to avoid others, may reduce susceptibility to disease and other welfare problems and promote harmonious social life. This preventive approach requires basic research into the principles affecting social organization under unconstrained free-ranging situations, and takes time to develop into practical and viable alternatives to existing systems. An alternative approach is to tackle specific welfare problems by altering existing social environments, either by providing opportunities for avoidance or by distributing resources in such a way that aggressive competition and social stress are minimized. We suggest that this trouble-shooting approach is most likely to be successful when the social history and social skills of group members are taken into account, and when the causes of problems, rather than their symptoms, are addressed. Other practical solutions to social problems in managed groups, such as genetic selection and social learning, are also discussed.

14.1 Introduction

From an animal welfare perspective, the social environment of managed animals presents us with a dilemma. On the one hand, social companionship is potentially the most effective way of enriching the lives of social species in captivity, a point clearly made by Humphrey (1976, p. 308):

> [The monkeys] live in social groups of about eight or nine animals in relatively large cages. But these cages are almost empty of objects, there is nothing to manipulate, nothing to explore; once a day the concrete floor is hosed down, food pellets are thrown in and that is about it. So I looked, and seeing this barren environment, thought of the stultifying effect it must have on the monkey's intellect. And then one day I looked again and saw a half-weaned infant pestering its mother, two adolescents engaged in a mock battle, an old male grooming a female whilst another female tried to sidle up to him, and I suddenly saw the scene with new eyes: forget about the absence of objects, these monkeys had each other to manipulate and explore.

The idea that relationships with other individuals can be a major source of comfort and entertainment to captive animals is supported by evidence from a variety of species. For example, captive primates show lower levels of abnormal behaviour in the presence of companions than when isolated (Bernstein, 1991), and horses (Feh and de Mazières, 1993), talapoin monkeys (Keverne *et al.*, 1989) and cattle (Galindo and Broom, 2002; Fig. 14.1) calm companions by grooming them.

Social motivation studies show that social attachment can be considered a biological need in social species (Estévez *et al.*, 2007), and epidemiological evidence reveals that the presence or absence of positive social experiences can have major effects on health and welfare (Carter and Keverne, 2002). For example, studies in rhesus macaques show that social enrichment not only results in increased time spent in species-typical activities and less time in abnormal behaviour, but also in a significantly enhanced cell-mediated immune response (Schapiro, 2002). Evidence is accumulating that a positive affective climate with abundant playful social activity is accompanied by the production of antidepressant neural growth factors (Watt and Panksepp, 2009). In contrast, social loss can lead to severe stress symptoms (Ruis *et al.*, 2001), and has been associated with spikes in pro-inflammatory

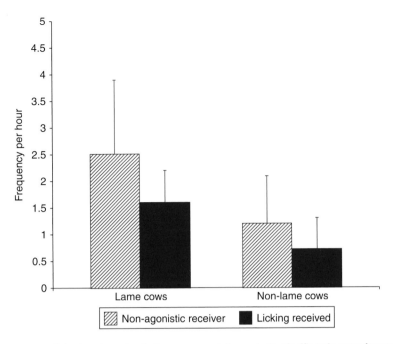

Fig. 14.1. Lame cows were licked and received other non-agonistic contacts significantly more frequently than non-lame cows. (From Galindo and Broom, 2002.)

cytokines implicated in depressive disease (Hennessy *et al.*, 2007; Miller *et al.*, 2009). These findings attest to the importance of social companionship.

On the other hand, damage that animals can inflict on each other by intimidation, overt aggression and redirected behaviours gives rise to serious animal welfare concerns. Although it is known that some of these problems occur when the social structure of animals is not well managed, for example as a result of overcrowding or constant regrouping, unfortunately, it is often difficult to predict the situations in which conspecifics will injure or otherwise damage each other (Visalberghi and Anderson, 1993). This lack of predictability creates a challenge to all those concerned with solving socially induced animal welfare problems.

Can we gather and use knowledge about social behaviour such that our manipulations of the social environment have more predictable and beneficial outcomes? Before addressing solutions, we first need to consider what the problems are and how they arise. We do this in the next two sections of the chapter by discussing the evolution of group living, and contrasting the structure and dynamics of groups living in their natural environment with those of captive groups.

14.2 Evolution of Group Living: the Dynamic Structure of Natural Social Groups

The evolution of social groups can best be considered in terms of the costs and benefits to individual group members. In natural environments, living in groups increases the chances for survival by diluting predation risk and enhancing abilities to detect predators. This is relevant when managing animals in captivity as they retain anti-predator behaviours. Other benefits of group living in terms of fitness and welfare include enhanced capacity to defend resources, to locate and catch food, to thermoregulate and resist disease through allogrooming, and improved immune function (Schapiro, 2002). Furthermore, living in a social group increases learning opportunities (Estévez *et al.*, 2007, p. 186):

> Through social learning and facilitation animals discover the location of food sources more easily, stimulating feeding and other functionally important behaviours such as dustbathing in hens. By living in groups, animals learn to avoid potentially dangerous situations and in young animals chances to interact with play mates stimulate motor and social skills which are basic components to secure a positive behavioural development and coping abilities.

Potential costs of living in a social group include increased risk of contracting parasites and contagious diseases, increased competition with group members for food and other resources, and increased aggression when these are limited. Some of these changes in behaviour may be dependent on group size and demography. For example, Bonaventura *et al.* (2008) mention that primates living in larger groups experience higher reproductive costs than animals in smaller groups.

For each individual, evolutionary theory predicts that the decision to join or leave a group depends upon the relative pay-off (benefits minus costs), in terms of inclusive fitness, of solitary or group living (Pulliam and Caraco, 1984). Vehrencamp (1983) has modelled the conditions under which different levels of cooperation or despotism should occur in animal groups when individuals are free to join and leave the group. Despotic behaviour results in increased variation or bias in the fitness of individuals within the group (Vehrencamp, 1983, p. 667):

> Selection acts simultaneously on the stronger, dominant members of the group to secure more resources for themselves at the expense of subordinates, and on subordinates to leave the group when excessively manipulated if they can do better elsewhere. When it is to the advantage of the dominant to maintain the group, the dominant will ultimately be limited in the degree of bias it can impose by the options available to subordinates outside the group.

Thus, in natural groups, we expect high-ranking individuals to moderate the amount of competitive pressure they exert on low rankers because low rankers can leave when options outside the group become relatively more beneficial. Decisions to join or leave groups are likely to be affected by a variety of factors, including predation pressure, resource availability and distribution, and the individual's ability to compete with others in the group.

In line with these predictions, the group structure of free-living animals is fluid and varies both within and between species, ranging from loose aggregations of individuals to long-term, cohesive groups. Members immigrate and emigrate, and there are seasonal and short-term changes in group size and composition, and fusion and fission of groups. For example, male feral domestic fowl switch between territoriality during the breeding season and a hierarchical group structure during winter. The females associate in groups with the males during winter and separate to rear their broods in summer (McBride *et al.*, 1969).

Natural social groups also exhibit polymorphism in personality types and social strategies that are consistent with theories regarding life history trade-offs between current and future reproduction. The former is predicted to favour rapidly growing individuals with bold and aggressive personalities whereas the latter is expected to favour slower growing individuals with a more risk-aversive personality type (Wolf *et al.*, 2007; Biro and Stamps, 2008). Owing to environmental fluctuations, these phenotypes can coexist within populations, with the relative proportion of each varying depending upon current conditions (e.g. predation pressure).

14.3 The Constrained Structure of Managed Social Environments

The social environment of laboratory and farm animals and, to a lesser extent, of zoo animals is often characterized more by human concerns such as effective use of space and uniformity of group membership than by consideration of evolved social structures and abilities. The resulting constraints on social behaviour may have detrimental welfare consequences. Clearly, there are dangers in arguing that the welfare of captive animals is good only if they show the same behavioural organization and repertoire as they would show in free-ranging conditions. It becomes even more problematic to relate the behaviour of domestic animals to that of their wild ancestors.

However, although domestication has altered the threshold and frequency at which some behaviour patterns are expressed (Price, 1999, 2004), domestic animals appear to retain the basic social characteristics of wild conspecifics. For example, the social behaviour of domestic pigs living in a semi-natural environment strongly resembles that of the European wild boar (Stolba and Wood-Gush, 1989). Feral cattle also show characteristics of habitat use and group size very similar to those of some species of wild ungulates (Hernández *et al.*, 1999). It thus seems likely that domestic animals are predisposed to deal most effectively with the range of group sizes and structures typical of their species, and that the freedom to make decisions about when to join or leave groups and which animals to associate with remains an important feature of their social environment.

Keeping individuals of group-living species in isolation, either temporarily or permanently, clearly affects their social behaviour. For example, rearing

F. Galindo *et al.*

calves in isolation places them at a competitive disadvantage when later grouped with more socially experienced animals (Le Neindre *et al.*, 1992). Involuntary separation from attachment figures elicits calling and vigorous attempts to reinstate contact (Colonnello *et al.*, 2010), and isolation often gives rise to increased emotional reactivity and a cascade of physiological stress responses (Ruis *et al.*, 2001; Weiss *et al.*, 2004). That social loss is an unpleasant experience for the animals is supported by the observation of parallels between the brain regions activated during mother–young separation in the guinea pig and those activated during the experience of sadness in humans (Panksepp, 2003). Furthermore, a single exposure to isolation stress in young animals is associated with changes in the expression of genes that may mediate both normal and pathological responses to subsequent stressors (Kanitz *et al.*, 2009). Changes in gene expression could explain why social isolation in early-weaned piglets is related to impairment in the ability to recognize familiar conspecifics (Souza and Zanella, 2008). Rearing females in isolation can also disrupt their future maternal behaviour (Berman, 1990). There is increasing recognition of the critical role played by the lack of a secure early social attachment in the intergenerational transmission of abusive parenting in primates (Maestripieri, 2005).

Not only does the disruption of social attachments create welfare problems, but problems also arise when groups are created by introducing unfamiliar individuals in a confined space and in unnatural aggregations. When unfamiliar individuals are suddenly forced together, the level of aggression can be much higher than that observed when unfamiliar individuals encounter each other under more natural conditions (Jensen, 1994). Even if there is no physical contact or injury, individuals may show a marked stress response, and can die if placed in the enclosure of an established group (von Holst, 2004). These effects may be related to the lack of opportunity for the animals to escape. They may also be related to the group composition. For example, in many species, creating groups of adult males when they would not naturally form such groups results in serious and sustained aggression (Love and Hammond, 1991). Placing young animals together in the absence of their mothers or other older individuals may also give rise to elevated agonistic behaviour (Le Neindre *et al.*, 1992), as dramatically illustrated by the violent behaviour of young male elephants following loss of adult family members from poaching (Bradshaw *et al.*, 2005). Furthermore, social rank appears to influence how animals will respond to unfamiliar groups. For example, an increased adrenocortical reaction at the beginning of the confrontation and a later bradycardia were observed particularly when former top-ranking pigs were confronted with an unfamiliar group as compared with a familiar group (Otten *et al.*, 1997).

Once groups are established, the design of housing systems usually prevents animals from emigrating in response to high levels of aggression. Therefore, high-ranking individuals may be able to exert more pressure on others than they would in free-living groups (Vehrencamp, 1983), with potentially damaging welfare consequences for low rankers (González *et al.*, 2003; Fig. 14.2). This problem is exacerbated when resources are limited and groups are housed in unstructured enclosures with no opportunities for avoidance. Rumbaugh *et al.* (1989, p. 360) showed that, for captive chimpanzees at least, there may be a simple solution to this problem:

> In the wild, chimpanzees vary their choice of companions as they move from place to place, yet most captive enclosures are designed so that chimpanzees are always together, either in a large arena, or in cages. Aggressive encounters diminish when chimpanzees have a series of areas they can go to, ideally furnished with doors they can shut behind them if they wish.

Other welfare problems that result from changes in the social environment of managed animals are those related to the early separation of mothers and young, a common condition in most domestic and exotic species reared in captivity. The impact of separation on the young is described by Newberry and Swanson (2008, p. 129):

> When abruptly and permanently separated from their mother at a young age, the offspring are faced with the sudden need to adapt from a milk diet and maternal care to a solid diet, a new housing environment, a new social environment, and loss of maternal contact, all of which can elicit behavioural signs suggestive of negative affect. Relocating weaned young and placing them in social isolation or mixing them with strangers also causes welfare concerns.

Furthermore, changes in social structure related to overcrowding, lack of individual recognition, and changes in the age and sex composition of the group are factors that can provoke

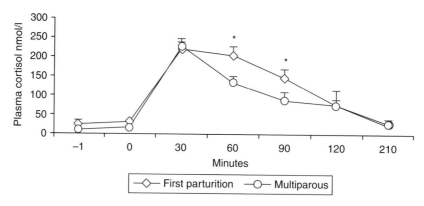

Fig. 14.2. Average (± SE) plasma cortisol levels of multiparous cows and lower ranking first-parturition cows when competing for resources at different times after the administration of ACTH (adrenocorticotrophic hormone) (* = $p < 0.05$). (From González *et al.*, 2003.)

aggression and welfare problems in a variety of species (Orihuela and Galindo, 2004; Verga *et al.*, 2007).

Evolved communication mechanisms between conspecifics are often overlooked in managed social environments, with potentially adverse welfare consequences. For example, Algers and Jensen (1991) observed that fan noise disrupted vocal communication between sows and piglets during nursing, and reduced milk intake by the piglets. In groups of male mice, Gray and Hurst (1995) found that the presence of male odours within the cage stimulated aggression when the mice were returned to the cage following removal for cage cleaning and other procedures. They recommended that the mice be returned to completely clean cages following handling to minimize aggression problems. In breeding rats, Burn and Mason (2008) noted that frequent cage cleaning was associated with elevated infanticide, presumably by interfering with maternal attachment to pups in the early postnatal period.

Although captive social environments can create problems, intra-group competition and destructive social behaviour are not always artefacts of captive housing. Socially induced stress, cannibalism, infanticide and fatal fights also occur in nature. We need to understand the environmental conditions that evoke these behaviour patterns so that we can avoid inadvertently creating situations in captivity where they are likely to occur. In the following sections we consider different approaches to tackling or pre-empting welfare problems in the captive social environment.

14.4 Designing Husbandry Systems Using Knowledge About the Social Organization of Species

The rationale of this approach has been to develop an understanding of the 'natural' social structure of a species as observed under conditions similar to those in which it evolved, and then to design a novel husbandry system which incorporates key features of this structure. The underlying assumption is that a 'natural social environment' promotes social stability and good welfare. This is a debatable point (Dawkins, 1990). Nevertheless, as argued earlier, knowledge of natural social behaviour can be used to enhance the welfare of captive animals.

As expressed by Špinka (2006, p. 126):

> Animals do not need 'natural behaviour' per se for good welfare. However, providing the opportunity for natural behaviour is often a very effective way to satisfy the needs and/or goals of the animals, to provide them with emotionally positive experience, and to stimulate their behavioural development in such a way that it brings long-term benefits. Therefore, natural behaviour of the species in question should be considered both when new housing systems are being developed and when solutions to particular problems in existing systems are being sought.

A seminal example of this approach is the work of Stolba and Wood-Gush (1984) on domestic pigs. Key social stimuli were identified by studying groups of pigs breeding and rearing young in semi-natural environments of varying size and complexity. Social features common to these environments were incorporated into a new housing system.

Characteristics of this 'family pen' system included stable core groups of similar size and composition to those observed in the semi-natural conditions, the provision of area subdivisions and enough space to allow individuals to avoid some group members and associate with others, natural weaning and a policy of avoiding the abrupt introduction of new group members. The system eliminated the need to separate and remix sows before and after farrowing, and thus avoided the intense aggression that occurs at this time (Edwards and Mauchline, 1993). Keeping sows and piglets in family groups also prevented the social trauma associated with early weaning and the abrupt mixing of unfamiliar, weaned piglets (Jensen, 1994).

A similar approach was used to develop a new group housing system for rabbits (Stauffacher, 1992). The provision of a relatively complex social environment for rabbits appeared to prevent the disturbed reproductive behaviour, infanticide and unstructured behaviour seen in singly caged individuals. Perhaps as a consequence, a breeding success rate of 89% was achieved compared with conception rates of only 30–70%, and pup mortality of 30%, in separately caged females.

These findings support the idea that systems designed in this way can reduce welfare problems in the social environment. However, uptake of these systems at the commercial level requires that they are easy to manage, productive and economically viable, or that they are imposed by legislation. In any case, when designing housing systems for captive animals there are multiple factors that affect social behaviour and welfare that must be taken into account. These include permanent characteristics of the buildings, such as space allowance, floor quality, feeding systems (Spoolder et al., 2009) and hiding areas that can be used as visual barriers (Kuhar, 2008), and also some more temporary aspects of the management system itself, such as substrates that may allow the modulation of social interactions. For example, a large bedding area (Nielsen et al., 1997), straw (Høøk Presto et al., 2009) and other enrichment materials (van de Weerd and Day, 2009) can help to reduce aggression.

The complex and dynamic interaction between the animal and its social environment represents a multifactorial system that is affected by evolutionary heritage and individual experience (Newberry and Estévez, 1997; Price and Stoinskia, 2007). In the context of animal welfare, the design of husbandry systems and social groupings in captivity should be carried out taking into account individual characteristics within a continually changing social environment because relationships between these factors can influence individual susceptibility to disease or other welfare problems.

14.5 Solving Socially Induced Welfare Problems in Existing Husbandry Systems

14.5.1 Treating symptoms

In contrast to the painstaking procedure of designing social environments from first principles, a more common approach has been to seek solutions to social problems by altering specific features of existing systems. Although a troubleshooting approach can be effective, constraints imposed by existing systems may result in a tendency to tackle symptoms rather than focusing on the underlying causes of the problem. The archetypal example of solving socially induced welfare problems by attempting to prevent the symptoms is partial amputation of body parts. For example, tail docking to control tail biting in pigs, and beak trimming to prevent feather pecking in laying hens, are relatively quick and simple procedures but they are often not 100% effective and have their own welfare consequences (e.g. pain). Although the causes of these damaging behaviours are undoubtedly complex, probably representing multiple phenomena controlled by different motivational mechanisms (e.g. Newberry et al., 2007; Taylor et al., 2010), a better understanding of them should lead to more effective and humane solutions.

Another example of treating symptoms without tackling underlying causes concerns methods used to address aggression when unfamiliar animals are mixed together to create new groups – or for other reasons, such as transportation. In domestic pigs, which show particularly vigorous aggression when encountering unfamiliar individuals, a number of researchers have attempted to reduce mixing-related aggression by administering sedating drugs, timing mixing to coincide with periods of low activity, or providing food or fresh straw to distract pigs from new group members. The rationale appears to be to suppress general activity, or engage the pigs in another behaviour to override the motivation to fight. These techniques often appear simply to postpone rather than reduce aggression.

For example, Luescher *et al.* (1990) showed that while administration of the drug azaperone (Stresnil) reduced aggression during the first 2 h following mixing of gilts relative to saline-treated control animals, further observations revealed that aggression was far higher in the azaperone-treated pigs during the following 4 h, such that no difference between the groups was evident after 6 h. Petherick and Blackshaw (1987, p. 609) noted similar findings in younger pigs:

> Azaperone reduced agonistic behaviour in mixed, weaned pigs during the period of sedation but, once the effect of the drug had worn off, the treated pigs showed the same level of agonistic behaviour as the untreated group. This might be expected as they still had to establish their dominance hierarchy.

Barnett *et al.* (1994) observed a similar delaying effect when mixing sows just before darkness at a time of low activity. Aggression was reduced immediately following grouping, but there was no effect on injury scores 3 days later, suggesting that aggression had simply been postponed until the next active period. Likewise, Arey and Franklin (1995) demonstrated that providing straw to distract newly mixed pigs from each other did not reduce levels of aggression.

14.5.2 Understanding causes of aggression when unfamiliar animals are introduced

Given the apparently limited success of tackling mixing-related aggression by treating symptoms, we now consider whether understanding its causes can suggest novel and more effective ways of minimizing this damaging behaviour. Aggression occurs when unfamiliar animals from social species are abruptly forced together, probably due to an evolved tendency to deter rapid intrusions into the (kin) group. For example, matrilineal groups of wild boar and feral pigs repel unfamiliar/unrelated animals, though they may accept the gradual integration into the group of a new member (Mendl, 1995).

Victory in fights may give individuals priority of access to resources and hence confer fitness advantages. However, because aggression can be a very costly activity, theorists propose that animals should avoid fighting unless they have a good chance of winning (Enquist and Leimar, 1983; Arnott and Elwood, 2009), especially if the probability of monopolizing resources is low (e.g. in large groups; Estévez *et al.*, 2002; Andersen *et al.*,

2004). Instead, they should assess each other's fighting abilities and settle disputes on this basis wherever possible. For example, red deer stags appear to assess each other's competitive abilities through roaring contests during disputes over harem ownership in the rutting season (Clutton-Brock and Albon, 1979).

Solving disputes by assessment, and by aggression, will occur more rapidly if there are clear asymmetries in the competitive abilities of group members, as is often the case in natural groups, which may contain animals of varying age and size. Some husbandry practices involve separating group members for a period of time (e.g. when sows leave a group to give birth), and then reuniting them. Aggression may occur when these previously familiar animals are brought together once more (Ewbank and Meese, 1971), and this is often the case for sows returning into groups following the weaning of their litters. This may be caused by animals failing to recognize or remember each other or needing to re-establish their relative status (Croney and Newberry, 2007).

The causes of mixing-related aggression listed above indicate that it could be reduced by: (i) enhancing asymmetries/differences between group members; (ii) minimizing opportunities for resource monopolization; (iii) facilitating assessment behaviour; and (iv) facilitating the recognition of previously familiar animals. We now consider these possibilities in turn, focusing primarily on the domestic pig.

Enhancing differences between group members

The greater the asymmetry in the competitive abilities of mixed animals, the more rapidly relative social status should be established. Support for this prediction was provided by Rushen (1987) who showed that if the weight difference between pigs was large enough the duration and severity of fighting following mixing was reduced, perhaps by facilitating the assessment of relative abilities. Andersen *et al.* (2000) observed similar results and also noted that fighting was particularly intense when there were small weight asymmetries between pigs and clumped resources to compete for.

Farmers may be reluctant to mix growing pigs of different weight together because this could decrease the chances of animals in a pen reaching slaughter weight at the same time (although it is often the case that pigs of similar weight at initial mixing do not reach slaughter weight simultaneously). Researchers have thus investigated the idea that asymmetries in other characteristics (e.g. behaviour, sex) may be used to minimize aggression at mixing. There has been particular interest in asymmetries in 'aggressiveness', which may be related to more general differences in 'coping style'. Hessing et al. (1994a) reported that when pigs categorized as highly (H) aggressive (proactive copers) or of low (L) aggressiveness (reactive copers) were mixed together, they rapidly developed a stable social order, helping to minimize aggression and maximize growth rates in newly formed groups. Similarly, Mendl and Erhard (1997) observed that fewer pairs of unfamiliar pigs fought following mixing in groups containing H and L pigs compared with those containing just H or just L pigs. Bolhuis et al. (2005) showed that L pigs may be particularly responsive to the abilities of other animals and moderate their behaviour accordingly, perhaps through some form of assessment, while H pigs tend to be more inflexible in their social tactics and persistent in showing aggression (see also D'Eath, 2002). Mixing pigs according to aggressiveness may thus help to reduce aggression. Measuring aggressiveness by aggression testing is unlikely to be feasible on most commercial farms, but using proxy measures such as response to being restrained in the supine position (given that highly aggressive animals struggle more in the 'back test'; Hessing et al., 1994b and Bolhuis et al., 2005; but see D'Eath and Burn, 2002) or levels of lesions (Turner et al., 2006) could be more practicable, especially on breeding units where selection indices might take into account aggressive behaviour (Turner et al., 2009).

A related and simpler approach is to introduce one or two clearly superior individuals into a group to suppress aggression in others. For example, Rushen (1987) noted that the presence of a dominant pig could inhibit aggression between subordinates. Adult boars are dominant in wild and feral groups and their presence helps to minimize aggression when domestic sows are mixed (Barnett et al., 1993 and Borberg and Hoy, 2009; but see Luescher et al., 1990). Similarly, McGlone (1990, p. 102) showed that the boar pheromone androstenone could decrease fighting among young pigs:

> I must conclude that androstenone does not operate as a relevant pheromone, but rather as a super-male odor. Prepuberal pigs are less likely to attack an animal that smells like an adult.

Similar phenomena have been observed in other species (Petit and Thierry, 1994). In rhesus monkeys, Reinhardt and colleagues (1989, p. 275) observed that:

> Young rhesus monkeys, 12 to 18 months old and of both sexes, tend to inhibit aggression in singly caged adults of both sexes.

In this case, a competitively inferior young companion could be used to help socialize isolated adults with little risk of aggression occurring.

Another method for minimizing aggression by mixing selected individuals is suggested by the finding of Colson et al. (2006) that levels of fighting were higher in mixed-sex groups than in single-sex groups of weaned pigs. The authors argued that the presence of female pigs enhanced aggressive behaviour in males, perhaps because females represent an important resource to be competed for (although the animals in this study were prepubertal). Whatever the exact reason, the study suggests that mixing animals into single-sex groups, out of contact with members of the opposite sex, may help to minimize aggression. A related approach has been explored in the management of captive primates. Gold and Maple (1994) subjectively rated selected behavioural characteristics of nearly 300 gorillas. Four factors describing aspects of individual behavioural style, 'extroverted', 'dominant', 'fearful' and 'understanding', were identified in subsequent analysis. On the basis of these a database was created which could be used to match animals that are moved between zoos (pp. 515–516):

> For example, suppose the [Gorilla Species Survival Plan] was considering the formation of an additional bachelor group and was deliberating over potential candidates ... It is known that males with a history of aggressiveness have trouble living in an all-male situation. Sociable males, on the other hand, tend to adapt easily to a bachelor group situation ... The [Gorilla Behaviour Index could be used] to look at the ranges of profile scores of males ... This analysis could lead to the selection of a male ... with a comparatively high extroverted score ... and a comparatively low dominant score.

Overall, an understanding of individual characteristics may be successfully used to reduce aggression when unfamiliar animals are mixed, including by increasing asymmetries between individuals.

Minimizing opportunities for resource monopolization

Minimizing the ability of newly mixed animals to monopolize resources may also help to decrease competitive aggression. Andersen *et al.* (2004) suggest that a simple way of doing this is to increase the size of the group into which animals are mixed because the probability of being able to monopolize resources diminishes as group size increases. They found that the number of fights per individual and the proportion of pigs engaging in aggression post-mixing was indeed significantly lower in groups of 24 newly mixed pigs compared with groups of six or 12 pigs. This would benefit pigs and farmers by decreasing the proportion of mixed individuals exposed to damaging aggressive behaviour. Further examples of this approach in the context of settled groups are provided in Section 14.6.2.

Another consideration when mixing animals in captivity is illustrated by the resident–intruder paradigm, whereby residents are likely to attack and defeat an unfamiliar individual suddenly placed into their territory (D'Eath, 2002; Nelson and Trainor, 2007). Aggression is especially likely if both resident and intruder are adult males and breeding females are present. Although this scenario creates a clear asymmetry, unless the (involuntary) intruder has very ample space to escape and modulate his exposure to the resident, he may be severely harmed despite displaying submissive postures. In this case, introducing the animals in a spacious neutral location, equally unfamiliar to all and in the absence of females and other highly valued resources, may be safer since it removes the emotionally laden, resource/territorial defence component of the interaction.

Facilitating assessment behaviour

The ability of animals to assess each other's competitive abilities may partly underlie some of the findings discussed above. Rushen (1990, p. 135) noted:

> In initial fights, young pigs seem unable to make such judgements. However, they appear to be able to judge the relative fighting ability of an opponent from events that occur during a fight ... The reduction in fighting ... occurs because the pigs accumulate information about relative fighting ability ... Thus, the initial fighting between unacquainted pigs is motivated by uncertainty about relative fighting abilities and can be considered a form of social exploration.

Jensen and Yngvesson (1998) found that pairs of pigs that were pre-exposed before mixing had shorter contests following mixing than did non-exposed pairs, suggesting that they may have been able to make some assessment of each other's relative abilities:

> It appears that the overt biting phase might be shortened by pre-exposure, and, since total contest length was significantly shortened by pre-exposure, that the fighting period following mixing will be shortened. Farmers may therefore be encouraged to find practical methods of exposing pigs to their contestants before mixing.

Although Rushen (1988) failed to find this effect in young pigs, Kennedy and Broom (1994) reported similar findings for older animals. Gilts pre-exposed to a group of sows received less aggression than non-exposed animals following mixing. Perhaps the pre-exposure period – which simulates the gradual integration of group members that occurs in natural environments (Stolba and Wood-Gush, 1989), allowed some assessment and familiarization which acted to decrease uncertainty and related fighting behaviour.

More generally, the ability of animals to observe social interactions between others and gather information on their competitive and social status relative to those that they watch – so-called 'eavesdropping' – has been demonstrated in a number of species (e.g. Oliveira *et al.*, 1998; Peake *et al.*, 2001; Paz-y-Miño *et al.*, 2004), including domestic chickens (Hogue *et al.*, 1996), and emphasizes the potential sophistication of social abilities that could be harnessed to help minimize aggression in captive environments. Conversely, it should be noted that offensive aggression may be elevated in animals able to view the aggressive behaviour of others before introduction (Oliveira *et al.*, 2001; Suzuki and Lucas, 2009). This may occur especially if such observations, coupled with prior experience of winning, lead them to anticipate winning future encounters, as indicated by their circulating androgen levels being elevated.

Social play experience when young is probably important for the learning of social assessment skills, as suggested by the more rapid resolution of social dominance among unfamiliar pigs with greater social play experience (Newberry *et al.*, 2000). More generally, there is support for the hypothesis that, through social play experience, animals develop the ability to regulate their emotional arousal, enabling them to cope adaptively

F. Galindo *et al.*

when confronted by strangers and to modify their social tactics according to the behaviour of opponents (Špinka *et al.*, 2001; Pellis *et al.*, 2010). If husbandry environments constrain opportunities to develop and use assessment abilities to sort out relative status, this may exacerbate levels of aggression when animals are mixed.

Facilitating recognition of previously familiar animals

When animals that have been separated are reunited, aggression may occur in order to re-establish social status or because individuals fail to recognize or remember their group mates. It is difficult to disentangle these two potential causes, but in both cases a better understanding of social recognition and memory processes (including memory of social status) may help to minimize these problems. Studies demonstrate that pigs can discriminate between familiar and unfamiliar individuals, that they can use olfactory, visual and probably auditory cues to do so, and that environmental pollutants (e.g. high levels of ammonia) may alter preferences for approaching familiar or unfamiliar individuals, but appear not to disrupt discrimination abilities completely (Kristensen *et al.*, 2001; McLeman *et al.*, 2005, 2008). The duration of social memory for a conspecific has been

investigated using aggressive behaviour as an indicator of failure to recognize another animal, and it appears that pigs can remember others 4 weeks after separation (Spoolder *et al.*, 1996), although memory may deteriorate relative to a 1-week separation (Hoy and Bauer, 2005) and may be different for animals of different social status (Ewbank and Meese, 1971). Increased investigation of unfamiliar rather than familiar social stimuli has also been used to assess the duration of social memory (Burman and Mendl, 2006; Fig. 14.3). Influences on social memory of factors such as the time that animals were together before being separated, the size of their original social group, the experiences that they had during the separation period and the context in which they are reunited all need to be investigated systematically (cf. Burman and Mendl, 2006).

14.6 Solving Problems in Established Social Groups

14.6.1 Providing cover and other opportunities for avoidance

In established groups, one important way of ensuring that individuals do not constrain the behaviour of fellow group members or adversely affect their welfare in other ways is to provide animals with opportunities to avoid each other. Avoidance of

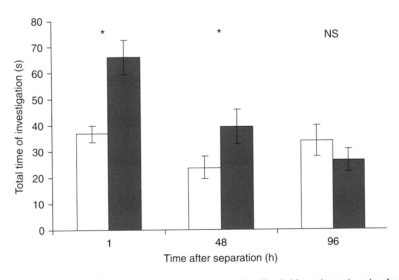

Fig. 14.3. Differences in the times of investigation directed towards familiar (white columns) and unfamiliar (blue columns) odour stimuli by laboratory rats, at different times after separation. The rats spent significantly more (*) time investigating the unfamiliar odour 1 h and 48 h after separation, but not significantly more (NS) time after 96 h of separation. (From Burman and Mendl, 2006.)

aggressive or cannibalistic individuals may be easier in larger than smaller enclosures. However, merely increasing the available space per individual may not be effective if that space is open and unstructured, thereby allowing individuals to pursue others unchecked. For example, in laying hen flocks, feather-pecked pariahs avoid the central litter area because they are relentlessly attacked by other females if they emerge into this area (Freire *et al.*, 2003). Likewise, in flocks of broiler breeders, males congregate in the litter area and multiple males may aggressively force matings upon females venturing there, resulting in injury to females. Consequently, females may avoid the litter area, with resultant low flock fertility (Leone and Estévez, 2008).

Such problems can be mitigated by breaking up the visual landscape with 'bushes', thus capitalizing on ancestral 'Jungle fowl' anti-predator behaviour. Newberry and Shackleton (1997) observed that chickens are attracted to short vertical panels that provide partial rather than solid visual cover (Fig. 14.4). They suggested that these panels have a Venetian blind effect, allowing the birds to hide but, at the same time, to monitor events on the other side of the barrier. Chickens were found to seek the proximity of these panels when engaging in vulnerable activities such as resting and preening, and to experience fewer disturbances from other chickens when resting near them (Newberry and Shackleton, 1997; Cornetto *et al.*, 2002). Furthermore, in adult broiler breeders, the provision of such cover panels in the litter area improved reproductive performance, most likely by

Fig. 14.4. Chickens seek proximity to discontinuous vertical cover for safety. (Newberry and Shackleton, 1997; Photo by Ruth Newberry.)

encouraging more females to enter this area. The resulting higher numbers of females available to the males may have altered male mating tactics, resulting in more courtship and fewer forcible matings (Leone and Estévez, 2008). These findings indicate that relatively simple solutions to serious social problems can be devised through an understanding of social behaviour and attention to the specific design features of protective structures.

An effect of visual cover in reducing aggression has been reported in a number of other species (Chamove and Grimmer, 1993; Whittington and Chamove, 1995; Honess and Marin 2006). In pigs, pop-holes in which the head can be hidden appear to be effective in terminating or avoiding aggressive attacks (McGlone and Curtis, 1985). In pigtail macaques, Erwin *et al.* (1976, p. 321) reported that:

> Decreased frequency of agonistic behaviour in the presence of the concrete cylinders, as contrasted with basic frequencies in the usual captive environment among stable groups, suggested that the barriers helped reduce aggression by providing cover for aggressed animals. Subjects began using the barriers immediately upon their introduction to sit on or in, and aggressed animals frequently avoided their aggressors by hiding in the cylinders.

Erwin *et al.* (1976, p. 322) also noted that the presence of cover was effective only for stable groups:

> The availability of cover was not sufficient to overcome the effects of typical macaque hostility towards strangers.

The provision of perches and elevated platforms can also provide harassed individuals with refuge, provided that the animals have obtained suitable experience in navigating in three-dimensional space, which may have a sensitive period in early life for optimal development (Gunnarsson *et al.*, 2000). However, in aviary housing systems for laying hens, where serious problems with cannibalism and feather pecking can occur (Lay *et al.*, 2011), long, continuous rows of closely spaced perches, although better than no perches, do not fully enable individuals to avoid harmful attacks. Tree-like structures would probably be more effective in this case.

14.6.2 Manipulating resource distribution

There are other ways to prevent individuals from adversely affecting the welfare of fellow group members. Ideally, a managed social environment will provide all group members with ways of obtaining resources without the need for competition. To achieve this, we need to be aware of the conditions that stimulate competition. Aggressive competition for a resource (such as food or females) tends to occur when the resource is defensible, as when food arrives in small quantities (Bryant and Grant, 1995) or is dispersed in a limited number of defendable patches (Pulliam and Caraco, 1984). Even if food is provided *ad libitum*, despotic individuals may be able to monopolize access to feeders, as in the case of computerized feeding stations for sows where low-ranking individuals may be intimidated by the presence of an aggressive sow lying at the entrance. In captive green iguanas, the ability to monopolize heat sources probably accounted for the faster growth of high-ranking individuals compared with their cage mates (Alberts, 1994). In contrast to aggressive competition, scramble competition occurs for resources that are not easily defendable, as when a supply of food arrives at a single location accessible to many animals (Milinski and Parker, 1991). In this case, faster eaters, and stronger animals able to withstand jostling at the food patch, will be able to eat more food than their weaker group mates (Fig. 14.5).

Understanding the relationship between resource distribution and agonistic behaviour is complicated by the intertwined effects of enclosure space, group size and stocking density. Thus, changes in stocking density or group size can have paradoxical effects on aggression owing to changes in the accessibility of resources. For example, in chickens kept in pens of a fixed size, the frequency of agonistic pecks and threats per bird decreases with increasing group size and stocking density (Estévez et al., 2003). This result may be a sequela to larger numbers of birds being located around the feeders, making resource defence uneconomical. As a result, birds with aggressive phenotypes may switch their competitive tactic from resource defence to scramble competition. In contrast, when the space allowance per bird is greater, there may be more fluctuation in numbers around feeders. Under these conditions, Estévez et al. (2002) found that declining numbers of hens around a food patch coincides with increased aggressive defence of the feed patch by the remaining hens. Al-Rawi and Craig (1975) also reported an increased frequency of agonistic behaviour with an increase in space allowance per individual.

Solutions to problems created by competition for resources include distributing resources to prevent monopolization by particular individuals and

Fig. 14.5. Chickens engage in scramble competition for a limited quantity of a favoured food (Estévez *et al.*, 2002; Photo by Ruth Newberry.)

desynchronizing activity to reduce competition. For example, from ideal free distribution theory, we can predict that when food is thinly dispersed over a large area and arrives simultaneously at all locations, animals will spread out evenly with a minimum of competition (Milinski and Parker, 1991). The uniformity of spread may be enhanced if animals can use an easily identifiable cue, such as trough length, as a reliable predictor of food availability. Alternatively, it may be possible to train individuals to feed at different times by signalling when they are allowed to feed. More generally, allowing lower ranking individuals alternative ways of gathering resources can help to ensure that differences in competitive ability do not lead to differences in welfare (Mendl and Deag, 1995).

14.7 Other Solutions to Social Problems

14.7.1 Genetic selection

Genetic selection can provide a long-term solution for reducing undesirable social behaviour. This approach takes advantage of heritable variation in behavioural characteristics such as aggressiveness, once again emphasizing the importance of taking individual differences in social behaviour into account. Particularly notable results have been achieved by applying a group selection strategy to the promotion of social harmony in domestic animal groups. With this approach, families are selected based on high reproductive output from the whole group, rather than accentuating individual differences in competitiveness as occurs when breeding stock are housed individually and the selection programme fails to emphasize social behaviour. For example, Craig and Muir (1993) demonstrated the effectiveness of group selection for high egg production and against mortality (primarily due to beak-inflicted injuries) as a means of reducing the incidence of social aggression, cannibalism and feather pecking in small groups of caged laying hens. As a consequence of the adoption of group selection by commercial poultry breeders, the need for beak trimming to control beak-inflicted injuries in caged hens has been greatly reduced, although problems remain in large flocks. Furthermore, whereas in caged groups it used to be common to observe dominant individuals aggressively defending the feeder, thus limiting the feed intake and egg production of low-ranking

hens, hens of at least one modern, group-selected strain, share limited feeder space without aggression or stress (Thogerson et al., 2009a, b).

These findings in poultry demonstrate the importance of aligning the selection pressures created by the social environment in which breeding animals are kept with those experienced by their offspring. Even when animals are group selected, differences in resource availability and distribution between the breeding environment and the environments to which the offspring are exposed can result in unwanted social consequences. This is especially a challenge when selecting for rapidly growing commercial offspring, given that, for long-term health, it is usually necessary to restrict the growth of the breeding stock. In fish, Ruzzante (1994) points out that artificial selection for growth rate could inadvertently provide a selective advantage for aggressiveness if selection is practised on breeding groups with a limited, monopolizable food supply (but see also Nielsen et al., 1995, who found no evidence for this effect in pigs). This problem can be avoided by conducting selection programmes under conditions that minimize aggressive competition for food.

14.7.2 Social learning

Learning from other group members may be used in a positive way to enhance animal welfare and production. For example, farm animals are often fed a series of diets at different stages of development, but setbacks in growth can occur at the time of a diet change owing to reluctance to try new foods. If certain group members are trained to eat the new food, the rest of the group may acquire a preference for this food quickly as a result of observational learning or social facilitation (Galef, 1993; Nicol and Pope, 1994). Similarly, social learning can enhance adjustment to new feeding and watering equipment, perches and other resources following a pen change. However, care must be taken to minimize opportunities for the social learning of undesirable habits such as cannibalism (Cloutier et al., 2002).

It is common to move animals from the group in which they were raised, either for production or breeding reasons. This process usually involves aggressive encounters (Wechsler and Lea, 2007). Whenever possible, animals that will be introduced to an unknown group should be given the opportunity to establish gradual contact before full immersion into the group to facilitate learning about group members and escape routes (González et al., 2003).

Nevertheless, close monitoring of introductions is recommended as, for example, captive adult male clouded leopards have been known to kill their designated mate when allowed to enter her pen for breeding purposes, despite gradual introduction (MacKinnon *et al.*, 2007), a bizarre reaction perhaps predicated by lack of appropriate early social experience. Furthermore, given that in large groups aggression tends to be directed in a frequency-dependent manner towards minorities that are 'different' (Dennis *et al.*, 2008), the provision of opportunities to gain experience in interacting with diverse phenotypes early in life may ease the future introduction of strangers and reduce the likelihood that particular individuals within a group will be bullied (Croney and Newberry, 2007).

14.8 Conclusions

- Many animal welfare problems in social groups are a result of not taking into account the social history and evolved social capabilities of the group members.
- Attempts to solve social problems in captivity have to focus on a deep understanding of social behaviour, including the role of individual differences within a dynamic social environment.
- Although the social environment, if badly managed, can be a constant source of trouble, it is also a potential tool to improve the welfare of individuals, as it is the most important source of stimulation, interest and comfort in the lives of many captive animals.
- For group-living species, it is important to devise ways of providing an appropriate social environment that can prevent health and welfare problems rather than resorting to social isolation to avoid problems.
- Epidemiological studies should take into account social behaviour as a risk factor affecting individual susceptibility to disease and other welfare problems.

References

Al-Rawi, B and Craig, J.V. (1975) Agonistic behaviour of caged chickens related to group size and area per bird. *Applied Animal Ethology* 2, 69–80.

Alberts, A.C. (1994) Dominance hierarchies in male lizards: implications for zoo management programs. *Zoo Biology* 13, 479–490.

Algers, B. and Jensen, P. (1991) Teat stimulation and milk production during early lactation in sows: effects of continuous noise. *Canadian Journal of Animal Science* 71, 51–60.

Andersen, I.L., Andenaes, H., Bøe, K.E., Jensen, P. and Bakken, M. (2000) The effects of weight asymmetry and resource distribution on aggression in groups of unacquainted pigs. *Applied Animal Behaviour Science* 68, 107–120.

Andersen, I.L., Naevdal, E., Bakken, M. and Bøe, K.E. (2004) Aggression and group size in domesticated pigs, *Sus scrofa*: 'when the winner takes it all and the loser is standing small'. *Animal Behaviour* 68, 965–975.

Arey, D.S. and Franklin, M.F. (1995) Effects of straw and unfamiliarity on fighting between newly mixed growing pigs. *Applied Animal Behaviour Science* 45, 23–30.

Arnott, G. and Elwood, R.W. (2009) Assessment of fighting ability in animal contests. *Animal Behaviour* 77, 991–1004.

Barnett, J.L., Cronin, G.M., McCallum, T.H. and Newman, E.A. (1993) Effects of "chemical intervention" techniques on aggression and injuries when grouping unfamiliar adult pigs. *Applied Animal Behaviour Science* 36, 135–148.

Barnett, J.L., Cronin, G.M., McCallum, T.H. and Newman, E.A. (1994) Effects of food and time of day on aggression when grouping unfamiliar adult pigs. *Applied Animal Behaviour Science* 39, 339–347.

Berman, C.M. (1990) Intergenerational transmission of maternal rejection rates among free-ranging rhesus monkeys. *Animal Behaviour* 39, 329–337.

Bernstein, I.S. (1991) Social housing of monkeys and apes: group formations. *Laboratory Animal Science* 41, 329–333.

Biro, P.A. and Stamps, J.A. (2008) Are animal personality traits linked to life-history productivity? *Trends in Ecology and Evolution* 23, 361–368.

Bolhuis, J.E., Schouten, W.G.P., Schrama, J.W. and Wiegant, V.M. (2005) Individual coping characteristics, aggressiveness and fighting strategies in pigs. *Animal Behaviour* 69, 1085–1091.

Bonaventura, M., De Bortoli Vozioli, A. and Eschino, G. (2008) Costs and benefits of group living in primates: group size effects on behaviour and demography. *Animal Behaviour* 2008, 76, 1235–1247.

Borberg, C. and Hoy, S. (2009) Mixing of sows with or without the presence of a boar. *Livestock Science* 125, 314–317.

Bradshaw, G.A., Schore, A.N., Brown, J.L., Poole, J.H. and Moss, C.J. (2005) Elephant breakdown. *Nature* 433, 807.

Bryant, M.J. and Grant, J.W.A. (1995) Resource defence, monopolization and variation of fitness in groups of female Japanese medaka depend on the synchrony of food arrival. *Animal Behaviour* 49, 1469–1479.

Burman, O.H.P. and Mendl, M. (2006) Long-term social memory in the laboratory rat (*Rattus norvegicus*). *Animal Welfare* 15, 379–382.

Burn, C.B. and Mason, G.J. (2008) Effects of cage-cleaning frequency on laboratory rat reproduction, cannibalism, and welfare. *Applied Animal Behaviour Science* 114, 235–247.

Carter, C.S. and Keverne, E.B. (2002) The neurobiology of social affiliation and pair bonding. In: Pfaff, D.W., Arnold, A.P., Etgen, A.M., Fahrbach, S.E. and Rubin, R.T. (eds) *Hormones, Brain and Behavior, Volume One.* Academic Press (an Imprint of Elsevier Science), San Diego, California, pp. 299–337.

Chamove, A.S. and Grimmer, B. (1993) Reduced visibility lowers bull aggression. *Proceedings of the New Zealand Society of Animal Production* 53, 207–208.

Cloutier, S., Newberry, R.C., Honda, K. and Alldredge, J.R. (2002) Cannibalistic behaviour spread by social learning. *Animal Behaviour* 63, 1153–1162.

Clutton-Brock, T.H. and Albon, S.D. (1979) The roaring of red deer and the evolution of honest signalling. *Behaviour* 69, 145–169.

Colonnello, V., Iacobucci, P. and Newberry, R.C. (2010) Vocal and locomotor responses of piglets to social isolation and reunion. *Developmental Psychobiology* 52, 1–12.

Colson, V., Orgeur, P., Courboulay, V., Dantec, S., Foury, A. and Mormède, P. (2006) Grouping piglets by sex at weaning reduces aggressive behaviour. *Applied Animal Behaviour Science* 97, 152–171.

Cornetto, T., Estévez, I. and Douglass, L.W. (2002) Using artificial cover to reduce aggression and disturbances in domestic fowl. *Applied Animal Behaviour Science* 75, 325–336.

Craig, J.V. and Muir, W.M. (1993) Selection for reduced beak-inflicted injuries among caged hens. *Poultry Science* 72, 411–420.

Croney, C. and Newberry, R.C. (2007) Group size and cognitive processes. *Applied Animal Behaviour Science* 103, 215–228.

Dawkins, M.S. (1990) From an animal's point of view: motivation, fitness, and animal welfare. *Behavioral and Brain Sciences* 13, 1–9, 54–61.

D'Eath, R.B. (2002) Individual aggressiveness measured in a resident-intruder test predicts the persistence of aggressive behaviour and weight gain of young pigs after mixing. *Applied Animal Behaviour Science* 77, 267–283.

D'Eath, R.B. and Burn, C.C. (2002) Individual differences in behaviour: a test of 'coping style' does not predict resident-intruder aggressiveness in pigs. *Behaviour* 139, 1175–1194.

Dennis, R., Newberry, R.C., Cheng, H.-W. and Estévez, I. (2008) Appearance matters: artificial marking alters aggression and stress. *Poultry Science* 87, 1939–1946.

Edwards, S.A. and Mauchline, S. (1993) Designing pens to minimise aggression when sows are mixed. *Farm Buildings Progress* 113, 20–23.

Enquist, M. and Leimar, O. (1983) Evolution of fighting behaviour: decision rules and assessment of relative strength. *Journal of Theoretical Biology* 102, 387–410.

Erwin, J., Anderson, B., Erwin, N., Lewis, L. and Flynn, D. (1976) Aggression in captive pigtail monkey groups: effects of provision of cover. *Perceptual and Motor Skills* 42, 319–324.

Estévez, I., Newberry, R.C. and Keeling, L.J. (2002) Dynamics of aggression in the domestic fowl. *Applied Animal Behaviour Science* 76, 307–325.

Estévez, I., Keeling, L.J. and Newberry, R.C. (2003) Decreasing aggression with increasing group size in young domestic fowl. *Applied Animal Behaviour Science* 84, 213–218.

Estévez, I., Andersen, I.-L. and Nævdal, E. (2007) Group size, density and social dynamics in farm animals. *Applied Animal Behaviour Science* 103, 185–204.

Ewbank, R.J. and Meese, G.B. (1971) Aggressive behaviour in groups of domesticated pigs on removal and return of individuals. *Animal Production* 13, 685–693.

Feh, C. and de Mazières, J. (1993) Grooming at a preferred grooming site reduces heart rate in horses. *Animal Behaviour* 46, 1191–1194.

Freire, R., Wilkins, L.J., Short, F. and Nicol, C.J. (2003) Behaviour and welfare of individual laying hens in a non-cage system. *British Poultry Science* 44, 22–29.

Galef, B.G. (1993) Functions of social learning about food: a causal analysis of effects of diet novelty on preference transmission. *Animal Behaviour* 46, 257–265.

Galindo, F. and Broom, D.M. (2002) The effects of lameness on social and individual behavior of dairy cows. *Journal of Applied Animal Welfare Science*, 5, 193–201.

Gold, K.C. and Maple, T.L. (1994) Personality assessment in the gorilla and its utility as a management tool. *Zoo Biology* 13, 509–522.

González, M., Yabuta, A.K. and Galindo, F. (2003) Behaviour and adrenal activity of first parturition and multiparous cows under a competitive situation. *Applied Animal Behaviour Science* 83, 259–266.

Gray, S. and Hurst, J.L. (1995) The effects of cage cleaning on aggression within groups of male laboratory mice. *Animal Behaviour* 49, 821–826.

Gunnarsson, S., Yngvesson, J., Keeling, L.K. and Forkman, B. (2000) Rearing without early access to perches impairs the spatial skills of laying hens. *Applied Animal Behaviour Science* 67, 217–228.

Hennessy, M.B., Schiml-Webb, P.A., Miller, E.E., Maken, D.S., Bullinger, K.L. and Deak, T. (2007) Anti-inflammatory agents attenuate the passive responses of guinea pig pups: evidence for stress-induced sickness behavior during maternal separation. *Psychoneuroendocrinology* 32, 508–515.

Hernández, L., Barral, H., Halffter, G. and Sánchez, C.S. (1999) A note on the behavior of feral cattle in the

Chihuahuan Desert of México. *Applied Animal Behaviour Science* 63, 259–267.

Hessing, M.J.C., Schouten, W.G.P., Wiepkema, P.R. and Tielen, M.J.M. (1994a) Implications of individual behavioural characteristics on performance in pigs. *Livestock Production Science* 40, 187–196.

Hessing, M.J.C. Hagelso, A.M., Schouten, W.G.P., Wiepkema, P.R. and Van Beek, J.A.M. (1994b) Individual behavioural and physiological strategies in pigs. *Physiology and Behavior* 55, 39–46.

Hogue, M.E., Beaugrand, J.P. and Laguë, P.C. (1996) Coherent use of information by hens observing their former dominant defeating or being defeated by a stranger. *Behavioural Processes* 38, 241–252.

Honess, P.E. and Marin, C.M. (2006) Enrichment and aggression in primates. *Neuroscience and Biobehavioral Reviews* 30, 413–436.

Høøk Presto, M., Algers, B., Persson, E. and Andersson, H.K. (2009) Different roughages to organic growing/finishing pigs – influence on activity behaviour and social interactions. *Livestock Science* 123, 55–62.

Hoy, S. and Bauer, J. (2005) Dominance relationships between sows dependent on the time interval between separation and reunion. *Applied Animal Behaviour Science* 90, 21–30.

Humphrey, N.K. (1976) The social function of intellect. In: Bateson, P.P.G. and Hinde, R.A. (eds) *Growing Points in Ethology*. Cambridge University Press, Cambridge, UK, pp. 303–317.

Jensen, P. (1994) Fighting between unacquainted pigs – effects of age and of individual reaction pattern. *Applied Animal Behaviour Science* 41, 37–52.

Jensen, P. and Yngvesson, J. (1998) Aggression between unacquainted pigs – sequential assessment and effects of familiarity and weight. *Applied Animal Behaviour Science* 58, 49–61.

Kanitz, E., Puppe, B., Tuchscherer, M., Heberer, M., Viergutz, T. and Tuchscherer, A. (2009) A single exposure to social isolation in domestic piglets activates behavioural arousal, neuroendocrine stress hormones, and stress-related gene expression in the brain. *Physiology and Behavior* 98, 176–185.

Kennedy, M.J. and Broom, D.M. (1994) A method of mixing gilts and sows which reduces aggression experienced by gilts. In: *Proceedings of the 28th International Congress of the ISAE [International Society for Applied Ethology]*, National Institute of Animal Science, Foulum, Denmark, p. 52 (abstract).

Keverne, E.B., Martenz, N.D. and Tuite, B. (1989) Beta-endorphin concentrations in cerebrospinal fluid of monkeys are influenced by grooming relationships. *Psychoneuroendocrinology* 14, 155–161.

Kristensen, H.H., Jones, R.B., Schofield, C., White, R.P. and Wathes, C.M. (2001) The use of olfactory and other cues for social recognition by juvenile pigs. *Applied Animal Behaviour Science* 72, 321–333.

Kuhar, C.W. (2008) Group differences in captive gorillas' reaction to large crowds. *Applied Animal Behaviour Science* 110, 377–385.

Lay, D.C. Jr, Fulton, R.M., Hester, P.Y., Karcher, D.M., Kjaer, J., Mench, J.A., Mullens, B.A., Newberry, R.C., Nicol, C.J., O'Sullivan, N.P. and Porter, R.E. (2011) Hen welfare in different housing systems. *Poultry Science* 90, 278–294.

Le Neindre, P., Veissier, I., Boissy, A. and Boivin, X. (1992) Effects of early environment on behaviour. In: Phillips, C. and Piggins, D. (eds) *Farm Animals and the Environment*. CAB International, Wallingford, UK, pp. 307–322.

Leone, E.H. and Estévez, I. (2008) Economic and welfare benefits of environmental enrichment for broiler breeders. *Poultry Science* 87, 14–21.

Love, J.A. and Hammond, K. (1991) Group housing rabbits. *Laboratory Animals* 20, 37–43.

Luescher, U.A., Friendship, R.M. and McKeown, D.B. (1990) Evaluation of methods to reduce fighting among regrouped gilts. *Canadian Journal of Animal Science* 70, 363–370.

MacKinnon, K.M., Newberry, R.C., Wielebnowski, N.C. and Pelican, K.M. (2007) Identifying early indicators for successful pairing of clouded leopards in captive breeding programs. In: Galindo, F. and Alvarez, L. (eds) *Proceedings of the 41st Congress of the ISAE [International Society for Applied Ethology]*, 30 July–3 August 2007, Merida, Mexico. ISAE, p. 24 (abstract).

Maestripieri, D. (2005) Early experience affects the intergenerational transmission of infant abuse in rhesus monkeys. *Proceedings of the National Academy of Science of the USA* 102, 9726–9729.

McBride, G., Parer, I.P. and Foenander, F. (1969) The social behaviour and organization of feral domestic fowl. *Animal Behaviour Monographs* 2, 125–181.

McGlone, J.J. (1990) Olfactory signals that modulate pig aggressive behavior. In: Zayan, R. and Dantzer, R. (eds) *Social Stress in Domestic Animals*. Kluwer Academic Publications, Dordrecht, The Netherlands, pp. 86–109.

McGlone, J.J. and Curtis, S.E. (1985) Behavior and performance of weanling pigs in pens equipped with hide areas. *Journal of Animal Science* 60, 20–24.

McLeman, M.A., Mendl, M., Jones, R.B. and Wathes, C.M. (2005) Discrimination of conspecifics by juvenile domestic pigs, *Sus scrofa*. *Animal Behaviour* 70, 451–461.

McLeman, M.A., Mendl, M.T., Jones, R.B. and Wathes, C.M. (2008) Social discrimination of familiar conspecifics by juvenile pigs, *Sus scrofa*: development of a non-invasive method to study the transmission of unimodal and bimodal cues between live stimuli. *Applied Animal Behaviour Science* 115, 123–137.

Mendl, M. (1995) The social behaviour of non-lactating sows and its implications for managing sow aggression. *Pig Veterinary Journal* 34, 9–20.

Mendl, M. and Deag, J. (1995) How useful are the concepts of alternative strategy and coping strategy in applied studies of social behaviour? *Applied Animal Behaviour Science* 44, 119–137.

Mendl, M. and Erhard, H.W. (1997) Social choices in farm animals: to fight or not to fight? In: Forbes, J.M., Lawrence, T.L.J., Rodway, R.G. and Varley, M.A. (eds) *Animal Choices*. British Society of Animal Science (BSAS) Edinburgh, UK, pp. 45–53.

Milinski, M. and Parker, G.A. (1991) Competition for resources. In: Krebs, J.R. and Davies, N.B. (eds) *Behavioural Ecology: An Evolutionary Approach*, 3rd edn. Blackwell Scientific Publications, Oxford, UK, pp. 137–168.

Miller, A.H., Maletic, V. and Raison, C.L. (2009) Inflammation and its discontents: the role of cytokines in the pathophysiology of major depression. *Biological Psychiatry* 65, 732–741.

Nelson, R.J. and Trainor, B.C. (2007). Neural mechanisms of aggression. *Nature Reviews Neuroscience* 8, 536–546.

Newberry, R.C. and Estévez, I. (1997) A dynamic approach to the study of environmental enrichment and animal welfare. *Applied Animal Behaviour Science* 54, 53–57.

Newberry, R.C. and Shackleton, D.M. (1997) Use of cover by domestic fowl: a Venetian blind effect? *Animal Behaviour* 54, 387–395.

Newberry, R.C. and Swanson, J.C. (2008) Implications of breaking mother–young social bonds. *Applied Animal Behaviour Science* 110, 3–23.

Newberry, R.C., Špinka, M. and Cloutier, S. (2000) Early social experience of piglets affects rate of adaptation to strangers after weaning. In: Ramos, A., Pinheiro Machado F., L.C. and Hötzel, M.J. (eds) *Proceedings of the 34th International Congress of the ISAE [International Society for Applied Ethology]*, 17–20 October 2000, Florianópolis. Federal University of Santa Catarina, Florianópolis, Brazil, p. 67 (abstract).

Newberry, R.C., Keeling, L.J., Estévez, I. and Bilčík, B. (2007) Behaviour when young as a predictor of severe feather pecking in adult laying hens: the redirected foraging hypothesis revisited. *Applied Animal Behaviour Science* 107, 262–274.

Nicol, C.J. and Pope S.J. (1994) Social learning in small flocks of hens. *Animal Behaviour* 47, 1289–1296.

Nielsen, B.L., Lawrence, A.B. and Whittemore, C.T. (1995) Effect of group size on feeding behaviour, social behaviour and performance of growing pigs using single-space feeders. *Livestock Production Science* 44, 73–85.

Nielsen, L.H., Mogensen, L., Krohn, C., Hindhede, J. and Sørensen, J.T. (1997) Resting and social behaviour of dairy heifers housed in slatted floor pens with different sized bedded lying areas. *Applied Animal Behaviour Science* 54, 307–316.

Oliveira, R.F., McGregor, P.K. and Latruffe, C. (1998) Know thine enemy: fighting fish gather information from observing conspecific interactions. *Proceedings of the Royal Society B – Biological Sciences* 265, 1045–1049.

Oliveira, R.F., Lopes, M., Carneiro, L.A. and Canário, A.V.M. (2001) Watching fights raises fish hormone levels. *Nature* 409, 475.

Orihuela, A. and Galindo, F. (2004) Etología aplicada en los bovinos. In: Galindo, F. and Orihuela, A. (eds) *Etología Aplicada*. Universidad Nacional Autónoma de México, México, pp. 89–131.

Otten, W., Puppe, B., Stabenow, B., Kanitz, E., Schijn, P.C., Briissow, K.P. and Niirnberg, G. (1997) Agonistic interactions and physiological reactions of top- and bottom-ranking pigs confronted with a familiar and an unfamiliar group: preliminary results. *Applied Animal Behaviour Science* 55, 79–90.

Panksepp, J. (2003) Feeling the pain of social loss. *Science* 302, 237–239.

Paz-y-Miño, G., Bond, A.B., Kamil, A.C. and Balda, R.P. (2004) Pinyon jays use transitive inference to predict social dominance. *Nature* 430, 778–781.

Peake, T.M., Terry, A.M.R., McGregor, P.K. and Dabelsteen, T. (2001) Male great tits eavesdrop on simulated male-to-male vocal interactions. *Proceedings of the Royal Society B – Biological Sciences* 268, 1183–1187.

Pellis, S.M., Pellis, V.C. and Bell, H.C. (2010) The function of play in the development of the social brain. *American Journal of Play* 2, 278–296.

Petherick, J.C. and Blackshaw, J.K. (1987) A review of the factors influencing the aggressive and agonistic behaviour of the domestic pig. *Australian Journal of Experimental Agriculture* 27, 605–611.

Petit, O. and Thierry, B. (1994) Aggressive and peaceful interventions in conflicts in Tonkean macaques. *Animal Behaviour* 48, 1427–1436.

Price, E.E. and Stoinskia, T.S. (2007) Group size: determinants in the wild and implications for the captive housing of wild mammals in zoos. *Applied Animal Behaviour Science* 103, 255–264.

Price, E.O. (1999) Behavioral development in animals undergoing domestication. *Applied Animal Behaviour Science* 65, 245–271.

Price, E.O. (2004) Efecto de la domesticación en la conducta animal. In: Galindo, F. and Orihuela, A. (eds) *Etología Aplicada*. Universidad Nacional Autónoma de México, México, pp. 29–50.

Pulliam, H.R. and Caraco, T. (1984) Living in groups: is there an optimal group size? In: Krebs, J.R. and Davies, N.B. (eds) *Behavioural Ecology: An Evolutionary Approach*, 3rd edn. Blackwell Scientific Publications, Oxford, UK, pp. 122–147.

Reinhardt, V., Houser, D., Cowley, D., Eisele, S. and Vertein, R. (1989) Alternatives to single caging of rhesus monkeys (*Macaca mulatta*) used in research. *Zeitschrift für Versuchstierskunde* 32, 275–279.

Ruis, M.A.W., te Brake, J.H.A., Engel, B., Buist, W.G., Blokhuis, H.J. and Koolhaas, J. M. (2001) Adaptation to social isolation: acute and long-term stress responses of growing gilts with different coping characteristics. *Physiology and Behavior* 73, 541–551.

F. Galindo *et al.*

Rumbaugh, D.M., Washburn, D. and Savage-Rumbaugh, E.S. (1989) On the care of captive chimpanzees: methods of enrichment. In: Segal, E. (ed.) *Housing, Care and Psychological Wellbeing of Captive and Laboratory Primates.* Noyes Publications, Park Ridge New Jersey, pp. 357–375.

Rushen, J. (1987) A difference in weight reduces fighting when unacquainted newly weaned pigs first meet. *Canadian Journal of Animal Science* 67, 951–960.

Rushen, J. (1988) Assessment of fighting ability or simple habituation – what causes young pigs (*Sus scrofa*) to stop fighting. *Aggressive Behavior* 14, 155–167.

Rushen, J. (1990) Social recognition, social dominance and the motivation of fighting by pigs. In: Zayan, R. and Dantzer, R. (eds) *Social Stress in Domestic Animals.* Kluwer Academic Publications, Dordrecht, The Netherlands, pp. 135–143.

Ruzzante, D.E. (1994) Domestication effects on aggressive and schooling behavior in fish. *Aquaculture* 120, 1–24.

Schapiro, S.J. (2002) Effects of social manipulations and environmental enrichment on behavior and cell-mediated immune responses in rhesus macaques. *Pharmacology, Biochemistry and Behavior* 73, 271–278.

Souza, A.S. and Zanella, A.J. (2008) Social isolation elicits deficits in the ability of newly weaned female piglets to recognise conspecifics. *Applied Animal Behaviour Science* 110, 182–188.

Špinka, M. (2006) How important is natural behaviour in animal farming systems? *Applied Animal Behaviour Science* 100, 117–128.

Špinka, M., Newberry, R.C. and Bekoff, M. (2001) Mammalian play: training for the unexpected. *The Quarterly Review of Biology* 76, 141–168.

Spoolder, H.A.M., Burbidge, J.A., Edwards, S.A., Lawrence, A.B. and Simmins, P.H. (1996) Social recognition in gilts mixed into a dynamic group of 30 sows. *Animal Science* 62, 630 (abstract).

Spoolder, H.A.M., Geudeke, M.J., Van der Peet-Schwering, C.M.C. and Soede, N.M. (2009) Group housing of sows in early pregnancy: a review of success and risk factors. *Livestock Science* 125, 1–14.

Stauffacher, M. (1992) Group housing and enrichment cages for breeding, fattening and laboratory rabbits. *Animal Welfare* 1, 105–125.

Stolba, A. and Wood-Gush, D.G.M. (1984) The identification of behavioural key features and their incorporation into a housing design for pigs. *Annales de Recherches Veterinaires* 15, 287–298.

Stolba, A. and Wood-Gush, D.G.M. (1989) The behaviour of pigs in a semi-natural environment. *Animal Production* 48, 419–425.

Suzuki, H. and Lucas, L.R. (2009) Chronic passive exposure to aggression escalates aggressiveness of rat observers. *Aggressive Behavior* 35, 1–13.

Taylor, N.R., Main, D.C.J., Mendl, M. and Edwards, S.A. (2010) Tail-biting: a new perspective. *The Veterinary Journal* 186, 137–147.

Thogerson, C.M., Hester, P.Y., Mench, J.A., Newberry, R.C., Pajor, E.A. and Garner, J.P. (2009a) The effect of feeder space allocation on behavior of Hy-line W-36 hens housed in conventional cages. *Poultry Science* 88, 1544–1552.

Thogerson, C.M., Hester, P.Y., Mench, J.A., Newberry, R.C., Okura, C.M., Pajor, E.A., Talaty, P.N. and Garner, J.P. (2009b) The effect of feeder space allocation on productivity and physiology of Hy-line W-36 hens housed in conventional cages. *Poultry Science* 88, 1793–1799.

Turner, S.P., Farnworth, M.J., White, I.M.S., Brotherstone, S., Mendl, M., Knap, P., Penny, P. and Lawrence, A.B. (2006) The accumulation of skin lesions and their use as a predictor of individual aggressiveness in pigs. *Applied Animal Behaviour Science* 96, 245–259.

Turner, S.P., Roehe, R., D'Eath, R.B., Ison, S.H., Farish, M., Jack, M.C., Lundeheim, N., Rydhmer, L. and Lawrence, A.B. (2009) Genetic validation of postmixing skin injuries in pigs as an indicator of aggressiveness and the relationship with injuries under more stable social conditions. *Journal of Animal Science* 87, 3076–3082.

van de Weerd, H.A. and Day, J.E.L. (2009) A review of environmental enrichment for pigs housed in intensive housing systems. *Applied Animal Behaviour Science* 116, 1–20.

Vehrencamp, S.L. (1983) A model for the evolution of despotic versus egalitarian societies. *Animal Behaviour* 31, 667–682.

Verga, M., Luzi, F. and Carenzi, C. (2007) Effects of husbandry and management systems on physiology and behaviour of farmed and laboratory rabbits. *Hormones and Behavior* 52, 122–129.

Visalberghi, E. and Anderson, J.R. (1993) Reasons and risks associated with manipulating captive primates' social environments. *Animal Welfare* 2, 3–15.

von Holst, D. (2004) Attachment and pair bonds in tree shrews: proximate causes and physiological consequences. *Journal of Psychosomatic Research* 56, 589.

Watt, D. and Panksepp, J. (2009) The depressive matrix: an evolutionarily conserved mechanism to terminate separation distress? A review of aminergic, peptidergic and neural network perspectives. *Neuropsychoanalysis* 11, 5–104.

Wechsler, B. and Lea, S.E.G. (2007) Adaptation by learning: its significance for farm animal husbandry. *Applied Animal Behaviour Science* 108, 197–214.

Weiss, I.C., Pryce, C.R., Jongen-Rêlo, A.L., Nanz-Bahr, N.I. and Feldon, J. (2004) Effect of social isolation on stress-related behavioural and neuroendocrine state in the rat. *Behavioural Brain Research* 152, 279–295.

Whittington, C.J. and Chamove, A.S. (1995) Effects of visual cover on farmed red deer behaviour. *Applied Animal Behaviour Science* 45, 309–314.

Wolf, M., van Doorn, G.S., Leimar, O. and Weissing, F.J. (2007) Life-history trade-offs favour the evolution of animal personalities. *Nature* 447, 581–584.

15 Human Contact

PAUL H. HEMSWORTH AND XAVIER BOIVIN

Abstract

Human–animal relationships can be viewed in various ways, but from an ethological perspective they can be conceptualized in terms of inter-individual relationships. This chapter will review the influence of human contact on the welfare of domestic animals. The model used will be the relationship between humans and farm animals, the subject of the majority of studies. The most studied aspect of the relationship from the perspective of farm animals has been the animal's fear responses to humans. Recently there has been increasing appreciation that animals may experience positive or pleasant emotions in the presence of humans that may arise from rewarding events and associations. There are three main lines of evidence concerning the implications for the welfare of farm animals: handling studies in controlled experimental conditions, observations in commercial settings and intervention studies in commercial settings. Although handling at an early age may be highly influential, subsequent handling is also influential and has the potential to modify early learning effects. Conditioning and habituation to humans, occurring both early and later in life, are probably the most influential factors affecting the behavioural responses of farm animals to humans. This review highlights the important role and responsibility of the human in the development of the human–animal relationship. The results of handling studies in the laboratory and intervention studies on farms on the relationship between stockperson attitudes, stockperson behaviour, animal behaviour and stress physiology provide evidence of causal relationships between these variables. Furthermore, this research provides a strong case for introducing stockperson training courses in the livestock industries which target stockperson attitudes and behaviour. The selection of stockpeople on the basis of attributes appropriate to the job provides another opportunity to improve animal welfare. This discussion demonstrates the important role and responsibility of the human in the development of human–animal relationships and thus underlines the need to understand not only these relationships but also the opportunities to improve them in order to safeguard animal welfare.

15.1 Introduction

This chapter explores the influence of human contact on the welfare of domestic animals by examining the close relationship between humans and domestic animals. The model for this review is the relationship between humans and farm animals, as there is a substantial body of research on the regulation of human–farm animal relationships and their implications, particularly for the animal. While the initial studies on human–farm animal relationships were driven by an interest in the effects on farm animal productivity, more recent research has been conducted because of the implications of this relationship for farm animal welfare.

Modern farm animals have undergone thousands of years of domestication (Serpell, 1986; Clutton-Brock, 1994; Price, 2002). Accompanying the ever-increasing use of domestic animals since the Neolithic period has been a developing discussion on the human–animal relationship, particularly its development, regulation and implication for both partners. Even after thousands of years of domestication, the subject of human–animal relationships still stimulates controversial views that directly question the current standards of welfare for domestic animals and the responsibility of humans in the use of these animals in society. Philosophical discussions on the inequity of the relationship for the partners and what is acceptable human behaviour in this relationship are still contentious and prominent (Digard, 1990; Hemsworth, 2007).

Human–animal relationships are viewed in various ways, with disciplines such as biology, philosophy, psychology, history, anthropology and sociology providing different perspectives. However, in this chapter, they will be viewed from

©CAB International 2011. *Animal Welfare*, 2nd Edition (eds M.C. Appleby *et al.*)

an ethological perspective in terms of inter-individual relationships, with the quality and frequency of interactions between two familiar individuals – as well as the context in which they occur – determining the nature or quality of the relationship between them.

15.2 Human Contact and Domestication

Most people have an intuitive view of what 'domestic animals' are, as these animals, as opposed to wild animals, are for the most part very familiar. Species that develop 'social' interactions with humans are often considered as domestic (Denis, 2004). Indeed, the word 'domestication' is often misused to indicate individual taming. However, it is now commonly recognized that domestication involves populations of animals, while taming can be defined as 'an experiential (learning) phenomenon occurring during the lifetime of an individual' (Price, 1999, p. 258). In their review of the early process of domestication, Zeder *et al.* (2006, p. 140) stated that:

> In contrast to crop plants, selection on animals initially brought under human management is most often directed towards modification of behaviour in the target species rather than towards morphological change. For example, both inadvertent and deliberate human efforts can select for increased tolerance of penning, sexual precocity and, above all, reduction of wariness and aggression.

However domestication is not simply a product of genetic selection; it is also achieved through environmental stimulation and experiences during an animal's lifetime (Price, 2002). Indeed, husbandry conditions and human contact have affected the responses of animals to humans since the Neolithic period and continue to do so. Even today, a small number of cattle live permanently in semi-wild conditions with African nomads, while millions of cattle live in large North American or Brazilian feedlots with mechanical feeding and reduced human contact. Small dog breeds are famous as favourite companion animals of celebrities, living in close contact with people in modern city dwellings, while some breeds work as herd protection dogs, living mostly among sheep and without close human contact. While rabbits and dogs may be sources of food for humans in some countries, they may also be used as companion or laboratory animals in the same or other countries. So the definition of domestication from Price

(2002) appears particularly relevant from a biological point of view, and especially from a human–animal relationship perspective. Price (2002, p. 11) defines domestication as a 'process by which a population of animals becomes adapted to man and to the captive environment by genetic changes occurring over generations and environmentally induced developmental events reoccurring during each generation'. Thus, animal populations are subjected to a perpetual domestication process through which they adapt to humans and the human environment and human needs. Among the many important environmental factors that have to be taken into account in the domestication process, it is essential to consider the human–animal interactions that occur during the lifetime of animals as assisting these animals to adapt to the constraints imposed by humans.

15.3 Interactions and Relationships Between Humans and Domestic Animals

It has been proposed by many authors that species-specific predisposed characteristics are essential in the process of domestication (Hale, 1969; Kretchmer and Fox, 1975; Scott, 1992; Price, 1999). Among these characteristics, social organization is an important behavioural trait that influenced the domestication of ungulate and galliform species (Stricklin and Mench, 1987). These species had the ability to live in relatively large groups without marked year-round territoriality. Also, stable dominance relationships are usually present, with the result that these animals can be kept in large groups and are less likely to injure one another or to expend large amounts of energy in competing for resources that may be limited, such as mates or food. This trait has been further emphasized through artificial selection during domestication.

As a consequence of humans keeping farm animals in relatively large groups in restricted areas, humans are able to interact regularly with their animals at several levels. Many interactions are associated with regular observation of the animals. Animals in most production systems have to be moved and, in addition to visual and auditory contact, stockpeople often use tactile interactions to move their animals. Human–animal interactions also occur in situations in which animals must be restrained and subjected to management or health procedures.

From an ethological perspective, the human–animal relationship can be conceptualized in terms of inter-individual relationships. In a chapter in the book *The Inevitable Bond: Examining Scientist–Animal Interactions*, Estep and Hetts (1992) proposed an ethological concept of the human–animal relationship. Utilizing the view of Hinde (1976) that inter-individual relationships are based on the history of regular interactions between two individuals, Estep and Hetts (1992) argued that human–animal relationships can be viewed in a similar manner and that studies of this relationship should be undertaken by investigating each partner's perception of the relationship. Each individual partner's perception of the relationship allows it to interpret and predict the other's interactions. Therefore, if animals are able to learn and anticipate future interactions, the concept of the relationship does not only exist for each partner of the relationship but also for an external observer.

One important methodological feature associated with this concept is the necessity of characterizing those interactions that have significance for the human and animal partners, so that the influence of the quantity and nature of these interactions on the human–animal relationship can be understood. This approach has been clearly illustrated by the research on stockpeople and farm animals which is described in this chapter. As discussed later, evidence from handling studies and observations on human–animal interactions in the livestock industries indicate that it is this history of interactions between humans and animals that leads to the development of a stimulus-specific response of farm animals to humans: through conditioning, a farm animal may associate humans with rewarding and punishing events that occur at the time of human–animal interactions, and thus develop conditioned responses to humans. Similarly, the stockperson's direct and indirect experiences with animals are influential determinants of the stockperson's attitudes and behaviour towards farm animals.

It is generally recognized that at least other mammals are likely to have affective experiences that resemble our own (Panksepp, 2005), and most researchers studying human–farm animal relationships have been specifically interested in the fear responses of animals to humans because of their implications for animal productivity and welfare (Hemsworth and Coleman, 1998). Nevertheless, the animal's perception of the relationship is likely to be determined not only by negative emotional states such as fear, but also by positive emotional states generated by interaction with humans. For example, Schmied et al. (2008) found that stroking the ventral region of the neck of dairy cattle encourages more neck stretching and lowers heart rate in these animals. In cattle, intraspecific social licking in this area is very common and these results suggest that stroking this preferred region may be perceived more positively than a neutral interaction by humans.

Extending the view of Aureli and Schaffner (2002) that social relationships allow animals to predict the actions and responses of their partners and therefore guide their own responses, Boivin et al. (2003) recommended that the range of emotions generated by the interaction with humans is likely to determine an animal's relationship with humans. Similarly, Waiblinger et al. (2006) suggest that different emotions and motivations are involved in the animal's perception of and reaction to humans, and that these are likely to determine the strength of an animal's relationship with humans, which may therefore vary from negative through neutral to positive (Fig. 15.1). Indeed, behavioural and physiological responses of animals to humans express such emotional states in the presence and absence of the human partner.

Furthermore, based on the cognitive theory of emotion, Désiré et al. (2002) proposed that emotions that are generated during interactions between humans and animals are likely to be determined by not only the properties of the other partner in the relationship but also by the perception and hence the interpretation of the whole situation. Emotions, therefore, are likely to be elicited by a combination of basic evaluations such as the suddenness, familiarity, valence, predictability and controllability of the interactions. For example, increases in activity and heart rate in farm animals are much more pronounced when novelty and suddenness are combined (Désiré et al., 2006).

The consequences of negative and positive emotional states arising from human interactions for the welfare of farm animals are obvious, particularly those human interactions leading to fear and thereby stress in animals. These emotional states may also affect the ease of handling of the animal and the safety of the handler as, for example, fear may lead to defensive behaviour such as aggression in large farm animals.

P.H. Hemsworth and X. Boivin

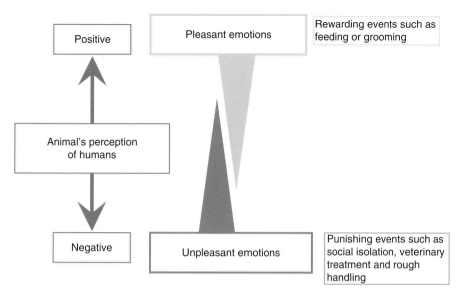

Fig. 15.1. Emotional dimensions affecting an animal's response to humans. (Modified from Waiblinger *et al.*, 2006.)

15.4 Assessment of the Human–Animal Relationship

The most studied aspect of the human–animal relationship from the perspective of the farm animal has been fear responses to humans, both behavioural and physiological (Fig. 15.2). As reviewed by Waiblinger *et al.* (2006), tests measuring animal responses to humans can be categorized into three main types: (i) responses to a stationary human; (ii) responses to a moving human; and (iii) responses to actual handling. These authors noted that possible confounding motivations or behavioural systems may differ between the test categories. In tests in which humans approach, fear responses may be easier to interpret than other motivations, such as curiosity. In contrast, when testing an animal's approach to a stationary human, latency in approaching and interaction by the animal may vary according to the levels of both curiosity and attraction to the human. In many of the tests measuring an animal's response to a stationary human, although the degree of novelty of the test arena may be reduced by the similarity of the arena to the animal's familiar environment, animals introduced into this new environment will be motivated to explore and familiarize themselves with the environment once the initial fear responses have waned. Therefore, although an animal may be motivated both to avoid and explore the arena and

the human stimulus, the animal's fear of humans will have a major influence on its approach to the human stimulus. This method of measuring an animal's response to a stationary human to assess fear is supported by the findings of behavioural and physiological correlates in the tests, together with findings that imposition of handling treatments designed to affect an animal's fear of humans differentially generally produces the expected variations in the behavioural responses of the animal to humans (Hemsworth and Coleman, 1998).

There are several sources of variation induced by non-standardized environmental conditions that can affect the accuracy of measurements on human–animal interactions in commercial farm settings. Hemsworth *et al.* (2009) have reviewed some of these in relation to studying the behaviour of stockpeople towards their animals. The presence of the observer, as well as the ability to observe stockperson behaviour under routine handling conditions, may affect the validity of these observations. de Passillé and Rushen (2005), Waiblinger *et al.* (2006) and Hemsworth *et al.* (2009) have also reviewed the main sources of variation that may affect the accuracy of measurements on the farm animal's response to humans. As with the human behaviour observations, the presence of an observer may affect the behaviour of the study animals. Furthermore, the validity of

Fig. 15.2. Scientists have assessed fear levels by measuring the amount of avoidance of farm animals to close approach by humans.

the measurements of fear responses may be affected by variation between farms (or studies) and between animals in relation to: the measures of fear studied; other motivations, such as curiosity and hunger; the context of the test setting, such as its novelty and its location relative to where routine handling or husbandry occurs; the properties of the human stimulus, such as the animal's familiarity with the human and the human's behaviour during testing; and pretest conditions that affect fear and other motivations. Therefore, standardized procedures are necessary to evaluate these human–animal interactions under both commercial and experimental settings.

Fear is not the only emotional state that has been assessed and that may influence the farm animal's response to humans. As discussed earlier, animals may experience positive or pleasant emotions in the presence of humans which may arise from associating humans with rewarding events. Through conditioning, farm animals associate humans with rewarding events such as feeding, and show increased attraction to humans on the basis of approach behaviour (see Hemsworth, 2003). Some forms of human contact also appear to elicit positive emotional responses in farm animals.

For example, stroking applied in a manner that is similar to intraspecific allogrooming has been shown to reduce heart rate and result in relaxed body postures and increased approach to humans in dogs (McMillan, 1999), horses (Fig. 15.3; Lynch et al., 1974; Feh and de Mazières, 1993; McBride et al., 2004), cattle (Schmied et al., 2008), lambs (Tallet et al., 2005, 2008) and foals (Ligout et al., 2008). There is some evidence that farm animals which experience positive emotional states in the presence of humans may have reduced stress responses in stressful situations: social isolation stress is reduced in dogs and sheep that show a strong affinity to humans (Boivin et al., 2000; Topál et al., 2005; Palmer and Custance, 2008; Tallet et al., 2008). Previous positive handling also reduces heart rate and salivary cortisol concentrations in lambs following tail docking (Tosi and Hemsworth, 2002), and reduces restlessness and heart rate in cows undergoing rectal palpation (Waiblinger et al., 2004). However, the field of positive emotions in farm animals is still largely unexplored (Waiblinger et al., 2006; Boissy et al., 2007) and clearly requires further study because of the implications for animal welfare and ease of handling.

P.H. Hemsworth and X. Boivin

Fig. 15.3. Some forms of human contact, such stroking applied in a manner that is similar to intraspecific allo-grooming, appear to elicit positive emotional responses in animals.

15.5 General Development of Animal–Human Relationships and Implications for Animal Welfare

Fear is considered a powerful emotional state that normally gives rise to defensive behaviour or escape. In concert with these behavioural effects, fear normally activates the autonomic nervous system and the neuroendocrine system, which both assist the animal to meet physical or emotional challenges through their effects on regulatory mechanisms such as energy availability and use, and cardiac and respiratory functions (Hemsworth and Coleman, 1998). Gray (1987) recognized that fear may be triggered by environmental stimuli that are novel, have high intensity – such as loud and large stimuli, have special evolutionary dangers – such as heights, isolation or darkness, arise from social interaction – such as

contagious learning, or have previously been paired with aversive experiences.

As discussed in more detail later in the chapter, there is substantial evidence that the fear responses of farm animals to humans are affected by the history of the animals' interactions with humans, particularly the nature and quantity of human contact. Habituation will occur over time as the animal's fear of humans is gradually reduced by repeated exposure to humans in a neutral context; that is, the presence of the human has neither rewarding nor punishing elements. Over time, young domesticated animals that have had limited experience with humans may habituate to the presence of humans and so may perceive them as part of the environment and without any particular significance. Even wild strains of rats and deer that are highly fearful of humans will habituate to humans over time (Galef, 1970; Matthews, 1993).

Furthermore, conditioned approach–avoidance responses develop as a consequence of associations between the stockperson and aversive and rewarding elements of the handling bouts. This has been demonstrated, for example, by studies examining the effects of a range of handling treatments on the behaviour of pigs (Gonyou *et al.*, 1986; Hemsworth *et al.*, 1981a, 1986, 1987, 1996a; Hemsworth and Barnett, 1991). Pigs that were slapped or shocked with a battery-operated prodder whenever they approached, or failed to avoid the experimenter during daily handling bouts lasting 15 to 30 s, learned to associate the presence of the handler with the punishment of the handling bouts. In contrast, pigs that received pats or strokes during brief daily handling bouts subsequently showed increased approach to humans (Fig. 15.4). Moreover, there is some evidence that pigs may associate the rewarding experience of feeding with the handler and that this conditioning results in pigs being less fearful of humans (Hemsworth *et al.*, 1996b). Although there is some controversy over the mechanism by which avoidance behaviour becomes conditioned by punishment (Walker, 1987), it is well established that animals learn to avoid conditioned stimuli that are paired with aversive events. Thus, through conditioning, the behavioural responses of animals to humans may be regulated by the nature of the experiences occurring around the time of their interactions with humans.

Other factors, such as age, social environment and genetics can also modulate an animal's responses to humans. There is evidence in some species that

Fig. 15.4. Positive handling reduces fear responses to humans in farm animals.

handling at an early age, subsequent handling is also influential and has the potential to modify such early learning effects. The literature on early handling of rodents is very extensive, and studies involving brief removal of pre-weaned animals from their home cages and the associated handling generally report increased growth and accelerated development, reduced activity and defecation in open-field tests, improved performance in learning tasks and physiological stress responses of lower magnitude to subsequent stressors (Dewsbury, 1992). The results of these early handling studies have often been interpreted as a consequence of either direct stimulation or acute stress advancing the rate of development of some behavioural and physiological processes (Schaefer, 1968). These results therefore suggest that early handling effects on farm animals, including effects on fear of humans, may not necessarily be solely due to handling per se, but may in part be a consequence of acute stress early in life associated with the separation and handling involved in the handling treatment, maternal care after handling and perhaps also early weaning in some studies. Indeed, most of the studies on farm animals showing persistent effects of early human contact on the animals' subsequent responses to humans have been done with artificially fed animals (see review by Boivin *et al.*, 2003). Consequently, although handling at an early age may be highly influential (for a more detailed discussion of this, see Rushen *et al.*, 2001), subsequent human contact can modify early learning effects. Furthermore, subsequent human contact is probably necessary to maintain early contact effects (Boivin *et al.*, 2000). The two types of learning, conditioning and habituation to humans, occurring both early and later in life, are probably the most influential factors affecting the behavioural responses of farm animals to humans (Hemsworth, 2004).

Social learning also appears to influence the development of an animal's relationship with humans. Lyons *et al.* (1988) observed that the approach behaviour of goat kids to humans was affected by the behaviour of other goats, particularly the dam. Recently, Boivin *et al.* (2002) and Krohn *et al.* (2003) demonstrated that the effects of patting and talking in association with feeding lambs and calves on reducing their later fear responses to humans (based on approach and avoidance behaviour) were less effective when the dam was present, even if the lambs and calves were

the age at which handling occurs is influential. The development of fear responses towards unfamiliar objects during early infancy has been well described, for example, in dogs (Scott and Fuller, 1965) and poultry (ducks and chickens; Hess, 1959). These studies suggest that, in contrast to later in life, animals in their infancy show low fear responses towards unfamiliar humans. Recent studies of horses show similar effects (Lansade *et al.*, 2007). Human contact of a positive nature during early life or at weaning has been shown to have persistent effects on fear of humans in many farm animals. For example, studies on cattle, pigs, sheep and silver foxes show that handling of a positive nature both early in life and after weaning reduces the subsequent fear responses of juveniles, based on their approach and avoidance behaviour to humans (Pedersen and Jeppesen, 1990; Hemsworth and Barnett, 1992; Markowitz *et al.*, 1998, Boivin *et al.*, 1992, 2000; Krohn *et al.*, 2001). However in some studies on cattle and pigs, early handling effects were not very persistent (Boissy and Bouissou, 1988; Hemsworth and Barnett, 1992), indicating that although farm animals may be sensitive to artificially fed. In a similar manner, Henry *et al.* (2005, 2007) observed that the behaviour of dams and their foals towards humans was correlated and that rewarding the dams in the presence of the foals reduced the flight distance of foals towards humans up to a year later. Boivin *et al.* (2009) observed that the docility score of beef calves – based on their behavioural responses to humans – was positively related with the docility of their dams. These studies indicate that the development of the young animal's relationship with humans is not only influenced by its experience with humans but is also affected by the dam's response to humans.

In addition, differences in the behavioural responses of livestock to humans may reflect inherent genetic differences (see review by Boissy *et al.*, 2005). For example, Murphey *et al.* (1981) reported marked differences in the flight distance of *Bos indicus* and *Bos taurus* breeds of cattle to humans. Hearnshaw *et al.* (1979) and Burrow and Corbet (2000) found marked differences in the behaviour of crossbred Brahman cattle and British breeds to handling, and reported that the behavioural response to restraint in a squeeze chute (or stall) in the close presence of humans (which some have considered a measure of cattle temperament; Petherick *et al.*, 2002), is moderately heritable in *B. indicus* cattle. Similar results on the behavioural response of *B. taurus* cattle when handling in a corral (called the docility test) have been reported by Le Neindre *et al.* (1995), Gauly *et al.* (2001) and Phocas *et al.* (2007). These differences may, in part, reflect differences in the animals' fear of unfamiliar stimuli (neophobia), as selection for neophobia will more likely affect the general fearfulness of naive animals rather than their responses to specific novel stimuli (Price, 1984). Thus, differences in neophobia will affect the initial responses of naive animals to novel stimuli, such as humans, but over time experience with humans should modify these responses to the extent that they become stimulus specific. For example, there is evidence that the behavioural response of relatively naive pigs to humans is moderately heritable, but that subsequent experience with humans modifies the responses (Hemsworth *et al.*, 1990). Murphy and Duncan (1977, 1978) studied two stocks of chickens, termed 'flighty' and 'docile' on the basis of their behavioural responses to humans, and found that early handling affected the behavioural responses of both of these stocks of birds to humans. However, the docile birds did not necessarily show less withdrawal response to novel

stimuli, such as a mechanical scraper and an inflating balloon (Murphy, 1976). What is more, a series of studies by Jones and colleagues (Jones, *et al.*, 1991; Jones and Waddington, 1992) showed that regular handling predominantly affected the behavioural responses of birds to humans, rather than to novel stimuli such as a blue light. These data indicate that experience with humans results in stimulus-specific effects rather than in effects on general fearfulness.

As a result of this developmental process, there are statements in the scientific literature that humans can be considered in certain circumstances to be the social partners of domestic animals. That is, humans can be perceived by animals as appropriate social partners and part of their social organizational system. Indeed, this view is probably widespread in many companion animal (pet) owners. A common and important characteristic of domestic animals is their sociality and, as with humans, farm animals demonstrate species-specific communication and social organizational systems. Several authors have described human–animal relationships as 'social' relationships (Estep and Hetts, 1992; Scott, 1992; Rushen et al., 2001), even though in the strictest sense a 'social' system is defined as intra-specific (McFarland, 1990). Attachment behaviour is considered important in organizing social relationships (Bowlby, 1958), and the integration of humans as a social partner in the social behaviour of dogs from an early age was called socialization by Scott and Fuller (1965). Scott (1992) defined socialization as the building of a social relationship with conspecifics as well as with humans, essentially based on an attachment process and an adjustment to the other members of the group, particularly during sensitive periods of the animal's life. Kraemer (1992) considered that an important aspect of the socialization process is that attachment is an emotional concept in which reassurance is provided by the presence of the attachment object and, in contrast, distress behaviour is elicited when the attachment object is removed. Lott and Hart (1979) viewed the interactions between Fulani African nomads and their cattle when living permanently together as a two-species social system. Educational programmes for dog and cattle handlers often recommend that the handlers should behave as friends and leaders towards their animals but be perceived as the dominant member of the social group (Grandin, 2000; Rooney *et al.*, 2000; Krueger, 2007) (see also Fig. 15.5).

Fig. 15.5. Some authors suggest that humans can be considered in certain circumstances to be the social partners of domestic animals. (Photograph courtesy of Wageningen UR Communication Services.)

However, what scientific evidence do we have that humans are perceived as a social partner and even possibly as a conspecific? Pioneering research by Lorenz (1935) with geese and by Scott and Fuller (1965) with dogs investigated the imprinting or attachment processes of animals to humans during sensitive periods of life. The research reported earlier in this section on the persistent effects of human contact in early life or at weaning provides evidence of sensitive periods for human contact on the behavioural response of farm animals to humans. Recent research has utilized human–animal interactions to study social cognition in domestic animals by examining, for example, attachment in dogs (Topál *et al.*, 1998; Palmer and Custance, 2008) and sheep (Boivin *et al.*, 2000; Tallet *et al.*, 2008), the effects of humans on sheep during social isolation (Price and Thos, 1980; Boivin *et al.*, 2000) and on cattle during husbandry procedures (Waiblinger *et al.*, 2004), and the use of human cues by dogs (Miklósi *et al.*, 1998; Udell *et al.*, 2008). As discussed in the previous section, there is evidence that farm and companion animals may experience positive or pleasant emotions when stroked by humans in a manner similar to intraspecific allogrooming. However, these results do not necessarily mean that the human is integrated into the animal's social organization, such as its social hierarchy, or perceived as the leader of the group. Social organization in the wolf has been used as a general model for dog social organization, and this model has been extended by some to include the human–dog relationship. However, Bradshaw *et al.* (2008) suggest that interactions between dogs may be influenced more by their specific experiences with each other rather than by the overall social structure or hierarchy of the group. Thus, applying a single 'social' model to interpret human–dog relationships from the dog's perspective may be misleading. In a similar manner, the so-called 'round pen technique' of training horses has been proposed by some as training the horse to perceive the handler as a dominant member of the group (Roberts, 2002). The technique basically involves forcing the horse to run in the round pen until it stops and shows 'signs of submission' towards the handler. However, Krueger (2007) did not observe a generalization of this behaviour to the social behaviour displayed by horses in groups on pasture.

The first part of this chapter has considered the human–animal relationship from a theoretical perspective. This discussion has highlighted the important role and responsibility of the human in the development of the human–animal relationship. Furthermore, it is obvious that while the animal's perspective and role are important in this relationship, this critical aspect is less recognized. Indeed the animal's perspective and role have implications for animal welfare, as both positive and negative emotions are generated by human–animal interactions that directly affect animal welfare (Fraser, 1995). The final part of the chapter will illustrate, using evidence from both commercial and experimental settings, the consequence of the quality of the human–animal relationship for animal welfare and how humans can practically and significantly improve both the relationship and animal welfare through changes in their behaviour.

15.6 Effects of Human Contact on Animal Welfare

There are three main lines of evidence that demonstrate the implications of human contact for the welfare of farm animals: handling studies under controlled conditions; observed relationships in the field; and intervention studies in the field that target human contact.

15.6.1 Evidence from handling studies

Handling studies, particularly of pigs and poultry, indicate that negative or aversive handling, imposed briefly but regularly, will increase fear of humans and reduce the growth, feed conversion efficiency, reproduction and health of these animals (see reviews by Hemsworth and Coleman, 1998; Waiblinger et al., 2006; Hemsworth et al., 2009). A chronic stress response has been implicated in these effects on productivity because in many of the pig-handling studies (see Hemsworth and Coleman, 1998), handling treatments that resulted in high fear levels also produced either a sustained elevation in the basal free cortisol concentrations or an enlargement of the adrenal glands. Therefore, there are likely to be concerns for the welfare of farm animals that are negatively handled. Fear is generally considered an undesirable emotional state of suffering in both humans and animals (Jones and Waddington, 1992), and one of the key recommendations proposed to the UK Parliament by the Brambell Committee (1965) was that intensively housed livestock should be free from fear. There are several reasons why fear of humans will reduce

the welfare of farm animals. Research has shown that farm animals that are both highly fearful of humans and in regular contact with humans are likely to experience not only an acute stress response in the presence of humans, but also a chronic stress response (Hemsworth and Coleman, 1998). Fearful animals are also more likely to sustain injuries trying to avoid humans during routine inspections and handling. Besides, as discussed later, in situations where human contact is negative, the stockperson's attitude towards the animal is likely to be poor and thus the stockperson's commitment to the surveillance of and the attendance to welfare (and production) problems facing the animal may be deficient.

There is also evidence from other farm animal species that aversive handling has both productivity and welfare effects. For example, handling studies in dairy cattle show that aversive handling may depress milk yield in cows (Rushen et al., 1999; Breuer, 2000; Breuer et al., 2003). The results of the study by Rushen et al. (1999) implicate the secretion of catecholamines under the influence of the autonomic nervous system as affecting milk letdown, while the study by Breuer et al. (2003) found evidence of chronic stress in negatively handled heifers.

15.6.2 Evidence from field observations

Consistent findings of negative inter-farm correlations between fear of humans (assessed on the basis of the behavioural response to humans) and the productivity of dairy cattle, pigs and poultry (see Table 15.1) have stimulated research in the livestock industries to identify stockperson characteristics associated with these fear responses of farm animals to humans. Coleman et al. (1998) and Hemsworth et al. (1989) found that the use of a high proportion of negative tactile behaviours by stockpeople, such as slaps and hits, was correlated with increased avoidance by breeding sows of an experimenter in a standard approach test used to assess fear. In studies on dairy cows housed outdoors all year round on pasture, Breuer et al. (2000) and Hemsworth et al. (2000) found that the use of a high proportion of negative tactile interactions, such as slaps, pushes and hits, was associated with increased avoidance by cows of an experimenter in a standard approach test. Similarly, in a study of dairy cows in indoor farms, Waiblinger et al. (2002, 2003) found that positive stockperson

Table 15.1. Correlations between fear and animal productivity in the livestock industries.

Species	Study	Inter-farm correlations between fear of humans and productivity[a]
Pig	Hemsworth *et al.* (1981b)	−0.51*
	Hemsworth *et al.* (1989)	−0.55*
	Hemsworth *et al.* (1994a)	−0.01
Dairy cow	Breuer *et al.* (2000)	−0.46*
	Hemsworth *et al.* (2000)	−0.27
Meat chicken	Hemsworth *et al.* (1994b)	−0.57**
	Cransberg *et al.* (2000)	−0.10
	Hemsworth *et al.* (1996b)	−0.49*
Laying hen	Barnett *et al.* (1992)	−0.58**

[a] Significant correlations at $P < 0.05$ (*) and < 0.01 (**), respectively.

behaviours, such as talking, petting and touching, and negative behaviours, such as forceful slaps and hits and shouting, were respectively negatively and positively associated with avoidance behaviour of cows to an approaching experimenter. Lensink *et al.* (2001) studied stockperson and calf behaviour at 50 veal calf units and found that the frequency of positive behaviour towards calves by the stockperson, such as touching, petting and allowing calves to suck the stockperson's fingers, was negatively associated with avoidance of an approaching experimenter. In studies on commercial meat chickens, Hemsworth *et al.* (1994b) and Cransberg *et al.* (2000) found that the speed of movement by the stockperson was positively correlated with avoidance of an approaching experimenter by meat chickens. Edwards (2009) studied caged laying hens and found that the incidence of noise made by stockpeople, such as shouting and cleaning with an air hose or leaf blower, was associated with greater avoidance by hens of an approaching experimenter, while the times that stockpeople spent standing stationary and spent close to the birds' cages were associated with less avoidance by hens of an approaching experimenter.

A number of these studies on stockperson and animal behaviour have shown that the attitudes of stockpeople towards interacting with their animals are predictive of the behaviour of the stockpeople towards their animals. Questionnaires were used to assess the attitudes of the stockpeople on the basis of their beliefs about their behaviour and the behaviour of their animals. In general, positive attitudes to the use of petting and the use of verbal and physical effort to handle animals were

negatively correlated with the use of negative tactile interactions such as slaps, pushes and hits in the dairy (Breuer *et al.*, 2000; Hemsworth *et al.*, 2000; Waiblinger *et al.*, 2002) and pig (Hemsworth *et al.*, 1989; Coleman *et al.*, 1998) industries. Lensink *et al.* (2000) also found that a positive attitude to the sensitivity of calves to human contact was predictive of the frequency of positive behaviour used by stockpeople towards veal calves, while Edwards (2009) found that negative attitudes to the sensitivity of hens to human contact as well as negative general beliefs about hens were associated with more noise, faster speed of movement and less time spent stationary near the hens.

15.6.3 Evidence from intervention studies in the field

Studies in the dairy and pork industries (Coleman *et al.*, 2000; Hemsworth *et al.*, 1994a, 2002) have shown that cognitive-behavioural training, in which the key attitudes and behaviour of stockpeople are targeted, can successfully improve the attitudes and behaviour of stockpeople towards their animals, with consequent beneficial effects on animal fear and productivity.

In these intervention studies in the dairy (Hemsworth *et al.*, 2002) and pig industries (Coleman *et al.*, 2000; Hemsworth *et al.*, 1994a), targeting the key attitudes and behaviour of stockpeople that were previously found to be correlated with fear responses of cows and pigs to humans resulted in reductions in these fear responses. Furthermore, concurrent improvements in animal productivity were observed: there were

P.H. Hemsworth and X. Boivin

improvements in the milk yield of dairy cows and a marked tendency for an improvement in the reproductive performance of sows.

15.7 Opportunities to Improve Human–Animal Relationships

The results of the intervention studies cited above, taken in conjunction with previous research on the relationship between stockperson attitudes, stockperson behaviour, animal fear and animal productivity, and research on handling farm animals, provide evidence of causal relationships between these stockperson and animal variables. Moreover, this research provides a strong case for introducing stockperson training courses in the livestock industries that target the attitudes and behaviour of the stockperson.

Cognitive-behavioural techniques basically involve retraining stockpeople's behaviour by targeting both the beliefs that underlie the behaviour (attitude) and the behaviour in question, and then by maintaining these changed beliefs and behaviours (Hemsworth and Coleman, 1998). The intervention studies cited above demonstrate that this training is practical and effective for a wide range of stockpeople working in a variety of situations. Commercial multimedia training programmes for pig stockpeople called 'ProHand Pigs' and for dairy cattle stockpeople called 'ProHand Dairy' (Animal Welfare Science Centre, 2008) have been developed and validated from these cognitive-behavioural intervention techniques, and are currently being used in Australia, New Zealand and the USA. In Europe, similar training programmes, called 'Quality Handling' for stockpeople working with pigs, laying hens and dairy and beef cattle, have been developed within the European Union 6th Framework Welfare Quality programme (e.g. Ruis *et al.*, 2009).

Stockpeople clearly require a basic knowledge of both the requirements and behaviour of farm animals, and also must possess a range of well-developed husbandry and management skills to care for and manage those animals effectively. Therefore, while cognitive-behavioural training addressing the key attitudes and behaviour of stockpeople that affect animal fear is important in improving animal welfare, knowledge and skills training are also fundamental to improving the welfare of commercial livestock.

Stockperson selection may provide another opportunity to improve animal welfare. The potential value of selecting stockpeople using screening aids is illustrated by a study of stockpersons in the Australian pig industry (Coleman, 2001; Carless *et al.*, 2007). At the commencement of their employment, a total of 144 inexperienced stockpeople completed a set of computerized questionnaires that included measures of personality, motivation, turnover potential, performance potential, and attitudes and empathy towards pigs. After 6 months of employment, their behaviour towards pigs, technical knowledge, and work motivation and commitment were directly assessed, and their performance and conscientiousness were measured by means of a supervisor's report. The main findings were that some measures of stockperson characteristics taken in the initial interview were correlated with the performance measures taken 6 months later. A positive attitude towards pigs was correlated with subsequent behaviour towards pigs and with technical skills and knowledge. Empathy towards animals was correlated with subsequent behaviour and technical knowledge of the stockperson. A pre-employment measure of work reliability and job satisfaction was also found to be a good predictor of the work motivation, behaviour towards pigs and technical knowledge. The results from this study suggest that measures of attitude, empathy, work reliability and job satisfaction may be useful in assisting the selection of stockpeople who will perform well in the ways studied here.

15.8 Conclusions

- Human–animal relationships can be studied by investigating each partner's perception of the other. The quality of this relationship from the perspective of the farm animal has been assessed by measuring fear, or lack of it, in response to a familiar stockperson or to humans in general.
- Evidence on human–animal interactions in the livestock industries indicates that the history of such interactions leads to a stimulus-specific response of farm animals to humans. This response is affected by several emotional states, including positive and negative emotions towards humans.
- This chapter has highlighted the important role and responsibility of the human in the development of the human–animal relationship. Characteristics of the animal, such as age, experience, social environment and genetics, also affect their response to humans.

- Furthermore, the animal's perception of the relationship has implications for its welfare, as emotions in animals are generated by human interactions which directly impact on the animal's welfare.
- Understanding stockperson–farm animal relationships has implications for improving farm animal welfare and productivity. Stockperson attitudes are amenable to change, so stockperson training can improve human–animal relationships in the livestock industries.
- While understanding stockperson attitudes to farm animals is essential in improving human interactions with farm animals (Coleman, 2004), understanding of the relationships between other stockperson characteristics (such as empathy towards animals and job satisfaction) and stockperson behaviour is also important in improving human–animal relationships in the livestock industries.

References

Animal Welfare Science Centre (2008) ProHand Pigs and ProHand Dairy. Available at: http://www.animalwelfare. net.au/educate/phcc.pdf (accessed 3 January 2011).

Aureli, F. and Schaffner, C.M. (2002) Relationship assessment through emotional mediation. *Behaviour* 139, 393–420.

Barnett, J.L., Hemsworth, P.H. and Newman, E.A. (1992) Fear of humans and its relationships with productivity in laying hens at commercial farms. *British Poultry Science* 33, 699–710.

Boissy, A. and Bouissou, M.F. (1988) Effects of early handling on heifer's subsequent reactivity to humans and to unfamiliar situations. *Applied Animal Behaviour Science* 20, 259–273.

Boissy, A., Bouix, J., Orgeur, P., Poindron, P., Bibé, B. and Le Neindre, P. (2005) Genetic analysis of emotional reactivity in sheep: effects of the genotypes of the lambs and of their dams. *Genetics Selection Evolution* 37, 381–401.

Boissy, A., Manteuffel, G., Jensen, M.B., Moe, R.O., Spruijt, B., Keeling, L.J., Winckler, C., Forkman, B., Dimitrov, I., Langbein, J., Bakken, M., Veissier, I. and Aubert, A. (2007) Assessment of positive emotions in animals to improve their welfare. *Physiology and Behavior* 92, 375–397.

Boivin, X., Le Neindre, P. and Chupin, J.M. (1992) Establishment of cattle–human relationships. *Applied Animal Behaviour Science* 32, 325–335.

Boivin, X., Tournadre, H. and Le Neindre, P. (2000) Hand-feeding and gentling influence early-weaned lambs' attachment responses to their stockperson. *Journal of Animal Science* 78, 879–884.

Boivin, X., Boissy, A., Nowak, R., Henry, C., Tournadre, H. and Le Neindre, P. (2002) Maternal presence limits the effects of early bottle feeding and petting on lambs' socialisation to the stockperson. *Applied Animal Behaviour Science* 77, 168–159.

Boivin, X., Lensink, J., Tallet, C. and Veissier, I. (2003) Stockmanship and farm animal welfare. *Animal Welfare* 12, 479–492.

Boivin, X., Gilard, F. and Egal, D. (2009) The effect of early human contact and the separation method from the dam on responses of beef calves to humans. *Applied Animal Behaviour Science* 120, 132–139.

Bowlby, J. (1958) The nature of the child's tie to his mother. *International Journal of Psychoanalysis* 39, 350–373.

Bradshaw, J.W.S., Blackwell, E.J. and Casey, R.A. (2008) Dominance in domestic dogs: useful construct or bad habit? In: Boyle, L., O'Connell, N. and Hanlon, A. (eds) *Proceedings of the 42nd Conference of the ISAE [International Society for Applied Ethology] 'Applied Ethology: Addressing Future Challenges in Animal Agriculture'*, University College, Dublin, Ireland, 5–9 August 2008. Wageningen Academic Publishers, Wageningen, The Netherlands, p. 4 (abstract).

Brambell Committee (1965) *Report of the Technical Committee to Enquire into the Welfare of Animal Kept under Intensive Livestock Husbandry Systems.* Command Paper 2836, Her Majesty's Stationery Office, London.

Breuer, K. (2000) Fear and productivity in dairy cattle. PhD thesis, Monash University, Victoria, Australia.

Breuer, K., Hemsworth, P.H., Barnett, J.L., Matthews, L.R. and Coleman, G.J. (2000) Behavioural response to humans and the productivity of commercial dairy cows. *Applied Animal Behaviour Science* 66, 273–288.

Breuer, K., Hemsworth, P.H. and Coleman, G.J. (2003) The effect of positive or negative handling on the behavioural responses of nonlactating heifers. *Applied Animal Behaviour Science* 84, 3–22.

Burrow, H.M. and Corbet, N.J. (2000) Genetic and environmental factors affecting temperament of zebu and zebu-derived beef cattle grazed at pasture in the tropics. *Australian Journal of Agricultural Research* 51, 155–162.

Carless, S.A., Fewings-Hall, S., Hall, M., Hay, M., Hemsworth, P. and Coleman, G.J. (2007) Selecting unskilled and semi-skilled blue-collar workers: the criterion-related validity of the PDI-Employment Inventory. *International Journal of Selection and Assessment as an Information Exchange* 15, 335–340.

Clutton-Brock, J. (1994) The unnatural world: behavioural aspects of humans and animals in the process of domestication. In: Manning, A. and Serpell, J. (eds) *Animals and Human Society: Changing Perspectives.* Routledge, London and New York, pp. 23–35.

Coleman, G.J. (2001) *Selection of stockpeople to improve productivity*. In: *Proceedings of the 4th Industrial and Organisational Psychology Conference*, Sydney, 21–24 June, p. 30.

Coleman, G.J. (2004) Personnel management in agricultural systems. In: Rollin, B.E. and Benson, J. (eds) *Maximizing Well-being and Minimizing Suffering in Farm Animals*. Iowa State University Press, Ames, Iowa, pp. 167–181.

Coleman, G.C., Hemsworth, P.H., Hay, M. and Cox, M. (1998) Predicting stockperson behaviour towards pigs from attitudinal and job-related variables and empathy. *Applied Animal Behaviour Science* 58, 63–75.

Coleman, G.J., Hemsworth, P.H., Hay, M. and Cox, M. (2000) Modifying stockperson attitudes and behaviour towards pigs at a large commercial farm. *Applied Animal Behaviour Science* 66, 11–20.

Cransberg, P.H., Hemsworth, P.H. and Coleman, G.J. (2000) Human factors affecting the behaviour and productivity of commercial broiler chickens. *British Poultry Science* 41, 272–279.

de Passillé, A.M.B. and Rushen, J. (2005) Can we measure human–animal interactions in on-farm welfare assessment? Some unresolved issues. *Applied Animal Behaviour Science* 92, 193–209.

Denis, B. (2004) Broadening concepts of domestication. *INRA Productions Animales* 17, 161–166.

Désiré, L., Boissy, A. and Veissier, I. (2002) Emotions in farm animals: a new approach to animal welfare in applied ethology. *Behavioural Processes* 60, 81–85.

Désiré, L., Veissier, I., Després, G., Delval, E., Toporenko, G. and Boissy, A. (2006) Appraisal process in sheep (*Ovis aries*): interactive effect of suddenness and unfamiliarity on cardiac and behavioral responses. *Journal of Comparative Psychology* 120, 280–287.

Dewsbury, D.A. (1992) Studies of rodent–human interactions in animal psychology. In: Davis, H. and Balfour, D. (eds) *The Inevitable Bond: Examining Scientist–Animal Interactions*. Cambridge University Press, Cambridge, UK, pp. 27–43.

Digard, J.-P. (1990) *L'Homme et les Animaux Domestiques, Anthropologie d'une Passion*. Collection: Le Temps des Sciences, Editions Fayard, Paris.

Edwards, L.E. (2009) The human–animal relationship in the laying hen. PhD thesis, University of Melbourne, Victoria, Australia.

Estep, D.Q. and Hetts, S. (1992) Interactions, relationships, and bonds: the conceptual basis for scientist–animal relations, In: Davis, H. and Balfour, A.D. (eds) *The Inevitable Bond: Examining Scientist–Animal Interactions*. Cambridge University Press, Cambridge, UK, pp. 6–26.

Feh, C. and de Mazières, J. (1993) Grooming at a preferred site reduces heart rate in horses. *Animal Behaviour* 46, 1191–1194.

Fraser, D. (1995) Science, values and animal welfare: exploring the inextricable connection. *Animal Welfare* 4, 103–117.

Galef, B.G. Jr (1970) Aggression and timidity: responses to novelty in feral Norway rats. *Journal of Comparative and Physiological Psychology* 70, 370–381.

Gauly, M., Mathiak, H., Hoffmann, K., Kraus, M. and Erhardt, G. (2001) Estimating genetic variability in temperamental traits in German Angus and Simmental cattle. *Applied Animal Behaviour Science* 74, 109–119.

Gonyou, H.W., Hemsworth, P.H. and Barnett, J.L. (1986) Effects of frequent interactions with humans on growing pigs. *Applied Animal Behaviour Science* 16, 269–278.

Grandin, T. (2000) Behavioural principles of handling cattle and other grazing animals under extensive conditions. In: Grandin, T. (ed.) *Livestock Handling and Transport*. CAB International, Wallingford, UK, pp. 63–85.

Gray J.A. (1987) *The Psychology of Fear and Stress*, 2nd edn. Cambridge University Press, Cambridge, UK.

Hale, E.B. (1969) Domestication and the evolution of behaviour. In: Hafez, E.S.E. (ed.) *The Behavior of Domestic Animals*. The Williams and Wilkins Company, Baltimore, Maryland, pp.23–42.

Hearnshaw, H., Barlow, R. and Want, G. (1979) Development of a "temperament" or "handling difficulty" score for cattle, *Proceedings of the Inaugural Conference of Australian Animal Breed Genetics* 1, 164–166.

Hemsworth, P.H. (2003) Human–animal interactions in livestock production. *Applied Animal Behaviour Science* 81, 185–198.

Hemsworth, P.H. (2004) Human–livestock interaction. In: Benson, G.J. and Rollin, B.E. (eds) *The Well-Being of Farm Animals, Challenges and Solutions*. Blackwell Publishing, Ames, Iowa, pp. 21–38.

Hemsworth, P.H. (2007) Ethical stockmanship. *Australian Veterinary Journal* 85, 194–200.

Hemsworth, P.H. and Barnett, J.L. (1991) The effects of aversively handling pigs, either individually or in groups, on their behaviour, growth and corticosteroids. *Applied Animal Behaviour Science* 30, 61–72.

Hemsworth, P.H. and Barnett, J.L. (1992) The effects of early contact with humans on the subsequent level of fear of humans by pigs. *Applied Animal Behaviour Science* 35, 83–90.

Hemsworth, P.H. and Coleman, G.J. (1998) *Human–Livestock Interactions: The Stockperson and the Productivity and Welfare of Intensively-farmed Animals*. CAB International, Wallingford, UK.

Hemsworth, P.H., Barnett, J.L. and Hansen, C. (1981a) The influence of handling by humans on the behaviour, growth and corticosteroids in the juvenile female pig. *Hormones and Behavior* 15, 396–403.

Hemsworth, P.H., Brand, A. and Willems, P.J. (1981b) The behavioural response of sows to the presence of human beings and their productivity. *Livestock Production* Science 8, 67–74.

Hemsworth, P.H., Barnett, J.L. and Hansen, C. (1986) The influence of handling by humans on the behaviour, reproduction and corticosteroids of male and female pigs. *Applied Animal Behaviour Science* 15, 303–314.

Hemsworth, P.H., Barnett, J.L. and Hansen, C. (1987) The influence of inconsistent handling by humans on the behaviour, growth and corticosteroids of young pigs. *Applied Animal Behaviour Science* 17, 245–252.

Hemsworth, P.H., Barnett, J.L., Coleman, G.J. and Hansen, C. (1989) A study of the relationships between the attitudinal and behavioural profiles of stockpersons and the level of fear of humans and reproductive performance of commercial pigs. *Applied Animal Behaviour Science* 23, 301–314.

Hemsworth, P.H., Barnett, J.L., Treacy, D. and Madgwick, P. (1990) The heritability of the trait fear of humans and the association between this trait and subsequent reproductive performance of gilts. *Applied Animal Behaviour Science* 25, 85–95.

Hemsworth, P.H., Coleman, G.J. and Barnett, J.L. (1994a) Improving the attitude and behaviour of stockpersons towards pigs and the consequences on the behaviour and reproductive performance of commercial pigs. *Applied Animal Behaviour Science* 39, 349–362.

Hemsworth, P.H., Coleman, G.J., Barnett, J.L. and Jones, R.B. (1994b) Fear of humans and the productivity of commercial broiler chickens. *Applied Animal Behaviour Science* 41, 101–114.

Hemsworth, P.H., Barnett, J.L. and Campbell, R.G. (1996a) A study of the relative aversiveness of a new daily injection procedure for pigs. *Applied Animal Behaviour Science* 49, 389–401.

Hemsworth, P.H., Coleman, G.C., Cransberg, P.H. and Barnett, J.L. (1996b) *Human Factors and the Productivity and Welfare of Commercial Broiler Chickens.* Research Report on Chicken Meat Research and Development Council Project, Attwood, Victoria, Australia.

Hemsworth, P.H., Coleman, G.J., Barnett, J.L and Borg, S. (2000) Relationships between human–animal interactions and productivity of commercial dairy cows. *Journal of Animal Science* 78, 2821–2831.

Hemsworth, P.H., Coleman, G.J., Barnett, J.L., Borg, S. and Dowling, S. (2002) The effects of cognitive behavioral intervention on the attitude and behavior of stockpersons and the behavior and productivity of commercial dairy cows. *Journal of Animal Science* 80, 68–78.

Hemsworth, P.H., Barnett, J.L. and Coleman, G.J. (2009) The integration of human–animal relations into animal welfare monitoring schemes. *Animal Welfare* 18, 335–345.

Henry, S., Hemery, D., Richard, M.-A. and Hausberger, M. (2005) Human–mare relationships and behaviour of foals toward humans. *Applied Animal Behaviour Science* 93, 341–362.

Henry, S., Briefer, S., Richard-Yris, M.A. and Hausberger, M. (2007) Are 6-month-old foals sensitive to dam's influence? *Developmental Psychobiology* 49, 514–521.

Hess, E.H. (1959) Imprinting: an effect of early experience, imprinting determines later social behavior in animals. *Science* 130, 133–141.

Hinde, R.A. (1976) Interactions, relationships and social structure. *Man* 11, 11–17.

Jones, R.B. and Waddington, D. (1992) Modification of fear in domestic chicks, *Gallus gallus domesticus* via regular handling and early environmental enrichment. *Animal Behaviour* 43, 1021–1033.

Jones, R.B., Mills, A.D. and Faure, J.M. (1991) Genetic and experimental manipulation of fear-related behaviour in Japanese quail chicks *(Coturnix coturnix japonica). Journal of Comparative Psychology* 105, 15–24.

Kraemer, G.W. (1992) A psychobiological theory of attachment. *Behavioural Brain Science* 15, 493–541.

Kretchmer, K.R. and Fox, M.W. (1975) Effects of domestication on animal behaviour. *Veterinary Record* 96, 102–108.

Krohn, C.C., Jago, J.G. and Boivin, X. (2001) The effect of early handling on the socialisation of young calves to humans. *Applied Animal Behaviour Science* 74, 121–133.

Krohn, C.C., Boivin, X. and Jago, J.G. (2003) The presence of the dam during handling prevents the socialization of young calves to humans. *Applied Animal Behaviour Science* 80, 230–237.

Krueger, K. (2007) Behaviour of horses in the "Round pen technique". *Applied Animal Behaviour Science* 104, 162–170.

Lansade, L., Bouissou, M.F. and Boivin, X. (2007) Temperament in preweaning horses: development of reactions to humans and novelty, and startle responses. *Developmental Psychobiology* 49, 501–513.

Le Neindre, P., Trillat, G., Sapa, J., Ménissier, F., Bonnet, J.N. and Chupin, J.M. (1995) Individual differences in docility in Limousin cattle. *Journal of Animal Science* 73, 2249–2253.

Lensink, J., Boissy, A. and Veissier, I. (2000) The relationship between farmers' attitude and behaviour towards calves, and productivity of veal units. *Annales de Zootechnie* 49, 313–327.

Lensink, B.J., Veissier, I. and Florland, L. (2001) The farmers' influence on calves' behaviour, health and production of a veal unit. *Animal Science* 72, 105–116.

Ligout, S., Bouissou, M.F. and Boivin, X. (2008) Comparison of the effects of two different handling methods on the subsequent behaviour of Anglo–Arabian foals toward humans and handling. *Applied Animal Behaviour Science* 113, 175–188.

Lorenz, K. (1935) Der Kumpan in der Umwelt des Vogels. *Zeitschrift Ornithology* 83, 289–413.

Lott, D.F. and Hart, B.L. (1979). Applied ethology in a nomadic cattle culture. *Applied Animal Ethology* 5, 309–319.

Lynch, J.J., Fregin, G.F., Mackie, J.B. and Monroe, R.R. Jr (1974) Heart rate changes in the horse to human contact. *Psychophysiology* 11, 472–478.

Lyons, D.M., Price, E.O. and Moberg, G.P. (1988) Social modulation of pituitary–adrenal responsiveness and individual differences in behavior of young domestic goats. *Physiology and Behavior* 43, 451–458.

Markowitz, T.M., Dally, M.R., Gursky, K. and Price, E.O. (1998) Early handling increases lamb affinity for humans. *Animal Behaviour* 55, 573–587.

Matthews, L.R. (1993) Deer handling and transport. In: Grandin, T. (ed.) *Livestock Handling and Transport*. CAB International, Wallingford, UK, pp. 253–272.

McBride, S.D., Hemmings, A. and Robinson, K. (2004) A preliminary study on the effect of massage to reduce stress in the horse. *Journal of Equine Veterinary Science* 24, 76–81.

McFarland, D. (1990) *The Oxford Companion to Animal Behaviour*. Oxford University Press, Oxford, UK.

McMillan, F.D. (1999) Effects of human contact on animal health and well-being. *Journal of the American Veterinary Medical Association* 215, 1592–1598.

Miklósi, Á., Polgárdi, R., Topál, J. and Csányi, V. (1998) Use of experimenter given cues in dogs. *Animal Cognition* 1, 113–121.

Murphey, R.M., Moura Duarte, F.A. and Torres Penendo, M.C. (1981) Responses of cattle to humans in open spaces: breed comparisons and approach–avoidance relationships. *Behaviour Genetics* 2, 37–47.

Murphy, L.B., (1976) A Study of the behavioural expression of fear and exploration in two stocks of domestic fowl. PhD Dissertation, Edinburgh University, UK.

Murphy, L.B. and Duncan, L.J.H. (1977) Attempts to modify the responses of domestic fowl towards human beings. 1. The association of human contact with a food reward. *Applied Animal Ethology* 3, 321–334.

Murphy, L.B. and Duncan, L.J.H. (1978) Attempts to modify the responses of domestic fowl towards human beings. II. The effect of early experience. *Applied Animal Ethology* 4, 5–12.

Palmer, R. and Custance, D. (2008) A counterbalanced version of Ainsworth's strange situation procedure reveals secure-base effects in dog–human relationships. *Applied Animal Behaviour Science* 109, 306–319.

Panksepp, J. (2005) Affective consciousness: core emotional feelings in animals and humans. *Consciousness and Cognition* 14, 30–80.

Pedersen, V. and Jeppesen, L.L. (1990) Effects of early handling on better behaviour and stress responses in the silver fox (*Vulpes vulpes*). *Applied Animal Behaviour Science* 26, 383–393.

Petherick, J.C., Holroyd, R.G., Doogan, V.J. and Venus, B.K. (2002) Productivity, carcass and meat quality of lot-fed *Bos indicus* cross steers grouped according to temperament. *Australian Journal of Experimental Agriculture* 42, 389–398.

Phocas, F., Boivin, X., Sapa, J., Trillat, G., Boissy, A. and Le Neindre, P. (2007) Genetic correlations between temperament and breeding traits in Limousin heifers. *Animal Science* 82, 805–811.

Price, E.O. (1984) Behavioral aspects of animal domestication. *The Quarterly Review of Biology* 59, 1–32.

Price, E.O. (1999) Behavioral development in animals undergoing domestication. *Applied Animal Behaviour Science* 65, 245–271.

Price, E.O. (2002) *Animal Domestication and Behavior*. CAB International, Wallingford, UK.

Price, E.O. and Thos, J. (1980) Behavioral responses to short-term social isolation in sheep and goats. *Applied Animal Ethology* 6, 331–339.

Roberts, M. (2002) *Horse Sense for People: The Man Who Listens to Horses Talks to People*. Penguin Group, New York.

Rooney, N.J., Bradshaw, J.W.S. and Robinson, I.H. (2000) A comparison of dog–dog and dog–human play behaviour. *Applied Animal Behaviour Science* 66, 235–248.

Ruis, M., Coleman, G.J., Waiblinger, S. and Boivin, X. (2009) A multimedia-based cognitive-behavioural intervention programme improves the attitude of stockpeople to handling pigs. In: Scientific Committee of the ISAE (eds) *Proceedings of 43rd Congress of the ISAE [International Society of Applied Ethology] 'Applied Ethnology for Contemporary Animal Issues'*, Cairns, Queensland, Australia, 6–10 July 2009. Organising Committee of the 43rd ISAE Congress, Wageningen, The Netherlands, p. 139 (abstract).

Rushen, J., de Passillé, A.M.B. and Munksgaard, L. (1999) Fear of people by cows and effects on milk yield, behaviour and heart rate at milking. *Journal of Dairy Science* 82, 720–727.

Rushen, J., de Passillé, A.M., Munksgaard, L. and Tanida, H. (2001) People as social actors in the world of farm animals. In: Keeling, L.J. and Gonyou, H.W. (eds) *Social Behaviour in Farm Animals*. CAB International, Wallingford, UK, pp. 353–372.

Schaefer, T. (1968) Some methodological implication of the research on "early handling" in the rat. In: Newton, G. and Levine, S. (eds) *Early Experience and Behaviour. The Psychobiology of Development*. Charles C. Thomas Publisher, Springfield, Illinois, pp. 102–141.

Schmied, C., Waiblinger, S., Scharl, T., Leisch, F. and Boivin, X. (2008) Stroking of different body regions by a human: effects on behaviour and heart rate of dairy cows. *Applied Animal Behaviour Science* 109, 25–38.

Scott, J.P. (1992) The phenomenon of attachment in human–non human relationships. In: Davis, H. and Balfour, A.D. (eds) *The Inevitable Bond: Examining Scientist–Animal Interactions*. Cambridge University Press, Cambridge, UK, pp. 72–92.

Scott, J.P. and Fuller, J.L. (1965) *Genetics and the Social Behavior of the Dog*. University of Chicago Press, Chicago, Illinois.

Serpell, J. (1986) *In the Company of Animals: A Study of Human–animal Relationships*. Blackwell, Oxford, UK.

Stricklin, W.R. and Mench, J.A. (1987) Social organization. In: Price, E.O. (ed.) *Farm Animal Behavior. The Veterinary Clinics of North America, Food Animal Practice, Volume 3(2)*.W.B. Saunders, Philadelphia, Pennsylvania, pp. 307–322.

Tallet, C., Veissier, I. and Boivin, X. (2005) Human contact and feeding as rewards for the lamb's affinity to their stockperson. *Applied Animal Behaviour Science* 94, 59–73.

Tallet, C., Veissier, I. and Boivin, X. (2008) Temporal association between food distribution and human caregiver presence and the development of affinity to humans in lambs. *Developmental Psychobiology* 50, 147–159.

Topál, J., Miklósi, A., Csányi, V. and Doka, A. (1998) Attachment behavior in dogs (*Canis familiaris*): a new application of Ainsworth's (1969) strange situation test. *Journal of Comparative Psychology* 112, 219–229.

Topál, J., Gácsi, M., Miklósi, A., Virányi, Z., Kubinyi, E. and Csányi, V. (2005) Attachment to humans: a comparative study on hand-reared wolves and differently socialized dog puppies. *Animal Behaviour* 70, 1367–1375.

Tosi, M.V. and Hemsworth, P.H. (2002) Stockperson–husbandry interactions and animal welfare in the extensive livestock industries. In: Koene, P. and the Scientific Committee of the 36th ISAE Congress (eds) *Proceedings of the 36th Congress of the ISAE [International Society for Applied Ethology]*, Egmond aan Zee, The Netherlands, 6–10 August 2002. Scientific Committee of the 36th ISAE Congress, Wageningen, The Netherlands, p. 129 (abstract).

Udell, M.A.R., Dorey N.R. and Wynne, C.D.L. (2008) Wolves outperform dogs in following human social cues. *Animal Behaviour* 76, 1767–1773.

Waiblinger, S., Menke, C. and Coleman, G. (2002) The relationship between attitudes, personal characteristics and behavior of stockpeople and subsequent behavior and production of dairy cows. *Applied Animal Behaviour Science* 79, 195–219.

Waiblinger, S., Menke, C. and Fölsch, D.W. (2003) Influences on the avoidance and approach behaviour of dairy cows towards 35 farms. *Applied Animal Behaviour Science* 84, 23–39.

Waiblinger, S., Menke, C., Korff, J. and Bucher, A. (2004) Previous handling and gentle interactions affect behaviour and heart rate of dairy cows during a veterinary procedure. *Applied Animal Behaviour Science* 85, 31–42.

Waiblinger, S., Boivin, X., Pedersen, V., Tosi, M.-V., Janczak, A.M., Visser, E.K. and Jones, R.B. (2006) Assessing the human–animal relationship in farmed species: a critical review. *Applied Animal Behaviour Science* 101, 185–242.

Walker, S. (1987) *Animal Learning: An Introduction*. Routledge and Kegan, London.

Zeder, M.A., Emshwiller, E., Smith, B.D. and Bradley, D.G. (2006) Documenting domestication: the intersection of genetics and archaeology. *Trends in Genetics* 22, 139–155.

16 Genetic Selection

PAUL M. HOCKING, RICHARD B. D'EATH AND JOERGEN B. KJAER

Abstract

Welfare is affected by the genetic inheritance that an animal receives and the environment in which it is kept. The focus of this chapter is on the contribution of genetics to animal welfare and how this may be harnessed to improve the welfare of animals. The genetic structure of a domestic breed is determined by its early history and the existence of a hierarchical breeding structure that combine to increase homozygosity. This may decrease health and welfare, as well as productivity, through inbreeding depression and by increasing the prevalence of genetic disease. In most farm animals, these processes are alleviated by the use of crossbreeding, which should be more widely practised in companion animals. Genetic selection may adversely affect welfare through inappropriate selection criteria, as in many breeds of dog, either inadvertently because of unexpected genetic correlations, or as a result of the neglect of important fitness traits. Many factors constrain selection by animal breeders – such as time, the number of potential traits, negative genetic correlations between economic and welfare traits, the costs of measuring and selecting animals and shifting market requirements. Examples of welfare problems with a genetic basis that have potential for genetic selection are considered under two headings: behavioural problems, and skeletal and physiological disease. New opportunities for genetic improvement of animal welfare based on DNA markers, new electronic measurement techniques and improved statistical procedures are likely to make selection for welfare traits more effective in the near future.

16.1 Introduction

All traits that are measured or characteristic of an animal are the result of the combined effects of the animal's genetic inheritance and the environment in which it is raised and kept, and traits affecting animal welfare are no exception. The focus of this chapter is on the structures and constraints in which animal breeders operate and the role that genetic selection has had in affecting animal welfare in domestic farm and companion species. Finally, the application of genetic principles to improve animal welfare is addressed with examples of how selection has and could be harnessed to improve the experiences of animals that are cared for by human beings. Although our discussion will consider genetic aspects of welfare in domestic animals in general, we do not address laboratory animals or the important area of transgenic animals. A transgenic animal carries a foreign gene that has either been deliberately inserted into its genome, or has inherited a transgene from its parents. Transgenic modification can produce unwanted changes that affect the welfare of animals because the gene may be harmful or cause undesirable side effects (Bruce and Bruce, 1998).

16.2 The Genetic Structure of Domestic Breeds

Individual members of a breed are obviously more closely related to each other than to those of any other breed. A consequence of this relatedness is a greater degree of homozygosity, or the likelihood that the genes inherited from an animal's sire and dam are identical to each other by descent from a common ancestor. This form of inbreeding is useful to distinguish a breed from other breeds in terms of coat colour or behavioural and production characteristics, but it also affects disease-causing genes and can lead to a phenomenon known as inbreeding depression. Some examples of both are given in Tables 16.1 and 16.2 respectively, and the effect of inbreeding depression on body weight in sheep is dramatically illustrated by the two ewes shown in Fig. 16.1.

Summers *et al.* (2010) summarized over 300 genetic diseases in British pedigree dogs caused by recessive (71%) or dominant (11%) autosomal and sex-linked (10%) genes that are all more prevalent in dogs that are inbred. Dog breeders sometimes mate a dog to his daughter or

Table 16.1. Consequences of inbreeding on the appearance of genetically recessive diseases (% occurrence) in 3006 inbred and 1693 outbred pigs from the same genetic base. Inbred lines were created by successive generations of brother–sister matings leading to inbreeding coefficients (i.e. predicted homozygosity) of 25–50%. (From Donald, 1955.)

Abnormality	Inbred	Outbred
Kinky tail	4.6	2.1
Skeletal deformities	1.1	0.4
Hernia	1.5	0.4
Cryptorchids	1.3	0.7
Other	0.8	0.9
Total	9.4	4.4

Table 16.2. The effects of a 10% increase in inbreeding (the probability of homozygosity) on phenotypic performance in farm animals (inbreeding depression). (From Falconer, 1964, p. 249.)

Trait	Decrease in units	% (mean)
Lactation yield (cattle)	−13.5 kg	3.2
Litter size (pigs)	−0.4 pigs	4.6
Weight (154 days, pigs)	−1.6 kg	2.7
Fleece weight (sheep)	−0.3 kg	5.5
Egg numbers (chickens)	−9.3 eggs	6.2
Hatchability (chickens)	−4.4%	6.4

granddaughter, resulting respectively in an increase in the inbreeding coefficient of 25% and 12.5%, ignoring other close relatives and historical inbreeding (see below). These proportions are estimates of the probability that a recessive gene in the ancestor is homozygous in the offspring and will result in genetic disease, and mating decisions like these compromise the future health and welfare of animals.

The deleterious genetic consequences of inbreeding occur even in the absence of mating between closely related individuals because of two factors that apply to all domestic breeds. First, there are relatively few common ancestors in the founding population of modern breeds in terms of the pedigree ancestors of contemporary animals. As a consequence the 'sample' of genes arising from the foundation of the breed may be very small. In thoroughbred horses, for example, 78% of genes can be traced to 27 stallions and three mares (Cunningham *et al.*, 2001). This represents a genetic 'bottleneck' that can only be corrected by crossing different breeds.

Secondly, pedigree breeding has a hierarchical structure in which a few popular breeders sell and exchange breeding stock among themselves, resulting in a breeding pyramid (Fig. 16.2). The effect of this phenomenon is well illustrated by an analysis of Finnish Kennel Club pedigree records, which showed that the parents of registered

Fig. 16.1. Inbreeding depression in Cheviot × Welsh Mountain sheep at 10 months of age. The two ewes were sired by the same ram; the one on the left had an inbreeding coefficient of 60% and was 40% of the body weight of the ewe on the right, which was not inbred. (From Wiener and Haytor, 1974.)

P.M. Hocking *et al.*

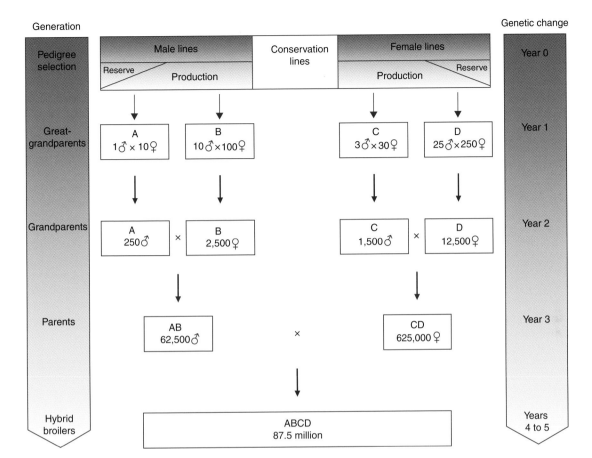

Fig. 16.2. Multiplication pyramid for a commercial broiler line. The box at the top of the figure corresponds to the pedigree selection level and is followed by the multiplication (crossing) phase that consists of a number of generations (left). The numbers in the boxes are the minimal numbers of birds in line A, and the corresponding numbers in lines B, C and D, that are required to generate hybrid broiler parent hens. The number of broilers that result from this process (lowest level) illustrates the power of the multiplication process to generate very large numbers of broilers from relatively few great-grandparents. The time required for (usually small, incremental) genetic changes at the pedigree level to appear in commercial flocks is indicated on the right of the figure. The numbers of male and female birds in each line and generation are based on the numbers of male and female chicks from male line, female line and parent broilers in commercial flocks (reported in Hocking and McCorquodale, 2008).

progeny were sired by only 10% of registered dogs (some with more than 300 registered offspring) and 20% of bitches (Mäki *et al.*, 2001). Moreover, animals that were more inbred had more hip and elbow dysplasia. An extreme example of this breeding pyramid is the Holstein–Friesian dairy cow: most dairy cattle worldwide trace their ancestry to the USA and are bred by artificial insemination (AI). The use of AI creates a very short loop from the top of the breeding pyramid to commercial herds of dairy cows, and

the extensive use of popular bulls leads to large numbers of related cows. The best bulls are also used to breed sons for subsequent evaluation (of female progeny) and, in time, the probability of a recessive gene appearing in homozygous form is increased. It is not surprising, therefore, that genetic defects occasionally appear in the Holstein population. Recently, a high incidence of malformed calves, abortions, stillbirths and poor fertility occurred in the US Holstein population, a syndrome collectively known as Complex

Vertebral Malformation (CVM). This was subsequently shown to be caused by a recessive gene. The inheritance of the gene was traced to a bull called Ivanhoe Bell and his sire Ivanhoe Star, which were widely used in AI. Paradoxically one of the advantages of the extensive use of relatively few bulls in AI is that such genes can be recognized and, if a genetic test is available, eliminated from the dairy herd. With the exception of the Holstein breed, the effective (genetic) size of commercial breeds of farm animals is at least as large as it was under traditional pedigree breeding systems because animal breeders have large breeding populations and avoid mating close relatives.

16.3   Undoing Genetic Inbreeding

Crossing two breeds maximizes heterozygosity and minimizes the chance of the animal suffering from a disease caused by a recessive gene in the homozygous state. Crossbreeding also has the opposite effect of inbreeding, increasing lifespan and reproductive function and having a positive effect on many production traits. In those farm species in which reproduction is a key component (beef or pork production and egg laying) impressive increases in overall productivity can be obtained, a phenomenon known as hybrid vigour or heterosis, and the majority of farm animals are crossbreds. For example, crosses of three breeds of beef cow of similar size and characteristics (Shorthorn, Angus and Hereford) in the USA led to an increase of 30% of weaned calf live weight, with most of the increased performance coming from the two-breed cross cow (Cundiff *et al.*, 1992). One caveat to this general rule has to be stated in that the relative advantage of a specific cross over pure-breds cannot be predicted and there may be no overall benefit. Crossbreeding is not generally practised in dairy cattle because the milk production of a crossbred is lower than that of the pure-bred Holstein and, in economic terms, the value of the milk compensates for the effects of the poor health and fertility of the breed. A recent report from an expert committee of scientists recommends that there should be increased emphasis on robustness in genetic selection to reduce the welfare insults caused by overemphasis on milk yield (Algers *et al.*, 2009).

Finally, we note that the advantages of crossbreeding are not routinely realized in companion animals because of the unhealthy pursuit of specific aesthetic appearance in different breeds of dogs and cats. From the perspective of improving animal welfare this neglect of the benefits of crossing breeds is unjustified.

16.4   Genetic Selection

The most extreme example of genetic selection by humans on changing a species is represented by the different breeds of dog. The pursuit of different breed characteristics in itself might be thought of as benign, but a recent report showed that the breed standard adversely affected welfare in each of the 50 most popular breeds in the UK (Asher *et al.*, 2009). For example, breeding King Charles cocker spaniels (Fig. 16.3) for small heads has led to a disorder in which the skull is too small for the brain, causing pain and fits; likewise, many bulldogs have been bred for narrow hips and cannot give birth naturally, leading to unnecessary Caesarean

Fig. 16.3. Selection for a small head in the King Charles cocker spaniel (pictured) has resulted in a skull that is too small for the brain. The brain protrudes out of its proper place and puts pressure on the spinal column, resulting in pain in the neck region. It may also cause scoliosis (twisted spine), limb weakness and ataxia. As many as half of all dogs of the breed suffer from this disorder and several other diseases related to line breeding to particular show winners (i.e. mating close relatives). The breed is an example of the problems associated with selection for a trait that has an unforeseen correlated response, and inbreeding that is exacerbated by a genetic 'bottleneck' at the founding of the breed. (Photograph courtesy of Dr K. Summers.)

P.M. Hocking *et al.*

sections. This type of selection is unacceptable and should lead to a radical re-evaluation of breed standards to reflect functional attributes that enhance rather than diminish health and welfare.

Genetic selection has been extensively used in farm animals from the middle of the 20th century to effect substantial changes in performance. When combined with improved nutrition and good husbandry, the benefits of these changes are economically and environmentally significant. The benefits of genetic selection in broiler chickens, for example, have resulted in less than half the food (and therefore half the land and fertilizer) required to produce the same weight of chicken in 2009 compared with 1959. However, while recognizing that housing and husbandry play a part, it is also widely believed that genetic selection has adversely affected the welfare of farm animals. In the following sections, we will examine some of these problems and suggest how genetic selection can be, and in some cases has been, used to provide solutions to welfare problems. Before that, we will explain briefly the constraints under which animal breeders operate and how these affect the way that animal breeders manage their selection programmes.

16.5 Factors Constraining Genetic Selection

The rate at which a trait such as broiler growth rate increases in a selection programme depends on how 'good' the selected animals are relative to the other chickens in the flock, and how much of this superior performance is passed on to the next generation. This depends on the heritability of the trait, which is the proportion of the overall variation in a trait that can be explained by genes that act additively (the remaining variation is due to environment and gene interactions). Trait measurement, mating and breeding the next generation take time, and the quicker the selection is performed, and the next generation is produced, the faster will be the annual rate of change. In a modern breeding programme, all sources of information from relatives will be used as this improves the accuracy of prediction of each animal's breeding value. In the extreme case, such as a sex-limited trait (e.g. egg production or milk yield) all the information for the selection of males comes from their daughters and female relatives. Finally, genetic change is 'diluted' if lots of traits are used in selection decisions. Breeders therefore try to

focus on relatively few traits in each population. The effects of negative correlations between commercially important traits on the rate of change are decreased by selecting for alternative objectives in different lines. This is usually the case for reproductive traits such as litter size in pigs or egg laying in chickens, and for measures of growth rate and meat yields. The different lines are then crossed to create commercial crossbreds that combine the genetic changes from the parent lines and capitalize on hybrid vigour for health and reproduction traits (Fig. 16.2). However, it is important to note that a negative genetic correlation does not mean that genetic selection for both traits cannot take place – merely that the rate of change in both traits will be relatively slow compared with single-trait selection.

Selection response is cumulative and permanent and is highly cost-effective, but it should not be confused with the short-term gain from crossbreeding, which is not permanent, does not accumulate over time and has to be 'recreated' in every generation. In practice, animal breeders utilize the breeding pyramid (Fig. 16.2) to cross different lines and produce large numbers of commercial offspring in a multiplication of the genetic change in their pedigree flocks and herds. Genetic selection (measurement and evaluation) is expensive and multiplication is essential to spread the cost over a large number of commercial animals. This has the further consequence that there is a time lag before genetic improvement in the pedigree flocks is passed on to commercial animals.

Genetic selection is a gradual process and the lag time between pedigree and commercial animals cannot be short circuited except in the case of AI in dairy bulls. Thus a change in selection criteria will begin to take effect 4 to 5 years later and will take many (perhaps 10 to 15) years to effect a substantial change in commercial animals. The breeder must successfully predict changes in the operating environment, such as the economics and systems of production, or develop alternative lines to insure against them. Such changes in the past include, for example, the introduction of payment for milk on the basis of fat or protein yield rather than volume and the current change in the European Union from battery cages to free range or aviary systems for laying hens.

Because genetic selection is both expensive and slow, with the benefits accruing over a long period of time, geneticists do not select for a trait if there are more immediate or easier routes to the same end. Historically, health and behaviour traits were not easily measured. Vaccination, or the inclusion of antibiotics in feed, were considered quicker and more effective routes to protect the health and welfare of farm animals than genetic selection. Changes in management (e.g. keeping laying hens in cages rather than on the floor to minimize feather pecking, feather removal and cannibalism) and husbandry (e.g. tail docking in pigs to minimize the risk of tail biting) decreased the value of incorporating these behavioural traits into selection programmes.

Two further constraints that animal breeders face are related to micro- and macroeconomics. Traits have to be recorded on large numbers of animals both cheaply and quickly: in general, there is little point in measuring a trait that either takes months to evaluate or is very expensive to record. Dairy cattle selection decisions, for example, are based on records of milk yield and quality that are used in the management of commercial dairy herds. The only additional costs for dairy bull evaluation are for the statistical analysis of these records and the costs of keeping the bulls to the completion of their progeny test. One of the challenges facing animal breeders is to develop and adopt techniques that are both feasible to conduct in large flocks and herds, and appropriate for traits that are problematic and difficult to record (some examples in farm animals are given below). To use hip and elbow dysplasia in dogs as an example, breeding decisions can be made on X-ray images taken by veterinarians at 1 year of age before the animals are bred. This is expensive and there is a debate about how effective it is. Nevertheless, a decline in hip dysplasia has been observed in those breeds of dog in which effective breeding programmes have been implemented (Table 16.3).

Table 16.3. Changes in the incidence of hip dysplasia in dogs in different countries where evaluation and selection of breeding stock has been made over several years. (From Nicholas, 1987.)

| Incidence | | | |
from	to	Number of years	Country
0.40	0.20	4	USA
0.44	0.12	7	Germany
0.41	0.28	14	Finland
0.50	0.27	5	Sweden

On the macroeconomic scale, commercial animals have to be profitable in a given national economy that is subject to market forces and competition, not only with farmers in the same country, but increasingly with those in other countries and even continents. An animal breeder therefore needs to select for those traits that are required for international competitiveness. As long as it is possible to sell products to a country when those products are produced in systems that are illegal in that country, there will be an inevitable tension between selection criteria for different markets, at least as far as selection for production versus welfare traits is concerned (see Chapter 17, Economics).

16.6 Problems Associated with Genetic Selection: Genetic Solutions

Problems resulting from selection for aesthetically desirable traits in dogs have already been discussed above (and see Fig. 16.3). In farm animals, genetic selection for production characteristics has led to unwanted negative side effects in a number of species. In many cases, breeders have responded by broadening breeding goals by adding additional traits to ameliorate these unwanted effects. Genetic selection in an environment that is unlike commercial production systems, however, may also lead to unexpected welfare consequences. For example, there is evidence that selection for egg production in laying hens housed in single-bird cages is associated with birds that are prone to peck each other when housed commercially in multiple bird cages or in non-battery systems. High biosecurity in the farms where pedigree selection takes place enhances genetic improvement in broilers and pigs by reducing the confounding effects of disease, but may result in animals that are more susceptible to disease challenges in the field.

Animals have evolved under natural selection so that their available resources are allocated optimally between important functions such as growth, reproduction, maintenance, immunity and so on (Rauw et al., 1998). Although artificial selection can result in an absolute increase in the resources available (e.g. through greater food intake and feed efficiency), it may also result in an imbalance in resource allocation between different functions, leading to welfare problems. For example, in dairy cattle, selection for high milk yield during the first lactation resulted in a reduction in longevity and fertility (cows need to become pregnant to

P.M. Hocking *et al.*

produce milk), and in health problems such as mastitis. Breeding goals in dairy cattle have now been broadened in a number of countries to reflect the negative economic impact of many of these 'functional' traits (fertility, mastitis and lameness), and may in time reduce the incidence, or at least reduce the rate of deterioration, of the traits (Lawrence *et al.*, 2004).

In pigs, selection for rapid growth and leanness gradually increased the proportion of pigs carrying a gene which produced these characteristics, but also (as an unwanted side effect) resulted in an enhanced sensitivity to stress. This resulted in sudden death or poor meat quality (Fig. 16.4). A single gene (*ryanodine receptor 1*) was identified as the cause and is now routinely genotyped by pig breeders to eliminate the susceptible allele from populations, or at least to avoid the damaging double recessive state.

Broiler parents are routinely fed on a limited quantity of feed to control body weight at the onset of lay because, if broiler breeders are fed *ad libitum*, the birds regularly ovulate two or more ova (egg yolks) per day, which interferes with the production of hatching eggs. Feed restriction during rearing is continued after sexual maturity, but at a much lower rate, and results in the production of many more eggs and decreased mortality (Table 16.4). These problems are associated with selection for high growth rates in the broiler offspring of the birds. Whereas the reduction in mortality is welcomed from a welfare perspective, feed-restricted birds are hungry and the degree of feed restriction practised in commercial flocks may compromise their welfare. A full description of this problem is

Table 16.4. Body weight, mortality, egg production, hatchability and feed intakes of broiler breeder females fed *ad libitum* or feed restricted from hatch to 60 weeks of age. (From Hocking *et al.*, 2002.)

Trait	*Ad libitum*	Restricted
Body weight (kg)	5.3	3.7
Mortality (%)	46	4
Egg production (no.)	58	157
Hatch of eggs set (%)	43	86
Feed intake (g/day)		
0–24 weeks	163	63
24–37 weeks	192	157
37–60 weeks	142	151

Fig. 16.4. Halothane anaesthesia was used to identify pigs inheriting a gene causing stress susceptibility in addition to increased lean muscle growth. The pig in the upper panel shows a normal relaxed response to anaesthesia and contrasts with the one in the lower panel, which shows a rigid extension of the limbs, identifying it as a homozygous carrier. The identification of the causative gene (*ryanodine receptor 1*) allowed the rapid development of a DNA test that avoided the need for this form of evaluation and allowed the identification of heterozygous carriers that did not show the full stress response. (From Webb *et al.*, 1986.)

outside the remit of this chapter but potential solutions will be discussed later (Section 16.8).

Some examples of specific welfare problems that may be addressed by genetic selection are discussed in detail below, although many others could have been considered.

16.7 Behavioural Problems

A variety of animal welfare problems that may not have been associated with genetic selection but are linked to the housing environment could also be addressed by appropriate genetic selection. Once the challenge of accurately measuring the problem has been overcome, estimates of heritability (the proportion of variation in a trait that is passed on to the next generation) would be needed to develop efficient selection procedures. Current evidence suggests that selection would be effective for many welfare traits, including a number of behavioural traits.

16.7.1 Damaging behaviours in pigs

Unfamiliar pigs fight when they are mixed, but there is considerable variation in aggressiveness between different pigs. Estimates suggest that around 30–40% of this variation in aggressiveness is genetic in origin. As aggressive behaviour is time-consuming to measure by observation, recording skin lesions after mixing can be used as a proxy measure. Skin lesions might appear to be a measure of the victims of aggression rather than of the aggressors, so this requires some explanation. Aggressive pigs engage in mutual fighting and also in one-sided attacks. Mutual fighting mainly damages the head and shoulders, while the victims of one-sided attacks receive skin lesions evenly all over the body. Thus, pigs with skin lesions at the front but few elsewhere can be identified as the perpetrators of aggression, making selection against this possible (Turner *et al.*, 2009).

Other damaging behaviours in pigs include tail, ear and flank biting. These are thought to result from redirected foraging behaviour and are reduced by access to foraging substrates. However, the majority of pigs are housed without adequate access to substrates and instead are routinely tail docked to reduce the consequences of tail biting. The farming industry in some countries is coming under increasing public pressure to limit the use of such mutilations, so selection to reduce tail biting

may have a role. Tail biting is heritable at a low level in some pig breeds (Breuer *et al.*, 2003), although it tends to occur in sporadic outbreaks, so regular and consistent selection is difficult. Development of proxy measures such as the amount that an artificial tail is chewed might present a valuable way of testing for likely perpetrators without waiting for an outbreak.

16.7.2 Feather pecking, removal and cannibalism in laying hens

Feather pecking is a behaviour that involves hens destroying the feathers of other hens, in some cases even plucking out feathers and eating them. In some severe cases, feather pecking can be followed by cannibalism, where hens eat the blood and tissue of other hens. Feather pecking, feather pulling and cannibalism are closely related complex behaviours influenced by a range of environmental and genetic factors. By controlling the physical environment (keeping lights very low and having small group sizes) it is possible to reduce the prevalence of these problems. This is achievable in cage systems but not in non-cage systems. Beak trimming (partial beak amputation) at a young age is routinely used to reduce the consequences of feather pecking and especially of cannibalism. The behaviour still persists after the operation but the damage to feathers and skin, like bloody wounds, is curbed. However, beak trimming is painful and may interfere with normal oral behaviour and should be avoided if the occurrence of damaging pecking could be prevented in another way.

That there are differences between breeds in the incidence of feather pecking and cannibalism demonstrates that there is a clear genetic basis for these behaviours, and makes breeding against them a potential option for poultry breeders. Lines of White Leghorn layer chickens have been developed by genetic selection that show low or high levels of feather pecking in relation to an unselected control line (Fig. 16.5). Studies of the selected lines have shown that the low feather pecking line has a higher egg mass output and better feed efficiency than the high pecking line (Su *et al.*, 2006). This better feed efficiency in the low pecking line was mainly the result of better feather cover, but might also partly be caused by a lower general activity. These results illustrate how genetic selection purely for welfare traits can also improve productivity and efficiency. If selection for low feather pecking

Fig. 16.5. Hens from two lines of laying hens selected for low (upper panel) or high (lower panel) feather pecking activity assessed by observation, and illustrating the potential of genetic selection to improve welfare related traits. (Further details can be found in Kjaer *et al.*, 2001; Su *et al.*, 2006.)

proves successful in commercial lines, there may be no further need for beak trimming, even in non-cage systems.

16.7.3 Inappropriate maternal and neonate behaviour in pigs and sheep

Some sows lie or step on their piglets in the first few days after they are born, crushing them to death. The farrowing crate reduces this by restricting sow movements, making rapid lying more difficult. However, the introduction of this engineering fix has meant that sows showing this poor maternal behaviour can still rear many piglets. Public pressure in some countries to improve welfare by doing away with this confinement system is likely to require pigs with a different genotype, one that is appropriate for them to perform well in a less confined system (Roehe *et al.*, 2009).

Similarly, many lowland breeds of sheep have been lambed inside with lots of human supervision and intervention. This has led to a relaxation of natural selection for appropriate maternal and neo-natal behaviour, because lambs that are slow to stand and suck or are not defended by their dams still survive (Dwyer and Lawrence, 2005). With a reduction in the economic value of reared lambs compared with the cost of labour, there is a growing interest in breeding sheep (or using existing breeds) that require far less assistance and supervision at lambing time.

16.8 Skeletal and Physiological Disease

16.8.1 Laying hens

The main skeletal problems in laying hens are associated with loss of bone mineral during the laying period, with poor bone quality and bone fracture being the main consequences. The loss of bone mineral may have one of two causes – osteomalacia, which has no direct genetic component, or osteoporosis. The latter is defined as a decrease in the amount of fully mineralized structural bone, leading to increased fragility and susceptibility to fracture. Fractures occur during the laying period, at depopulation, during transport and during the slaughter process (Gregory and Wilkins, 1992). Osteoporosis is probably a product of genetic selection for persistently high rates of lay that require a high degree of calcium mobilization from bones for eggshells.

A significant genetic component in osteoporosis was found in a White Leghorn population studied by Bishop *et al.* (2000), and selection for improved bone strength proved to be possible. Tibial strength, humeral strength and keel radiographic density were found to be moderately to strongly inherited, and these measurements were combined in a Bone Index which was used as a basis for selection. All bone characteristics used in the Bone Index responded rapidly to divergent selection for high or low values of the index. In the last (sixth) generation, the lines differed by 25% for tibial strength, 13% for humeral strength and 19% for keel radiographic density.

16.8.2 Meat type poultry

Skeletal defects in meat type chickens and turkeys are related to selection for rapid body weight gain and may be caused by a combination of bones, cartilage, tendons and ligaments of poor structural quality and low tensile strength, and high body weight resulting in mechanical stress. Rapid growth may be the main factor in defects occurring up to about 4 weeks of age, after which the high body weight produces stress on the bones, tendons and ligaments, particularly in turkeys. The pathogenesis of some rapid growth problems may be related to the high requirement of specific nutrients such as vitamin D_3, as deficiency may develop in the fastest growing birds in a group even if an adequate general level is present in the diet (Thorp et al., 1993).

A subjective score of broiler walking ability (gait score, GS) was developed in order to give an overall evaluation related to the welfare of broilers (Chapter 12; Kestin et al., 1992). The gait score is 1 for birds with no problems and 5 for birds that cannot walk. This score evaluates reduced walking ability that can arise from one or more of many different problems, such as tibial dyschondroplasia (TD), valgus-varus deformity (VVD; bowed or splayed legs), angular bone deformity, twisted legs, crooked toes, epiphyseal ischaemic necrosis or foot pad dermatitis (FPD; see below). The heritability of GS was found to be 0.20 in a female line parent of commercial broilers. Large differences were found in walking ability between four crosses of commercial broiler lines (Kestin et al., 1999). Selection for rapid growth rate in broilers has been accompanied by a decrease in walking ability and a highly unfavourable phenotypic correlation of 0.8 has been found between body weight and overall walking ability (Kestin et al., 2001).

TD is a specific form of growth-plate abnormality in which transitional chondrocytes accumulate, forming a mass of avascular cartilage underlying the layer of proliferative chondrocytes. Lesions develop at about 2–3 weeks old in broilers and at 9–12 weeks old in turkeys. TD has been the subject of a number of genetic studies. Ducro and Sørensen (1992) reported a heritability of 0.33 using a portable X-ray machine (Lixiscope) to measure TD at 4 weeks of age in a divergent selection programme. Given the high additive genetic variability and availability of the Lixiscope technique for scoring TD in live birds, commercial breeding companies have selected successfully against TD. A survey in Denmark revealed that the level of broilers with general leg problems has been reduced to around 15%, and only 2% of these had TD (Sørensen, 2004, personal communication).

'Twisted legs' is a term that covers VVD, angular bone deformity and other rotational effects. VVD is defined as lateral (valgus) or medial deviation of the distal tibeotarsus, resulting in a 'hocks-in/feet-out' (valgus) stance or a 'hocks-out/feet-in' (varus) stance. The pathogenesis of VVD is not known, but the deformity in broilers is related to fast growth under continuous light. VVD is one of the most frequent causes of leg problems, with reported incidences as high as 30–50% in commercial broilers in Europe (Sanotra et al., 2003), and has been subject to several successful selection experiments. The heritability of VVD has been estimated to be moderate to high and only small-to-moderate genetic correlations were found between varus and valgus scores, indicting two distinct causes of these deformities.

16.8.3 Foot pad dermatitis

Foot pad dermatitis, or FPD, is a type of contact dermatitis affecting the plantar region of the feet in broilers and turkeys. The lesions are commonly named 'ammonia burns', and are thought to be caused by a combination of moisture, high ammonia content and other not-yet specified chemical factors in the litter, although recent research in turkeys has shown that high litter moisture alone is sufficient to cause FPD. The FPD condition is an important problem for poultry welfare and, in severe cases, can cause pain resulting in unsteady walking. Closely related to FPD are 'hock burns' (HBs) where the skin of the hock becomes dark brown, and in severe cases scabs are observed.

In order to gain more specific knowledge on genetic effects on FPD, pedigreed broilers from a fast- and a slow-growing strain were studied by Kjaer et al. (2006). No FPD lesions and very few low-grade HB lesions were found in chickens from the slow-growing strain. In the fast-growing strain, the first signs of FPD and HB were seen at 2 weeks of age, and the incidence of both types of lesions increased thereafter. Heritabilities were estimated to be moderately high for FPD but low for HB, with low genetic correlations between the traits. The relatively high heritability of FPD, with a low genetic correlation to body weight, suggests that genetic selection against susceptibility to FPD should be possible without negative effects on body weight gain.

16.8.4 Reproduction in broiler parents

There is evidence that there is a genetic correlation between selection for high growth rate and the incidence of multiple ovulation (see Hocking, 2009). Multiple ovulation is associated with low egg production through internal ovulation, and the formation of double-yolk eggs or eggs with poor shells that cannot be incubated. Controlling body weight in broiler and duck female parents by feed restriction restores the single ovulation state and normal egg production. The degree of feed restriction necessary to control ovulation results in hungry birds and causes concerns about their ability to cope with the limited supply of feed. The welfare of feed-restricted birds is clearly open to question, although it is difficult to assess objectively, and selection to decrease ovulation rate would enable farmers to provide sufficient feed to optimize the bird's welfare without compromising egg production. Unfortunately, it is not feasible to measure ovulation rate in live birds. The degree of feed restriction expressed as a proportion of the body weight of unrestricted birds has increased over the years, and the development of alternative methods of decreasing ovulation rate is urgently required. One promising option is the use of genetic (DNA) markers that are linked to multiple ovulation that would facilitate genetic selection to reduce ovulation rate.

16.9 New Opportunities for Genetic Improvement of Animal Welfare

16.9.1 DNA markers for single-gene defects

The development of rapid genotyping methods has facilitated the identification of an increasing number of recessive genes causing developmental disorders and disease. Once the gene has been discovered, the causative mutation can be identified and simple tests developed to test animals before selection. A brief list of such tests is given in Table 16.5, and will doubtless be added to in the coming years. DNA tests have already led to the elimination of the mutation or genetic management (by only breeding heterozygotes) of pigs inheriting the *ryr1* gene that causes stress susceptibility (Fig. 16.4) and low-quality meat (pale, soft and exudative, or PSE, meat).

16.9.2 Whole genome selection

Recent advances in technology have resulted in the sequencing of the complete DNA code of individuals representing several species, including dogs, chickens, pigs and cattle (Ensembl, 2009). Within the DNA of a species are many single nucleotide changes or polymorphisms (SNPs) at unique locations along the genetic code. Using another technology (SNP chips) to genotype animals at many SNP loci at the same time, these differences can be used to map traits to specific SNPs that are then used in marker assisted selection (MAS). Alternatively, if several hundreds of thousands of SNPs are available, animals could be selected wholly or partly on the basis of previously identified associations between SNPs and the desired traits (whole genome selection

Table 16.5. A selection of current DNA tests for inherited diseases in several species available from commercial laboratories. (From Laboklin, 2009; Schmutz, 2009; vetGen, 2009.)

Disease	Gene action[a]	Species	Breeds affected
Alpha-mannosidosis	R	Cattle	Angus, Galloway
Bovine lymphocyte adhesion deficiency	R	Cattle	Holstein
Complex vertebral malformation (CVM)	R	Cattle	Holstein
Platelet bleeding disorder	R	Cattle	Simmental
Protoporphyri	R	Cattle	Limousin
Centronuclar myopathy	R	Dog	Labrador
Pyruvate kinase deficiency	R	Dog	West Highland White Terrier
Renal cystadenocarcinoma nodular dermatofibrosis	D	Dog	German Shepherd
Cyclic neutropenia	R	Dog	Collie
Haemophilia B	X	Dog	Bull terrier, Lhasa Apso
Feline polycystic kidney disease	D	Cat	Persian
Gangliosidosis GM1	R	Cat	Korat, Siam
Severe combined immunodeficiency	R	Horse	Arabian
Lethal white foal syndrome	R	Horse	Paint Horse, Quarter Horse

[a] R = autosomal recessive; D = autosomal dominant; X = sex linked.

or genomic selection). This technology has already been used to select dairy bull calves for progeny testing (to compare the potential merit of full brothers, for example) (Seidel, 2010). The method is ideally suited to selection for other traits that involve measurements which are difficult to make, expensive or cause harm to the animal, e.g. testing for disease resistance or the avoidance of a behavioural problem. Many welfare traits come into this category, and the adoption of these techniques could lead to substantial improvements in traits that would be very difficult to improve by conventional methods of selection.

16.9.3 New measurement techniques

The cost of measuring many traits is not insignificant. Examples of traits that are expensive to measure include food conversion efficiency and carcass composition in pigs and poultry. Animal breeders are therefore always looking for quick, economical and efficient measures of significant economic or welfare traits. Imaging techniques are constantly improving and becoming more widely available at lower costs and have been adopted for assessing skeletal disease, for example. In hens, one reason for recording egg production in cages is to allow individual egg recording, but the development of electronic identification and nest boxes where an egg can be attributed to the hen that laid it have facilitated the recording of individual egg production in group-housed hens. This leads to improved bird welfare during selection in a group housing system which is like those that will be used for commercial flocks in the future. Methods for individual identification and recording the feed intake of broiler chickens have also obviated the need for individual caging, which has led to better welfare; the new environment also represents a more relevant selection environment, because broiler chickens are not kept in cages. These methods have also made it possible to measure activity (by recording the number of feeding bouts or time spent on the nest) and improved the prospect of selecting for more active broilers and laying hens. More active adult laying hens may lead to improved leg and skeletal health (but might also increase feather pecking and cannibalism; Kjaer, 2009), and to layers that tend always to use the nest box rather than the floor to lay their eggs. Techniques such as these will continue to become available and contribute to improved animal welfare.

16.9.4 Improved methods of conventional selection

Feather damage, feather removal and cannibalism in chickens, and aggression in pigs, are difficult to measure and are affected by the members of the group, i.e. the cage or pen mates, which suggests that these traits should be recorded in the groups in which the animals are housed. This approach has been examined both in laying hens and in pigs. In laying hens, the number of hen days without beak-inflicted injuries, a trait that combined measurements of feather pecking and cannibalism leading to severe injury or death, was used in a study of group selection (Craig and Muir, 1996). Group selection was very effective in reducing the incidence of these beak-inflicted injuries. Each sire family was held in a multiple-bird cage and was exposed to very bright light. Selection was on the basis of group survival. After two generations of selection, realized family heritability was estimated to be large (0.65). Mortality decreased from an initial unacceptable level of 68% to 9% in generation 3, indicating that a major gene may have been involved in these behaviours in this population. Recently, Muir (2005) and Bijma *et al.* (2007a,b) presented a quantitative genetic framework for the prediction of response to selection and for statistical analyses of traits affected by social interactions. In pigs, social genetic effects for finishing traits, such as growth rate, were found to be very large in one population (Bergsma *et al.*, 2008), but very low in another (Chen *et al.*, 2009). The modelling of these social genetic effects is undergoing further development and debate, and represents a promising route to improve welfare traits that are affected by the social environment.

16.9.5 Selection for resistance to disease

It has become clear in the last decade that heritable variation for resistance to disease is commonplace, with heritabilities similar to those of other fitness traits. Disease has a major impact on the welfare and protection of farm animals, and has become an increasingly important consideration with the reduced use of in-feed antibiotics and other drugs, the emergence of new diseases, the increasing virulence of existing disease organisms and the subsequent ineffectiveness of vaccination strategies to maintain animal health and welfare. The new genetic techniques based on DNA technologies will make selection for disease resistance practical (Biscarini *et al.*, 2010), and are likely to become increasingly important.

16.10 Negative Aspects of Genetic Selection for Improved Animal Welfare

We have outlined many instances of welfare problems that either may have a genetic basis or can potentially be solved using genetic selection. Nevertheless, we note that the problems of negative correlated responses and unexpected consequences of selection apply for welfare as much as for production traits. In addition, the warning from the first edition of this book (Mills *et al.*, 1997, p. 225) that 'great care must be used in defining the traits to be selected because "what you select is precisely what you get" Beilharz *et al.* (1993)' is very relevant to the use of genetic selection to improve animal welfare.

Most of our examples in farm animals are in poultry and pigs; this is partly because these are more easily studied than ruminants and partly because these species have higher reproductive rates and faster generation turnover, with the result that genetic selection has been more effective in these than in other farm animals. None the less, genetic selection to improve welfare traits is not without its detractors and there are potential dangers. Whereas genetic selection to improve the skeletal condition of laying hens may be successful, some have argued that this masks an inappropriate production environment. In this specific case, the evidence suggests that the cause is genetic selection for egg production that ignored skeletal strength, a difficult trait to measure and an additional selection criterion that will, therefore, negatively affect genetic progress for egg production. Similarly, selection to reduce foot pad dermatitis, if successful, could simply mask a very poor environment that might continue to have negative effects on other welfare measures. The latter scenario is indeed possible, but in practice is unlikely to happen because such poor welfare will affect other welfare and performance traits.

Selection against behavioural traits such as damaging feather pecking has been criticized on the same grounds, but also because of the possibility that, whereas the behaviour may be changed, the underlying psychological 'disease' may become worse or simply not be observed. There is some evidence that this can occur where there is a mental attribute of emotion or feeling, such as fear. Adult laying hens are flighty and generally assumed to be more fearful than broilers, which are relatively phlegmatic. However, both breeds have been shown to display similar responses to tests of fear that did not involve an active response (Keer-Keer *et al.*, 1996). Furthermore, earlier research had shown that the heart rate of medium body weight (brown) layers remained high longer than did that of lighter, more flighty (white) layers in response to a fearful event (Duncan and Filshie, 1980). Both lines of evidence suggest that the breeds had different strategies for responding to fearful stimuli, and that apparently docile birds may be as frightened as flighty birds in physiological terms.

Another criticism of selection for behavioural attributes is that it may lead to animals that are mere zombies, unresponsive and psychologically 'inert'. These animals would lose their 'naturalness' or species integrity to some degree. Leaving aside the fact that domestic animals are behaviourally very different from their wild progenitors, there is no evidence that such fundamental changes have occurred or are even possible: domestication has itself produced remarkable genetic changes in adaptability to humans and confinement.

In conclusion, we suggest that it is imperative that genetic selection for behavioural changes should be conducted with proper evaluation of the potential physiological and psychological consequences to guard against any undesirable responses to selection.

16.11 Conclusions

- The genetic structure of domestic breeds is affected by their history and breeding structures, which increase levels of inbreeding and the risk of genetic disease. Whereas crossbreeding is widely practised in most farm animals and leads to improved reproductive fitness and health, it is not popular in companion animals where the benefits of specific breed crosses should be investigated.

- Genetic selection can adversely affect welfare traits directly, as in many dog breeds, inadvertently as a result of unexpected correlated responses to selection, or as a consequence of undue selection pressure for narrow production goals that ignore fitness traits.

- Selection for improved welfare is also possible and there are now many examples of welfare-related traits that have been shown to have a genetic component that could respond to selection.

- The assessment of welfare traits in the environment in which the animals will be kept is particularly important to avoid ineffective selection in the presence of genotype by environment interactions.
- A number of constraints on breeders restrict the introduction of welfare traits to selection programmes including: the difficulty or cost of measuring traits, the fact that increasing the number of traits in an index reduces progress on each, negative genetic correlations between traits and difficulties in placing an economic value on a welfare trait.
- Recent developments in selection theory and DNA technology, and the potential of whole genome selection, suggest that genetic improvement of welfare without compromising production efficiency in farm livestock will be increasingly adopted and adapted to improving the genetic status and welfare of companion animals.

Acknowledgements

The work of the first author is supported by a core strategic grant from the BBSRC to the Roslin Institute. The Scottish Agricultural College (where the second author works) is supported by the Scottish Government.

References

Algers, B., Blokhuis, H.J., Botner, A., Broom, D.M., Costa, P., Domingo, M., Greiner, M., Hartung, J., Koenen, F., Müller-Graf, C., Mohan, R., Morton, D.B., Osterhaus, A., Pfeiffer, D.U., Roberts, R., Sanaa, M., Salman, M., Sharp, J.M., Vannier, P. and Wierup, M. (2009) Scientific opinion of the panel on animal health and welfare on a request from european commission on the overall effects of farming systems on dairy cow welfare and disease. *The EFSA Journal* 1143, 1–38.

Asher, L., Diesel, G., Summers, J.F., McGreevy, P.D. and Collins, L.M. (2009) Inherited defects in pedigree dogs: 1. Disorders related to breed standards. *The Veterinary Journal*, 182, 402–422.

Beilharz, R.G., Luxford, B.G. and Wilkinson, J.L. (1993) Quantitative genetics and evolution: is our understanding of genetics sufficient to explain evolution? *Journal of Animal Breeding and Genetics* 110, 161–170.

Bergsma, R., Kanis, E., Knol, E.F. and Bijma, P. (2008) The contribution of social effects to heritable variation in finishing traits of domestic pigs (*Sus scrofa*). *Genetics* 178, 1559–1570.

Bijma, P., Muir, W.A. and Van Arendonk, J.A.M. (2007a) Multilevel selection 1: Quantitative genetics of inheritance and response to selection. *Genetics* 175, 277–288.

Bijma, P., Muir, W.M., Ellen, E.D., Wolf, J.B. and Van Arendonk, J.A.M. (2007b) Multilevel selection 2: Estimating the genetic parameters determining inheritance and response to selection. *Genetics* 175, 289–299.

Biscarini, F., Bovenhuis, H., Van Arendonk, J.A.M., Parmentier, H.K., Jungerius, A.P. and Van Der Poel, J.J. (2010) Across-line SNP association study of innate and adaptive immune response in laying hens. *Animal Genetics* 41, 26–38.

Bishop, S.C., Fleming, R.H., McCormack, H.A., Flock, D.K. and Whitehead, C.C. (2000) Inheritance of bone characteristics affecting osteoporosis in laying hens. *British Poultry Science* 41, 33–40.

Breuer, K., Sutcliffe, M.E.M., Mercer, J.T., Rance, K.A., O'Connell, N.E., Sneddon, I.A. and Edwards, S.A. (2003) Heritability of clinical tail-biting and its relation to performance traits. In: van der Honing, Y. (ed.) *Book of Abstracts of the 54th Annual Meeting of the European Association for Animal Production*, Rome, Italy, 31 August–3 September 2003. Wageningen Academic Publishers, Wageningen, The Netherlands, pp. 87–94.

Bruce, D.M. and Bruce, A. (eds) (1998) *Engineering Genesis: Ethics of Genetic Engineering in Non-Human Species*. Earthscan, London.

Chen, C.Y., Johnson, R.K., Newman, S., Kachman, S.D. and Van Vleck, L.D. (2009) Effects of social interactions on empirical responses to selection for average daily gain of boars. *Journal of Animal Science* 87, 844–849.

Craig, J.V. and Muir, W.M. (1996) Group selection for adaptation to multiple-hen cages: beak-related mortality, feathering, and body weight responses. *Poultry Science* 75, 294–302.

Cundiff, L.V., Nunezdominguez, R., Dickerson, G.E., Gregory, K.E. and Koch, R.M. (1992) Heterosis for lifetime production in Hereford, Angus, Shorthorn, and crossbred cows. *Journal of Animal Science* 70, 2397–2410.

Cunningham, E.P., Dooley, J.J., Splan, R.K. and Bradley, D.G. (2001) Microsatellite diversity, pedigree relatedness and the contributions of founder lineages to thoroughbred horses. *Animal Genetics* 32, 360–364.

Donald, H.P. (1955) Controlled heterozygosity in livestock. *Proceedings of the Royal Society of London Series B – Biological Sciences* 144, 192–203.

Ducro, B.J. and Sørensen, P. (1992) Evaluation of a selection experiment on tibial dyschondroplasia in broiler chickens. In: *Proceedings of the XIX World's Poultry Congress*, Amsterdam, The Netherlands, 20–24 September 1993, Vol. 2, p 386–389.

P.M. Hocking *et al.*

Duncan, I.J.H. and Filshie, J.H. (1980) The use of radio telemetry devices to measure temperature and heart rate in domestic fowl. In: Amlaner, C.J. and MacDonald, D.W. (eds) *A Handbook on Biotelemetry and Radio Tracking*. Pergamon Press, Oxford, UK, pp. 579–588.

Dwyer, C.M. and Lawrence, A.B. (2005) A review of the behavioural and physiological adaptations of hill and lowland breeds of sheep that favour lamb survival. *Applied Animal Behaviour Science* 92(3), 235–260.

Ensembl (2009) Browse a Genome – The Ensembl project produces genome databases for vertebrates and other eukaryotic species, and makes this information freely available online. Available at: http://www.ensembl.org/index.html (accessed 13 October 2009).

Falconer, D.S. (1964) *Introduction to Quantitative Genetics*. Oliver and Boyd, Edinburgh, UK.

Gregory, N.G. and Wilkins, L.J. (1992) Skeletal damage and bone defects during catching and processing. In: Whitehead, C.C. (ed.) *Bone Biology and Skeletal Disorders in Poultry: Poultry Science Symposium No. 23, 1992*. Carfax Publishing, Abingdon, UK, pp. 313–328.

Hocking, P.M. (2009) Feed restriction. In: Hocking, P.M. (ed.) *Biology of Breeding Poultry*. Poultry Science Symposium Series, CAB International, Wallingford, UK, p 307–330.

Hocking, P.M. and McCorquodale, C.C. (2008) Similar improvements in reproductive performance of male line, female line and parent stock broiler breeders genetically selected in the UK or in South America. *British Poultry Science* 49, 282–289.

Hocking, P.M., Bernard, R. and Robertson, G.W. (2002) Effects of low dietary protein and different allocations of food during rearing and restricted feeding after peak rate of lay on egg production, fertility and hatchability in female broiler breeders. *British Poultry Science* 43, 94–103.

Keer-Keer, S., Hughes, B.O., Hocking, P. and Jones, R. (1996) Behavioural comparison of layer and broiler fowl: measuring fear responses. *Applied Animal Behaviour Science* 49, 321–333.

Kestin, S.C., Knowles, T.G., Tinch, A.E. and Gregory, N.G. (1992) Prevalence of leg weakness in broiler chickens and its relationship with genotype. *Veterinary Record* 131, 190–194.

Kestin, S.C., Su, G. and Sørensen, P. (1999) Different commercial broiler crosses have different susceptibilities to leg weakness. *Poultry Science* 78, 1085–1090.

Kestin, S.C., Gordon, S., Su, G. and Sørensen, P. (2001) Relationships in broiler chickens between lameness, liveweight, growth rate and age. *Veterinary Record* 148, 195–197.

Kjaer, J.B. (2009) Feather pecking in domestic fowl is genetically related to locomotor activity levels: implications for a hyperactivity disorder model of feather pecking. *Behavior Genetics* 39, 564–570.

Kjaer, J.B., Sørensen, P. and Su, G. (2001) Divergent selection of feather pecking behaviour in laying hens (*Gallus gallus domesticus*). *Applied Animal Behaviour Science* 71, 229–239.

Kjaer, J.B., Su, G., Nielsen, B.L. and Sørensen, P. (2006) Foot pad dermatitis and hock burn in broiler chickens and degree of inheritance. *Poultry Science* 85, 1342–1348.

Laboklin (2009) Genetic Diseases [animal genetic testing]. Available at: http://www.laboklin.co.uk/laboklin/GeneticDiseases.jsp (accessed 13 October 2009).

Lawrence, A.B., Conington, J. and Simm, G. (2004) Breeding and animal welfare: practical and theoretical advantages of multi-trait selection. *Animal Welfare* 13 (Supplement 1), 191–196.

Mäki, K., Groen, A.F., Liinamo, A.E. and Ojala, M. (2001) Population structure, inbreeding trend and their association with hip and elbow dysplasia in dogs. *Animal Science* 73, 217–228.

Mills, A.D., Beilharz, R. and Hocking, P.M. (1997) Genetic selection. In: Appleby, M.C. and Hughes, B.O. (eds) *Animal Welfare*. CAB International, Wallingford, UK, pp. 219–231.

Muir, W.M. (2005) Incorporation of competitive effects in forest tree or animal breeding programs. *Genetics* 170, 1247–1259.

Nicholas, F.W. (1987) *Veterinary Genetics*. Oxford University Press, Oxford, UK.

Rauw, W.M., Kanis, E., Noordhuizen-Stassen, E.N. and Grommers, F.J. (1998) Undesirable side effects of selection for high production efficiency in farm animals: a review. *Livestock Production Science* 56, 15–33.

Roehe, R., Shrestha, N.P., Mekkawy, W., Baxter, E.M., Knap, P.W., Smurthwaite, K.M., Jarvis, S., Lawrence, A.B. and Edwards, S.A. (2009) Genetic analyses of piglet survival and individual birth weight on first generation data of a selection experiment for piglet survival under outdoor conditions. *Livestock Science* 121, 173–181.

Sanotra, G.S., Berg, C. and Lund, J.D. (2003) A comparison between leg problems in Danish and Swedish broiler production. *Animal Welfare* 12, 677–683.

Schmutz, S. (2009) DNA tests for Cattle. Available at: http://homepage.usask.ca/~schmutz/tests.html#disease%20tests (accessed 13 October 2009).

Seidel, G.E. (2010) Brief introduction to whole-genome selection in cattle using single nucleotide polymorphisms. *Reproduction Fertility and Development* 22, 138–144.

Su, G., Kjaer, J.B. and Sørensen, P. (2006) Divergent selection on feather pecking behavior in laying hens has caused differences between lines in egg production, egg quality, and feed efficiency. *Poultry Science* 85, 191–197.

Summers, J.F., Diesel, G., Asher, L., McGreevy, P.D. and Collins, L.M. (2010) Inherited defects in pedigree dogs. Part 2: Disorders that are not related to breed standards. *The Veterinary Journal* 183, 39–45.

Thorp, B.H., Ducro, B., Whitehead, C.C., Farquharson, C. and Sørensen, P. (1993) Avian tibial dyschondroplasia – the interaction of genetic selection and dietary 1,25-dihydroxycholecalciferol. *Avian Pathology* 22, 311–324.

Turner, S.P., Roehe, R., D'Eath, R.B., Ison, S.H., Farish, M., Jack, M.C., Lundeheim, N., Rydhmer, L. and Lawrence, A.B. (2009) Genetic validation of postmixing skin injuries in pigs as an indicator of aggressiveness and the relationship with injuries under more stable social conditions. *Journal of Animal Science* 87, 3076–3082.

vetGen (2009) CNM-Centronuclear Myopathy. Available at: http://www.vetgen.com/canine-centronuclear-myopathy.html (accessed 13 October 2009).

Webb, A.J., Cameron, N.D. and Haley, C.S. (1986) Genetic research for pig improvement. In: *ABRO Report 1986*, Animal Breeding Research Organisation, Edinburgh, UK, pp. 6–8.

Wiener, G. and Haytor, S. (1974) Crossbreeding and inbreeding in sheep. *ABRO Report 1974*, Animal Breeding Research Organisation, Edinburgh, UK, pp. 19–26.

17 Economics

RICHARD BENNETT AND PAUL THOMPSON

Abstract

This chapter discusses the relevance of economics to the study of animal welfare. It considers the relationship between ethics and economics, and emphasizes that economics is about the attainment of human well-being. It considers both conventional and alternative economic thinking, the relevance of economics to policy decisions about animal welfare, and how economics can offer insights into the costs and benefits of animal exploitation. Economic frameworks for analysis are presented, together with some assessment of their use and the implications for policy, which can help to clarify the problems that a purely pragmatic approach may generate. Policy options and instruments for addressing questions relating to animal welfare in society are identified. A case study of an economic survey is reported which suggests that people have a substantial willingness to pay to improve the welfare of animals used for food. However, surveys may not always accurately predict actual purchasing or other economic behaviour. Economic considerations are central to the animal welfare debate and are integral (and inescapable) aspects of issues concerning the use of animals and of any interdisciplinary inquiry into animal welfare; they should help us to make better decisions concerning human use of, and obligations towards, animals.

17.1 Introduction

In many people's minds economics is perhaps synonymous with accountancy and with the importance of monetary considerations. Indeed, when asking students new to the study of economics what they think economics is about, they usually mention money at the outset. However, money is merely employed by economists as a useful measuring rod of people's preferences, and accountancy is primarily concerned with the presentation of financial information. In contrast, economics, which grew out of the study of moral philosophy, tries to address much broader issues that are central concerns of society. Indeed, Sen (1987) states that economics can help us to address the important question of 'how we should live'. More expansively, Alfred Marshall wrote in 1890 (Marshall, 1947, pp. 1, 22, 39) that:

> Economics is a study of mankind in the ordinary business of life; it examines that part of individual and social action which is most closely connected with the attainment and with the use of the material requisites of wellbeing … Money is a means towards ends … and is sought as a means to all kinds of ends, high as well as low, spiritual as well as material … Thus though it is true that 'money' or 'general purchasing power' or 'command over material wealth', is the centre around

which economic science clusters; this is so, not because money or material wealth is regarded as the main aim of human effort, nor even as affording the main subject matter for the study of the economist, but because in this world of ours it is the one convenient means of measuring human motive on a large scale … But with careful precautions money affords a fairly good measure of the moving force of a great part of the motives by which men's lives are fashioned … Economics has a great and an increasing concern in motives connected with … the collective pursuit of important aims. Economics has then as its purpose … to throw light on practical issues.

Among these practical issues that Marshall raises is that of 'those who suffer the evil, but do not reap the good' and poses the question 'how far is it right that they should suffer for the benefit of others?' Although Marshall was referring to human society, this question is surely central to the economic study of animal welfare.

However, although the ethical and scientific aspects of animal welfare and animal rights have been extensively debated over the last nearly 50 years (Harrison, 1964; Singer, 1975, 1980; Regan, 1982, 1984), the relevance of economic considerations has received relatively little attention. This is despite some early economic thinking regarding the economic relationship between animals and humans.

In his 'Theory of Moral Sentiments', Adam Smith (1790, Part II, Sec. III, Chap. 1, para. 4), the 'founding father' of economics, succinctly noted the economic relationship between animals and humans by writing 'Animals are not only the causes of pleasure and pain, but are also capable of feeling those sensations'. The relevance of the latter to economics – which is fundamentally concerned with human welfare and not with that of other species – becomes clearer given that Smith (1790, Part I, Sec. I, Chap. 1, para. 1) also wrote:

> How selfish soever man may be supposed, there are evidently some principles in his nature which interest him in the fortune of others and render their happiness necessary to him though he derives nothing from it except the pleasure of seeing it.

Bentham (1789) explicitly included human benevolence and sympathy to animals among his categorization of (human) 'pleasures'. This utilitarian ethic of weighing up pleasures and pain (benefit and cost) underlies the discipline of economics. This is perhaps most tangibly evident in cost–benefit analysis, a cornerstone of applied economics, which itself has been described as the 'economic ethic' (Boulding, 1969).

Although there have been some specific studies over the last decade and a half, economists have not fully come to terms with animal welfare as a topic for theoretical or applied research. One explanation for this may be found in the assumption that animal welfare and how humans treat animals are moral issues. Although what is good or bad for animals may involve matters of fact concerning relative states of veterinary health or cognitive well-being, economic considerations remain secondary or even irrelevant. Given this conceptualization of animal welfare, some find the thought that monetary aspects should be taken into consideration antithetical to the goal of acting in an ethical manner: how can we put a price on animal suffering? At the same time, if animal welfare is defined as an ethical problem, scientifically inclined economists will be quickly dissuaded from undertaking any research on it at all.

However, regulations on human use of animals have obvious economic implications. It is relatively straightforward (though as we will discuss, potentially controversial) to apply standard concepts in economic theory to quantify certain elements of these implications. Thus, there have been several studies to estimate the economic impact of regulations intended to improve farm animal welfare. For example, Bennett (1997) and Bennett and Blaney (2003) estimated people's willingness to pay for legislation to ban the use of cages in egg production in Europe and weighed the results obtained against the higher industry costs of production, while Carlsson et al. (2007) undertook a similar exercise in relation to livestock slaughter facilities in Sweden. Sumner et al. (2008) estimated how animal-welfare promoting regulations in California would affect the cost of production for producers, and concluded that Californian producers would no longer be competitive with producers from other states. The study went on to estimate the effects on the costs of animal products for Californian producers if production was displaced to neighbouring states, as well as the overall impact on the Californian economy associated with the loss of employment and tax revenue. A similar study published in 2009 estimated the costs of shifting all egg production in the USA to non-cage systems (PROMAR International, 2009). It would certainly be possible to conduct similar studies on the impact of animal welfare regulations on medical research, zoos, the keeping of companion animals and even wildlife conservation.

Such studies are generally intended as decision aids. The cost and benefit estimates derived are put forward in order to help voters or political decision makers make decisions about using resources to achieve the things that we want. The central problem for society, as economics conceives it, is that we all want a multitude of different things, but the resources that can be deployed in satisfying those wants are limited. As a society, we cannot have everything that we want. This leaves us with three important and interrelated decisions: (i) What social outcomes should be pursued in allocating resources (i.e. What do we want)?; (ii) What allocation of resources produces the best or optimal mix of possible outcomes (i.e. What should we do)?; and finally, (iii) How are the costs and benefits of these outcomes distributed (i.e. Who wins, and who loses)? These questions have their roots in ethics, and the discipline of economics has evolved through a persistent attempt to clarify the questions and to bring data and rigorous analysis to bear on how to answer them.

17.2 Perspectives on Economic Analysis

One philosophical assumption followed by many economists sees each of the above questions as straightforward issues of fact, and construes economic science as an empirical inquiry that succeeds to the extent that it produces an accurate description or measure of the factual matters implicit in each question. So although the question of what we *should* want (for example, in terms of the welfare of animals) may be inherently ethical, the question of what agents in society (e.g. consumers) do in fact want (or prefer) might be inferred by careful observation of their behaviour. Although the question of which policy choices should be made is inherently political, it may be possible to estimate which policies most closely approximate outcomes consistent with the preferences of individuals as evidenced in their behaviour. In addition, it may be possible to provide accounting of winners and losers by identifying which parties' preferences are satisfied by the policy outcome. Given this orientation, economics can be understood as the development of methods that model these social facts (e.g. preferences with regards to animal welfare) through the study of economic behaviour (e.g. people's expenditure on 'animal welfare friendly' products, or how much they donate to animal welfare causes).

Economists who have conducted their researches under the influence of this philosophical paradigm have been careful to note limitations in their ability to provide adequate models of facts observable from economic behaviour. Early theoretical models made a number of implausible assumptions on the way to inferring that actual choice behaviour reveals the preference of an economic agent: agents were assumed to act rationally and with full information, for example; for an emotive issue such as animal welfare with substantial information deficiencies/asymmetries, this is unlikely to be the case. Despite these acknowledged limitations, a number of factors have combined to make this conceptualization of economics into a seductive one. For politicians in democracies, the prospect of aligning policy with outcomes that maximize the number of individuals enjoying benefits was especially attractive. Also, the scientific and empirical orientation toward these questions allowed economists themselves to assume a position of impersonal authority. The combined effect of these factors is that admittedly incomplete economic models have had substantial influence on the resolution of inherently ethical and political issues without seeming to take on subjective or contestable ethical positions.

Yet in addition to implausible assumptions, studies of economic behaviour generally involve a number of additional value commitments, often implicit within the property rights, legal rules and history of accumulation associated with the status quo (e.g. the prevailing status of different species in society). Thus pertinent to studies that calculate the economic impact of regulations to promote the welfare of animals, these estimates are based on economic behaviour that presupposes virtually no legal rights or protection for animal interests. Similar studies might well have accurately predicted significant (negative) economic impacts from laws banning the practice of human slavery. It is therefore possible to question whether putatively scientific studies based on economic behaviour ever transcend the normative commitments implied by the accretion of custom, law and accumulated wealth or influence inherent in any given characterization of the status quo. In other words, in the context of animal welfare, for example, conventional economic analyses of animal welfare policy, and the results that flow from them, are themselves largely a result of the status quo (e.g. regarding current human–animal relations, institutional arrangements, etc.). For example, prices are largely determined by institutional arrangements, so that if it is legal to give no consideration to the welfare of animals then the cost of animal production and the price of animal products will be lower than if there are laws safeguarding animal welfare.

It is in recognition of this situation that Daniel Bromley has proposed a more pragmatic interpretation of the capabilities and orientation of economics. For Bromley, the status quo is always shaped by both formal and informal economic institutions: the legal rules that govern exchange and contracts, and behavioural tendencies reflective of custom and existing distributions of status, wealth and social influence. In this connection, Bromley notes that any given status quo will be diversely regarded as satisfactory or problematic by diverse interests within society. For example, certain current livestock production methods may be generally accepted by farmers but rejected by some consumers. Economics can provide a service by helping individuals and social decision makers to gain a clearer grasp of the likely consequences of political action or policy change. Nevertheless, Bromley argues, it can be quite misleading to present these outcomes within a larger framework of social costs or benefits. To characterize an outcome as a cost or benefit is to imply an ethical value judgement that may well be contestable (Bromley, 2006).

17.3 Economic Perspectives on Animal Welfare

Both conventional economic science and a pragmatic approach to economic institutions can be extended to the issue of animal welfare. From the mainstream approach, the task of economics is to model facts derived from a conceptualization of economic behaviour. From this perspective, it is logically possible to imagine the health and cognitive well-being of non-humans as being one aspect of the outcomes flowing from the collective behaviour of economic agents transacting for personal gain. The utilitarian ethics of Peter Singer, for example, approximate a conventional economic approach by postulating the relative satisfaction or suffering of animals as being among the outcomes given consideration in computing the optimal mix or allocation of satisfaction of benefit and cost (Singer, 1993). However, it has been far more typical for economists to presume that these animals do not themselves evince any specifically economic behaviour. As such, economists presume that any economic analysis of animal interests must be reflected in the economic behaviour of human beings. McInerney (1994, pp. 13–14), for example, writes:

> Animal welfare is just a subset of man's perception of his own welfare, and only indirectly to do with what is good for animals. There should be no surprise, therefore, that the welfare standards a society pursues are a coincidental outcome of its primary concern – the pursuit of human welfare … In economic terms animals are no more than resources employed in economic processes which generate benefits for people.

This perspective also receives philosophical support from Jeremy Bentham's assertion (1789, Chap. XVII, para. 6) that animals 'stand degraded into the class of things'. From this perspective, the fact that animals may suffer in the process of producing goods and services for humans becomes a (regrettable) side effect of the production system. Economists often refer to such indirect effects as 'externalities' (environmental pollution is another example of such an externality). Within this framework, animal welfare matters only because of human sensibilities towards animal suffering that affects human welfare. Thus, those that feel empathy with animals may feel pain (or pleasure) associated with their perception that animals are unnecessarily suffering (or experiencing well-being), which reduces (or increases) their human welfare (utility) and so imposes a cost (or benefit) on them and, where a number of individuals are so affected, on society as a whole.

Arguably, concerns over poor animal welfare are determined entirely by people's perceptions of animal suffering and how they interpret what they see or measure in terms of animal behaviour, changes in physiological processes, etc. People's perceptions are likely, therefore, to depend on the degree of anthropomorphism or reference to human suffering that is involved (Sandøe and Simonsen, 1992; Mason and Mendl, 1993). Perceptions are related to what people consider to be the necessary or unnecessary, or acceptable or unacceptable uses of animals, which depend, in turn, on the perceived feasible alternatives (alternative production systems, products etc.) that are available. For example, many people may feel that the slaughter of farm animals is necessary and acceptable (although they may recognize that some animal suffering will be involved even with 'humane' methods) because they see no other alternative if they are to continue to eat meat. However, these same individuals may consider the caged production of eggs unacceptable because they see a more acceptable alternative in the form of free-range systems. Others may feel that all farm animal production systems are unacceptable and find acceptable non-animal alternative products to consume. It is clear that these perceptions are influenced by a host of factors ranging from cultural practices to aspects of lifestyle, and revolve around people's awareness about current uses of animals and alternatives to them.

A pragmatic or institutional approach would begin with the observation that animal welfare is problematic in multiple ways. First, many human uses of animals can be understood as causing problems for the animals themselves, though as above, economics is not in itself a reliable source of insight into these problems. However, having some sense in which current use causes problems for animals will be critical to any analysis that is attempting to determine whether changed economic institutions result in an improvement in the condition of animals. Secondly, political action by animal advocates provides direct articulation of respects in which they view the status quo to be problematic; they advocate specific institutional changes. Finally, the proposal for institutional change itself becomes a potentially problematic dimension of the status quo for other humans who use animals, either directly as in the case of researchers, livestock producers and

R. Bennett and P. Thompson

other keepers of animals, or through the consumption of animal products. One potentially significant implication of the pragmatic approach is that it illustrates the potential for divergence between the actual impact of an outcome for animals themselves and its impact on the preferences of human beings. It is at least logically possible for an outcome that even animal advocates find satisfying to make the situation for animals themselves worse. A US ban on horse slaughter on ethical grounds, for example, arguably created economic incentives that resulted in both the abandonment of unwanted horses and the long-distance shipment of horses to slaughtering facilities in Canada and Mexico. An economic analysis focused solely on human feelings as surrogate for animal interests might have failed to deploy analytic methods that detected this possibility.

17.4 Economic Analysis and Animal Welfare – Weighing Costs and Benefits

Economists use markets and market prices as a basis of values for both benefits and costs. However, they recognize that this can be problematic because markets do not necessarily reflect the true values of things to society. This is because: (i) market prices only reflect the value of things that are exchanged through those markets and only reflect the preferences of people participating in markets, so there may be indirect costs and benefits ('externalities') that are not represented by market prices; (ii) market prices may be distorted in some way – for example, by government subsidies or taxation; and (iii) market prices reflect marginal values not total values. Therefore, we cannot rely solely on market prices for information on the value that we place on something in society. This means that economists must sometimes use other means of valuing benefits and costs.

The negative animal welfare externalities (animal suffering) associated with the production of animal products and services do not explicitly feature in the markets for animal products and so remain as hidden costs. An important aspect of these externalities is that the preferences of people *not* buying goods in markets are not considered. For example, consider the market for veal produced by keeping calves in crates. The preferences of people that consume veal are recorded in terms of the quantity they are willing to buy and the price they are willing to pay. But there may be people in society who experience considerable disquiet about the use of calves in this way, which reduces their (human) welfare and so imposes a cost on them (and hence, within a cost–benefit analysis framework, on society). These people have no way of expressing their wants through the market system. Even people who consume veal may not be entirely happy with how it is produced and would prefer that an alternative method were used, but they can only express their feelings by not buying veal, assuming that they are not offered a more 'calf-friendly' alternative. Because markets only take account of the preferences of a subset (perhaps a minority) of people in society, they fail to allocate resources in a socially optimal way that maximizes the net benefit to society as a whole. Again, the economic argument is that the costs associated with such animal welfare externalities should be taken into account in making decisions about the use of animals. A formal economic analysis of how economics might address this issue is provided by Bennett (1995), and only the main argument of that analysis is presented here.

The above considerations have some important implications for policies toward animal welfare. A conventional approach in economics raises two major questions. How can we estimate the optimum level of animal exploitation and how might we best ensure that this level is actually achieved? Figure 17.1 shows a likely relationship between the production of animal goods and services (food products, companionship, research benefits, etc.) and animal welfare. The relationship in Fig. 17.1 assumes that up to a point (i.e. from A to B) animals and humans may derive mutual benefit from their association (although some may question this assumption – even for companion animals). Thus point B marks a relationship of maximum welfare for animals with some important benefits for the human species. However, this does not maximize the output of animal products for human benefit. This would be achieved at point D, but at a cost to the welfare of animals. If humans exploited animals beyond this point their welfare would be so affected that they would no longer be efficient producers of food and other products and services, and hence operating beyond D would be highly inefficient. The decision for society is where on the curve between B and D should we be? No doubt animals would prefer point B, whereas a human society uncaring about the welfare of animals (unless it affects production) would prefer point D. Arguably, within society at present, we might be operating at a point such as C. This implicitly gives a relative value to the welfare of animals (shown by the slope of the tangent at C which is $-pAP/pAW$, where pAP is the implicit price of animal products and pAW the implicit price of animal welfare), because it implies that, as

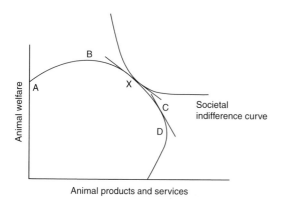

Fig. 17.1. The relationship between the production of animal products and services and animal welfare. From A to B animals and humans derive mutual benefit. B = max. animal welfare and D = max. animal product/service output. X and C denote higher and lower implicit societal values placed on animal welfare; X maximizes the net benefit to society. (From Bennett, 1995.)

a society, we would be unwilling to forgo a unit of animal product to gain an additional unit of animal welfare (or vice versa). This unit of animal product (for example, milk or eggs in the case of farmed livestock) may be traded in markets and have a market price attached to it. Effectively then, society is placing an implicit (money) value on animal suffering. However, we may feel that point C does not accurately reflect people's concerns for animal welfare and that perhaps if we knew society's true preferences – shown by the societal indifference curve – then we should rather be at point X, which gives a greater implicit value to animal welfare relative to animal products than the value given at point C (i.e. a lower level of output of animal products but higher level of animal welfare). At point X, the ratio of the marginal cost of production of animal goods to the marginal cost of animal welfare is equal to the ratio of the marginal utility (i.e. the benefit to society at this point) of production of animal goods to society's marginal utility of animal welfare. From a theoretical economic viewpoint, this is an optimal position (of production of animal products and of levels of animal welfare) that maximizes the net benefit to society.

Figure 17.2 shows a further orthodox economic theoretical analysis of the optimal use of animals in society. The curve labelled TSB shows the total social benefits (from consumption) of increasing

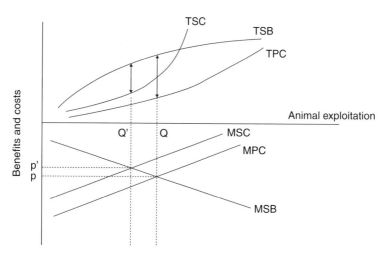

Fig. 17.2. Private versus social costs and benefits of animal exploitation. TSC = total social cost; TSB = total social benefits; TPC = total private cost; MSC = marginal social cost; MPC = marginal private cost; MSB = marginal social benefit; Q = level of animal exploitation when only private costs (not social costs) are taken into account; Q' = socially optimum level of animal exploitation, where social costs are taken into account; p = market price at level of animal exploitation Q; p' = market price at level of animal exploitation Q'. (Derived from Bennett, 1995.)

R. Bennett and P. Thompson

our use of animals to produce food, etc. (labelled as 'animal exploitation'). As we increase our use of animals and so produce, and consume, more food and other animal products, so our benefit increases. The benefits of consumption of animal products can be estimated from information concerning people's willingness to pay for those products in markets. The line TPC denotes the total private cost of production/consumption; in other words, how much it costs (in terms of the money value of resources – as estimated in the studies discussed in the very first section of the paper) to produce a particular quantity of animal products. The direct costs of production of animal products can also be estimated according to the value of resources needed to produce them. The difference between the two is the net benefit to be derived from consumption – which is maximized at level Q of animal exploitation (i.e. the greatest difference between TSB and TPC). The line labelled MSB is the marginal social benefit associated with the TSB. It is, in fact, a consumer demand function which shows how much consumers are willing to pay, in aggregate, for different quantities of animal products denoted by the level of animal exploitation. The line MPC is the marginal private cost associated with production/consumption and forms the producer supply function. It can be seen that maximum net benefit is achieved where MSB = MPC which is where the two lines cross and where demand equals supply and the market is said to be in equilibrium (i.e. where the quantity of animal products produced equals the quantity bought by consumers). This equilibrium would be achieved in the market at price p. However, point Q is not the optimum level of animal exploitation for society. This is because the externality costs of animal use have not been taken into account. If these were to be adequately taken into account then the true costs of animal exploitation would be represented by line TSC (the total social cost) instead of TPC. This means that a lower level of animal exploitation/production of animal products – shown by point Q' – now maximizes net benefit, and this is the true optimum level of animal exploitation. The corresponding marginal cost curve of MSC (marginal social cost) shows that this level of animal exploitation would be achieved in the market at price p' (which is higher than the former price, p). There are two main implications that follow from this simple analysis. First, unless we take account of the animal welfare externalities associated with

animal use we will overexploit animals and reduce the net benefit for society that could be achieved. Secondly, to achieve this optimum level of animal use we may need to ensure that animal use and animal products have a higher explicit market price which reflects their true cost to society. Of course, there are other means of achieving the optimum level of animal exploitation – for example, by regulation that imposes level Q'. Policy instruments for improving animal welfare are considered further in Section 17.5.

For a pragmatic or institutional approach, the point of economic analysis is ultimately to inform an ethical and political discussion concerning institutional change. It is not to measure putatively optimal distributions, and an economist working in this tradition would be more sceptical about the suggestion that economics *can* identify the optimum level of animal exploitation. Existing consumer preferences reflect long-standing customs, not to mention considerable ignorance about the circumstances under which animals are actually kept. This ignorance is itself complex and consists of a wide range of errors. For example, many consumers may idealize the conditions currently existing in livestock production, failing to recognize a dissonance with their own values. Others may overestimate the suffering of animals as they are currently kept or alternatively overestimate the relative benefits of alternative means of keeping animals. In either case, current consumer preferences merely reflect property rules and customary practices; they might be quite different under different rules or different histories. This does *not* imply that the study of latent consumer demand for animal welfare is unimportant from a pragmatic perspective. If a given configuration of market rules and practices provides little opportunity for consumers who would pay for more humanely produced animal products, the discovery and documentation of this latent demand contributes to a more thorough characterization of the problem definition. What is more, such studies provide a basis for more accurately (if still imperfectly) predicting the likely consequences of a policy change that empowered choice based on these latent preferences.

Whether one takes a conventional or an institutional view of why one is doing the economic analysis, the procedures that can be used to estimate the values of non-market goods, such as the externalities of animal welfare, are the same (Mitchell and Carson, 1989). One of the most obvious procedures is to ask people what something (usually referred to as a 'good' in economics) is worth to them. This is the basis of methods such as contingent valuation and choice experiments, sophisticated survey techniques which present people with hypothetical market systems involving the good in question, and then elicit their willingness to pay for the good. A conventional economist would elicit information from people on what they would be willing to pay for animals to be exploited at different levels (thereby obtaining an estimate of the value they assign to animal welfare externalities) in order to determine the optimum level of animal exploitation for society. An institutional economist would interpret the same information not as evidence of the overall *value* to be assigned to animal welfare, but as a basis for predicting and understanding behaviour. A case study of animal welfare valuation is presented in Section 17.6.

The institutional approach runs into an immediate problem in this respect, for it is well known that the response given to surveys seldom matches actual behaviour in the marketplace. The amount of willingness to pay for animal welfare measured in surveys is almost never observed when putatively welfare-friendly products are actually made available in markets. There are several possible explanations. It may be that people lack confidence in labels proclaiming animal welfare-friendly production methods, for example. However, it may be more plausible to interpret these survey results as evidence for economic behaviour in the domain of political economy, rather than consumer choice. That is, responses to surveys more accurately reflect how people will behave in elections or other political situations than how they will make purchases in the grocery store. The phenomenon of 'binding' occurs when someone engages in behaviour that limits the choices they can make in other situations (Elster, 1979, 2002). Under this scenario, people support institutional changes through political action that run *contrary* to the choices that they would make in consumer situations, if given the opportunity. Recent work suggests that some individuals may go so far as to engage in consumer choice with the *intention* of influencing political outcomes rather than of satisfying personal preference (Lusk *et al.*, 2007). This result is almost wholly inconsistent with the assumptions of the conventional economic paradigm, and suggests that even certain economic behaviour has a symbolic function that is not fully captured in the conventional understanding of cost–benefit analysis.

17.5   Animal Welfare Policy Instruments

The economic framework presented above helps to consider possible policy options for reducing the social costs associated with animal exploitation. In relation to farm animals, Bennett (1995, p. 58) wrote:

> There are three main policy options for trying to achieve a balance between the production of livestock products and farm animal welfare which optimizes the welfare of society. The first is to use the market mechanism and allow people to make an informed choice about the products they consume. For this to work, consumers need to be well-informed about the animal welfare characteristics of the products they purchase and about alternative production practices and products. This would no doubt require government intervention in the supply of information. However, even given such a high level of awareness on the part of consumers ... the market mechanism would still fail to adequately capture the negative animal welfare externalities, particularly the public bad *[negative externality that imposes a cost on people generally]* aspects of livestock product consumption. In order to try to address these aspects, government could intervene in two main ways. First, it could regulate the production of livestock products through legislation or codes of practice to ensure that the preferences of citizens for animal welfare friendly production practices are serviced. Secondly, it could tax those producing negative animal welfare externalities (the polluter pays principle) and/or subsidise those producing animal welfare goods, where those goods are not valued by the market.

In relation to Fig. 17.2 presented above, it is clear that the market mechanism could be used, in theory, to achieve a level of animal exploitation closer to the optimum by appropriately pricing the exploitation of animals. This could take the form of either a subsidy for producers/products which are perceived as resulting in good animal welfare or a tax on producers/products which are perceived as resulting in animal suffering. Alternatively, government could intervene through legislation and ensure that only the level of animal exploitation takes place that it believes would be optimal.

In practice, there is a host of different policy instruments that might be used to help protect or improve the welfare of animals. Table 17.1 provides a categorization of these policy instruments, with examples of their application to animal health and welfare, and some assessment of their relative merits and limitations (see FAWC, 2008, for more detail).

These policy instruments can be used singly or in combination with one another. Historically, policy makers have tended to use legislation as the main policy instrument to protect or improve animal welfare (this aspect is covered in some detail in Chapter 18, Incentives and Enforcement). Furthermore, in terms of a theoretical optimum use of animals, data have not been available to gain a clear idea of where the optimum level of animal exploitation might lie. Policy makers and others have rather put forward proposals concerning the level that they consider appropriate, based on scientific evidence, moral beliefs and the political process.

McInerney (1994, p. 18) wrote:

> If economic analysis is to extend beyond the drawing of diagrams which set out the conceptual basis for

decisions on animal welfare standards, the task of research is to identify the structure and relative weight of those preferences in society if our notion of a social optimum is to be pursued. On the other hand, imposing a welfare standard institutionally by the establishment and enforcement of particular codes raises the question as to *whose* value function it claims to reflect.

This emphasizes the need, in democratic society, to know people's preferences for animal welfare in order for government to set standards or intervene with other policy instruments to protect and improve animal welfare on society's behalf.

The next section presents a case study of research aimed at eliciting people's willingness to pay for farm animal welfare improvements.

Table 17.1. Categories of policy instrument for protecting or improving animal welfare. (Adapted from FAWC, 2008.)

Type of policy instrument	Example applied to animal welfare and health	Strengths	Weaknesses
1. Legal rights and liabilities	EU (European Union) Protocol on Animal Welfare	Self-help	May not prevent events resulting from accidents or irrational behaviour
2. Command and control	Minimum space rules for poultry	Force of law Minimum standards set Transparent	Costly Inflexible
3. Direct action (by government)	Welfare inspections Border controls	Can separate infrastructure from operation	Danger of being perceived as 'heavy handed'
4. Public compensation/ social insurance	Compensation for animals slaughtered for welfare reasons	Insurance provides economic incentives	May provide adverse incentives Can be costly to tax payers
5. Incentives and taxes	Cross compliance Pillar II monies for farm animal welfare improvements (see Chapter 18)	Low regulator discretion Low cost application Economic pressure to behave acceptably	Rules required Predicting outcomes from incentives difficult Can be inflexible
6. Institutional arrangements	EFSA (European Food Safety Authority) Sub-committee on Animal Welfare	Specialist function Accountability	Potential for narrow focus of responsibility
7. Disclosure of information	Reporting of notifiable diseases Labelling	Low intervention	Information users may make mistakes
8. Education and training	Animal welfare in veterinary education, national school curriculum	Ensures education and skills required by society	Can be too prescriptive and inflexible
9. Research	Funding for animal welfare research (e.g. Welfare Quality® project)	Provide information to policy	May duplicate or displace private sector activities

Continued

Table 17.1. Continued.

Type of policy instrument	Example applied to animal welfare and health	Strengths	Weaknesses
10. Promoting private markets			
(a) Competition laws	Market power of companies in the food supply chain, and prices to farmers to meet production costs	Economies of scale in use of general rules Low level of intervention	No expert agency to solve technical/commercial problems in the industry Uncertainties and transaction costs
(b) Franchising and licensing	Veterinary drugs/treatments Animal husbandry equipment	Low cost (to public) of enforcement	May create monopoly power
(c) Contracting	Hire of private vets to provide public services	Combines control with service provision	Confusion of regulatory and service roles
(d) Tradable permits	Permits for intensive livestock production systems (e.g. The Netherlands)	Permits allocated to greatest wealth creators	Require administration and monitoring
11. Self-regulation			
(a) Private	(a) Farm assurance schemes, veterinary profession, industry codes of practice	High commitment Low cost to government Flexible	(a) Self-serving; monitoring and enforcement may be weak
(b) Enforced	(b) Member State enforcement of EU legislation	Enforcement ensures greater compliance	(b) Enforcement may be variable

17.6 People's 'Willingness To Pay' for Animal Welfare: a Case Study

A questionnaire was designed and a telephone survey of 300 people in Great Britain was recently undertaken to elicit their attitudes to and willingness to pay for farm animal welfare. The sample was selected randomly but stratified to represent the demographic and socio-economic characteristics of the population. The questionnaire was structured into four main sections. The first section contained questions on people's attitudes to farm animal welfare and questions concerning their consumption of livestock products. The second section provided them with information about methods by which the welfare of farm animals throughout their life could be assessed and measured on an index scale of 0–100, where 40 would denote the current legal minimum standard and 100 would denote the highest possible levels of welfare for the animal. Thus, for example, a welfare score of 60 would denote an animal with a substantially higher level of welfare than legal minimum standards but with potential

for these standards to be higher still. The welfare assessment method and score was compatible with those developed by the EU (European Union)-funded Welfare Quality® project (Welfare Quality®, 2010). The next section of the questionnaire presented respondents with a series of willingness to pay questions. A sample of these is shown in Box 17.1. The levels of welfare for which respondents were asked for their willingness to pay were in two pairs of 60 and 80, and 70 and 90. Description of the animal welfare assessment and scoring method and the willingness to pay questions were sent by post to respondents before the telephone interviews so that they could read the material beforehand and have it in front of them during the interview. Following these willingness to pay questions, there was an open-ended question asking people to explain briefly the reasoning behind their responses to the willingness to pay questions, and some further attitudinal questions regarding animal welfare and the scenario presented to them. The final section asked some socio-economic questions of the

Assume that in your usual food store there are meat and meat products with high welfare scores. If you buy this meat with a welfare score above the legal minimum of 40 your monthly food bill will rise.

If the meat had a welfare score of 60 would you buy it for the extra cost of …?

	Definitely yes	Probably yes	Don't know	Probably no	Definitely no
£5 per month					
£11 per month					
£22 per month					

If the meat had a welfare score of 80 would you buy it for the extra cost of …?

	Definitely yes	Probably yes	Don't know	Probably no	Definitely no
£8 per month					
£16 per month					
£32 per month					

respondents, such as age, income, family size and level of education. Analysis of willingness to pay responses was undertaken using a Bayesian Ordered Probit model.

Results of the analysis showed a mean willingness to pay for meat with a welfare score of 60 of £19.31 per month, or nearly £232 per year. For meat with a welfare score of 80, people's stated willingness to pay was £23.63 per month, or nearly £284 per year. Over the 60–80 welfare score range, this gives a marginal willingness to pay of £2.59 per unit of welfare score per year. In contrast, people's marginal willingness to pay over the 70–90 welfare score range was £1.36 per unit, suggesting that people give a higher value to raising animal welfare from 60 to 80 than from 70 to 90. Such diminishing marginal willingness to pay is to be expected beyond a certain point and, indeed, is predicted by conventional economic theory (the well-known 'law of diminishing marginal utility' as more of a product is consumed).

The levels of willingness to pay elicited from people appeared reasonable and rational in relation to their incomes and their responses to attitudinal questions. The survey highlights that not only do the vast majority of people say that they care about animal welfare but most also state that they have a willingness to pay to improve it. This is a strong indication of people's preferences for high levels of animal welfare but, for reasons discussed previously, may not accurately predict or be reflected by people's actual food purchases or other economic behaviour.

The recommendation of the Farm Animal Welfare Council (FAWC) of Great Britain is that legal minimum standards of farm animal welfare should result in animals having 'a life worth living' (FAWC, 2009). This would equate to the current legal minimum standard welfare score of 40 (or above, if current standards are thought not to give farm animals a life worth living), as used in the survey. In addition, FAWC states that a policy aim should be that an increasing number of animals also have 'a good life'. This would equate to a welfare score of substantially beyond that of legal minimum standards, such as a score of 60 or more.

17.7 Conclusions

- Animal welfare and human welfare are inextricably linked. Economic analysis explicitly acknowledges this and considers the relationship between them.

- It is important that economics is perceived not merely as some financial accounting exercise, but as a discipline capable of incorporating different ethical considerations and using information which can help to make better decisions concerning human use of, and obligations to, animals.
- Economic analysis can help decision makers regarding choice of different policy options and instruments intended to protect and improve the welfare of animals.
- Economic considerations are central to the animal welfare debate and are integral, inescapable aspects of issues concerning the use of animals and of any interdisciplinary inquiry into animal welfare, alongside ethics, veterinary science and other disciplines.

References

Bennett, R.M. (1995) The value of farm animal welfare. *Journal of Agricultural Economics* 46, 46–60.

Bennett, R.M. (1997) Farm animal welfare and food policy. *Food Policy* 22, 281–288.

Bennett, R.M. and Blaney, R.J.P. (2003) Estimating the benefits of farm animal welfare legislation using the contingent valuation method. *Agricultural Economics* 28, 265–278.

Bentham, J. (1789) *Introduction to the Principles of Morals and Legislation*, 1996 Imprint. Clarendon Press, Oxford, UK.

Bromley, D. (2006) *Sufficient Reason: Volitional Pragmatism and the Meaning of Economic Institutions*. Princeton University Press, Princeton, New Jersey.

Boulding, K.E. (1969) Economics as a moral science. *The American Economic Review* 59, 1–12.

Carlsson, F., Frykblom, P. and Lagerkvist, C.J. (2007) Consumer willingness to pay for farm animal welfare – transportation of farm animals to slaughter versus the use of mobile abattoirs, *European Review of Agricultural Economics* 34, 321–344.

Elster, J. (1979) *Ulysses and the Sirens*. Cambridge University Press, Cambridge, UK.

Elster, J. (2002) *Ulysses Unbound: Studies in Rationality, Precommitment and Constraint*. Cambridge University Press, Cambridge, UK.

FAWC (Farm Animal Welfare Council) (2008) *Opinion on Policy Instruments for Protecting and Improving Farm Animal Welfare*. FAWC, London.

FAWC (2009) *Farm Animal Welfare in Great Britain: Past, Present and Future*. FAWC, London.

Harrison, R. (1964) *Animal Machines*. Vincent Stuart, London.

Lusk, J.L., Nilsson, T. and Foster, K. (2007) Public preferences and private choices: effect of altruism and free riding on demand for environmentally certified pork. *Environmental and Resource Economics* 36, 499–521.

Marshall, A. (1947) *Principles of Economics*, 8th edn reprint. Macmillan, London.

Mason, G.J. and Mendl, M. (1993) Why is there no simple way of measuring animal welfare? *Animal Welfare* 2, 301–319.

McInerney, J. (1994) Animal welfare: an economic perspective. In: Bennett, R.M. (ed.) *Valuing Farm Animal Welfare*. University of Reading, Reading, UK, pp. 9–25.

Mitchell, R.C. and Carson, R.T. (1989) *Using Surveys to Value Public Goods. The Contingent Valuation Method*. Resources for the Future, Washington, DC.

PROMAR International (2009) *Impacts of Banning Cage Egg Production in the United States*. PROMAR International, Washington, DC.

Regan, T. (1982) *All That Dwell Therein: Animal Rights and Environmental Ethics*. University of California Press, Berkeley and Los Angeles, California.

Regan, T. (1984) *The Case for Animal Rights*. Routledge, London.

Sandøe, P. and Simonsen, H.B. (1992) Assessing animal welfare: where does science end and philosophy begin? *Animal Welfare* 1, 257–267.

Sen, A. (1987) *On Ethics and Economics*. Basil Blackwell, Oxford, UK.

Singer, P. (1975) *Animal Liberation*. Jonathan Cape, London.

Singer, P. (1980) Animals and the value of life. In: Regan, T. (ed.) *Matters of Life and Death: New Introductory Essays in Moral Philosophy*. Random House, New York.

Singer, P. (1993) *Practical Ethics*, 2nd edn. Cambridge University Press, Cambridge, UK.

Smith, A. (1790) *The Theory of Moral Sentiments*, revised edn. T. Cadell, London. Republished in 1975 by Oxford University Press, Oxford, UK.

Sumner, D.A., Rosen-Molina, J.T., Matthews, W.A., Mench, J.A. and Richter, K.R. (2008) *Economic Effect of Proposed Restrictions on Egg-Laying Housing in California*. University of California Agricultural Issues Center, Davis, California.

Welfare Quality® (2010) Research results. Available at: http://www.welfarequality.net/everyone/34056/5/0/22 (accessed 5 January 2011).

18 Incentives and Enforcement

UTE KNIERIM, EDMOND A. PAJOR, WILLIAM T. JACKSON
AND ANDREAS STEIGER

Abstract

The application of welfare principles can, in practice, be achieved by several different means. The first part of the chapter deals with one of these: legislation. The main legal instruments involved are intergovernmental agreements, supranational statute law, and national law and case law. Increasingly, international harmonization is taking place to engender common ethical attitudes and to remove barriers to trade and distortion of competition. Limitations in terms of the efficacy of legislation are discussed. They relate, for example, to difficulties and differences in implementation and to a relatively low flexibility in response to public demands or development of technological advances. The second part of the chapter describes examples of measures to raise animal welfare standards that work either as incentives to improve animal welfare or work on a voluntary basis. These measures may be developed and applied by public institutions or by private organizations. Examples are private activities resulting from an ethical commitment, non-governmental standards for biomedical research and the incorporation of aspects of farm animal welfare into supply chain management in the private sector. Public incentives may include direct payments or other rural development support measures subject to the fulfilment of either legal standards or extra standards, public information campaigns and the incorporation of animal welfare aspects into teaching and training curricula. Legislation, public incentives and voluntary measures complement each other. Their appropriate combination is important for the long-term improvement of animal welfare.

18.1 Introduction

People who keep animals or interact with them may do this in an animal welfare-friendly way out of intrinsic motivation. Yet often there are competing motivations – such as economic interests or forces, human health or safety interests, aspects of convenience or tradition, or inadequate knowledge – which conflict with the interests of the animals. The weighing up of the differing interests will result in very different conclusions for different individual people, for example according to their cultural, religious or professional background.

However, the weighing up of various interests is also taking place at a societal level, and when found important enough, this leads to the setting of standards in order to safeguard a certain level of animal welfare. Very often this takes the form of legislation (statute law) or court case decisions (case law). For example, as early as the third century BC, Asoka, King of Maghada in North India, who was a Buddhist, implemented legislation that abolished sacrificial slaughter, banned the Royal Hunt and planted trees to give shade to humans and animals (Brown, 1974). More than 1800 years later, a body of case law concerning the treatment of animals gradually began to arise in various parts of the world. Attempts at the establishment of statute law in Europe began in England and Wales only in 1800, and it took until 1822 before a law was passed to establish the legal obligation to treat animals humanely; the law was 'to prevent the cruel and improper treatment of cattle'. This was followed by numerous pieces of statute all over the world, for example within the criminal law of the German kingdom of Saxony of 1838, the French 'Loi Grammont' of 1850 and the 1866 New York Cruelty to Animals laws.

Historically, it was only an offence to mistreat animals in public. The goal was to protect the public from witnessing cruel behaviour. However, increasingly, animals became legally protected for their own sake as sentient living beings, in line with ethical views expressed by, for example, the 18th century utilitarian philosopher and lawyer Jeremy Bentham, who argued that animals deserve moral and legal protection because they can suffer (see Chapter 1).

In the last century, pioneers giving worldwide impetus for improved animal welfare were, in particular, Ruth Harrison with regard to farm animals, and Russell and Burch in the field of animal experimentation. In her book *Animal Machines*, Harrison (1964) criticized the intensive housing of laying hens, calves and pigs, and initiated a worldwide public discussion on animal welfare. This resulted in the UK Ministry of Agriculture forming an expert committee to investigate the welfare of farm animals under intensive husbandry systems, known as the Brambell Committee. In 1965, this committee published the Brambell Report, in which many principles of good husbandry and the foundation for the development of the 'Five Freedoms' (FAWC, 2009) were articulated (Brambell Committee, 1965). Russell and Burch (1959) developed the well-known 3R principles of animal protection in animal experimentation: replacement (no use of animals at all), reduction (less animals per experiment) and refinement (less suffering of the animal).

Legal minimum requirements concerning the use of animals are not the only means to raise or safeguard welfare standards. Although countries vary considerably, the past few decades have seen a general increase in concern about animals, reflecting a change in societal attitudes towards them. More specifically, a new social ethic has emerged that moves away from traditional emphasis on animal cruelty towards concern about their welfare (Rollin, 2004). In line with this, over time, a large number of further measures have developed that predominantly work as incentives to improve animal welfare, while others work on a voluntary basis. These measures may be developed and applied by public institutions or by private organizations. Examples of such measures will be discussed below, in addition to an outline of various legal devices. We will not attempt to give a complete overview of the measures in force or employed worldwide. This would be futile considering the abundance of measures, and especially because legislation, label programmes, etc., evolve and change quite rapidly. Instead we will explain the basic principles, using measures currently in force for purposes of illustration. Being from Europe and North America, we may be accused of a limited view and of neglecting measures used in other parts of the world. This is true, and the same applies for a certain emphasis on farm animals. However, the use of enforcement and initiatives to enhance animal welfare are currently mostly discussed and applied in these regions, and with regard to those animals. We recognize that this may well change over time, as is already visible, particularly in regions such as South America and Asia.

18.2 Legislation

Legislation in democratic countries is shaped by what is legally and politically possible. Animal welfare measures may not unduly violate personal human rights (for example concerning freedom of religious practices or scientific work), and while they should be based as far as possible on scientific knowledge and established experience, they always represent some kind of compromise between differing views of the different stakeholders involved, including parliaments, governments and authorities, if applicable.

There are a number of different ways by which animal welfare measures can be implemented or influenced by legislation. The main legal methods are intergovernmental agreements (treaties, conventions), supranational statute law (e.g. European Community (EC) Directives or EC Regulations), national statute law and case law (as authority in itself or used to interpret statute law).

18.2.1 Intergovernmental agreements

Increasing worldwide trade and the removal of borders have a considerable impact on the use of animals. Traffic in animals (farm animals for breeding or slaughter, wild animals for zoos and private owners, experimental animals and companion animals for breeding or selling) expands, thereby increasing competition between farmers, industry and scientists of the different countries. Standards which impose restrictions on production or research and impair competitiveness will more readily be accepted and also be more effective if they must be equally observed by all competitors. Thus, for an effective protection of animals, there is a need for international harmonization, although it is often much easier to reach agreement only at the national level, and progress can be considerably slowed down. Additionally, for the sake of compromise, standards in those agreements are often formulated in a way that leaves ample room for interpretation. The wider the range of countries with differing attitudes towards animal welfare is, the more general and vague the wordings of recommendations or conventions have perforce to be. However, such international standards also largely help to promote the idea of animal welfare in countries which have as yet either minimal animal welfare legislation or none.

International harmonization may take place at very different levels. One possible level is that certain international institutions, set up on the basis of a convention (which is a special form of treaty between two or more sovereign states), become active in the field of animal welfare, although this is not their primary mission. The activities of the World Organisation for Animal Health (OIE, originally Office International des Epizooties) or the Organisation for Economic Co-operation and Development (OECD) are examples of this. The activities of the latter organization are, moreover, an example of rather indirect effects on animal welfare (see below). Conventions dealing explicitly with animal welfare are another possible level, which can be found for example within the Council of Europe. The most direct influence is given when, based on a treaty, states form a supranational body which, among other activities, issues animal welfare legislation that is binding to all member states. This applies to the European Union (EU). More detail on all these examples will be presented in the following sections.

World Organisation for Animal Health (OIE)

The OIE was initially created in 1924 in Paris by 28 countries. Its main mission is to improve animal health worldwide (OIE, 2010a). Over its history, the OIE has grown to 175 member countries in 2010. It is a very influential organization, and its standards are used as the international reference in the field of animal diseases for the World Trade Organization (WTO). In 2002, recognizing the link between animal diseases and the suffering and welfare of animals, the OIE received a mandate to develop guidelines in animal welfare that could be used for international trade and serve as a foundation for legislation in countries that currently do not have animal welfare legislation. The OIE insists that its guidelines be science based and its efforts are guided by eight principles. These include the Five Freedoms, the 3Rs, the recognition of value assumptions as being part of animal welfare and the use of animal-based criteria rather than design criteria as the basis for comparing

guidelines (OIE, 2010b). The *OIE Guiding Principles on Animal Welfare* were included in the *OIE Terrestrial Animal Health Code (Terrestrial Code)* in 2004; since 2005, the OIE has developed six animal welfare guidelines in this Code. These guidelines cover: (i) the transport of animals by land; (ii) the transport of animals by sea; (iii) the transport of animals by air; (iv) the slaughter of animals for human consumption; (v) the killing of animals for disease control purposes; and (vi) the control of stray dog populations. This was the first time that a global governance organization had provided guidance on the issue of animal welfare (Ransom, 2007). The OIE's future activities include developing guidelines for the housing and management of farm animals, as well as laboratory animals (OIE, 2009a). While the OIE's development of broad guiding principles and specific guidelines is a step forward for animal welfare worldwide, there are also associated limitations and risks. In order to gain agreement within such a large international organization which, moreover, traditionally concentrates on sanitary measures, the OIE's welfare guidelines address basic needs and are set relatively low compared with numerous other intergovernmental agreements, national standards, or even industry guidelines or codes of practice. The OIE's guidelines may provide obtainable goals for countries that are currently developing animal welfare standards, but they may also lead to inertia among other countries or organizations that already meet the guidelines. An issue of greater concern is that standard-setting organizations themselves are likely to advocate for the general acceptance of their own standards and actively work against the setting of any higher welfare standards. For example, the OIE claims that it has become 'the leading international organisation for animal welfare and for publication of standards and guidelines in this field', and names as one important objective the avoidance of unjustified trade barriers (OIE, 2009b) that may emerge through the setting of higher welfare standards in certain countries or regions.

Organisation for Economic Co-operation and Development (OECD)

The OECD was founded in 1961 in Paris. The organization has, at present (June 2010), 31 member states (19 EU Member States, Australia, Canada, Chile, Iceland, Japan, Korea, Mexico,

New Zealand, Norway, Switzerland, Turkey and the USA). According to its mission statement (OECD, 2010):

> OECD brings together the governments of countries committed to democracy and the market economy from around the world to support sustainable economic growth, boost employment, raise living standards, maintain financial stability, assist other countries' economic development and contribute to growth in world trade.

The organization aims to coordinate domestic and international policies (OECD, 2010). Among its different and very broad activities, the OECD has adopted guidelines on the testing of chemical substances before they are brought on to the market. The guidelines on the one hand impose certain animal testing, and on the other hand ensure that such tests will be acceptable in all OECD states to obviate the need for duplicate testing. At the same time, there is a slow but significant movement to replace or refine methods that involve severe animal suffering. The standard methods are defined in the 'Good Laboratory Practice' guidelines, which harmonize toxicological testing methods of industrial chemicals, pesticides, food and articles of everyday use.

Animal welfare conventions of the Council of Europe

The Council of Europe is another intergovernmental organization, and must not be confused with the EU itself (or the Council of the European Union or the European Council). The Council of Europe has been in existence since 1949, and has 47 member states (as of June 2010). In particular, it undertakes action in the legal field with a view to harmonizing national laws. In addition to its well-known work in the field of human rights, and other activities, the Council of Europe has set up five Conventions on animal welfare; these concern the international transport of animals (1968, revised 2003), animals kept for farming purposes (1976), slaughter (1979), animals used for experimental or other scientific purposes (1986) and pet animals (1987). These five Conventions are based on ethical concepts common to all the participating countries and aim to avoid unnecessary suffering or injury to animals, and to provide conditions in accordance with the specific needs of animals. All of the Conventions (Council of Europe, 2011a) have mechanisms to elaborate within their frameworks

more detailed recommendations, for example for different species. Contracting parties are bound to bring their rules into line with the Conventions through legislation or administrative practice, although they may maintain or apply more stringent provisions in their own territories. However, instruments of control and strict enforcement are lacking at this level. Thus, the obligation is largely a moral and political one. Moreover, interpretations of the provisions of the Conventions may differ between parties. For instance, Article 6 of the *Convention for the Protection of Animals kept for Farming Purposes* states that 'no animal shall be provided with food or liquid in a manner, nor shall such food or liquid contain any substance, which may cause unnecessary suffering or injury'. Whereas it is the interpretation of many contracting parties that force feeding of ducks or geese (to produce foie gras) must be forbidden following this article, in other countries this practice is continued, with the parties claiming that the treatment does not cause the animals any suffering or injury.

Despite the weakness of enforcement, the Council of Europe Conventions and their Recommendations are important as a foundation of legislation. For instance, at the time of writing detailed standards for the keeping of 13 different farm animal categories including farmed fish and ratites have been adopted, more than are covered by most national law (Council of Europe, 2011b). Moreover, the EU is a contracting party to all of the Conventions except that for companion animals, and EU legislation on animal welfare is to a large extent based on the Conventions and Recommendations of the Council of Europe.

18.2.2 Statute law in the European Union

The EU, which currently (as of June 2010) has 27 Member States, was founded in 1992 by the Treaty on European Union (the Treaty of Maastricht), but it built on the European Communities, which had been established by the Treaty of Rome in 1957. The next step in the unification process was the Treaty of Lisbon, signed by EU leaders in 2007, which came into force in 2009. The Treaty of Lisbon amends the current EU and EC treaties, without replacing them. The now consolidated version of the Treaty on the functioning of the EU states in Article 13:

> In formulating and implementing the Union's agriculture, fisheries, transport, internal market, research and technological development and space policies,

the Union and the Member States shall, since animals are sentient beings, pay full regard to the welfare requirements of animals, while respecting the legislative or administrative provisions and customs of the Member States relating in particular to religious rites, cultural traditions and regional heritage.

Certain principles of EU policy place certain limits on actions in the field of animal welfare legislation. One is the principle of subsidiarity, which here means that any matter that can be regulated on a national level must be left to the Member States. The other principle, on the contrary, imposes restrictions on possible national measures for animal protection. An integral and most important part of the EU Treaty is to ensure freedom of trade and to prevent distortion of competition. Hence, a national ban on, say, the importation of animals or animal products treated or produced in a way not in accordance with national animal welfare provisions is in general against the Treaty. However, this does not prevent Member States from prohibiting certain treatments or production methods as long as it is only their own inhabitants who are saddled with the prohibition. An example is the production of foie gras, which is prohibited in many EU Member States, but whose importation from (say) France, Belgium or Hungary cannot be prohibited or limited. The same applies to, for example, the keeping of fur animals, which is prohibited in a few Member States.

Legislation in the EU is set by the Council of the European Union, which is composed of the national ministers responsible for the specific topic of legislation, one per country. The Council partly shares the legislative power with the European Parliament. The European Commission is responsible for drafting law, and certain legislative power can be delegated to the Commission. Two main types of binding EU legislation are important in the field of animal welfare: EU Directives and EU Regulations. Directives are binding as to objectives only and require national legislation or administrative arrangements in each country to become effective. Explicitly or implicitly, the form of a Directive says that Member States may, within the limits of the Treaty (see above), maintain or apply more stringent provisions than those contained in the Directive. Thus, a Directive allows Member States some flexibility in their decisions concerning national measures. This is the reason why originally exclusively EC animal welfare legislation took the form of a Directive (e.g. Directive 86/609/EEC or Directive 1999/74/EC).

An EU Regulation is binding in its entirety and is directly applicable in all member states. Regulations are used when it is especially important to safeguard a uniform implementation, such as a marketing standards for eggs (Regulation (EC) No. 589/2008) or for organic products (834/2007 and 889/2008). As a recent development, there is a tendency to issue Regulations on animal welfare where European traffic plays a major role. Rules on animal transport as well as on slaughter or killing are now laid down in Regulations (e.g. 1/2005, 1099/2009). Other examples of areas which are controlled by EC Regulations and affect animal welfare are fishing with drift nets (e.g. 809/2007) or the trapping methods of wild fur animals (3254/91).

EU legislation partly contains provisions requiring countries outside the EU to certify that animals or meat that they export to the EU received treatment at least equivalent to that guaranteed by EC legislation (Directives 91/629/EEC and 91/630/EEC on calves and pigs, Regulation (EC) No. 1099/2009 on slaughter). Such mechanisms also exist in the USA, where the requirements of the Humane Methods of Slaughter Act of 1978 apply not only to all federal and state livestock (excluding poultry) slaughter facilities, but also to facilities in foreign countries exporting to the USA (Thaler, 1999). However, it is, in general, questionable whether such provisions are in concordance with the WTO's Agriculture Agreement (WTO, 2010); in fact, it appears that they are not, or they are not fully enforced. All EU law is freely accessible at EUR-Lex (2010).

18.2.3 National statute and case law

Before explaining legal measures at the national level, it is necessary to discuss briefly two different systems of law so as to make differences between some countries easier to understand. The relationship between case law and statute law (as well as procedures) differs depending upon whether countries have common law histories (the UK, the USA, and other countries which have historical links with the UK) or Romano–Germanic law histories (most European countries and those historically linked with them).

Romano–Germanic law is bound up with the law of Ancient Rome and was essentially developed by universities and jurists. Law was brought together in unified codes, a process called codification. The main weight here is on legal texts, which have more authority than cases.

Common law is largely based on unwritten but commonly agreed principles promulgated primarily by judges. It is still being developed further, with judges deciding what the law is in a particular case. Sometimes people misinterpret this process as changes to the law, but that is not so. The judges merely apply well-understood rules of common law rules to a new situation. Some very important criminal offences (e.g. murder) are common law offences. Case law has considerable importance and can have effects on statute law. The decisions of certain courts are binding. For example, in the UK, decisions are binding from the Crown Court or above, i.e. the Queen's Bench Divisional Court, the Court of Appeal or the Supreme Court.

Statutes

A statute is an Act of Parliament. Many jurisdictions have mechanisms requiring the consent of both upper and lower Houses of Parliament, as in the UK (the Lords and the Commons), and especially in those countries with a federal structure. Here, usually one house is the Federal Parliament, and the other is the representative body of the federal states or regions. Examples are the US Congress, which consists of the House of Representatives and the Senate, and the German Bundestag and Bundesrat. Individual federal states may also have a significant level of autonomy with regard to the regulation of the housing, care, transport and slaughter of animals within their borders, as for instance in the USA (Mench, 2008). US state legislatures can also pass statute law. Some states allow constitutional amendments on specific propositions through voter referendums. This process has been used to ban specific on-farm practices, such as the use of gestation crates for sows in Florida, the use of gestation crates and veal crates in Arizona, as well as the use of gestation crates, veal crates and battery cages for laying hens in California. The overwhelming voter support of such propositions and their associated costs has resulted in the passing of similar legislation, voluntarily, in other states, such as Oregon, Colorado and Michigan.

In many countries, statutory provisions relating to animal welfare can be found in specific animal welfare Acts, or in codes or other measures covering a wider range of areas. For example, in France, animal welfare provisions are contained in the *Code Rural*, with penalties being specified in the *Code Pénal*, while in the UK, animal welfare provisions

are made in the *Protection of Animals Acts 1911–1988* and the Agriculture (Miscellaneous Provisions) Act of 1968. In the USA, the US Congress passed its *Animal Welfare Act* in 1966; this focuses on the welfare of animals used in research, although many species are excluded (Mench, 2008). Federal statutes in the USA address farm animal welfare issues around humane slaughter and transportation.

Statutes may be both regulatory and 'enabling', i.e. delegating powers to ministers. The latter method is used where matters are too technical to be decided there and then, or where the precise circumstances under which a need might arise are unpredictable. By this means, the law can be made flexible and hence more effective. Legislation enacted under an enabling Act is called delegated legislation. It can, for instance, be a Regulation, a Statutory Instrument, an Order in Council or a Welfare Code. Although this kind of legislation also passes through Parliament, the procedure is usually easier. For example, in Germany, Regulations need the agreement of only the Bundesrat. In contrast, in the UK, Statutory Rules and Orders have only to be placed in the library of the House of Commons for 3 weeks; if nobody objects, they become law. However, UK Welfare Codes, prepared or revised after consultations, must be approved by both Houses of Parliament.

Welfare Codes should not be confused with legal codes as used in mainland European jurisdictions (e.g. in France). Welfare codes as used in countries including New Zealand or the UK are not mandatory. The UK *Agriculture (Miscellaneous Provisions) Act 1968* provides that the breach of a Code provision, while not an offence in itself, can nevertheless be used in evidence as tending to establish the guilt of anyone accused of causing suffering under the Act (Section 3(4)). As an illustration, the UK 'Highway Code', although established by law, is not itself law. Although this is so, a failure to follow its provisions is normally difficult to justify. If, for example, a driver elects to drive on the wrong side of the road, that is not illegal but if a head-on collision with another driver is the result of such an action, the highway code provision will assist the prosecution of any driver flouting that part of the code.

Case law

Because it is, in practice, rarely possible to frame statute law in terms leaving no room for argument, it is a basic task of courts to ascertain the intention of the legislator. The interpretation of animal welfare statute law by the court concentrates on a particular case. However, it can sometimes be very important for decisions in similar cases. The significance of decisions differs between countries according to their legal system.

In countries with common law systems, the Doctrine of Precedent applies, i.e. a court is bound by decisions of a court above itself in the hierarchy and usually also by a court of equivalent standing. Decisions of foreign courts, although not binding, can act as valuable persuasive authority. In the UK, many cases in Commonwealth or former Commonwealth countries have been quoted and, increasingly, cases from other European countries are being used. The precedent rules are not applied with inappropriate automatism; where necessary, judges have always been adept at 'distinguishing' cases, i.e. finding slight differences of fact, because it is only when facts are exactly the same that decisions based on similar cases are binding.

In contrast, in Romano–Germanic law systems, for example in Germany, courts are formally independent in their decisions from previously decided cases (an exception is the decisions of the Federal Constitutional Court). Even though courts may make use of lines of reasoning in other courts, they may well come to a different decision based on the same set of facts.

A major problem that courts face with regard to animal welfare legislation is the controversial basis for decisions on the needs of animals and what constitutes suffering. Many countries make use of expert opinions on these matters. Expert evidence sometimes differs, but the court decides.

Correspondences and differences in the contents of animal welfare legislation in different countries

Each of an increasing number of countries that have taken legal measures designed for the protection of animals has adopted its own approach. In many statutes, reference is made to the biological needs of animals that must be supplied to spare the animals suffering or harm. A scientific basis is generally regarded as necessary to define such needs and thereby to make sensible decisions on animal welfare.

The kind of animals that are protected differs between countries, depending upon whether the animals are (warm-blooded) vertebrates or invertebrates, and wild, feral or captive animals. In some countries, legislation is mostly limited to companion animals, and experimental and farm animals (e.g. Spain); in others (e.g. the USA in the *Animal Welfare Act* of 1966) it is limited to only warm-blooded animals bred for commercial sale, used in research (excluding rodents and birds), transported commercially, or exhibited to the public. Sometimes, welfare legislation in general applies to all vertebrates and to some specified invertebrates (e.g. Norway and Switzerland) or to the whole animal kingdom (e.g. Austria, Germany). However, independently of the scope of application, the majority of provisions refer to animals used as experimental, farm or pet animals.

In general, there are two ways to regulate. One is to allow anything that not explicitly forbidden (a 'negative list' approach); the other is to forbid anything not explicitly allowed (a 'positive list' approach). For instance, the *Dutch Animal Health and Welfare Act* of 1992 generally forbids the keeping or killing of animals unless permitted under Order in Council. Sometimes both approaches can be found in one Act, and there are good arguments for and against either. The development of a positive list is time-consuming but there is the advantage that it has the effect of stimulating thorough debate. However, even in a positive list, expressions included to provide flexibility to allow for differing situations tend to leave the matter somewhat open to interpretation, and although a positive list might be more restrictive than a negative list, this need not necessarily be the case, depending on the generality of the terms used.

Some countries have identified certain areas as in need of special control and, to this end, have introduced the obligation to obtain official approval of matters in those areas. This sometimes applies to the keeping of wild animal species (e.g. Switzerland, the UK, Austria), to the installation or reconstruction of farm animal holdings (e.g. Sweden), to the commercial breeding and trading of certain animals or to the running of riding establishments (e.g. Germany, Sweden, the UK), and to other areas. In Switzerland and Sweden, new husbandry equipment or technologies must be tested before they can be brought on to the market or be used; in Austria, Germany and the Netherlands there is a long-standing discussion about the possible introduction of similar measures. With regard to animals used for experimental or other scientific purposes, most European and North American countries have established procedures whereby experiments that may subject the animals to suffering may only be performed if authorized. The level of authorization differs between an authorization of each experiment (e.g. in Germany and Switzerland) and a general authorization with the obligation to notify experiments to the competent authority (e.g. Belgium and the UK).

Animal Welfare Acts in Austria, Belgium, Germany and Luxembourg also protect, in principle, the animal's life itself. It is an offence to kill an animal without reasonable excuse, which at least stimulates discussion about reasonable excuses, especially in regard to 'unwanted animals'.

Measures to enforce legislation are as important as the content of the provisions themselves. Between countries, competent authorities differ in their organization as well as in the number and the qualifications of their personnel; however, veterinarians are usually involved at some point. Sometimes, provision is made for advisory committees or boards associated with the competent authority; for example, in Norway, Animal Welfare Boards of the local authorities that consist of three to five members and may seek information on animals kept in the district, and even do inspections without any warning.

Limitations to legislation

For legislation to be a useful strategy for improving animal welfare in the long term, it needs to be effective, enforceable and economically feasible (Mench *et al.*, 2008). In fact, proposed legislative measures that would ban certain practices considered to have a negative welfare impact, or to set certain minimum standards, are often not adopted because they have been evaluated as economically unfeasible; for example, it is feared that producers will go out of business or production sites will be relocated to other states or countries.

In some cases, sufficiently long transition periods laid down in legislation can facilitate gradual change, but in others, animal welfare improvements carry such a cost burden that only by other measures such as additional public payments or label programmes allowing the achievement of higher prices, can sustainable progress be made. In addition, private initiatives may bear the advantage in comparison with legislative standards in that they can be more flexible and quicker to respond to public demands or in the development of technological advances in husbandry practices.

For instance, in response to the public discussion on male pig castration without anaesthesia and analgesia, retailers, especially in the Netherlands, but also in other European countries, decided to discontinue the selling of meat from pigs castrated in this way well before any respective amendment of legislation will come into force (Fredriksen *et al.*, 2009). Another benefit of this approach may be an increased acceptance by stakeholders, facilitating cooperative relationships, while adversarial relationships often develop in the context of legislative measures. In situations where market forces are limited, legislation may still be necessary to improve animal welfare. However, its effectiveness is likely to vary from country to country, or trading block to trading block, and is likely to be dependent on cultural differences regarding the use of legislation to address issues of animal use, as well as to logistic difficulties in implementation. In general, regulation without enforcement would fail to assure the public that the regulations were being followed. In many countries, the development of the infrastructure to enforce legislation may be a greater challenge both economically and politically than the passing of the legislation. For example, when legislation is applied to individual farms, the cost of the number of qualified inspectors and the additional administration required for enforcement is rather high. In the USA, there is currently a debate on how to address the cost of enforcement. Various models exist, ranging from government support to passing costs directly to producers and possibly, consequently, to consumers. However, consumers are likely to choose products predominantly according to price and so in the case of regulation at the state or local level an uncompetitive economic situation will arise.

Another important aspect regarding the effectiveness of legislation is whether standards are set concerning resources (e.g. dimensions, certain equipment), management (e.g. feeding, stocking densities) or outcomes in terms of animal welfare (e.g. lameness rates, ability to perform certain behaviours). Historically, there is an emphasis on the first two approaches, among other reasons because the control of these is much easier than is that of animal-based welfare outcomes. However, owing to complex interactions between all the different environmental factors and the animals themselves, it is by no means certain that the keeping of a number of resource or management standards guarantees that the desired welfare outcome is reached. Moreover, different housing systems or practices usually have both advantages and disadvantages with regard to the different dimensions of animal welfare. These are the reasons why, increasingly, there are calls for a stronger emphasis on animal-based measures in animal welfare assessment (e.g. Blokhuis et al., 2003). A large European research project, Welfare Quality® (2011), developed this approach further from 2004 to 2009, although the emphasis was not on legislation, but on an information system (see below). It is likely that, in future, European legislation will increasingly include limits concerning animal-based measures, while continued work will be necessary on methodological challenges (e.g. Knierim and Winckler, 2009). Already now there are single examples of the welfare outcome approach in the EU. For example, a Commission Decision (97/182/EC) on the protection of calves contains a provision that food shall contain sufficient iron to ensure an average blood haemoglobin level of at least 4.5 mmol/l, and the Directive on the protection of broilers (2007/43/EG) allows increased stocking densities only if certain mortality rates are not exceeded.

In general, one of the tasks of animal welfare scientists is to provide a sound scientific basis for decision making about whether and which kind of legislative measures are necessary or justified. However, it must not be forgotten that during the evolution of legislation, interest groups such as animal welfare organizations and professional or hobby animal users apply pressure for aims that are often opposing. While keeping in mind the objects of the measure, efforts are made by the legislators to offend as few people as possible. Therefore, legislation always represents a compromise.

18.3 Incentives and Voluntary Measures

As indicated above, state measures other than legislation, and non-state measures and forces, are important for the long-term improvement of animal welfare. Incentives to change or safeguard welfare-relevant practices in the use of animals that are set by public bodies are often of an economic nature, while voluntary measures may often be intrinsically motivated and adopted in association with an ethical commitment. Examples of the latter include measures by non-governmental organizations and individuals dedicated to responsible consumption behaviour, to improved housing and management of one's own farm, companion, zoo or laboratory animals, or to improvements of one's own animal experiments. However, ethical commitment does not exclude other possibilities, such as that companies profitably incorporate consumer concerns in their commercial strategies (Roe and Buller, 2008), and, in fact, voluntary measures can be adopted by industry just for economic reasons.

From an impressive number of different voluntary and governmentally steered activities in the field of animal welfare, selected examples will now be presented, organized into voluntary measures on the one hand and public incentives on the other.

18.3.1 Voluntary measures

Innumerable and engaged activities of local, national and international animal welfare organizations contribute considerably to progress in animal welfare, including the management of animal shelters and kennels, the performance of castration in stray cats and dogs, or the running of political campaigns, publicity work and information activities. The annual reports of many animal welfare societies are evidence of the impressive work done by these organizations.

Furthermore, many veterinary surgeons play an important role as consultants and as independent and competent welfare experts. Some veterinary organizations have developed their own ethical guidelines regarding animal welfare. A number of pet shops improve housing conditions above the minimal requirements of legislation, sometimes connected to a specific welfare labelling. Canine organizations are often active in programmes with courses for dog owners on dog education, and puppy play groups to ensure proper behavioural development. Modern zoos often have large and enriched enclosures for zoo animals, considerably above the minimal requirements of legislation. Private organizations offer educational courses for people that care for animals in pet shops, zoos, animal shelters and laboratory animal institutions. Other organizations offer educational courses for personnel who transport and slaughter animals or in other areas of animal husbandry. Institutions such as universities and the pharmaceutical industry organize courses for animal experimenters and laboratory personnel. They often install internal ethics committees for the assessment of their own animal experimental projects, and sometimes draw up ethical guidelines on animal experimentation on a level above the minimum legislative level.

With regard to non-governmental standards for biomedical research, accreditation from the Association for Assessment and Accreditation of Laboratory Animal Care International (AAALAC International) is highly regarded for research facilities. Accreditation is intended to ensure compliance with the US standards in the Guide for the Care and Use of Laboratory Animals (National Research Council, 1996), as well as with any other national, state, or local laws on animal welfare.

While some of the examples listed above are also economically motivated, for the marketing of animal products, this motivation is a particularly important aspect deserving some discussion. The relationship between market forces and animal welfare legislation, private standards and quality appreciation is affected by broad structural changes in the food supply chain (Thompson et al., 2007). Consolidation of food production companies, packing plants and retailers has altered the power relationship between these actors. Regulation of the industry is moving towards greater private control, and the power of retailers has dramatically increased. In addition, the consolidation of the food supply chain creates a situation where targeted campaigns – for example, by activist groups – can result in de facto changes to an industry (Schweikhardt and Browne, 2001). This fosters the increasing incorporation of aspects of farm animal welfare into supply chain management, either as part of companies' corporate social responsibility, thereby providing retailer differentiation, or in relation to special products or brands. Those can be premium products requiring, among other criteria, higher welfare standards that are not identified at point of sale, or specifically labelled products, often bundled with other production process qualities such as 'free range', 'grass fed' or 'organic' (Buller, 2009).

There is also a limited number of examples of specifically labelled products that focus on animal welfare (e.g. 'Certified Humane' in the USA, which is based on the 'Freedom Food' programme in the UK). Many of the labels used require certification according to a specific set of standards which may be either private or public or a combination of these. For instance, for organic production legal standards apply (e.g. National Organic Standards in the USA, Regulations (EC) No. 834/2007 and 889/2008 in the EU), but private standards of organic associations, which often include higher welfare requirements, are added if the product also carries the label of this association. The methods used to verify production processes vary from label to label, and increasingly require third-party auditing (Buller, 2009), whereas others simply require producer affidavits. The development of private standards is more pronounced in countries where legislation is less likely to be used to regulate production, whereas in other countries (e.g. Norway) reliance is more on robust welfare legislation (Buller, 2009). Nevertheless, in the USA, animal welfare labelling has attracted only a few customers (Mench et al., 2008), and also in Europe animal welfare is rarely a stand-alone selling point for food. However, welfare concerns and welfare standards are increasingly prevalent within supply chains (Buller, 2009).

An example of producers having influenced the development of animal welfare standards and significantly changed animal production practices is United Egg Producers (UEP) in the USA. They established an animal welfare scientific committee in the late 1990s that was charged with reviewing the existing scientific literature and making recommendations to improve the welfare of laying birds. One of the major recommendations was to decrease the stocking density of birds in cages. These recommendations were then used to develop a set of animal welfare guidelines to be used by UEP members. In addition, UEP established a third-party auditing programme to encourage compliance with its guidelines. Passing the audit allows producers to display a UEP-certified label on their egg cartons.

Another example of the incorporation of welfare standards into the supply chain is McDonald's. They established an animal welfare committee in 1999 and began to establish on-farm animal welfare standards for their suppliers. Numerous other retailers followed suit; currently many have incorporated animal welfare standards into their buying specifications (Mench, 2003, 2008). Consequently, there are strong economic incentives for producer organizations to consider adopting standards or changing on-farm practices in order to obtain access to this market. For example, when Smithfield, the largest producer of pork in the world, announced that it would ban the use of gestation stalls in the USA, the company denied that pressure from activists or the success of legislative initiatives had influenced its decision, and said that it was responding to concerns raised by its customers, such as McDonald's and others.

These are but a few examples of various voluntary initiatives by the private sector to address animal welfare issues directly or indirectly. Programmes range from those aimed at quality assurance to specific production methods (free range, organic) or aspects of welfare (Veissier *et al.*, 2008). The various voluntary programmes at supranational, national, regional and private sector (retail and producer) levels result in a diverse number and type of claims made regarding the welfare of animals, as well as other issues, such as animal health, environmental issues and product quality or taste (Roe *et al.*, 2005; Thompson *et al.*, 2007). Careful evaluation of the validity of these claims will become increasingly important in developing robust standards and potential legislation.

18.3.2 Public incentives

Public incentives may include, in particular, binding direct payments or other rural development support measures to the fulfilment of either legal standards or extra standards, and the setting of legal minimum standards which have to be fulfilled for the labelling of products that allow consumer choice, whereby this labelling can either be mandatory or voluntary.

As an example of incentives from the state, direct payments to farmers for especially animal-friendly farm housing systems were introduced in Switzerland in 1993 and 1996 on a voluntary basis. The housing standards of two programmes for ruminants, pigs and poultry are on a higher level than the minimum requirements of the legislation on animal protection. They include, for example, daily outside access (pasture or special outside area), the provision of litter and a ban of tethering. The great success of these programmes is reflected by an increase of farms with such animal-friendly systems by 2008: from 19% to 73% of all Swiss farms getting direct payments in one programme (daily outside access), and from 9% to 43% in the other programme (animal-friendly housing) (Federal Office for Agriculture, 2009).

Within the EU, support for rural development can be granted, subject to national programmes applied, if, among others, it is the goal to improve animal welfare (Regulation (EC) No 1698/2005). Possible measures with regard to animal welfare comprise: aid for modernization of agricultural holdings that improve their animal welfare status; support for investments in the processing and marketing sector,

opening new market opportunities for agricultural products by putting emphasis on animal welfare; incentive payments for participation of farmers in food-quality schemes; and support for information and promotion activities regarding these schemes. Moreover, measures can be taken to help farmers to adapt to demanding EU welfare standards. This support was used in Germany, for instance, to provide loans to laying hen farmers for the change from battery cages to alternative systems before the end of the legal transition period; 1.3 million laying hen places were built on this basis (Deutscher Bundestag, 2007). While these measures aim at a clear improvement of the current welfare level, there is also another EU mechanism: 'cross-compliance' (Regulation (EC) No 796/2004). This makes direct payments to farmers subject to compliance with basic standards concerning, among other topics, animal welfare. Cross-compliance represents the baseline or reference level for agri-environment measures for which the costs have to be borne by the farmer, and is intended to ensure that the support granted contributes to promoting sustainable agriculture and, thereby, responds positively to concerns of citizens at large (European Commission, 2009).

It was mentioned in the section on voluntary measures that for the labelling of animal products linked to specific welfare standards, these standards may be set (and controlled) by public bodies, even if their application is voluntary. An example of this is the marketing of organic products in many countries. However, mandatory use of such labelling is also possible, as applied in the EU for the marketing of table eggs. There are four permitted and obligatory production system labels: eggs from caged hens, barn eggs, free-range eggs and organic eggs. The requirements for these production systems are laid down in a Commission Regulation (No. 589/2008). The label gives consumers information that they may interpret as an indicator of animal welfare. Since the implementation of the legislation, the percentage of non-caged egg production has increased significantly in nearly all Member States (EC Commission, 2009).

Moreover, public bodies may play a role with respect to public information. They may launch campaigns to raise awareness of European consumers in order to increase the market share of welfare-friendly products, as has been done in the EU for organic products (EC Commission, 2009). In addition, in their responsibility for teaching and training curricula, public bodies may include animal welfare aspects in teaching and training programmes for farmers, and for those responsible for animal care in zoos, pet shops, animal refuges and animal laboratories, as well as for veterinarians, technicians, experimenters, people transporting, slaughtering or killing animals, etc. Also, by funding of research and education activities or the provision of information and exchange platforms such as the Gateway to Farm Animal Welfare by the Animal Production and Health Division of FAO (Food and Agriculture Organization of the United Nations, 2010), public bodies can contribute to animal welfare improvements.

18.4 Conclusions

- A remarkable body of animal welfare legislation has been built up nationally, supranationally and internationally. The level of animal protection and enforcement of these standards differs considerably between countries and regions.
- Increasingly, international harmonization is taking place to engender common ethical attitudes and to remove barriers to trade and the distortion of competition.
- For various reasons, legal measures have limitations in their effectiveness.
- State measures other than legislation, and non-state measures and forces, are important for the long-term improvement of animal welfare. Legislation, public incentives and voluntary measures complement each other.

References

Blokhuis, H.J., Jones, R.B., Geers, R., Miele, M. and Veissier, I. (2003) Measuring and monitoring animal welfare: transparency in the food product quality chain. *Animal Welfare* 12, 445–455.

Brambell Committee (1965) *Report of the Technical Committee to Enquire into the Welfare of Animal Kept under Intensive Livestock Husbandry Systems.* Command Paper 2836, Her Majesty's Stationery Office, London.

Brown (1974) *Who Cares for Animals?* Heinemann, London.

Buller, H. (2009) What can we tell consumers and retailers? In: Butterworth, A., Blockhuis, H., Jones, B. and Veissier, I. (eds) *Proceedings Conference, Delivering Animal Welfare and Quality: Transparency in the Food Production Chain, Including the final results of the Welfare Quality® project*, 8–9 October 2009, Uppsala, Sweden, pp. 43–46. Available at: http://www.welfarequality.net/downloadattachment/43160/20099/Def_ProceedingsTotal20091012_plus%20annex.pdf (accessed 6 January 2011).

Council of Europe (2011a) Human Rights and Legal Affairs. Biological safety – use of animals by humans. Available at: http://www.coe.int/t/e/legal_affairs/legal_co-operation/biological_safety_and_use_of_animals/default.asp (accessed 25 January 2011).

Council of Europe (2011b) Human Rights and Legal Affairs. Biological safety – use of animals by humans. Texts and Documents. Available at: http://www.coe.int/t/e/legal_affairs/legal_co-operation/biological_safety_and_use_of_animals/farming/A_texts_documents.asp#TopOfPage (accessed 25 January 2011).

Deutscher Bundestag (2007) *Tierschutzbericht 2007.* Available at: http://www.bmelv.de/cae/servlet/contentblob/383104/publicationFile/22248/Tierschutzbericht_2007.pdf (accessed 23 June 2010).

EC Commission (2009) *Report from the Commission to the European Parliament, the Council, the European Economic and Social Committee and the Committee of the Regions: Options for Animal Welfare Labelling and the Establishment of a European Network of Reference Centres for the Protection and Welfare of Animals.* Brussels, 28.10.2009, COM(2009) 584 final. Available at: http://ec.europa.eu/food/animal/welfare/farm/options_animal_welfare_labelling_report_en.pdf (accessed 6 January 2011).

EUR-Lex (2010) EUR-Lex.europa.eu. Available at: http://www.eur-lex.europa.eu/ (accessed 23 June 2010).

European Commission (2009) Agriculture and Rural Development. Agriculture and the Environment. Cross Compliance. Available at: http://www.ec.europa.eu/agriculture/envir/cross_com/index_en.htm (accessed 23 June 2010).

Federal Office for Agriculture (2009) *Swiss Agriculture on the Move: The New Agricultural Act Ten Years On.* Swiss Confederation, Federal Office for Agriculture. Available at: http://www.blw.admin.ch/dokumentation/00018/00498/index.html?lang=en&download=NHzLpZeg7t,lnp6I0NTU042l2Z6ln1ad1IZn4Z2qZpnO2Yuq2Z6gpJCEdIJ9gGym162epYbg2c_JjKbNoKSn6A– (accessed 23 June 2010).

FAWC (Farm Animal Welfare Council) (2009) *Five Freedoms.* Available at: http://www.fawc.org.uk/freedoms.htm (accessed July 2010).

Food and Agriculture Organization of the United Nations (2010) Gateway to Farm Animal Welfare. Available at: http://www.fao.org/ag/againfo/programmes/animal-welfare/aw-abthegat/aw-whaistgate/en/ (accessed 23 June 2010).

Fredriksen, B., Font i Furnols, M., Lundström, K., Migdal, W., Prunier, A., Tuyttens, F.A.M. and Bonneau, M. (2009) Practice on castration of piglets in Europe. *Animal* 3, 1480–1487.

Harrison, R. (1964) *Animal Machines: The New Factory Industry.* Vincent Stuart, London.

U. Knierim *et al.*

Knierim, U. and Winckler, C. (2009) On-farm welfare assessment in cattle – validity, reliability and feasibility issues and future perspectives with special regard to the Welfare Quality® approach. *Animal Welfare* 18, 451–458.

Mench, J.A. (2003) Assessing animal welfare at the farm and group level: a United States perspective. *Animal Welfare* 12, 493–503.

Mench, J.A. (2008) Farm animal welfare in the U.S. – farming practices, research, education, regulation and assurance programs. *Applied Animal Behaviour Science* 113, 298–312.

Mench, J.A., James, H., Pajor, E.A. and Thompson, P.B. (2008) The welfare of animals in concentrated animal feeding operations. In*: Report to the Pew Commission on Industrial Farm Animal Production. Pew Commission on Industrial Farm Animal Production*, Washington, DC.

National Research Council (1996) *Guide for the Care and Use of Laboratory Animals*. National Academy Press, Washington, DC.

OECD (2010) About OECD. Available at: http://www.oecd.org (accessed 23 June 2010).

OIE (2009a) The OIE's objectives and achievements in animal welfare. Available at: http://www.oie.int/Eng/bien_etre/en_introduction.htm (accessed 23 June 2010).

OIE (2009b) Statement from the World Organisation for Animal Health (OIE). In: Butterworth, A., Blockhuis, H., Jones, B. and Veissier, I. (eds) *Proceedings Conference, Delivering Animal Welfare and Quality: Transparency in the Food Production Chain, Including the final results of the Welfare Quality® project*, 8–9 October 2009, Uppsala, Sweden, pp. 59–60. Available at: http://www.welfarequality.net/downloadattachment/43160/20099/Def_ProceedingsTotal20091012_plus%20annex.pdf (accessed 6 January 2011).

OIE (2010a) The World Organisation for Animal Health (OIE). Available at: www.oie.int (accessed 23 June 2010).

OIE (2010b) *Animal Welfare*. Available at: http://www.oie.int/eng/ressources/AW_EN.pdf (accessed 23 June 2010).

Ransom, E. (2007) The rise of agricultural animal welfare standards as understood through a neo-institutional lens. *International Journal of Sociology of Food and Agriculture* 15, 26–44.

Roe, E. and Buller, H. (2008) *Marketing Farm Animal Welfare*. Welfare Quality® Fact sheet. Welfare Quality®, Lelystad, The Netherlands. Available at: http://www.welfarequality.net/downloadattachment/41858/19515/Fact%20sheet%20Marketing%20Farm%20animal%20welfare%20final.pdf (accessed 23 June 2010).

Roe, E., Murdoch, J. and Marsden, T. (2005) The retail of welfare-friendly products: a comparative assessment of the nature of the market for welfare-friendly products in six European Countries. In: Butterworth, A. (ed.) *Science and Society Improving Animal Welfare: Welfare Quality Conference Proceedings*, 17–18 November 2005, Brussels, Belgium. Welfare Quality®, Lelystad, The Netherlands/European Economic and Social Committee, 7 pp. (unnumbered). Available at: http://www.welfarequality.net/downloadattachment/31550/15865/proceedings%20WQ%20conference%2017-18%20November%202005.pdf (accessed 23 June 2010).

Rollin, B.E. (2004) Annual Meeting Keynote Address: Animal agriculture and emerging social ethics for animals. *Journal of Animal Science* 82, 955–964.

Russell, W.M.S. and Burch, R.L. (1959) *The Principles of Humane Experimental Technique*. Methuen, London (New edition 1992, special edition of first edition of 1959. Universities Federation for Animal Welfare (UFAW), Wheathampstead, UK).

Schweikhardt, D.B., and Browne, W.P. (2001) Politics by other means: the emergence of a new politics of food in the United States. *Review of Agricultural Economics* [now *Applied Economic Perspectives and Policy*] 23, 302–318.

Thaler, A.M. (1999) The United States perspective towards poultry slaughter. *Poultry Science* 78, 298–301.

Thompson, P., Harris, C., Holt, D. and Pajor, E.A. (2007) Livestock welfare product claim: the emerging social context. *Journal of Animal Science* 85, 2354–2360.

Veissier, I., Butterworth, A., Bock, B. and Roe, E. (2008) European approaches to ensure good animal welfare. *Applied Animal Behaviour Science* 113, 279–297.

Welfare Quality® (2011) Welfare Quality®: Science and society improving animal welfare in the food quality chain EU funded project FOOD-CT-2004-506508. Available at: http://www.welfarequality.net/everyone/26536/5/0/22 (accessed 6 January 2011).

World Trade Organization (2010) Agriculture. Available at: http://www.wto.org/english/tratop_e/agric_e/agric_e.htm (accessed 23 June 2010).

19 International Issues

MICHAEL C. APPLEBY AND STELLA MARIS HUERTAS

Abstract

Globalization and global issues affect all animals, particularly those kept primarily for monetary reasons. This contributes to internationalism – activity by individuals or groups on an international basis, including sharing of information. We discuss the impacts on animal welfare of increasing trade in animals and animal products, including competition for lower costs but also the initiation of global welfare standards and some tendency towards the 'levelling up' of animal treatment. Two aspects of treatment that pose major problems for the welfare of huge and increasing numbers of animals are transport and killing, including slaughter for food production. However, here too there is growing awareness of the benefits to both animals and people of humane animal treatment and, therefore, the implementation of improved techniques in many countries. Both increased communication about animal welfare and greater involvement of stakeholders (including civil society) in international decision making are resulting in positive outcomes for welfare, although many problems remain to be addressed, for vast numbers of animals worldwide.

19.1 Introduction

We are all world citizens now. Consider the effects of China joining the World Trade Organization (WTO) in 2002. China has about one fifth of the world's human population but nearly half of the world's pigs. It exports pork, and would doubtless be keen to export more, but cannot do so because of the prevalence of diseases such as foot-and-mouth and classical swine fever. On the contrary, increasing meat consumption in China is being supplied by increased imports from other major producers, such as the European Union (EU), USA and Canada (Fuller *et al.*, 2003). That implies increased competition within both those countries and others for cheap pork production and exports, with consequent effects on grain prices, water supplies and so on. Such changes will affect everyone, and arguably all animals.

The most acute effects of the global economy on animal welfare are on the interactions of humans with animals kept primarily for monetary reasons, notably farm animals. However, the global economy affects relationships between people and all animals (Appleby, 2005). Interactions with wild animals are affected both directly, for example through ecotourism, hunting, capture and trade in bushmeat, and indirectly, through impacts on habitat. Relationships with companion animals are

affected by diverse factors, such as the availability of exotic species, the cost of pet food and the risk of diseases brought from abroad. Treatment of laboratory animals is influenced by international standards – or lack of these – in the testing of pharmaceuticals, food products and so on.

Much information about and understanding of animal welfare issues has always been international, but in recent years consideration of the international context has become much more important and explicit. Reasons include the following overlapping aspects:

- Increases in trade, including trade in food, medicines, wildlife and products from wildlife.
- Concerns about disease, disasters, climate change and other issues not restricted by borders.
- Attempts to promote or regulate trade on an international basis and to address its problems (such as the risk of disease spread).
- Attention to the differences between developed and developing countries.
- The rapid development of the internet and other factors, including travel, promoting information transfer.

These factors interact in complex ways to produce some positive and some negative effects on animal welfare.

19.2 Internationalism

The previous chapters outlined some of the ways in which economic and political decisions affecting animals are made, primarily within countries. Within at least democratic countries, decision making – such as the passing of legislation – must take public opinion into account. In recent decades, the consideration of public opinion has increasingly attempted to involve all relevant stakeholders, such as producers, retailers and users of animals or animal products, animal welfare scientists, veterinarians, legislators, the media and other people active on welfare issues. However, the relationships between international decision making and public opinion have generally been more tenuous.

This chapter is concerned with the international context for decisions affecting animals. When trade representatives of the member countries of the WTO, or national veterinary officers who make up the constituency of the World Organisation for Animal Health (OIE), meet, they represent the citizens of their countries to some extent, but the government, industries and companies probably to a greater extent. So given the diversity of those countries, the influence of public opinion, about animal welfare among other matters, on their discussions has in the past been weak at most. That influence is now increasingly strong – or to put it another way, the involvement of all relevant stakeholders is now more complete, more similar to that within countries – because of the growth of internationalism.

By internationalism, we mean activity by individuals or groups on an international basis, taking a multinational or global viewpoint, interacting with other international and intergovernmental organizations (IGOs) and utilizing or influencing international connections, networks and agreements. Such activity is apparent in many different groups such as the following:

- Producers with operations in more than one country.
- Producer groups such as the International Egg Commission and the International Federation of Organic Agriculture Movements.
- Multinational companies like the chain restaurant company McDonald's.
- International non-governmental organizations (NGOs).
- Veterinary and other scientific associations and committees, such as the International Council for Laboratory Animal Science.

- Groups of countries, such as the Council of Europe and the European Union.
- The United Nations, including its Food and Agriculture Organization (FAO), and the WTO, OIE and other IGOs.

This is particularly so because there has also been growth in the attention paid to international issues, in requirements for accountability, and hence in communication about both such activity and stakeholder involvement.

Yet, not surprisingly, these different groups take different approaches to international issues depending on how they perceive the interests of their own stakeholders. Here are three examples. First, the EU may be expected to act on behalf of its citizens, partly as individuals and partly as represented by their national governments. It maintained a balance between individual concerns for animal welfare and established government positions in the 1997 Treaty of Amsterdam (EUR-Lex, 1997):

THE HIGH CONTRACTING PARTIES, DESIRING to ensure improved protection and respect for the welfare of animals as sentient beings, HAVE AGREED UPON the following provision to the Treaty establishing the European Community, In formulating and implementing the Community's agriculture, transport, internal market and research policies, the Community and the Member States shall pay full regard to the welfare requirements of animals, while respecting the legislative or administrative provisions and customs of the Member States.

The second example, the McDonald's company, naturally takes a commercial approach. Their Corporate Responsibility policy (McDonald's, 2009) includes statements that assert the importance of animal welfare while allowing their managers to manage the business, including making different decisions in different countries:

McDonald's cares about the humane treatment of animals, and we recognize that being a responsible purchaser of food products means working with our suppliers to ensure industry-leading animal husbandry practices. Our approach is based on our Animal Welfare Guiding Principles, which express our commitment to ensuring that animals are 'free from cruelty, abuse and neglect' ... In general, McDonald's local supply chain leaders have the flexibility to implement country-specific requirements while working within the global Guiding Principles. They take into account such factors as scientific evidence, availability, and local customer preferences.

The third example may also be expected to give regard to commercial factors, but across the whole sector of farming. The International Federation of Agricultural Producers (IFAP, 2008) states that:

> Good animal welfare practices reward farmers with good productivity. Animal welfare must be safeguarded in the production of farm animals, in the breeding process, when designing housing, in feeding and in production systems, as well as during transport and slaughter … Farmers realize that animal welfare has also become a global concern in a context of increasing market globalization. They recognize that the adoption of and respect for internationally harmonized minimum standards for animal welfare requirements are necessary to maintain consumer confidence in livestock products. Indeed, the high costs for animal welfare compliance will be rewarded with better opportunities for international trade.

The financial factors and pressure for harmonized standards introduced in that statement are applicable to three of the most pervasive international issues concerning animal welfare, to which we shall now turn: trade, then transport and slaughter.

19.3 Trade

In the livestock agriculture of developed countries, a predominant tendency over the last 60 years has been a drive for efficiency. One result has been the development of intensive farming and other practices that can cause problems for animal welfare, a situation that is also now occurring in many developing countries. Pressure for cheap production has been more variable for other categories of animals, such as companion and laboratory animals, but where it existed it has had similar effects, for example in development of 'puppy mills', producing large numbers of pedigree dogs for sale, often in very poor conditions. Meanwhile, there has been increasing concern regarding animal welfare in many countries. This has restricted practices deleterious to welfare or, conversely, has led to standards for the protection or promotion of welfare, albeit very unevenly. But fear is often expressed that such standards are put under further pressure by the growth of international trade in animal products and the increased competitiveness that this produces.

Trade in agricultural products is certainly increasing, and this is clearly regarded as desirable by governments and the agricultural industry. The goal of discussions in the WTO (1995) to bring agriculture under its remit is to accelerate this growth. Yet so-called free trade does not occur in a vacuum: it could be argued that other changes in the context within which production occurs are equally important, such as information exchange. Increased communication about animal welfare is maintaining the upward trend in international awareness. People concerned for animals hope, with some justification, that the positive effects of such awareness of animal welfare and welfare protection standards are accelerating (Turner and D'Silva, 2006).

Trade is then not wholly free, and this is particularly true of international trade. A country's own animal products, produced to certain welfare standards, may be promoted by its authorities (by advertising, use of tariffs, etc.) and favoured by its citizens both because of those standards themselves and for other reasons, such as local food security. There are also practical factors supporting in-country production, notably transport costs and hygiene controls – although these certainly do not always overwhelm other factors. Thus, egg production in Europe is subject to competition from imports (partly because of costs in Europe associated with more stringent animal welfare standards, but more because other costs such as labour and feed are greater than in competitor countries). This will probably have little impact on sales of whole eggs produced in Europe, but more on those of liquid and dried egg products, which are both more easily transported and sterilized (Fisher and Bowles, 2002).

Where restrictions on international trade exist, not directly warranted by the WTO, they may be challenged under that authority, and the precedents for welfare safeguards being permitted are not good. However, no challenges on welfare grounds have yet been made, and one possible defence increasingly mentioned is Article XX of the General Agreement on Tariffs and Trade (WTO, 2007), which says:

> Nothing in this agreement shall be construed to prevent the adoption or enforcement by any contracting party of measures: (a) necessary to protect public morals; (b) necessary to protect human, animal or plant health[.]

Nations faced with imports from countries with less extensive welfare standards could therefore potentially argue that protecting the welfare of their animals was important for public morals and both human and animal health.

It remains true that the pressure of increased trade, both domestic and international, makes creation, strengthening or even maintenance of animal welfare

M.C. Appleby and S. Maris Huertas

standards even more difficult than they would be in a less competitive market. Some existing standards may prove incompatible with an increasingly free market, although both the factors discussed so far in this section and the strong public support for the establishment of those standards suggest that these will be few. Perhaps the strongest effect of freer trade will be a reduction in the creation of new standards, or a weakening of any that are newly created. As one example, the EU passed a Directive for protection of broiler chickens in 2007 (Council of the European Union, 2007). This was weaker than earlier drafts, apparently because of concerns over the pressure of imports of chicken meat and the threat to the industry of avian influenza. Accordingly, international trade rules can constitute obstacles to the adoption of improved animal welfare standards by producers because of the possibility or perception that animal welfare policies and legislation of countries with more stringent standards might act as a non-tariff barrier to imports from other countries.

However, there are two overlapping ways in which such negative effects of trade pressure on animal welfare are being offset by positive developments. First, there is consumer demand within certain countries for 'high-welfare' products (food, clothing, cosmetics and so on) and for other niche products that consumers may perceive to be associated with improved welfare, such as organic food. This demand has produced a market for the export of such products from other countries, including developing countries (Bowles *et al.*, 2005). This market is small, but growing, and its potential has been recognized by the International Finance Corporation (IFC), part of the World Bank Group, for example in its publication *Creating Business Opportunity Through Improved Animal Welfare* (IFC, 2006a, p. 1):

> The sustainability of your business depends, among other things, on you responding positively to marketplace trends and grasping new opportunities. Consumers globally are increasing their demand for animal welfare assurances in their food supply. Meeting these demands is not only good for the animals involved, but also greatly enhances animal production and business efficiency.

Probably the most important area of improvement resulting to date has been in treatment of animals before and during slaughter (see below).

Secondly, increasing trade in animal products has been a major factor stimulating proposals that there should be global animal welfare standards, recognized by all countries, as in the quotation above from the IFAP. This task has been taken on by the OIE (2009a), because it recognized that animal health is affected by other aspects of animal welfare. In 2005, the first standards on transport and slaughter were agreed unanimously by OIE's member countries (which, at the time of writing, number 174); the OIE is now working both on standards for the rearing of farm animals and on standards for the treatment of street dogs and laboratory animals. All of these standards may be more basic than those of some member countries: for example, than EU legislation on farm animals, which is the strongest in the world. Nevertheless, their implementation (which is starting in a number of countries) will lead to improved animal welfare in the majority of OIE member countries, and there is no indication that any reduction in standards is likely in countries that already have safeguards for welfare. There is, therefore, strong reason to believe that this increased attention to animal welfare worldwide is leading not to a 'levelling down' effect, as was feared, but to a 'levelling up'. An additional process that might contribute will follow if the WTO comes to accept – either implicitly or explicitly – the validity of countries considering animal welfare in their trade decisions. The WTO already recognizes OIE standards on animal health as a consideration in trade disputes.

Levelling up is actively promoted by the EU and FAO, among others. The EU has funded a major project called 'Welfare Quality®: Science and society improving animal welfare in the food quality chain' (Welfare Quality®, 2009) involving a number of European countries and also four Latin American countries. The FAO has led discussions on capacity building for developing countries to implement good animal welfare practices (Fraser *et al.*, 2009), and has launched a web portal (FAO, 2009) for dissemination of information on the subject.

19.4 Transport

While trade in animal products has effects on welfare, the trading and other transport of live animals has more direct effects, and the scale on which animals are transported worldwide is huge. This includes companion, laboratory, sport and zoo animals, and high-value breeding animals, all of which are often moved by air (a procedure governed by specific regulations; see IATA, 2009). But by far the largest numbers of animals that are transported are production animals, with more than 60 billion farm animals yearly transported at least once, to slaughter, by land or sea. This is often over long distances, despite the fact that it would frequently be possible to slaughter them nearer to the point of production and transport carcasses instead (Box 19.1).

The fact that the OIE's first animal welfare standards were on transport (by land and sea, together with standards for slaughter, considered below) may reflect the severity and frequency of welfare problems, and the association of these problems with disease. It is widely agreed (Knowles and Warriss, 2000, p. 385) that:

> Transport is generally an exceptionally stressful episode in the life of the animal and one which is sometimes far removed from an idealized picture of animal welfare.

Stress may have negative effects on the immune system, and this can result in increased susceptibility to infection and increased infectiousness. Furthermore, transport augments the intensity and frequency of contacts between animals and this can result in diseases being spread (Manteca, 2008).

Most of the OIE's recommendations concerning transport are based on the behaviour and health of animals, the design of loading and unloading facilities (Fig. 19.1), the responsibility and competence of people involved in animal transport, and the planning and duration of the journey (OIE, 2009b). The OIE's emphasis on safeguarding welfare during transport may also have been because this subject should be relatively uncontroversial. Protecting an animal during a journey should also protect its value on arrival, whether for further use alive or for slaughter. Indeed, producers and owners of livestock sometimes claim that this overlap between owners' and animals' interests proves that concerns about welfare – in transport and at other stages of animal production – must be groundless, or at least that any problems that exist must be unavoidable. However, their primary concern is with group performance. From the perspective of the animals themselves, it is the individuals that matter. Furthermore, the owner's decisions must be affected by financial considerations: for example, some modifications to transport methods may reduce weight loss in animals and

---

Box 19.1. Transport of live farm animals versus carcasses.

In February 1882, the sailing ship Dunedin left New Zealand on a voyage to London, UK. It was carrying frozen sheep carcasses for the newly formed Bell-Coleman Mechanical Refrigeration Company. When the meat reached London 12,000 miles and 98 days later, *The Times* newspaper reported (1882):

> To-day we have to record such a triumph over physical difficulties as would have been unimaginable very few years ago. New Zealand has sent into our London market five thousand dead sheep in as good condition as if they had been slaughtered in some suburban abattoir.

In 1890, Samuel Plimsoll – now known mainly for the Plimsoll Line on ships, which indicates the depth to which they can be safely loaded – published the book *Cattle Ships*, describing the conditions during live transport as 'prolonged torture' (p. 54). In view of this and other major problems, such as the spread of infectious diseases and disasters at sea, from fires to shipwreck, he asked (p. 4):

> why cattle for food are imported alive at all, seeing that great quantities of beef are imported in a refrigerated state from America, Australia and New Zealand.

Furthermore, the chief inspector at Smithfield Market, London's main meat market, told Plimsoll that the best beef to eat was that which came over as dead meat, because the animals brought over alive were injured on the journey, reducing the quality of the carcass. Plimsoll's manifesto also complained that unscrupulous salesmen were passing off live imports as locally reared animals – an issue still current in the 21st century.

In the 21st century, more than 60 billion farm animals yearly – plus many others – are transported at least once, to slaughter. That includes many exported live from one country to another, for example 6 million sheep per year transported from Australia to the Middle East (Fisher and Jones, 2008). Reasons include preferences for fresh meat, financial arrangements by the operators, requirements for religious slaughter and availability of slaughterhouses. There is considerable scientific evidence that long-distance transport causes many welfare problems for farm animals. The European Food Safety Authority is one of an increasing number of organizations that takes such evidence seriously, and says (2004, p. 1) that:

> Transport should therefore be avoided wherever possible and journeys should be as short as possible.

M.C. Appleby and S. Maris Huertas

therefore increase the price received from their sale but, nevertheless, be considered too expensive to implement. So the interests of owners and animals do not overlap completely. As such, applying scientific approaches directly to animal welfare may produce different conclusions to those of traditional animal production science. In fact, the conventional approach, which emphasizes financial efficiency, has not always identified the best methods even to achieve its own aims. Thus it took an alternative approach, aimed at reducing problems for the animals concerned, to identify the fact that understanding animal behaviour can improve the design of handling systems and hence the efficient use of labour in handling livestock (Grandin, 2007; Fig. 19.2).

This difference in approaches is perhaps most clearly demonstrated by the fact that records of welfare problems in the handling and transport of farm animals have been sparse in many countries, even for unequivocal problems such as mortality. This reflects the commercial assumption that the prevention of such problems must be impossible or financially prohibitive. Yet record keeping is a basic requirement for scientific understanding of a problem, and for conclusions on how to address it. This is now better understood, and record keeping is required for most on-farm assurance schemes, although it is still frequently underemphasized for transport. Records of mortality in commercial or experimental conditions, both during and after transport, have now shown, for example, that it is increased by high or low temperatures (Knowles *et al.*, 1997; Knowles and Warriss, 2000), by long journey times (Warriss *et al.*, 1992) and by transporting very young animals (Staples and Haugse, 1974; Knowles, 1995).

Precursors to mortality are also, of course, important welfare problems in themselves. The incidence and severity of injury and disease are directly measurable, and aspects of transport that affect these have been reviewed elsewhere (Appleby *et al.*, 2008). In addition, a considerable amount is known about the causes and effects of injury and disease, at both an anatomical and a physiological level (Flecknell and Molony, 1997; Hughes and Curtis, 1997), together with their implications for welfare. For example, Flecknell and Molony (1997, p. 63) say that:

> Injury is of concern both because of the consequent pain which is likely to arise from traumatized tissues, and also because of its incapacitating effect on the animal. This incapacity can lead to other problems such as hunger, thirst and inability to find shelter.

Fig. 19.1. Some of the worst welfare problems associated with transport occur during loading and unloading. Good facilities for these procedures allow animals to enter or leave the transport on the level or by a gently sloping ramp with good footing, and with races and barriers that encourage quiet movement. Such facilities need not be expensive to construct.

Fig. 19.2. Minimizing welfare problems while moving animals requires understanding of their behaviour: for example, that cattle generally react much more calmly to a human on horseback than to one on foot. This understanding improves efficiency in use of labour and, when applied to pre-slaughter handling, increases meat yield and quality, food safety, disease control, worker safety and profit.

Long-distance transport of animals for slaughter occurs in all regions of the world and also between regions – most notably from Australia to the Middle East, a trade that currently includes 6 million sheep per year (Fisher and Jones, 2008). Variation in practices is loosely associated with the degree of development of a particular country. In developed countries, there is often more legislation protecting the welfare of transported animals; however, these countries also tend to have good infrastructure, such as roads, which enables more systematic and often larger scale transport of animals over long distances. Developing countries have fewer structures in place for legislation or for the supervision of animal treatment. Their transport systems are generally less advanced, so animals are not often moved over such long distances; however, it is more common for unsatisfactory vehicles and other procedures to be used (Gallo, 2008; Appleby *et al.*, 2008) (see also Box 19.2 and Table 19.1).

Despite the scale of animal transport worldwide, and the many welfare problems caused, this is an area where there is reason to believe that progress is being made in the prevention of such problems, and will continue to be made. As more information becomes available, the economic advantage of considering welfare becomes clearer. For example, short-term costs in slaughtering animals closer to the farm where they are produced may be covered by the long-term benefits of avoiding disease spread or reduced meat quality.

19.5 Killing and Prior Management

As for transport, the argument is gaining ground that humane management of animals before and while they are killed is advantageous not just for animal welfare but also for human benefit.

If dogs or other companion animals do have to be killed, gentle handling and careful euthanasia

M.C. Appleby and S. Maris Huertas

Box 19.2. Livestock transport in Latin America.

Countries in the central and northern part of Latin America (Bolivia, Colombia, Ecuador, Peru, Venezuela) are less developed and give less priority to animal welfare than those in the south (Argentina, Brazil, Chile, Paraguay, Uruguay). The region includes some countries among the world's most important beef exporters (Brazil, Argentina) and others where this business is small but important (Chile, Uruguay). Extreme variation in country size, socio-economic and cultural diversity, climate and geography also contribute to variation in welfare during transport across Latin America (Table 19.1).

Journey durations generally range from 1 to 12 h, but sometimes reach 60 h, mostly as a result of bad weather or poor road conditions (Gallo, 2007). Countries exporting animal products (mostly to Europe) have government welfare guidelines and requirements from consumers, so welfare of animals is taken into account. In other countries, animals including cattle, sheep and goats are transported in varied ways (by foot, trucks and occasionally boats), and welfare can be severely compromised. Common problems found in most countries are excessive stocking density to reduce transport costs, and

poor handling during loading and unloading (Gallo and Tadich, 2008).

As a specific example, Uruguay is the 7th largest beef exporter in the world, with only 3 million people, but 13 million cattle; and animal welfare and meat quality are therefore increasingly important here. Several institutions, led by the Veterinary Faculty of the University of Uruguay, carried out research on welfare problems in cattle transport. During 2002 and 2003, trucks transporting steers (average weight 450 kg) to slaughter plants travelled a mean of 214 km in 5 h. However, 50% of carcasses had bruises, and more than 2 kg of meat was lost per animal during the dressing process, totalling at least 4000 t from the 2 million animals slaughtered each year. Trucks were old but well maintained. Methods used to move animals were mostly prods and sticks, as well as dogs and shouts (Huertas *et al.*, 2003). A 2 year programme of training was carried out all over the country, sponsored by the Ministry of Livestock, the Uruguayan meat board, the producers' association and academia, and by 2008 carcass bruising had decreased by more than half (INAC, 2009).

Table 19.1. Cattle transport in Latin America. Under 'Welfare legislation', 'No' means that there are no specific regulations on animal welfare but guidelines from private or ministry agencies; 'Partial' means that there are regulations to avoid suffering during transport and slaughter; 'Yes' means that there are specific laws on animal welfare. Under 'Road conditions', 'Regular' means that most roads are not paved; 'Good' means that most roads are paved. Countries included in 'Others' are mostly in Central America. (From Gallo, 2007.)

Country	Area (km²)	Type of production	Cattle (million)/ Usual species	Meat consumption (kg/person p.a.)	Welfare legislation	Mean transport time (h)	Road conditions	Personnel training courses
Paraguay	406,752	Mostly extensive	10/*Bos indicus*	46	Partial	36	Regular	No
Uruguay	175,215	Mostly extensive	13/*Bos taurus*	66	Yes	5	Good	Yes
Argentina	3,761,274	Mostly extensive	55/*Bos taurus*	63	Partial	5–12	Regular	Yes
Brazil	8,511,965	Mostly extensive	170/*Bos indicus*	30	No	12–24	Regular to acceptable	Yes
Chile	756,623	Mixed	5/*Bos taurus*	23	Partial	8–20	Regular	Yes
Others	–	–	58	20	Partial	–	Regular to acceptable	No

will be more acceptable and safer for the workers involved than are inhumane methods. Guidelines on the euthanasia of dogs and cats are available (AVMA, 2007; Tasker, 2008).

However, sometimes good management involves a decision not to take the obvious course of killing animals and instead to treat them. For example, over 55,000 people are killed by rabies worldwide every year, mainly infected by dog bites (WHO, 2005); a common reaction by municipal authorities is to attempt to kill street dogs, for example by shooting (Windiyaningsih *et al.*, 2004) or beating. Yet many such attempts are not just inhumane but also ineffective, for reasons that include increased breeding and mobility among survivors. Since 1996, an NGO called Help in Suffering has been trying an alternative approach in the city of Jaipur, India. Over 50% of dogs in an area of 8 × 14 km have been captured humanely, sterilized, vaccinated against rabies, treated for any other health problems and released (Reece, 2007). Not only the dogs benefit: cases of human rabies reported to hospitals in the area fell from ten in 1993 to none in 2001 and 2002, while those outside the area continued to increase (Reece and Chawla, 2006). Consideration of the needs of animals led to a positive outcome for both the animals and the people involved.

Conversely, consideration of welfare may promote acceptance that humane killing is appropriate to terminate avoidable suffering. Discussion among countries in the Organisation for Economic Co-operation and Development (OECD) has led to agreement by many countries that laboratory tests for the safety of chemicals do not require death of the animals within the test as an end point. Clinical signs of toxicity and other safety problems have been agreed, allowing animals to be humanely killed rather than continue to suffer. Furthermore, Demers *et al.* (2006, p. 700), reviewing progress on the harmonization of laboratory animal use, report that:

> These instances of collaboration have reduced unnecessary duplication of studies involving animals by developing internationally accepted common methods for chemical testing.

With farm animals, it is increasingly recognized that humane handling and slaughter improve yield of saleable meat, food safety, disease control, worker safety and, therefore, profit. As just one example, moving animals at the slaughterhouse by beating them with sticks causes bruising, whereas using flags or rattles avoids this problem for both welfare and meat quality (Grandin, 2007; Fig. 19.3). As with transport, the fact that the OIE chose slaughter as an early topic for welfare standards may have been partly because of this recognition of potential 'win–win' opportunities and partly because of severe

Fig. 19.3. Moving animals at the slaughterhouse by beating them with sticks or other violent methods causes bruising, whereas using flags or rattles avoids this problem for both welfare and meat quality. Well-designed facilities at slaughterhouses such as this curved race also benefit operators, while improving animal welfare. Global communication about animal welfare is increasing the implementation of such ideas in many countries. The slaughterhouse shown is in Uruguay.

M.C. Appleby and S. Maris Huertas

welfare problems that occurred and continue to occur in connection with slaughter, sometimes constituting abuse. In one instance, the Humane Society of the United States (2008) obtained video in a California slaughterhouse of workers trying to force sick or injured cows to walk to slaughter by kicking them, ramming them with the blades of a forklift, jabbing them in the eyes and applying electric shocks. It is to be hoped that such behaviour is rare in most countries, but welfare problems during pre-slaughter handling and slaughter are not.

A recent development in the design and management of slaughter systems for poultry, for which such multiple benefits have been claimed, is Controlled Atmosphere Killing. This is carried out by passing birds in their transport crates through a chamber containing gas, usually argon or carbon dioxide or both, mixed with air. The welfare problems associated with other killing methods are described by Raj (1998) when he points out that gas killing can eliminate:

> stress and trauma associated with removing conscious birds from their transport containers, in particular, under the bird handling systems which require tipping or dumping of live poultry on conveyors; the inevitable stress, pain and trauma associated with shackling the conscious birds, i.e. compression of birds' hock bones by metal shackles; the stress and pain associated with conveying conscious birds hanging upside down on a shackle line which is a physiologically abnormal posture for birds; the pain experienced by some conscious birds that receive an electric shock before being stunned (pre-stun shocks); ... the pain and distress experienced by some conscious birds which miss being stunned adequately (due to wing flapping at the entrance to the water bath stunners) and then pass through the neck cutting procedure; [and] the pain and distress associated with the recovery of consciousness during bleeding due to inadequate stunning and/or inappropriate neck cutting procedure.

To that list must be added the pain and distress of some birds that are still conscious when they enter the scalding tanks for feather removal and die by scalding or drowning (Duncan, 1997). By contrast, Duncan (1997, p. 9) says of Controlled Atmosphere Killing that:

> In my opinion, this is the most stress-free, humane method of killing poultry ever developed. The birds are quiet throughout the operation. They remain in the transport crate until dead and the killing procedure itself is fast, painless, and efficient. There is no risk of recovery from unconsciousness.

In 2004, Deans Foods, one of the largest processors of end-of-lay hens and breeders in the UK, adopted Controlled Atmosphere Killing, and reports major advantages for bird welfare, carcass quality, plant efficiency and working conditions (Castaldo, 2004). However, other producers and commentators are not convinced, and discussions continue on the advantages and disadvantages of different poultry stunning and slaughter methods.

In the international context, the most important factor in acceptance that humane slaughter methods are desirable has again been international trade. In recent years, the USA and the EU have together accounted for over a third of the world's agricultural trade (Kelch and Normile, 2004), with the EU as the largest agricultural importer and the USA not far behind with annual imports of animals and animal products worth over US$10 billion (Brooks, 2004). In these and other developed countries, there are stringent requirements for food hygiene and quality, and hence for the conditions in which animals are slaughtered for food and handled beforehand. In addition, there are increasing requirements applied by purchasers such as supermarket companies for those conditions to be humane, especially for organic and other niche markets. As such requirements are applied at exporting slaughterhouses, it proves that careful attention to the design and management of handling and slaughter processes, primarily intended to improve humaneness and hygiene, also improve working efficiency, worker safety, meat yield and quality, and hence contribute in many ways to profitability (Grandin, 2007) – ways that are additional to the main aim of securing the intended export market.

This approach is spreading in developing as well as developed countries. These include large meat-exporting countries, such as Brazil and Argentina, and other large meat producers, such as China. For example, the World Society for the Protection of Animals is working with government authorities in both Brazil and China to promote humane slaughter by making recommendations on the design of slaughterhouses and facilities, implementing 'train the trainer' programmes (these reached over 3000 slaughterhouse workers in China in 18 months) and advising on legislation (Kolesar et al., 2011). A similar approach is also found in some smaller countries, such as Namibia (IFC, 2006b, p. 8):

The Namibian beef industry has a strong reputation for superior beef. This is due partly to a national assurance scheme which addresses animal health and welfare, transport and handling, and slaughter, and partly to its guaranteed hormone-free status. The scheme gives this beef industry advantages over its competitors and the country is the largest exporter to the UK of beef from the African continent. Over 100,000 tonnes of beef are produced each year, of which about 80% is exported.

However, many welfare problems continue in pre-slaughter handling and slaughter in many countries, perhaps particularly those where even limited short-term expenditure to improve facilities or training is difficult. Problems also persist in the slaughter of animals that have little economic value, such as end-of-lay hens and cull sows in some countries, and animals that are killed for disease control. Standards for the humane killing of animals both for human consumption and for disease control have been agreed by all member countries of the OIE (OIE, 2009a), but implementation of these in some countries is likely to be slow.

19.6 Conclusions

- Globalization and changes in socio-economic and cultural patterns in most countries of the world have made animal welfare an international issue. Increases in trade, concern about transnational problems (such as disease, disasters and climate change) and burgeoning communication have considerable influences on animal welfare – some positive, some negative – in both developed and developing countries.

- Growth of international trade in animal products has increased competitiveness, with some negative effects on welfare, and trade rules can be obstacles to the adoption of improved welfare standards. However, consumer demand in some countries for high-welfare products has produced a market for export from other countries, including developing countries. Increasing trade has also been a factor in the initiation of global animal welfare standards.

- Transport of animals occurs on a vast scale worldwide and causes many welfare problems. Considerable progress is being made in preventing such problems and is likely to continue. A controversial practice is live transport for slaughter; this should be replaced where possible with the transport of carcasses.

- Good practices in pre-slaughter handling, slaughter and other killing of animals are being adopted in many countries, benefiting both human and animal welfare, partly because of information sharing and partly because of legal, advisory and financial incentives. However, many welfare problems associated with the killing of animals continue, particularly in countries where expenditure for improvements is difficult.

- In many cases, the welfare of humans and the welfare of animals are closely linked. It remains true that in many countries the financial benefits of considering welfare have more impact than ethical arguments.

- More research is needed on the transport, management and killing of animals, including the training of people involved in animal handling, and on economic measures that can promote animal welfare while allowing trade between countries to provide development and benefits for people worldwide.

References

Appleby, M.C. (2005) Human–animal relationships in a global economy. In: de Jonge, F. and van den Bos, R. (eds) *The Human–Animal Relationship*. Van Gorcum, Assen, The Netherland, pp. 279–287.

Appleby, M.C., Cussen, V., Garcés, L., Lambert, L.A. and Turner, J. (eds) (2008) *Long Distance Transport and Welfare of Farm Animals*. CAB International, Wallingford, UK.

AVMA (American Veterinary Medical Association) (2007) *AVMA Guidelines on Euthanasia (Formerly Report of the AVMA Panel on Euthanasia)*, June 2007. Available at: http://www.avma.org/issues/animal_welfare/euthanasia.pdf (accessed November 2009).

Bowles, D., Paskin, R., Gutiérrez, M. and Kasterine, A. (2005) Animal welfare and developing countries: opportunities for trade in high-welfare products from developing countries. *Scientific and Technical Review of the OIE* 24, 783–790.

Brooks, N. (2004) *U.S. Agricultural Trade Update*, FAU-95, Nov. 12, 2004. Available at: http://usda.mannlib.cornell.edu/usda/ers/FAU//2000s/2004/FAU-11-12-2004.pdf (accessed August 2009).

Castaldo, D. (2004) Stunning advice: U.K. processor Deans Foods takes a new step in its animal welfare policy. Cited in: The Humane Society of the United States (HSUS) (2004) *Controlled Atmosphere Killing for Chickens and Turkeys*, September 2004 (Available at http://www.humanesociety.org/assets/pdfs/Gas_killing.pdf) as [then] available at: http://www.meatnews.com/index.cfm?fuseaction=Article&artNum=8111 MeatNews.com (accessed 13 September 2004).

Council of the European Union (2007) Council Directive 2007/43/EC of 28 June 2007 laying down minimum rules for the protection of chickens kept for meat production. *Official Journal of the European Union* 12.7.2007, EN, L 182/19–28.

Demers, G., Griffin, G., De Vroey, G., Haywood, J.R., Zurlo, J. and Bédard M. (2006) Harmonization of animal care and use guidance. *Science* 312, 700–701.

Duncan, I.J.H. (1997) *Killing Methods for Poultry: A Report on the Use of Gas in the U.K. to Render Birds Unconscious Prior to Slaughter.* Campbell Centre for the Study of Animal Welfare, Guelph, Ontario.

EFSA (European Food Safety Authority) (2004) Opinion of the Scientific Panel on Animal Health and Welfare on a request from the Commission related to the welfare of animals during transport. *The EFSA Journal* 44 (The welfare of animals during transport), 1–36. Available at: http://www.efsa.europa.eu/en/efsajournal/doc/44.pdf (accessed 7 January 2011).

EUR-Lex (1997) Treaty of Amsterdam Amending The Treaty on European Union, The Treaties Establishing The European Communities and Related Acts: Protocol on Protection and Welfare of Animals. Available at: http://eur-lex.europa.eu/en/treaties/dat/11997D/htm/11997D.html#0110010013 (accessed 7 January 2011).

FAO (Food and Agriculture Organization of the United Nations) (2009) Gateway to Farm Animal Welfare. Available at: http://www.fao.org/ag/againfo/programmes/animal-welfare/en/ (accessed July 2010).

Fisher, C. and Bowles, D. (2002) *Hard-Boiled Reality: Animal Welfare-Friendly Egg Production in a Global Market.* RSPCA, Horsham, UK.

Fisher, M.W. and Jones, B.S. (2008) Australia and New Zealand. In: Appleby, M.C., Cussen, V., Garcés, L., Lambert, L.A. and Turner J. (eds) *Long Distance Transport and Welfare of Farm Animals.* CAB International, Wallingford, UK, pp. 324–354.

Flecknell, P.A. and Molony, V. (1997) Pain and injury. In: Appleby, M.C. and Hughes, B.O. (eds) *Animal Welfare.* CAB International, Wallingford, UK, pp. 63–73.

Fraser, D., Kharb, R.M., McCrindle, C., Mench, J., Paranhos da Costa, M., Promchan, K., Sundrum, A., Thornber, P., Whittington, P. and Song, W. (2009) *Capacity Building to Implement Good Animal Welfare Practices.* Report of the FAO Expert Meeting, FAO Headquarters (Rome), 30 September – 3 October 2008. Available at: ftp://ftp.fao.org/docrep/fao/011/i0483e/i0483e00.pdf (accessed July 2010).

Fuller, F., Beghin, J., de Cara, S., Fabiosa, J., Fang, C. and Matthey, H. (2003) China's accession to the World Trade Organization: What is at stake for agricultural markets? *Review of Agricultural Economics* [now *Applied Economic Perspectives and Policy*] 25, 399–414.

Gallo, C.B. (2007) Animal welfare in the Americas, Working Document, Technical Item 1. In: *1st Interamerican Meeting on Animal Welfare,* Panama city, Panama. Available at: http://www.rr-americas.oie.int/in/Novedades/bienestar_animal/documentos/WORKING%20DOCUMENT%202006_ENG.doc (accessed 7 January 2011).

Gallo, C.B. (2008) Using scientific evidence to inform public policy on the long distance transportation of animals in South America. *Veterinaria Italiana* 44, 113–120.

Gallo, C.B. and Tadich, T.A. (2008) South America. In: Appleby, M.C., Cussen, V., Garcés, L., Lambert, L.A. and Turner, J. (eds) *Long Distance Transport and Welfare of Farm Animals.* CAB International, Wallingford, UK, pp. 261–287.

Grandin, T. (ed.) (2007) *Livestock Handling and Transport,* 3rd edn. CAB International, Wallingford, UK.

Huertas, S., Gil, A., Suanes, A., Cernicchiaro, N., Zaffaroni, R., de Freitas, J. and Invernizzi, I. (2003) Estudio de los factores asociados a la presencia de lesiones traumaticas en carcasas de bovinos faenados en Uruguay. *Jornadas Chilenas de Buiatria* (Pucón, Chile) 6, 117–118.

Hughes, B.O. and Curtis, P.E. (1997) Health and disease. In: Appleby, M.C. and Hughes, B.O. (eds) *Animal Welfare.* CAB International, Wallingford, UK, pp. 109–125.

HSUS (Humane Society of the United States) (2008) Rampant Animal Cruelty at California Slaughter Plant: Undercover investigation reveals rampant animal cruelty at California slaughter plant – a major beef supplier to America's school lunch program. Available at: http://www.humanesociety.org/news/news/2008/01/undercover_investigation_013008.html (accessed June 2009).

IATA (International Air Transport Association) (2009) Live Animals Regulations (LAR). Available at: http://www.iata.org/ps/publications/live-animals.htm (accessed August 2009).

IFAP (International Federation of Agricultural Producers) (2008) *Policy Brief "Animal Welfare: Maintaining Consumer Confidence in Livestock Products is a Responsibility of Farmers," July 2008.* IFAP, Paris.

IFC (International Finance Corporation) (2006a) *Quick Note: Creating Business Opportunity through Improved Animal Welfare.* April 2006. Available at: http://www.ifc.org/ifcext/enviro.nsf/AttachmentsByTitle/p_AnimalWelfare_QuickNote/$FILE/Animal+Welfare+QN.pdf (accessed July 2009).

IFC (2006b) *Good Practice Note: Animal Welfare in Livestock Operations.* October 2006. Available at: http://www.ifc.org/ifcext/enviro.nsf/AttachmentsByTitle/p_AnimalWelfare_GPN/$FILE/AnimalWelfare_GPN.pdf (accessed August 2009).

INAC (Instituto Nacional de Carnes) (2009) Uruguay: Auditorias de la Carne. Available at: http://www.inac.gub.uy/innovaportal/v/4775/1/innova.net/auditorias_de_la_carne (accessed November 2009).

Kelch, D. and Normile, M.A. (2004) European Union adopts significant farm reform. Available at: http://www.ers.usda.gov/Amberwaves/september04/Features/europeanunion.htm (accessed August 2009).

Knowles, T.G. (1995) A review of post transport mortality among younger calves. *Veterinary Record* 137, 406–407.

Knowles, T.G. and Warriss, P.D. (2000) Stress physiology of animals during transport. In: Grandin, T. (ed.) *Livestock Handling and Transport*, 2nd edn. CAB International, Wallingford, UK, pp. 385–407.

Knowles, T.G., Warriss, P.D., Brown, S.N., Edwards, J.E., Watkins, P.E. and Phillips, A.J. (1997) Effects on calves less than one month old of feeding or not feeding them during road transport of up to 24 hours. *Veterinary Record* 140, 116–124.

Kolesar, R., Lanier, J. and Appleby, M.C. (2011) Implementing OIE animal welfare standards: WSPA's humane slaughter training programme in Brazil and China. *Scientific and Technical Review of the OIE*, in press.

Manteca, X. (2008) Physiology and disease. In: Appleby, M.C., Cussen, V., Garcés, L., Lambert, L.A. and Turner, J. (eds) *Long Distance Transport and Welfare of Farm Animals*. CAB International, Wallingford, UK, pp. 69–76.

McDonald's (2009) McDonald's Corporate Responsibility Policy. Available at: http://www.crmcdonalds.com/publish/csr/home/report/sustainable_supply_chain/animal_welfare.html (accessed August 2009).

OIE (World Organisation for Animal Health) (2009a) The OIE's objectives and achievements in animal welfare. Available at: http://www.oie.int/eng/bien_etre/en_introduction.htm (accessed August 2009).

OIE (2009b) Terrestrial Animal Health Code. Available at: http://www.oie.int/eng/normes/mcode/en_sommaire.htm (accessed November 2009).

Plimsoll, S. (1890) *Cattle Ships*. Kegan Paul, Trench and Trubner, London.

Raj, A.B.M. (1998) Untitled. In: *Proceedings, 'Inert Gas: A Workshop to Discuss the Advantages of Using Inert Gas for Stunning and Killing of Poultry*, 30 March 1998, University of Guelph, Ontario. Cited in: Humane Society of the United States (HSUS) (undated) *An HSUS Report: The Economics of Adopting Alternative Production Practices to Electrical Stunning Slaughter of Poultry*. HSUS, Washington, DC, p. 1. Available at: http://www.abolitionistapproach.com/media/pdf/econ_elecstun.pdf (accessed 7 January 2011).

Reece, J.F. (2007) Rabies in India: an ABC approach to combating the disease in street dogs. *Veterinary Record* 161, 292–293.

Reece, J.F. and Chawla, S.K. (2006) Control of rabies in Jaipur, India, by the sterilisation and vaccination of neighbourhood dogs. *Veterinary Record* 159, 379–383.

Staples, G.E. and Haugse, C.N. (1974) Losses in young calves after transportation. *British Veterinary Journal* 130, 374–378.

Tasker, L. (2008) Methods for the euthanasia of dogs and cats: comparison and recommendations. World Society for the Protection of Animals, London. Available at: http://www.icam-coalition.org/downloads/Methods%20for%20the%20euthanasia%20of%20dogs%20and%20cats-%20English.pdf (accessed 7 January 2011).

The Times (1882) To-day we have to record.… *The Times*, 27 May 1882, London.

Turner, J. and D'Silva, J. (eds) (2006) *Animals, Ethics and Trade: The Challenge of Animal Sentience*. Earthscan, London.

Warriss, P.D., Bevis, E.A., Brown, S.N. and Edwards, J.E. (1992) Longer journeys to processing plants are associated with higher mortality in broiler chickens. *British Poultry Science* 33, 201–206.

Welfare Quality® (2009) Welfare Quality®: Science and society improving animal welfare in the food quality chain. Available at: http://www.welfarequality.net/everyone (accessed November 2009).

WHO (World Health Organization) (2005) *WHO Expert Consultation on Rabies: First Report*. WHO Technical Report Series 931, Geneva, Switzerland. Available at: http://www.who.int/rabies/trs931_%2006_05.pdf (accessed 7 January 2011).

Windiyaningsih, C., Wilde, H., Meslin, F.X., Suroso, T. and Widarso, H.S. (2004) The rabies epidemic on Flores Island, Indonesia (1998–2003). *Journal of the Medical Association of Thailand* 87, 1389–1393.

WTO (World Trade Organization) (1995) Agreement on Agriculture. WTO, Geneva, Switzerland. Available at: http://www.wto.org/english/docs_e/legal_e/14-ag.pdf (accessed 7 January 2011).

WTO (2007) WTO Analytical Index: GATT 1994. General Agreement on Tariffs and Trade 1994, Article XX. Available at: http://www.wto.org/english/res_e/booksp_e/analytic_index_e/gatt1994_07_e.htm#article20 (accessed 7 January 2011).

M.C. Appleby and S. Maris Huertas

Index

Page numbers in **bold** refer to figures and tables

aversions **185**
   see also aversive experiences; avoidance
aversive experiences 83–84, 89, 122–123, 171, **185**,
      221–222
   see also handling; transport
avoidance
   ability 221–222
   contact **250**, 255, 256
   events 86–87, 89
   learning **65**
   measures 192–193
   opportunities provision 237–**238**
   reactions 23, 79, 80, 83, 229
   responses 72, **102**, 150
   states induction 112

bacteraemia 123, 124, 205
bedding 190
behaviour
   abnormal 144–149, 150–151
   approaches 151
   assessment 138
   baseline in intensive systems 139
   changes 52, 56–57, 64, 72, 141–143
   constraints 230
   diversity reduction 34–35
   dysfunction expression 145
   evolution 24
   flow disruption 35
   health and welfare guide 138–151
   impaired control 147–148
   inappropriate 271
   individual differences 81
   inflexibility 147–148
   luxury 55, 144
      see also exploration; play
   measuring and testing 149–150
   normal **65**, 138–144, 150
   observer interpretations 150
   oro-nasal 109
   patterns identification 99
   performance importance **24**, 105–106, 215–216
   perserverative 147, 148
   priorities 217
   problems 270–271
   redirected 50, 229
   restriction 21, 98–114
   retention 21
   suppression 143
   tests 149–150, 171
   types accommodation **22**, **139**, 194
   versatility 35
   see also activity; motivation; natural behaviour;
      stereotypies; vocalizations
benefits and costs 52, 106–112, 131, 280, 283–286
Bentham, Jeremy 2, 3, 280, 282, 291

bioenergetics **165**
biological approaches 155–158, **161–166**, 223
biomarkers 70
biosecurity 126, 268
birds **73**, **99**, 125, 139, **172**, 254
   see also poultry
blood
   plasma 50–**51**, 123, **167**, 169, **176**, **232**
   pressure **102**, **161**
   sampling 160
   strain tests **162**, **163**
bone diseases 68, 127, 128, **268**, 271
boredom **34**, 35, 99, 112–114
brain 64, **65**, 71, 170
   see also cognition
Brambell Committee
   behaviour importance for welfare assessment 138
   cage wire recommendation 184
   disease or injury suffering 120
   formation 15–16, 292
   freedom from fear recommendation 255
   housing systems behavioural restriction and
      frustration 21
   on mental well-being 78
   report on intensive farming 98
   welfare importance of disease recognition 19
   welfare importance of feelings
      acknowledgement 18
breeders 219–220, 225
breeding 10, 67, 264, **265**, 267
   see also breeds; genetics selection
breeds 129, 263–266, **268**
broilers *see* poultry

cages
   battery **17**
   cleaning effects 232
   design 224–225
   flooring debate 184
   layer fatigue 127
   use legislation 280, 296
calls 110
   see also vocalizations
caloric restriction 53
calves **17**, **47**, 108, 173, 220, 231
cannibalism 146–147, 217, 270, 274
capability 28
   see also abilities
capacities 7
captivity conditions 98
cardiovascular responses 70
care
   duty of 1–11
   ethics of 8
   guides 139, 206, 300
   manuals 206

imaging techniques 170, 274
immobility effects 122
immune system 123, **163**, 175–176, 228
imports 306–307
    see also trade
imprinting 254
    see also attachment
inactivity induction 146
inbreeding 263, **264**–265, 266
incapacity 309
    see also injury
incentives 291, 292, 299, 300, 301–302
infanticide 23
infections 124, 132, 175–176
    see also disease
inflammation 67, 123, 128
information
    exchange 306
    gathering 187
    provision 203, 205–206, 208–209, 212
    public 301–302
    systems 299
    update regularity 28
injury 19, 20, 132, 217, 309
inspection 28, 130
insulin 175
integrity 38, 138–140
intensive systems **17**, 98, 99–**100**, 139
intentionality 38
interactions
    human–animal 247–248
    see also relationships
International Academy of Veterinary Pain
        Management 74
International Federation of Agricultural Producers
        (IFAP) 306
International Finance Corporation (IFC) 307
internationalism 304, 305–306
intervention point, veterinarian 126
intervention studies, evidence 256–257
isolation 80, 88, 107, 144, 217–219, 231
    see also separation

judgement bias 87, 88, 149

killing 4–5, 6, 112, 192–193, 310–314
    see also euthanasia; slaughter
knowledge 29, 232
    see also information; learning

labelling 299, 300, 301
    see also certification
laboratory animals
    analgesic use limiting factors 74

behaviour, preferences link 183
guidelines 139, 293, 294, 300
humane endpoints 211
pain relief 67–68
testing use 304
lambs **86, 87**
lameness
    assessment 141, 201
    avoidance 75
    causes 127–128
    clinical 20
    dairy cattle percentage, USA 19
    effects 122, 127–128
    flooring link 216
    measures 203–206
    prediction 202
    recovery aid 141, **142**
    reduction 206
    scores 72–74
laws 200, 291, 294–295, 296–297
    see also legislation
LD_{50} (Lethal Dose 50) 211
learning
    associative 28
    capability 28
    eavesdropping 236
    experiential (learning) phenomenon 247
    instrumental 32
    modification through human contact 252
    observation 29
    play role 236
    social 229–230, 240–241, 252–253
    see also habituation
leg health **142**, 272
    see also lameness
legislation
    advice 313
    alternative diets 53
    cage use 280, 296
    compliance checking 201–**202**
    cost 298
    crates 296
    delegated 296
    European Union 125, 224, 294, 295
    limitations 298–299
    management risk factors control 125
    methods 294
    practices prohibition 16
    pregnant sow confinement 224
    slaughter 280
    standards setting 291
    transport **311**
    variation 297–298
    see also incentives
leptin 171, 175, **176**
lesions 128, 270
Lethal Dose 50 (LD_{50}) 211

Critical
Thinking Skills for your
Nursing
Degree

**CRITICAL
STUDY SKILLS**

Critical Study Skills for Nursing Students

Our new series of study skills texts for nursing and other health professional students has four key titles to help you succeed in your degree:

Studying for your Nursing Degree

Academic Writing and Referencing for your Nursing Degree

Critical Thinking Skills for your Nursing Degree

Communication Skills for your Nursing Degree

Register with **Critical Publishing** to:

- be the first to know about forthcoming nursing titles;
- find out more about our new series;
- sign up for our regular newsletter for special offers, discount codes and more.

Visit our website at:  **www.criticalpublishing.com**

Our titles are also available in a range of electronic formats. To order please go to our website www.criticalpublishing.com or contact our distributor NBN International by telephoning 01752 202301 or emailing orders@nbninternational.com.